AN INTRODUCTION TO ORDINARY DIFFERENTIAL EQUATIONS

Garret J. Etgen
William L. Morris
University of Houston

AN INTRODUCTION TO ORDINARY DIFFERENTIAL EQUATIONS

with difference equations, numerical methods, and applications

HARPER & ROW, PUBLISHERS
New York, Hagerstown, San Francisco, London

Sponsoring Editor: Charlie Dresser
Project Editor: Eleanor Castellano
Designer: Emily Harste
Production Supervisor: Stefania J. Taflinska
Compositor: Syntax International Pte. Ltd.
Printer and Binder: The Maple Press Company
Art Studio: Vantage Art, Inc.

**AN INTRODUCTION
TO ORDINARY
DIFFERENTIAL
EQUATIONS
with difference
equations,
numerical
methods,
and applications**

Library of Congress Cataloging in Publication Data

Etgen, Garret J   1937–
    An introduction to ordinary differential
equations.

    Includes index.
    1. Differential equations.   2. Difference
equations.   I. Morris, William L., 1931–
joint author.   II. Title.
QA372.E8   515'.352   76-54685
ISBN 0-06-041913-X

Contents

4

THE FINITE CALCULUS   144

5

LINEAR DIFFERENCE EQUATIONS   178

6

ANALYSIS OF NUMERICAL METHODS   210

7

SERIES SOLUTIONS OF DIFFERENTIAL EQUATIONS   247

8

THE LAPLACE TRANSFORM   316

SYSTEMS OF LINEAR EQUATIONS   360

FIRST-ORDER DIFFERENTIAL EQUATIONS   428

APPLICATIONS   478

Preface

This text contains an introductory treatment of certain kinds of functional equations, that is, equations in which the "unknown" is a function. The particular kinds of functional equations that are studied are known as *differential equations* and *difference equations*.

The text is intended to provide a reasonable survey of those topics which meet the needs of most students. For the engineering and science students, the book emphasizes various approaches to studying the classes of equations that occur most frequently in their mathematical models. For the mathematics majors, the book can be a source of numerous examples of abstract concepts that are in their immediate future. The choice of topics has been influenced by the effects of the computer revolution, as well as by the increased use of analytical methods in the behavioral sciences.

The presentation in the text presupposes that the reader is familiar with the main ideas treated in a beginning calculus course. In addition, some familiarity with the concepts of elementary linear algebra would be helpful, but this is not essential. Chapter 1 contains a summary of some of the more frequently used topics from algebra and calculus that are needed in the text. It is not expected that the reader have a complete understanding of all of the material in Chapter 1.

The main body of the text is contained in Chapters 2 through 10. The material in these chapters offers students an opportunity to review and reinforce all their prior mathematical training while learning many new and useful concepts. The results that are derived are stated as theorems and are justified by proofs. This is a time-honored manner of mathematical exposition that seems to have more advantages than disadvantages. All the results and methods developed in the text are useful in a wide variety of applications. In this regard, Chapter 11 contains a number of detailed illustrations of applications in engineering, science, and commerce.

The book can be used as the text for a number of one-semester courses, and there is enough material for a two-semester course. Chapter 1 is not designed for classroom presentation. It is intended to be used as a beginning reading assignment in order to introduce the symbolism and terminology within the text. The items in this chapter are referenced throughout the text, so that additional opportunities for discussion of the material in the chapter arise during a course. Chapters 2 and 3 form the core of the text. Each chapter begins with a brief statement that indicates its dependence on other chapters. It is not necessary, or even recommended, that a course consist of consecutively numbered chapters. Some illustrations of one-semester courses are:

(A) Chapters 2, 3, 10 (except Section 47), 7 and 8 can be used for a traditional course for engineering and science majors.
(B) Chapters 2, 3, 4, 5, and 6 can be used for a course that emphasizes computational mathematics.
(C) Chapters 2, 3, 4, 5, and 9 (except Section 42) can be used for a course in which applications of linear algebra are emphasized.

Although the text is separated into chapters by topic, the primary organizational feature is the section. The sections are numbered consecutively with the exception of Chapter 11, which consists of numbered examples. In each section, important items, that is, definitions, theorems, examples, and so on, are numbered consecutively, and these numbers are used for reference throughout the section.

We gratefully acknowledge the help that we have received in editing and class testing the material in this book. In particular, we are indebted to our colleagues Professors Richard Dowell Byrd and Tom Wannamaker. Finally, we are indebted to George J. Telecki and Eleanor Castellano at Harper & Row.

Garret J. Etgen
William L. Morris

1

PRELIMINARIES

Section 1 Introduction

The purpose of this chapter is to present, in summary form, a survey of the material which serves as the mathematical background for most of the main body of the text. This chapter is not necessarily intended for classroom presentation, but rather it, like Chapter 11, is recommended to the reader for frequent reference and review as he or she progresses through the other chapters. However, an initial cursory reading of this chapter is advised, since it provides an introduction to the symbolism and terminology which is used throughout the text.

A study of differential equations presupposes that the reader is familiar with the basic concepts and techniques of calculus. The most frequently used concepts from calculus are reviewed in Sections 2 and 6. The presentation also requires some knowledge of the theory of equations. The essential concepts in this area are covered in the brief review of complex numbers and polynomials that is contained in Section 3.

The major portion of this text deals with linear equations of various kinds. The unifying principles that seem to be most helpful in this regard are the concepts of linear algebra. Therefore, a brief survey of linear algebra and related topics is included in Sections 4 and 5. It is not essential that the reader be familiar with this material, as the approach taken in the beginning chapters of the text is to interpret the properties of differential equations in terms of the concepts of linear algebra. This approach will make more sense after reading Chapters 2 and 3.

Because of the usual limitations of space, the brief treatment of pre-requisites in this chapter is bound to be inadequate, and so an occasional reference to calculus and linear algebra texts is certainly recommended.

Section 2 Functions and Calculus

Since some assumptions must be made, it is assumed that the reader is very familiar with the concept of a function. By and large our interest in this text will be in the type of functions which are familiar to you from calculus, namely, real-valued functions of a real variable. The first six chapters are devoted almost exclusively to two such types of functions, those whose domain is the set of real numbers on an interval, and those whose domain is the set of nonnegative integers. A function of the latter type is most often called a sequence. In addition to these familiar types of functions, however, we shall also be concerned with functions whose domain and range are vector spaces, especially vector spaces of functions. Such functions are usually called transformations, or operators, and our interest in functions of this type will be indicated in Section 4.

A basic definition connected with the function concept which will be required often in the work which follows is that of equality of functions. This notion is often overlooked in elementary calculus, and so we will state it here. Two functions f and g are *equal*, written $f = g$, if and only if they have the same domain, say X, and $f(x) = g(x)$ for all $x \in X$.

The set of all real numbers will be denoted by $\mathscr{R}$. If x and y are members of $\mathscr{R}$ and $x < y$, then the (open) interval consisting of all real numbers that are greater than x and less than y is denoted by (x, y). In more formal set notation,

$$(x, y) = \{r \in \mathscr{R} : x < r < y\}$$

The other types of (bounded, or finite) intervals are:

$$(x, y] = \{r \in \mathscr{R} : x < r \le y\}$$
$$[x, y) = \{r \in \mathscr{R} : x \le r < y\}$$
$$[x, y] = \{r \in \mathscr{R} : x \le r \le y\}$$

The last interval is said to be a closed interval. Similarly, the unbounded intervals are specified by using the symbols ∞ and $-\infty$. For example,

$$(-\infty, y) = \{r \in \mathscr{R} : r < y\}$$

is an unbounded open interval.

Let J be an open interval and let f be a function whose domain is J. If $c \in J$, then $\lim_{x \to c} f(x) = L$ means that given $\varepsilon > 0$, there is a number $\delta > 0$ such that if $x \in J$ and $0 < |x - c| < \delta$, then $|f(x) - L| < \varepsilon$.

The function f is *continuous* at the point $c \in J$ if $\lim_{x \to c} f(x) = f(c)$, and f is *continuous* on J if f is continuous at each point of J.

The function f is *differentiable* at the point $c \in J$ if there is a number m, called the *derivative of f at c*, such that

$$\lim_{x \to c} \frac{f(x) - f(c)}{x - c} = m$$

The derivative concept also has a useful geometric interpretation. In particular, if f is differentiable at the point $c \in J$, then the derivative of f at c is the slope of the line T which is tangent to the graph of f at the point $(c, f(c))$. See the figure here.

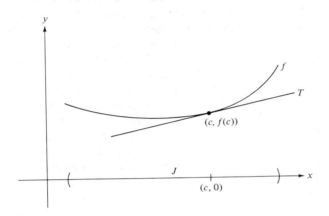

As with continuity, we say that f is *differentiable on J* if f is differentiable at each point of J. Functions which are differentiable on an open interval are studied in some detail in the sequel.

The relationship between continuity and differentiability is given by the following important result.

1 THEOREM

Let J be an open interval and let f be a function whose domain is J. If f is differentiable at the point $c \in J$, then f is continuous at the point c.

The converse of this theorem, namely, if f is continuous at the point $c \in J$, then f is differentiable at the point c, is false. The reader should be able to supply an example.

Of the many symbolic conventions which are used to denote the derivative of f at c, we prefer $Df(c)$. If f is differentiable on J, then we can determine the new function Df whose domain is J by calculating the derivative of f at each point of J. For example, if f is defined by $f(x) = x^2 + 3x - 1$ on the interval $(-\infty, \infty)$, then f is differentiable on this interval and $Df(x) = 2x + 3$ on $(-\infty, \infty)$.

Suppose that f is differentiable on the interval J. Then Df is a function whose domain is J, and it makes sense to inquire as to whether this function is continuous or differentiable. In particular, if Df is continuous, then f is said to be *continuously differentiable on J*. If Df is differentiable on J, then its derivative $D(Df)$ is called the *second derivative of f* and is denoted by D^2f. Thus, in the above example, Df is differentiable on $(-\infty, \infty)$ and D^2f is given by $D^2f(x) = 2$. Of course, we could continue successively in this manner and define the higher derivatives of f. Specifically, the nth derivative of f is denoted by D^nf, with the understanding that $f, Df, D^2f, \ldots, D^{n-1}f$ are each differentiable on the interval J, and that $D^nf = D(D^{n-1}f)$ on J. The function f is said to be *n-times differentiable on J* if D^nf is a function whose domain is J. Similarly, f is *n-times continuously differentiable on J* if D^nf is continuous on J. Finally, for convenience, we define $D^0f = f$ for each function f, and we make no distinction between D^1 and D.

It will become apparent as our work progresses that our interest is not in specific functions defined on some interval J, but rather we will be concerned with sets (collections, families) of functions defined on J. In particular, let J be an open interval. The set of all functions which are continuous on J will be denoted by $\mathscr{C}(J)$, and the set of all n-times continuously differentiable functions on J, n a positive integer, will be denoted by $\mathscr{C}^n(J)$. In addition, if f is n-times differentiable on J for each positive integer n, then f is said to be *infinitely differentiable on J*. Many of the functions which are studied in detail in calculus are infinitely differentiable on some open interval J. For example, the polynomials, the trig functions, $\sin(x)$ and $\cos(x)$, and the exponential function, $\exp(x)$, are each infinitely differentiable on the interval $(-\infty, \infty)$. The set of all infinitely differentiable functions on the interval J is denoted by $\mathscr{C}^\infty(J)$.

There is a relationship between the sets of functions defined above which can be obtained from Theorem 1. In particular, since a differentiable function on an interval J is continuous on J, it follows that $\mathscr{C}^n(J) \subset \mathscr{C}^{n-1}(J)$ for all positive integers $n \geq 2$, $\mathscr{C}^1(J) \subset \mathscr{C}(J)$, and $\mathscr{C}^\infty(J) \subset \mathscr{C}^n(J)$ for all positive integers n.

Members of $\mathscr{C}^n(J)$, n a positive integer, satisfy the hypotheses of Taylor's theorem. A statement of this theorem which is suitable for our purposes is as follows.

2 THEOREM

Taylor's theorem: If J is an open interval, $[a, b] \subset J$, and $f \in \mathscr{C}^n(J)$, then there exists a number $c \in (a, b)$ such that

$$f(b) = f(a) + Df(a)(b - a)$$

$$+ \frac{D^2f(a)}{2!}(b - a)^2 + \cdots + \frac{D^{n-1}f(a)}{(n-1)!}(b - a)^{n-1}$$

$$+ \frac{D^nf(c)}{n!}(b - a)^n$$

The special case $n = 1$ in the theorem yields the mean value theorem, which should be familiar to the reader as one of the most important and useful results in elementary calculus. With this observation, Taylor's theorem could also be called the *extended mean value theorem*.

Every constant function, that is, a function whose range consists of a single number, is a member of $\mathscr{C}^\infty(J)$ for every open interval J, and no symbolic distinction is made between a constant and a constant function. It is an easy consequence of the definition of the derivative and the mean value theorem that $Df(x) = 0$ for all x on an interval J if and only if f is a constant function on J. Equivalently, the functions F and G have the property $DF(x) = DG(x)$ for all $x \in J$ if and only if $G(x) = F(x) + c$ on J, where c is a constant.

The familiar rules concerning the derivative of the sum of two functions and the derivative of a constant times a function are:

$$D(f + g) = Df + Dg$$

and

$$D(cf) = c\, Df$$

where c is a constant. Each of the equations above expresses the equality of two functions and thus, by the definition, the equations mean

$$D(f + g)(x) = Df(x) + Dg(x)$$

and

$$D(cf)(x) = c\, Df(x)$$

for all x in J, where J is the domain of the functions f and g. The general versions of the rules above are:

(3) $\qquad D^n(f + g) = D^n f + D^n g$

(4) $\qquad D^n(cf) = c\, D^n f$

where it is assumed that each of f and g is n-times differentiable on some interval J, and where c is a constant.

Since a complete and motivated definition of the definite integral of a function is a lengthy process, we assume that the reader has some familiarity with the meaning of

$$\int_a^b f(t)\, dt$$

Recall that the variable t in this expression has no special significance, and that $\int_a^b f(t)\, dt$, $\int_a^b f(u)\, du$, $\int_a^b f(s)\, ds$, and so forth, all denote the same number.

The important relationship between D and $\int$ is expressed in the following theorem.

5 THEOREM

Let J be an open interval. If $f \in \mathscr{C}(J)$, $a \in J$, and

$$F(x) = \int_a^x f(t)\, dt, \qquad x \in J$$

then

$$F \in \mathscr{C}^1(J) \quad \text{and} \quad DF = f$$

This theorem provides a means of obtaining a function F such that $DF = f$. As noted above, if G is any other function whose derivative is f, then G differs from F by a constant; that is, $G(x) = F(x) + c$ for some constant c. Equivalently, the collection of all functions of the form $F(x) + c$, where c is any constant, represents all functions whose derivative is f. The collection of functions $F(x) + c$ is often called a *one-parameter family*. A consequence of Theorem 5 and these observations is the fundamental theorem of calculus.

6 THEOREM

Let J be an open interval and let $f \in \mathscr{C}(J)$. If $a, b \in J$, $a < b$, and if G is any function such that $DG = f$ on J, then

$$\int_a^b f(t)\, dt = G(b) - G(a)$$

For the most part, the function F of Theorem 5 has no better representation than that given. In special cases, of course, there might be more convenient representations of F. For example, if $f(x) = \cos x$ on $(-\infty, \infty)$, then

$$F_1(x) = \int_0^x \cos(t)\, dt = \sin x$$

is one example of a function whose derivative is f. However, the function F_2 given by

$$F_2(x) = \int_1^x \cos(t)\, dt = \sin x - \sin 1$$

is also a function whose derivative is f. Of course, $F_2 - F_1$ is a constant. In contrast to the function $f(x) = \cos x$, consider the function $g(x) = \sin x^2$. The function G given by

$$G(x) = \int_0^x \sin(t^2)\, dt$$

has no other convenient representation. The reader is urged to try some "methods of integration" to find another representation of G. Thus, in addition to generating functions in $\mathscr{C}^1(J)$, Theorem 5 indicates that there are many more members in this set than those few functions studied in calculus.

The following properties of $\int$ are derived from corresponding properties of D:

$$\int_a^b [f(x) + g(x)]\, dx = \int_a^b f(x)\, dx + \int_a^b g(x)\, dx$$
$$\int_a^b cf(x)\, dx = c \int_a^b f(x)\, dx$$

An additional property of $\int$, which follows either from Theorem 6 or from the geometric interpretation of the integral of f as being the area bounded by the graph of f and the x-axis, is:

$$\int_a^c f(x)\, dx = \int_a^b f(x)\, dx + \int_b^c f(x)\, dx$$

for any $a, b, c \in J$.

We assume that the reader is familiar with the fact that the concepts of continuity, differentiation, and integration, as discussed briefly here, can be extended to functions of more than one variable. Since this text requires only a limited amount of background information from the multivariable calculus, we will not go through a corresponding review of the basic concepts for functions of several variables. One concept which will arise, however, is that of a partial derivative of a function of several variables, and so we give an appropriate definition and the notation. Let f be a function of two variables, say x and y, whose domain is some region A in the x-y plane. Let $(x_0, y_0) \in A$. Then the *partial derivative of f with respect to x at the point* (x_0, y_0) is given by

$$\lim_{x \to x_0} \frac{f(x, y_0) - f(x_0, y_0)}{x - x_0}$$

provided this limit exists. Similarly, *the partial derivative of f with respect to y at the point* (x_0, y_0) is given by

$$\lim_{y \to y_0} \frac{f(x_0, y) - f(x_0, y_0)}{y - y_0}$$

provided this limit exists. The function f is *differentiable on A* if each of the partial derivatives of f exists at each point $(x, y) \in A$. If f is differentiable on A, then the notation

$$\frac{\partial f}{\partial x} \quad \text{and} \quad \frac{\partial f}{\partial y}$$

is used to denote the partial derivatives of f with respect to x and y, respectively. For additional information concerning multivariable calculus, the reader is urged to consult a calculus text.

The preceding discussion is a brief summary of some of the basic concepts and principal results of calculus. We shall assume throughout the remainder of this text that the reader is familiar with this basic material. We conclude this section by indicating some additional concepts and facts from calculus which might not be as familiar as those above, but which will be useful in work which follows.

Let J be an interval and let f be a function whose domain is J. Then f is *bounded on J* if there exists a positive number M such that

$$|f(x)| \le M$$

for all $x \in J$. The following theorem relates the concepts of continuity and boundedness.

7 THEOREM

If f is continuous on the finite, closed interval $J = [a, b]$, then f is bounded on J.

The familiar rule for differentiating the product of two functions is:

$$D(fg) = f\, Dg + g\, Df$$

If f and g are elements of $\mathscr{C}^2(J)$, then by differentiating this equation and using (3), we have

$$D^2(fg) = D[D(fg)] = D[f\, Dg] + D[g\, Df]$$
$$= f\, D^2g + 2(Df)(Dg) + g\, D^2f$$

The reader should recognize that this result is analogous to the expansion

$$(\alpha + \beta)^2 = \alpha^2 + 2\alpha\beta + \beta^2$$

Proceeding by induction, we can obtain a formula for the nth derivative of the product $f \cdot g$. This formula is the analog of the binomial theorem, that is, the expansion of $(\alpha + \beta)^n$, and it is known as *Leibnitz's rule*.

8 THEOREM

Leibnitz's rule: Let J be an open interval. If $f, g \in \mathscr{C}^n(J)$, then

$$D^n(fg) = \sum_{i=0}^{n} \binom{n}{i} (D^i f)(D^{n-i} g)$$

The symbol

$$\binom{n}{i}$$

is a *binomial number*, and it is defined by

$$\binom{n}{i} = \frac{n!}{i!(n-i)!}$$

Also, by our convention indicated previously, $D^0 f = f$.

It is often necessary to evaluate a limit of the form

$$\lim_{x \to a} \frac{f(x)}{g(x)}, \qquad a \in \mathscr{R}, \qquad \text{or} \qquad a = \pm\infty$$

where either

$$\lim_{x \to a} f(x) = 0 = \lim_{x \to a} g(x)$$

or

$$\lim_{x \to a} f(x) = \infty = \lim_{x \to a} g(x)$$

Such limits are called *indeterminate forms* of type $0/0$ or ∞/∞.

The following theorem, known as *L'Hôpital's rule*, provides a method for treating limits of quotients having an indeterminate form.

9 THEOREM

L'Hôpital's rule: Let f and g be differentiable functions. If

$$\lim_{x \to a} f(x) = 0 = \lim_{x \to a} g(x)$$

and if

$$\lim_{x \to a} \frac{Df(x)}{Dg(x)} = L$$

then

$$\lim_{x \to a} \frac{f(x)}{g(x)} = L$$

This conclusion also holds for the case

$$\lim_{x \to a} f(x) = \infty = \lim_{x \to a} g(x)$$

10 Examples

a. The limit

$$\lim_{x \to 0} \frac{\sin x}{x}$$

has the indeterminate form $0/0$. Since $D[\sin x] = \cos x$, $D[x] = 1$, and

$$\lim_{x \to 0} \frac{D[\sin x]}{D[x]} = \lim_{x \to 0} \frac{\cos x}{1} = 1$$

we have, by L'Hôpital's rule,

$$\lim_{x \to 0} \frac{\sin x}{x} = 1$$

b. Consider

$$\lim_{x \to \infty} \frac{x^2}{e^x}$$

This limit has the indeterminate form ∞/∞. The derivatives of x^2 and e^x are $2x$ and e^x, respectively, and

$$\lim_{x \to \infty} \frac{2x}{e^x}$$

also has the form ∞/∞. However, since

$$\lim_{x \to \infty} \frac{D[2x]}{D[e^x]} = \lim_{x \to \infty} \frac{2}{e^x} = 0, \qquad \lim_{x \to \infty} \frac{2x}{e^x} = 0$$

so that

$$\lim_{x \to \infty} \frac{x^2}{e^x} = 0$$

In the definition of the definite integral, $\int_a^b f(x)\,dx$, it is assumed that $[a, b]$ is a finite closed interval and that f is defined and bounded on $[a, b]$. The concept of the definite integral can be extended to include the following types of integrals:

$$\int_a^\infty f(x)\,dx \qquad \int_{-\infty}^b f(x)\,dx \qquad \int_{-\infty}^\infty f(x)\,dx$$

These integrals are called *improper integrals*.

Let a be a real number and let f be defined and integrable on $[a, c]$ for each number $c \geq a$. Then

(11) $$\int_a^\infty f(x)\,dx = \lim_{c \to \infty} \int_a^c f(x)\,dx$$

If this limit exists and has the value M, then the improper integral is said to *converge*, and the limit M is assigned as its value. If the limit does not exist, then the improper integral is said to *diverge*.

12 Examples

a.

$$\int_1^\infty \frac{1}{x}\,dx = \lim_{c \to \infty} \int_1^c \frac{1}{x}\,dx = \lim_{c \to \infty} \ln(c)$$

This limit does not exist and the improper integral diverges.

b.

$$\int_1^\infty \frac{1}{x^2}\,dx = \lim_{c \to \infty} \int_1^c \frac{1}{x^2}\,dx = \lim_{c \to \infty} \left[-\frac{1}{c} + 1 \right] = 1$$

This improper integral converges and

$$\int_1^\infty \frac{1}{x^2}\,dx = 1$$

c.

$$\int_0^\infty \sin(x)\,dx = \lim_{c \to \infty} \int_0^c \sin x\,dx = \lim_{c \to \infty} [1 - \cos(c)]$$

This limit does not exist, and so the improper integral diverges.

The improper integrals of the type (11) are the only ones needed in the text.

Section 3 Complex Numbers and Real Polynomials

A need for extending, or enlarging, the set of real numbers is evident when considering solutions of simple equations. For example, the equation

$$x^2 + 1 = 0$$

has no real solutions, for if x is any real number, then $x^2 \geq 0$ and thus $x^2 + 1 \geq 1$.

It is possible to extend the real number system to the system of so-called complex numbers in such a way that the arithmetic of the real numbers is preserved. Since this process is straightforward, only the important details of this construction are presented here.

A *complex number* is an ordered pair of real numbers (a, b). If $z_1 = (a_1, b_1)$ and $z_2 = (a_2, b_2)$ are complex numbers, then *equality* is defined by $z_1 = z_2$ if and only if $a_1 = a_2$ and $b_1 = b_2$. The set of all complex numbers will be denoted by $\mathfrak{C}$.

The arithmetic operations of addition and multiplication are defined on $\mathfrak{C}$ as follows: if $z_1 = (a_1, b_1)$ and $z_2 = (a_2, b_2)$ are any two complex numbers, then

(i) $z_1 + z_2 = (a_1 + a_2, b_1 + b_2)$

(ii) $z_1 \cdot z_2 = (a_1 a_2 - b_1 b_2, a_1 b_2 + a_2 b_1)$

It is easy to verify that addition and multiplication are associative and commutative, and that the distributive property holds. Also, the following arithmetic properties hold.

If $z = (a, b)$, then:

(1)

(i) $z + (0, 0) = (0, 0) + z = z$

(ii) $z + (-a, -b) = (-a, -b) + z = (0, 0)$

(iii) $(1, 0)z = z(1, 0) = z$

(iv) $\dfrac{1}{z} = \left(\dfrac{a}{a^2 + b^2}, \dfrac{b}{a^2 + b^2} \right)$ provided $z \neq (0, 0)$

Thus the ordered pair $(0, 0)$ plays the role of the zero element in the arithmetic of complex numbers. Since the complex number $(-a, -b)$ is the additive inverse of z, it is properly denoted $-z$. The complex number $(1, 0)$ is the multiplicative identity. Finally, it is easy to verify that $z(1/z) = (1, 0)$, so that $1/z$ is the multiplicative inverse of z and is denoted z^{-1}.

Notice that $(a, 0) + (b, 0) = (a + b, 0)$ and $(a, 0) \cdot (b, 0) = (ab, 0)$. Therefore, the set of complex numbers of the form $(x, 0)$ behaves, with respect to addition and multiplication, in exactly the same way as the set of real numbers. Clearly, there is a one-to-one correspondence between the set of real numbers and the set of complex numbers of the form $(x, 0)$. In this sense, then, the complex numbers contain the set of real numbers, and for convenience we denote $(x, 0)$ as simply x.

With this convention and the fact that $(a, b) = (a, 0) + (0, 1) \cdot (b, 0)$, we can denote this complex number as $(a, b) = a + (0, 1) \cdot b$. Finally, if the complex number $(0, 1)$ is denoted by i, then each complex number (a, b) can be expressed as $a + ib$. Note that $i^2 = (0, 1) \cdot (0, 1) = (-1, 0)$, which, by our convention, is -1.

If $z = a + ib$, then a is called the *real part* of z and b is called the *imaginary part* of z. This is abbreviated by $a = Re(z)$ and $b = Im(z)$. The *conjugate* of z is denoted by $\bar{z}$ and is defined by $\bar{z} = a - ib$. Some of the facts about the conjugate are:

(2)

(i) $z + \bar{z} = 2\, Re(z)$

(ii) $\bar{\bar{z}} = z$

(iii) $z \cdot \bar{z} = [Re(z)]^2 + [Im(z)]^2 \geq 0$

(iv) $\overline{(z^{-1})} = (\bar{z})^{-1}$

(v) $\overline{(z_1 + z_2)} = \bar{z}_1 + \bar{z}_2$

(vi) $\overline{(z_1 z_2)} = \bar{z}_1 \cdot \bar{z}_2$

The *modulus of* z is denoted by $|z|$ and defined by $|z| = (z \cdot \bar{z})^{1/2}$. Notice that in the case where z is a real number, $|z|$ is the same as the absolute value of the real number z. Also, if $|z| = 1$, then $[Re(z)]^2 + [Im(z)]^2 = 1$ and so it is possible to choose a number θ so that $Re(z) = \cos \theta$ and $Im(z) = \sin \theta$.

If $z = a + ib \neq 0$, then

$$z = |z| \left[\frac{a}{|z|} + i \frac{b}{|z|} \right]$$

and since $a/|z| + ib/|z|$ is a complex number of modulus 1, it is possible to represent z by

$$z = |z|(\cos \theta + i \sin \theta)$$

using our remarks above. This result is known as the *polar form of* z. An immediate justification for deriving the polar form of a complex number is the simplification of the operation of multiplication. In particular, if

$$z = |z|(\cos \theta + i \sin \theta) \quad \text{and} \quad w = |w|(\cos \varphi + i \sin \varphi)$$

are two complex numbers in polar form, then by the trigonometric addition formulas, we have

$$zw = |z|\,|w|[\cos(\theta + \varphi) + i \sin(\theta + \varphi)]$$

At this point it is useful to consider the extension of the exponential function $\exp(x)$ to a complex-valued function of a complex variable. A complete discussion of some of the deeper mathematical points involved in this extension is beyond the scope of this book. If, however, we define

(3) $\exp(x + iy) = \exp(x)(\cos y + i \sin y)$

where x and y are real numbers, then this new function agrees with the real exponential function when $y = 0$. This amounts to defining

(4) $\exp(i\theta) = \cos \theta + i \sin \theta$

for all real numbers θ, and it implies that the polar form of a complex number z can be written as

$$z = |z| \exp(i\theta)$$

Another observation which tends to make definition (4) plausible is that the formal power series expansion for $\exp(i\theta)$ is:

(5) $\exp(i\theta) = 1 + i\theta + \dfrac{(i\theta)^2}{2!} + \dfrac{(i\theta)^3}{3!} + \cdots$

By using the facts that $i^2 = -1; i^3 = -i; i^4 = 1$; and so on, it can be shown that simplification of (5) leads to the power series for $\cos \theta + i \sin \theta$ as predicted in (4).

This survey of the properties of the complex numbers is intentionally brief. However, the few facts presented here allow a detailed discussion of a family of functions that play an important role in the study of differential and difference equations.

By a *polynomial* we mean a function p defined by

$$p(x) = a_0 x^n + a_1 x^{n-1} + \cdots + a_{n-1} x + a_n$$

where n is a nonnegative integer and $a_0, a_1, \ldots, a_n$ are numbers called the *coefficients* of p. If $a_0 \neq 0$, then p has *degree* n. Unless otherwise specified, we consider only real numbers as coefficients.

The polynomials of degree zero are the nonzero constant functions. It is convenient to consider the constant function 0 to be a polynomial, but we do not assign any degree to this polynomial. The polynomials of degree one have the form

$$p(x) = a_0 x + a_1$$

where $a_0 \neq 0$. These polynomials are called *linear polynomials* and their graphs are straight lines.

A number r (real or complex) is called a *root* (or *zero*) of a polynomial p if $p(r) = 0$. Obviously, the polynomials of degree zero do not have any roots, while a root of the linear polynomial $p(x) = a_0 x + a_1$ is $-a_1/a_0$. As implied by our discussion of the polynomial $p(x) = x^2 + 1$ at the beginning of this section, a polynomial with real coefficients does not necessarily have a real root. Of course, the complex number i is a root of this polynomial.

Two central results concerning polynomials are contained in the following theorem.

6 THEOREM

If p is a polynomial of degree $n \geq 1$, and if the coefficients of p are complex numbers, then

(i) there exists a root r of p, and

(ii) $p(r) = 0$ if and only if $p(x) = (x - r)q(x)$, where q is a polynomial of degree $n - 1$.

A proof of Theorem 6 requires a deeper understanding of the set of complex numbers than that presented here. However, an obvious consequence of Theorem 6 is the next theorem.

7 THEOREM

If p is a (real) polynomial of degree $n \geq 1$, that is, if

$$p(x) = a_0 x^n + a_1 x^{n-1} + \cdots + a_n$$

then p has exactly n (complex) roots, $r_1, r_2, \ldots, r_n$ (not necessarily all distinct), and p can be written in the *factored form*

$$p(x) = a_0(x - r_1)(x - r_2) \cdots (x - r_n)$$

By using the properties of the conjugate of a complex number, we can obtain the following important result.

8 THEOREM

If p is a (real) polynomial, then for each complex number $z = a + ib$, $p(\bar{z}) = \overline{p(z)}$. In particular, if z is a complex root of p, then the conjugate $\bar{z} = a - ib$ of z is also a root of p.

Some information concerning the roots of a polynomial with real coefficients is now evident. For example, the complex roots of the polynomial p must occur in conjugate pairs, from which it follows that if p has odd degree, then p must have at least one real root. As observed above, the (real) number $-a_1/a_0$ is the root of the linear polynomial $p(x) = a_0 x + a_1$. Polynomials of degree two are called *quadratics*. The two roots r_1 and r_2 of the quadratic polynomial

$$p(x) = a_0 x^2 + a_1 x + a_2$$

are given by the well-known quadratic formula

$$r_1 = \frac{-a_1 + \sqrt{a_1{}^2 - 4a_0 a_2}}{2a_0}$$

$$r_2 = \frac{-a_1 - \sqrt{a_1{}^2 - 4a_0 a_2}}{2a_0}$$

The expression $a_1{}^2 - 4a_0a_2$ determines the character of the roots r_1 and r_2, that is, complex conjugates, real and equal, real and unequal, and it is called the *discriminant*. In general, determining the roots of a polynomial p whose degree is greater than two is difficult.

If $p(x) = (x - r)^k q(x)$, where $q(r) \neq 0$, then the number r is called a *root of p of multiplicity k*. This situation can be characterized in terms of the derivatives of p as follows:

9 THEOREM

If p is a (real) polynomial, then r is a root of multiplicity k if and only if $p(r) = Dp(r) = \cdots = D^{k-1}p(r) = 0$ and $D^k p(r) \neq 0$.

A proof of this result can be based on Leibnitz's rule, Theorem 8 in Section 2, and Taylor's theorem, Theorem 2 in Section 2, and the reader should be able to supply the details.

Section 4 Linear Algebra

The basic concepts of linear algebra provide a natural setting for the study of an important class of functional equations. As should be suspected from the terminology, the study of linear mathematics is in a state of much higher development than nonlinear mathematics. It is for this reason that linear algebra plays such a central role in much of mathematics and science.

A *real vector space* consists of a nonempty set $\mathscr{V}$ together with two binary operations, vector addition and scalar multiplication, which satisfy the following axioms. If $\alpha, \beta, \gamma \in \mathscr{V}$ and $x, y \in \mathscr{R}$, then

(i)	$(\alpha + \beta) \in \mathscr{V}$ and $(x \cdot \alpha) \in \mathscr{V}$	(closure)
(ii)	$\alpha + \beta = \beta + \alpha$	(commutativity)
(iii)	$\alpha + (\beta + \gamma) = (\alpha + \beta) + \gamma$	(associativity)
(iv)	$x(y \cdot \alpha) = (xy) \cdot \alpha$	(associativity)
(v)	$(x + y) \cdot \alpha = x \cdot \alpha + y \cdot \alpha$	(distributivity)
(vi)	$x \cdot (\alpha + \beta) = x \cdot \alpha + x \cdot \beta$	(distributivity)

(vii) there is a unique vector $\varnothing \in \mathscr{V}$ such that $\alpha + \varnothing = \varnothing + \alpha = \alpha$

(viii) to each α there corresponds a unique vector $-\alpha$ such that $\alpha + (-\alpha) = (-\alpha) + \alpha = \varnothing$

(ix) $1 \cdot \alpha = \alpha$

If the numbers x and y in this definition are allowed to be complex, then the concept is called a *complex vector space*. For almost all applications, we consider only real vector spaces and this is to be understood unless otherwise stated.

1 Example

Let $\mathscr{V}_n = \{(a_1, a_2, \ldots, a_n) : a_i \in \mathscr{R}\}$ with

$$(a_1, a_2, \ldots, a_n) + (b_1, b_2, \ldots, b_n) = (a_1 + b_1, a_2 + b_2, \ldots, a_n + b_n)$$

and

$$r \cdot (a_1, a_2, \ldots, a_n) = (ra_1, ra_2, \ldots, ra_n)$$

as the definitions of vector addition and scalar multiplication. This vector space is usually called Euclidean n-space. The use of horizontal arrays of real numbers, or row vectors, in defining $\mathscr{V}_n$ is merely a convenience, as so-called column vectors would do just as well. For $n = 1, 2,$ or 3, $\mathscr{V}_n$ has the usual geometric interpretation. Members of $\mathscr{V}_n$ can also be interpreted as real-valued functions whose domains are the first n positive integers. The zero vector is $\varnothing = (0, 0, \ldots, 0)$.

2 Example

Let $\mathscr{V}_\infty = \{(a_1, a_2, \ldots, a_n, \ldots) : a_i \in \mathscr{R}\}$ with

$$(a_1, a_2, \ldots, a_n, \ldots) + (b_1, b_2, \ldots, b_n, \ldots)$$
$$= (a_1 + b_1, a_2 + b_2, \ldots, a_n + b_n, \ldots)$$

and

$$r(a_1, a_2, \ldots, a_n, \ldots) = (ra_1, ra_2, \ldots, ra_n, \ldots)$$

This example is the vector space of all real sequences, or, equivalently, the vector space of all real-valued functions whose domains are the set of positive integers. The zero vector in this space is the zero sequence $\varnothing = (0, 0, \ldots, 0, \ldots)$.

3 Example

If J is an open interval, then $\mathscr{C}(J)$, as defined in Section 2, is a real vector space. Addition is the usual addition of functions and scalar multiplication is the usual multiplication of a function by a constant, that is,

$$(f + g)(x) = f(x) + g(x)$$
$$(rf)(x) = rf(x)$$

for each $x \in J$. Since we recall from calculus that the sum of continuous functions is continuous and the product of a continuous function with a real number yields a continuous function, the verification that $\mathscr{C}(J)$ is a vector space is straightforward. The zero vector in this vector space is the constant function $\varnothing(x) = 0$ for all

$x \in J$. Since $\mathscr{C}(J)$ is a vector space, it is not uncommon to refer to $f \in \mathscr{C}(J)$ as a vector, even though f is, more particularly, a continuous function.

4 Example

If addition and multiplication are defined as in Example 3, then $\mathscr{C}^k(J)$ is a vector space for each positive integer k. Property (i) in the definition of vector space holds because of equations (3) and (4) in Section 2. In fact, $\mathscr{C}^\infty(J)$ is also a vector space. As remarked in Section 2, the following relation between these vector spaces of functions holds:

$$\mathscr{C}^\infty(J) \subset \mathscr{C}^k(J) \subset \mathscr{C}(J)$$

If $\mathscr{V}$ is a vector space and $\mathscr{S} \subset \mathscr{V}$ ($\mathscr{S}$ is a subset of $\mathscr{V}$), then $\mathscr{S}$, together with the vector addition and scalar multiplication of $\mathscr{V}$, may or may not be a vector space. If $\mathscr{S}$ with the operations of $\mathscr{V}$ is a vector space, then it is called a *subspace* of $\mathscr{V}$. In verifying that a subset $\mathscr{S}$ is also a subspace, it is not necessary to check all nine properties in the definition of vector space. The following theorem is easy to verify.

5 THEOREM

If $\mathscr{V}$ is a vector space and $\mathscr{S} \subset \mathscr{V}$, then $\mathscr{S}$ is a subspace of $\mathscr{V}$ if and only if

(i) $\alpha, \beta \in \mathscr{S}$ implies that $(\alpha + \beta) \in \mathscr{S}$, and

(ii) $r \in \mathscr{R}$ and $\alpha \in \mathscr{S}$ implies that $(r \cdot \alpha) \in \mathscr{S}$.

6 Example

The vector spaces of Example 4 are subspaces of $\mathscr{C}(J)$. In fact, if $m > n > 0$ are two integers, then $\mathscr{C}^m(J)$ is a subspace of $\mathscr{C}^n(J)$, which is a subspace of $\mathscr{C}(J)$.

If $\mathscr{V}$ is a vector space, k is a positive integer, $\alpha_1, \alpha_2, \ldots, \alpha_k \in \mathscr{V}$, and $c_1, c_2, \ldots, c_k \in \mathscr{R}$, then

$$c_1 \cdot \alpha_1 + c_2 \cdot \alpha_2 + \cdots + c_k \cdot \alpha_k = \sum_{i=1}^{k} c_i \cdot \alpha_i$$

is called a *linear combination* of the vectors $\alpha_1, \alpha_2, \ldots, \alpha_k$. Notice that any linear combination of vectors in $\mathscr{V}$ is also a vector in $\mathscr{V}$. The numbers $c_1, c_2, \ldots, c_k$ are called the *coefficients* in the linear combination. If at least one of the coefficients is nonzero, the linear combination is called *nontrivial*. The linear combination in which all the coefficients are zero is the *trivial linear combination*.

7 Example

If $\mathcal{V}$ is a vector space and $\mathcal{S}$ is a subset of $\mathcal{V}$, then the set of all linear combinations of vectors in $\mathcal{S}$ is a subspace of $\mathcal{V}$. It is called the *subspace generated by* $\mathcal{S}$ and is denoted by $[\mathcal{S}]$.

A finite set of vectors, $\alpha_1, \alpha_2, \ldots, \alpha_k$, in a vector space $\mathcal{V}$ is said to be *linearly dependent* if a nontrivial linear combination of them is equal to the zero vector; otherwise the vectors are called *linearly independent*. Both concepts concern the equation

(8) $$c_1 \cdot \alpha_1 + c_2 \cdot \alpha_2 + \cdots + c_k \cdot \alpha_k = \varnothing$$

where $\alpha_1, \alpha_2, \ldots, \alpha_k$ is the given set of vectors from $\mathcal{V}$. Obviously, $c_1 = c_2 = \cdots = c_k = 0$ is always a solution of (8), and if this is the only solution, then the vectors are linearly independent; otherwise, the vectors are linearly dependent.

Frequently, a determination that a given set of vectors is, or is not, linearly independent involves the solution of a system of linear equations in which the unknowns are the coefficients in a linear combination. However, different approaches to this problem are possible depending upon the specific nature of the vector space under consideration.

9 Example

The vectors $(1, 2)$ and $(-\frac{1}{2}, -1)$ are linearly dependent in $\mathcal{V}_2$ since $(\frac{3}{2}) \cdot (1, 2) + 3 \cdot (-\frac{1}{2}, -1) = (0, 0) = \varnothing$. On the other hand, $(1, 0)$ and $(0, 1)$ are linearly independent since

$$c_1 \cdot (1, 0) + c_2 \cdot (0, 1) = (c_1, c_2) = (0, 0) = \varnothing$$

if and only if $c_1 = c_2 = 0$.

10 Example

The functions defined by $\alpha_1(x) = 1$, $\alpha_2(x) = x$, $\alpha_3(x) = 2x - 1$ are vectors in the vector space $\mathcal{C}(-\infty, \infty)$. This set of vectors is linearly dependent since

$$1 \cdot \alpha_1 - 2 \cdot \alpha_2 + 1 \cdot \alpha_3 = \varnothing$$

where $\varnothing(x) = 0$ is the zero function.

11 Example

The functions defined by $\alpha_1(x) = \sin x$ and $\alpha_2(x) = \cos x$ are vectors in $\mathcal{C}^\infty(-\infty, \infty)$. They are linearly independent, since if

$$c_1 \sin x + c_2 \cos x = \varnothing(x)$$

then

$$c_1 \sin 0 + c_2 \cos 0 = \varnothing(0) = 0$$

or

$$c_2 = 0$$

and

$$c_1 \sin\left(\frac{\pi}{2}\right) + c_2 \cos\left(\frac{\pi}{2}\right) = \varnothing\left(\frac{\pi}{2}\right) = 0$$

or

$$c_1 = 0$$

There is no special significance in the two particular numbers, 0 and $\pi/2$, used in Example 11. It was an obvious convenience to choose these two members of the domain of α_1 and α_2 to show that $c_1 = c_2 = 0$.

12 Example

The functions defined by $\alpha_1(x) = 1$, $\alpha_2(x) = x, \ldots, \alpha_k(x) = x^{k-1}$, $\alpha_{k+1}(x) = x^k$, where k is a positive integer, are all members of $\mathscr{C}^\infty(-\infty, \infty)$. This set of vectors is linearly independent since a nontrivial linear combination of these vectors is a polynomial of some degree n, $n \le k$. From Theorem 7 in Section 3, we know that a polynomial of degree n has exactly n zeros (some of which may be complex numbers). Hence, only the trivial linear combination of these vectors could be equal to the zero function. Here we are making use of the particular properties of these vectors.

The notion of linear independence can be extended to arbitrary, that is, not necessarily finite, subsets of a vector space $\mathscr{V}$ in the following manner: if $\mathscr{S} \subset \mathscr{V}$, then $\mathscr{S}$ is *linearly independent* if every finite subset of $\mathscr{S}$ is linearly independent. A subset $\mathscr{B} \subset \mathscr{V}$ is a *basis* for $\mathscr{V}$ if $\mathscr{B}$ is linearly independent and $[\mathscr{B}] = \mathscr{V}$.

An existence theorem of fundamental importance in the study of vector spaces, the proof of which is beyond the scope of this brief summary, is as follows.

13 THEOREM

Every vector space has a basis.

The next examples illustrate that a vector space need not have a unique basis. In fact, many questions in linear algebra are settled most conveniently by selecting an appropriate basis.

14 Example

The vectors $(1, 0)$ and $(0, 1)$ are a basis for $\mathscr{V}_2$. It was shown in Example 9 that these vectors are linearly independent, and since

$$(a, b) = a(1, 0) + b(0, 1)$$

it follows that $[(1, 0), (0, 1)] = \mathscr{V}_2$. Another basis for $\mathscr{V}_2$ is, for example, $\{(1, 1), (-1, 3)\}$. It is easy to verify that this pair of vectors is linearly independent and that

$$(a, b) = \frac{3a + b}{4}(1, 1) + \frac{b - a}{4}(-1, 3)$$

15 Example

It can be verified that both

$$\mathscr{B}_1 = \{(1, 0, \ldots, 0), (0, 1, \ldots, 0), \ldots, (0, 0, \ldots, 1)\}$$

and

$$\mathscr{B}_2 = \{(1, 1, \ldots, 1), (0, 1, 1, \ldots, 1), \ldots, (0, 0, 0, \ldots, 1)\}$$

are bases for $\mathscr{V}_n$.

16 Example

If $\mathscr{B} = \{1, x, x^2, \ldots, x^n, \ldots\}$, then $\mathscr{B} \subset \mathscr{C}(J)$ for any interval J. Since any nontrivial linear combination of members of $\mathscr{B}$ is a polynomial, it follows, as in Example 12, that $\mathscr{B}$ is a linearly independent set of vectors. Also, $[\mathscr{B}]$ is a subspace of $\mathscr{C}(J)$ and, in fact, $[\mathscr{B}]$ is the set of all polynomials along with the zero function. It is also true that $[\mathscr{B}]$ is a subspace of $\mathscr{C}^\infty(J)$. The vector space $[\mathscr{B}]$ is denoted by $\mathscr{P}$ and is called the vector space of all polynomials.

While a vector space need not have a unique basis, the number of vectors in a basis is invariant. This fact is the content of the next theorem.

17 THEOREM

If $\mathscr{V}$ is a vector space, n is a positive integer, and $\mathscr{D} = \{\beta_1, \beta_2, \ldots, \beta_n\}$ is a basis for $\mathscr{V}$, then every basis for $\mathscr{V}$ has n vectors.

If the vector space $\mathscr{V}$ is as given in Theorem 17, then $\mathscr{V}$ is said to be a *finite-dimensional vector space*, or a *vector space of dimension n*.

18 Example

In Example 14, $\mathscr{V}_2$ has dimension 2, and in Example 15, $\mathscr{V}_n$ has dimension n. Example 16 shows that the vector spaces $\mathscr{C}^\infty(J), \mathscr{C}^k(J)$, and $\mathscr{C}(J)$ are not finite-dimensional since

$$\mathscr{B} \subset \mathscr{P} \subset \mathscr{C}^{\infty}(J) \subset \mathscr{C}^{k}(J) \subset \mathscr{C}(J)$$

and the set $\mathscr{B}$, which is a basis for $\mathscr{P}$, is not a finite set.

19 Example

If $\mathscr{B}_k = \{1, x, x^2, \ldots, x^k\}$, then $[\mathscr{B}_k]$ is a vector space and, according to Example 12, $\mathscr{B}_k$ is a basis for $[\mathscr{B}_k]$. This vector space is denoted by $\mathscr{P}_k$ and consists of all polynomials of degree k or less, together with the zero function. The vector space $\mathscr{P}_k$ has dimension $k + 1$ and

$$\mathscr{P}_0 \subset \mathscr{P}_k \subset \mathscr{C}^{\infty}(J) \subset \mathscr{C}^{n}(J) \subset \mathscr{C}(J)$$

where k and n are any pair of positive integers.

The final topic in this brief review of linear algebra involves a special class of functions whose domain and range are vector spaces. Such functions are usually called *transformations*, and much of linear algebra is a detailed study of a particular type of transformation. If T is a transformation with domain the vector space $\mathscr{V}$ and range a subset of the vector space $\mathscr{W}$, then T is a *linear transformation* provided that for each α, $\beta \in \mathscr{V}$ and $r \in \mathscr{R}$,

(i) $T(\alpha + \beta) = T(\alpha) + T(\beta)$, and

(ii) $T(r \cdot \alpha) = r \cdot T(\alpha)$.

The symbolism that is used to denote a transformation T with domain the vector space $\mathscr{V}$ and range the vector space $\mathscr{W}$ is $T : \mathscr{V} \to \mathscr{W}$.

20 Example

Let $T : \mathscr{R} \to \mathscr{R}$ be defined by $T(x) = mx$, where m is a constant. The geometric interpretation of T is a straight line through the origin with slope m. It is easy to verify that T is a linear transformation. If $S : \mathscr{R} \to \mathscr{R}$ is defined by $S(x) = 3x - 7$, then S is not a linear transformation.

21 Example

If $T : \mathscr{V}_2 \to \mathscr{V}_2$ is defined by $T(a, b) = (a, 0)$, then T is a linear transformation.

22 Example

Since $\mathscr{C}^{n}(J)$ and $\mathscr{C}(J)$, where J is an open interval and n is a positive integer, are vector spaces, and since D^n has properties (i) and (ii) of the definition [see equations (3) and (4) in Section 2], then $D^n : \mathscr{C}^{n}(J) \to \mathscr{C}(J)$ is a linear transformation.

A linear transformation whose domain is a vector space of functions, such as in Example 22, is commonly called a *linear operator*. Thus D^n is a linear operator for $n = 1, 2, 3, \ldots$. In Section 2, we defined D^0 by $D^0 f = f$ for each function f. This operator is called the *identity operator* and it is often denoted by I.

The basic properties of the definite integral familiar from calculus yield another example of a linear operator.

23 Example

Let $a, b \in \mathcal{R}$, $a < b$. Define $T : \mathscr{C}(-\infty, \infty) \to \mathcal{R}$ by

$$T(f) = \int_a^b f(x)\, dx$$

Then referring to the discussion of the properties of $\int$ in Section 2, T is a linear operator.

24 Example

Let J be an open interval and let $f \in \mathscr{C}(J)$. Define the operator $f \cdot D$ by $(f \cdot D)(g) = f \cdot Dg$, where $g \in \mathscr{C}^1(J)$. Clearly, $f \cdot D : \mathscr{C}^1(J) \to \mathscr{C}(J)$, and $f \cdot D$ is a linear operator. Similarly, $f \cdot D^n : \mathscr{C}^n(J) \to \mathscr{C}(J)$ is a linear operator for each positive integer n.

25 Example

Let $f_i \in \mathscr{C}(J)$, where J is an open interval and $i = 0, 1, \ldots, n$. Define the operator L by

$$L[g] = f_0 \cdot D^n g + f_1 \cdot D^{n-1} g + \cdots + f_{n-1} \cdot Dg + f_n \cdot D^0 g$$

for all $g \in \mathscr{C}^n(J)$. The linear operators defined in Example 24 are special cases of L. Evidently $L : \mathscr{C}^n(J) \to \mathscr{C}(J)$, and it is easily verified using equations (3) and (4) of Section 2 that L is a linear operator. Operators of this form are called linear differential operators and they play an important role in the study of differential equations.

26 Example

Let $J = (-\infty, \infty)$ and let $f_0(x) = e^x$, $f_1(x) = x^2$, $f_2(x) = \sin x$. Then $L = e^x D^2 + x^2 D + \sin(x) D^0$ is a linear operator and

$$L[e^{-x}] = 1 - x^2 e^{-x} + \sin(x) e^{-x}$$

If $T : \mathscr{V} \to \mathscr{W}$ is a linear transformation, then

$$\mathscr{N}(T) = \{\alpha \in \mathscr{V} : T(\alpha) = \varnothing\}$$

is called the *null space of* T. The following theorem is easy to verify.

27 THEOREM

Let $\mathscr{V}$ and $\mathscr{W}$ be vector spaces. If $T:\mathscr{V} \to \mathscr{W}$ is a linear transformation, then the null space, $\mathscr{N}(T)$, of T is a subspace of $\mathscr{V}$.

It will be shown in Chapter 2 that the problem of solving certain differential equations is equivalent to determining the null space of a linear operator of the type in Example 25.

28 Example

As noted in Example 22, $D:\mathscr{C}^1(J) \to \mathscr{C}(J)$, where J is an open interval, is a linear operator. The null space of D is $\mathscr{N}(D) = \{f \in \mathscr{C}^1(J):Df = \varnothing$, that is, the zero function on $J\}$. An application of the mean value theorem, Theorem 2 in Section 2, using the case $n = 1$, shows that $f \in \mathscr{N}(D)$ if and only if f is a constant function. Thus $\mathscr{N}(D) = \mathscr{P}_0$, where $\mathscr{P}_0$ is defined in Example 19.

Section 5 Linear Equations and Matrices

A vector space that plays a central role in the study of systems of linear algebraic equations is $\mathscr{V}_n$ of Example 1 in Section 4. Here, however, it is more convenient to represent members of $\mathscr{V}_n$ as vertical arrays of real numbers rather than the horizontal arrays displayed in Example 1, Section 4. Thus, in this section, $\zeta \in \mathscr{V}_n$ means

$$(1) \qquad \zeta = \begin{bmatrix} x_1 \\ x_2 \\ \vdots \\ x_n \end{bmatrix}$$

where each x_i is a real number. Addition and scalar multiplication are defined component-wise just as in Example 1, Section 4, and the members of $\mathscr{V}_n$ are referred to as vectors, or more explicitly, column vectors.

A *matrix* is a rectangular array of scalars and is denoted by an uppercase letter. For example,

$$(2) \qquad A = \begin{bmatrix} a_{11} & a_{12} & \cdots & a_{1n} \\ a_{21} & a_{22} & \cdots & a_{2n} \\ \vdots & \vdots & & \vdots \\ a_{k1} & a_{k2} & \cdots & a_{kn} \end{bmatrix}$$

is a matrix with k rows and n columns, or a $k \times n$ matrix. The real numbers a_{ij}, $1 \le i \le k$, $1 \le j \le n$, are called the *entries*, or *components*, of A. The matrix A can also be interpreted as an array formed by n column vectors; that is,

$$(3) \qquad A = (\alpha_1, \alpha_2, \ldots, \alpha_n)$$

where

$$\alpha_i = \begin{bmatrix} a_{1i} \\ a_{2i} \\ \vdots \\ a_{ki} \end{bmatrix}$$

is the ith column of A.

The definitions of addition and scalar multiplication for members of $\mathscr{V}_n$ can be used to define corresponding operations on the set of $k \times n$ matrices. In particular, if

$$A = \begin{bmatrix} a_{11} & a_{12} & \cdots & a_{1n} \\ a_{21} & a_{22} & \cdots & a_{2n} \\ \vdots & \vdots & & \vdots \\ a_{k1} & a_{k2} & \cdots & a_{kn} \end{bmatrix} \quad \text{and} \quad B = \begin{bmatrix} b_{11} & b_{12} & \cdots & b_{1n} \\ b_{21} & b_{22} & \cdots & b_{2n} \\ \vdots & \vdots & & \vdots \\ b_{k1} & b_{k2} & \cdots & b_{kn} \end{bmatrix}$$

then

$$A + B = \begin{bmatrix} a_{11} + b_{11} & a_{12} + b_{12} & \cdots & a_{1n} + b_{1n} \\ a_{21} + b_{21} & a_{22} + b_{22} & \cdots & a_{2n} + b_{2n} \\ \vdots & \vdots & & \vdots \\ a_{k1} + b_{k1} & a_{k2} + b_{k2} & \cdots & a_{kn} + b_{kn} \end{bmatrix}$$

and

$$cA = \begin{bmatrix} ca_{11} & ca_{12} & \cdots & ca_{1n} \\ ca_{12} & ca_{22} & \cdots & ca_{2n} \\ \vdots & \vdots & & \vdots \\ ca_{k1} & ca_{k2} & \cdots & ca_{kn} \end{bmatrix}$$

where c is any real number. In terms of the representation (3), these definitions become

$$A + B = (\alpha_1 + \beta_1, \alpha_2 + \beta_2, \ldots, \alpha_n + \beta_n)$$
$$cA = (c\alpha_1, c\alpha_2, \ldots, c\alpha_n)$$

where $A = (\alpha_1, \alpha_2, \ldots, \alpha_n)$ and $B = (\beta_1, \beta_2, \ldots, \beta_n)$. Notice that the set of $k \times n$ matrices with these arithmetic operations is another example of a vector space. The zero vector in this space is the $k \times n$ matrix $\varnothing$ all of whose entries are 0.

The general form of a system of k linear equations in n unknowns is:

(4)
$$\begin{aligned} a_{11}x_1 + a_{12}x_2 + \cdots + a_{1n}x_n &= c_1 \\ a_{21}x_2 + a_{22}x_2 + \cdots + a_{2n}x_n &= c_2 \\ \vdots \qquad \vdots \qquad\quad \vdots \qquad \vdots \\ a_{k1}x_1 + a_{k2}x_2 + \cdots + a_{kn}x_n &= c_k \end{aligned}$$

It is this system of equations that provides the motivation for the definition of the product of a matrix and a column vector. If A is the matrix in (2) and

ζ is the n-component column vector in (1), then $A\zeta$ is the k-component column vector γ whose ith component c_i is given by

(5) $\qquad c_i = a_{i1}x_1 + a_{i2}x_2 + \cdots + a_{in}x_n$

$i = 1, 2, \ldots, k$. Note that this definition requires the number of components of the vector ζ to be exactly the same as the number of columns of the matrix A. Note, also, that (5) is the ith equation in the system (4), and so this system can be represented more compactly using the vector-matrix form

(6) $\qquad A\zeta = \gamma$

An equivalent means of defining the product $A\zeta$ makes use of the representation (3) of A. Using (3),

(7) $\qquad A\zeta = x_1\alpha_1 + x_2\alpha_2 + \cdots + x_n\alpha_n$

To give some indication of the relationship between the concepts being presented in this section and those treated in Section 4, notice that if A is a $k \times n$ matrix and ζ is any n-component vector, that is, a member of $\mathscr{V}_n$, then $A\zeta = \gamma$ is a k-component vector, that is, a member of $\mathscr{V}_k$. Thus the matrix A can be thought of as a function whose domain is $\mathscr{V}_n$ and whose range is in $\mathscr{V}_k$. It is easy to verify that for any number r and for any vectors $\alpha, \beta \in \mathscr{V}_n$,

$$A(\alpha + \beta) = A\alpha + A\beta$$

$$A(r \cdot \alpha) = r(A\alpha)$$

Therefore, $A: \mathscr{V}_n \to \mathscr{V}_k$ is an example of a linear transformation.

Equation (6), representing the system of equations (4), is known as the *general linear algebraic equation*. It is understood that A and γ are given and that ζ is the "unknown" vector. These linear equations are classified into two types: if $\gamma = \varnothing$, then equation (6) is called *homogeneous*; and if $\gamma \neq \varnothing$, then (6) is called *nonhomogeneous*.

A homogeneous equation always has a solution, namely $\zeta = \varnothing$, and this is called the *trivial solution*. If a given homogeneous equation has solutions in addition to the trivial solution, then they are said to be *nontrivial solutions*. If the trivial solution is the only solution of the equation $A\zeta = \varnothing$, then the columns of A are linearly independent vectors. This fact follows immediately from the definition of linear independence in Section 4 and the representation (7) for the product $A\zeta$.

8 Examples

a. Consider the system of equations

$$3x_1 + x_2 + x_3 = -4$$
$$3x_1 + 2x_2 - x_3 = 4$$

The vector-matrix representation of this system is:

$$\begin{bmatrix} 3 & 1 & 1 \\ 3 & 2 & -1 \end{bmatrix} \begin{bmatrix} x_1 \\ x_2 \\ x_3 \end{bmatrix} = \begin{bmatrix} -4 \\ 4 \end{bmatrix}$$

It is easy to verify that

$$\zeta = \begin{bmatrix} 1 \\ -2 \\ -5 \end{bmatrix}$$

is a solution.

b. If

$$\begin{bmatrix} 1 & 1 & 1 \\ 0 & 1 & 1 \\ 0 & 0 & 1 \end{bmatrix} \begin{bmatrix} x_1 \\ x_2 \\ x_3 \end{bmatrix} = \begin{bmatrix} 0 \\ 0 \\ 0 \end{bmatrix}$$

then

$$x_1 \begin{bmatrix} 1 \\ 0 \\ 0 \end{bmatrix} + x_2 \begin{bmatrix} 1 \\ 1 \\ 0 \end{bmatrix} + x_3 \begin{bmatrix} 1 \\ 1 \\ 1 \end{bmatrix} = \begin{bmatrix} 0 \\ 0 \\ 0 \end{bmatrix}$$

and, clearly,

$$\zeta = \varnothing = \begin{bmatrix} 0 \\ 0 \\ 0 \end{bmatrix}$$

is the only solution of this equation.

The definition of the product of a matrix and a vector can be used to define the product of two matrices. In particular, if A is a $k \times n$ matrix and B is an $n \times p$ matrix, then the product AB is the $k \times p$ matrix whose entry c_{ij} is:

(9) $c_{ij} = a_{i1}b_{1j} + a_{i2}b_{2j} + \cdots + a_{in}b_{nj}$

Note that if $p = 1$, that is, if B had only one column, then (9) agrees with (5). In general, AB can be thought of as the matrix whose columns are the vectors formed by calculating the products $A\beta_i$, $i = 1, 2, \ldots, p$, where β_i is the ith column of B. Note, also, that in order to form the product AB, the number of rows of B must be the same as the number of columns of A. In contrast to multiplication of real numbers, observe that matrix multiplication is not commutative. Indeed, if A is $k \times n$ and B is $n \times p$, then BA is not even defined if $p \neq k$.

10 Example

Let

$$A = \begin{bmatrix} 5 & 4 & 1 \\ 2 & 0 & 1 \end{bmatrix} \qquad B = \begin{bmatrix} 1 & 0 & 1 \\ 0 & 2 & -1 \\ 0 & 0 & 4 \end{bmatrix} \qquad C = \begin{bmatrix} 1 & -1 & 2 \\ -1 & 3 & 1 \\ 2 & 0 & 0 \end{bmatrix}$$

Then

$$AB = \begin{bmatrix} 5 & 8 & 5 \\ 2 & 0 & 6 \end{bmatrix}$$

BA is not defined,

$$BC = \begin{bmatrix} 3 & -1 & 2 \\ -4 & 6 & 2 \\ 8 & 0 & 0 \end{bmatrix} \qquad CB = \begin{bmatrix} 1 & -2 & 10 \\ -1 & 6 & 0 \\ 2 & 0 & 2 \end{bmatrix}$$

The $k \times n$ matrix A specified in (2) is said to be a *square matrix* if $k = n$, and in this case the matrix is said to have *order n*. Linear equations of the form (6) that are most common in studying functional equations involve square matrices, and so for the the remainder of this summary, we shall restrict our attention to this set of matrices.

If A and B are any two matrices of order n, then each of the products AB and BA is defined, and each yields an $n \times n$ matrix. In general, $AB \neq BA$ [see Example 10], so that matrix multiplication in the set of $n \times n$ matrices is not commutative. It can be shown, however, that matrix multiplication is associative. The *identity matrix* of order n is denoted by I and is given by

$$I = \begin{bmatrix} 1 & 0 & 0 & \cdots & 0 \\ 0 & 1 & 0 & \cdots & 0 \\ \vdots & \vdots & \vdots & & \vdots \\ 0 & 0 & 0 & \cdots & 1 \end{bmatrix}$$

It is easy to verify that $AI = IA = A$ for all matrices A of order n. If A is a matrix of order n, and if there is a matrix B of order n such that $AB = BA = I$, then B is called the *inverse of A*, and it is denoted by A^{-1}. The fact that not every $n \times n$ matrix A, $A \neq \emptyset$, has an inverse is illustrated by the following example.

11 Example

Let

$$A = \begin{bmatrix} 2 & 1 \\ 4 & 2 \end{bmatrix} \qquad \text{and} \qquad B = \begin{bmatrix} 0 & -1 \\ 0 & 2 \end{bmatrix}$$

A simple calculation shows

$$\begin{bmatrix} 2 & 1 \\ 4 & 2 \end{bmatrix} \begin{bmatrix} 0 & -1 \\ 0 & 2 \end{bmatrix} = \begin{bmatrix} 0 & 0 \\ 0 & 0 \end{bmatrix} = \varnothing$$

Suppose A had an inverse. Then

$$B = IB = (A^{-1}A)B = A^{-1}(AB) = A^{-1}\varnothing = \varnothing$$

and this is a contradiction. Thus A cannot have an inverse. The same argument can be used to show that B does not have an inverse.

Associated with each matrix A of order n is a real number called the *determinant of A* and denoted $\det(A)$. Although it is not feasible to include a complete discussion of the determinant function in this brief treatment, we shall indicate how to calculate the determinants of 2×2 and 3×3 matrices. It will be matrices of these orders which will occur most frequently in the text. Consider the 2×2 matrix

$$A = \begin{bmatrix} a & b \\ c & d \end{bmatrix}$$

The determinant of this matrix is:

(12) $\det(A) = ad - bc$

Consider the 3×3 matrix

$$A = \begin{bmatrix} a_{11} & a_{12} & a_{13} \\ a_{21} & a_{22} & a_{23} \\ a_{31} & a_{32} & a_{33} \end{bmatrix}$$

The determinant of this matrix can be expressed in terms of determinants of 2×2 matrices in a variety of equivalent ways. The following expression is called the expansion of $\det(A)$ by the first column:

(13)
$$\det(A) = a_{11} \det \begin{bmatrix} a_{22} & a_{23} \\ a_{32} & a_{33} \end{bmatrix} - a_{21} \det \begin{bmatrix} a_{12} & a_{13} \\ a_{32} & a_{33} \end{bmatrix} + a_{31} \det \begin{bmatrix} a_{12} & a_{13} \\ a_{22} & a_{23} \end{bmatrix}$$

so that

$$\det(A) = a_{11}(a_{22}a_{33} - a_{23}a_{32}) - a_{21}(a_{12}a_{33} - a_{13}a_{32}) + a_{31}(a_{12}a_{23} - a_{13}a_{22})$$

The expansion of $\det(A)$ by the first row is:

(14)
$$\det(A) = a_{11} \det \begin{bmatrix} a_{22} & a_{23} \\ a_{32} & a_{33} \end{bmatrix} - a_{12} \det \begin{bmatrix} a_{21} & a_{23} \\ a_{31} & a_{33} \end{bmatrix} + a_{13} \det \begin{bmatrix} a_{21} & a_{22} \\ a_{31} & a_{32} \end{bmatrix}$$

The reader should verify that the right sides of (13) and (14) are identical. Note that in each of the expansions (13) and (14), $\det(A)$ is expressed as a sum of three terms of the form

$$a_{ij} \det(A_{ij})$$

where A_{ij} is the 2×2 matrix which remains after deleting the ith row and jth column of A. Also note the signs of the terms which appear in each of the expansions. Associated with each position ij in the matrix A is an algebraic sign: $+$ if $i + j$ is even, and $-$ if $i + j$ is odd. These signs are independent of the sign of the number which actually appears in the position. The signs associated with each position in a 3×3 matrix are indicated by the following array:

$$\begin{bmatrix} + & - & + \\ - & + & - \\ + & - & + \end{bmatrix}$$

It should be clear from this discussion that $\det(A)$ can be calculated by an expansion using any row or column of A. In particular, the expansion of $\det(A)$ by the jth column, $1 \le j \le 3$, is:

$$\det(A) = (-1)^{1+j} a_{1j} \det(A_{1j}) + (-1)^{2+j} a_{2j} \det(A_{2j})$$
$$+ (-1)^{3+j} a_{3j} \det(A_{3j})$$

and a similar expression for $\det(A)$ can be formed using the ith row of A, $1 \le i \le 3$.

15 Example

Let

$$A = \begin{bmatrix} 2 & 1 \\ 4 & 2 \end{bmatrix} \qquad B = \begin{bmatrix} 1 & 2 \\ -1 & 7 \end{bmatrix} \qquad C = \begin{bmatrix} 1 & -2 & -1 \\ -1 & 0 & 3 \\ 2 & 0 & 1 \end{bmatrix}$$

Then $\det(A) = 2 \cdot 2 - 1 \cdot 4 = 0$, and $\det(B) = 1 \cdot 7 - 2(-1) = 9$. Calculating $\det(C)$ using the expansion by the first row gives

$$\det(C) = 1 \cdot \det \begin{bmatrix} 0 & 3 \\ 0 & 1 \end{bmatrix} - (-2) \cdot \det \begin{bmatrix} -1 & 3 \\ 2 & 1 \end{bmatrix}$$

$$+ (-1) \cdot \det \begin{bmatrix} -1 & 0 \\ 2 & 0 \end{bmatrix} = 0 + 2(-7) + 0 = -14$$

The most efficient calculation of $\det(C)$ takes advantage of the 0's in the second column. In particular, expansion by the second column yields

$$\det(C) = -(-2) \det \begin{bmatrix} -1 & 3 \\ 2 & 1 \end{bmatrix} = 2(-7) = -14$$

The procedure outlined for calculating the determinant of a 3×3 matrix can also be used for calculating the determinant of a higher-order matrix. If A is an $n \times n$ matrix and $\det(A) \neq 0$, then A is said to be *nonsingular*; if $\det(A) = 0$, then A is said to be *singular*. The relationship between the determinant, the inverse of a matrix, and the existence of solutions of a linear equation of the form (6) is given in the following important theorem.

16 THEOREM

If A is an $n \times n$ matrix, then the following are equivalent.

(i) $\det(A) \neq 0$, that is, A is nonsingular.

(ii) A^{-1} exists.

(iii) The columns of A are linearly independent vectors.

(iv) The only solution of the equation $A\zeta = \varnothing$ is the trivial solution.

(v)   The equation $A\zeta = \gamma$ has a unique solution.

Let A be an $n \times n$ matrix. A number r (which may be real or complex) is an *eigenvalue of A* if there exists a vector β, $\beta \neq \varnothing$, such that

$$(17) \qquad A\beta = r\beta$$

The nonzero vector β in (17) is called an *eigenvector of A associated with the eigenvalue r*. Note that an eigenvector of A is a vector which A transforms into a multiple of itself.

If r is an eigenvalue of A and β is an associated eigenvector, then equation (17) can be written equivalently as

$$(18) \qquad (rI - A)\beta = \varnothing$$

Since $\beta \neq \varnothing$, we can conclude from Theorem 16 that $\det(rI - A) = 0$. Conversely, if $\det(rI - A) = 0$, then there exists a nonzero vector β such that

$$(rI - A)\beta = \varnothing \qquad \text{or} \qquad A\beta = r\beta$$

19 THEOREM

Let A be an $n \times n$ matrix. The number r is an eigenvalue of A if and only if

$$\det(rI - A) = 0$$

It is easy to verify that the expansion of $\det(rI - A)$ results in a polynomial in r whose degree is n, the order of A. This polynomial is called the *characteristic polynomial of A*. Referring to the discussion of polynomials in Section 3, the nth degree polynomial $\det(rI - A)$ has exactly n roots, $r_1, r_2, \ldots, r_n$ (not necessarily distinct). Thus the $n \times n$ matrix A has exactly n eigenvalues, the n roots of $\det(rI - A)$.

20 Examples

a. Let

$$A = \begin{bmatrix} 2 & -6 \\ -1 & -3 \end{bmatrix}$$

Then

$$rI - A = \begin{bmatrix} r-2 & 6 \\ 1 & r+3 \end{bmatrix}$$

and

$$\det(rI - A) = (r-2)(r+3) - 6 = r^2 + r - 12 = (r-3)(r+4)$$

Thus the eigenvalues of A are $r_1 = 3, r_2 = -4$.

b. Let

$$B = \begin{bmatrix} -2 & 1 & 0 \\ -5 & 2 & 1 \\ -2 & -2 & 1 \end{bmatrix}$$

Then

$$rI - B = \begin{bmatrix} r+2 & -1 & 0 \\ 5 & r-2 & -1 \\ 2 & 2 & r-1 \end{bmatrix}$$

and

$$\det(rI - B) = (r+2)[(r-2)(r-1) + 2] - (-1)[5(r-1) + 2]$$
$$= r^3 - r^2 + 3r + 5$$

(Expansion by the first row.) The roots of this polynomial are $r_1 = -1, r_2 = 1 + 2i$, and $r_3 = 1 - 2i$.

Example 20b illustrates that although A is a real matrix and $\det(rI - A)$ is a polynomial with real coefficients, the roots of $\det(rI - A)$, that is, the eigenvalues of A, can be complex numbers. Of course, from Theorem 8, Section 3, the complex eigenvalues of A must occur in conjugate pairs. For a complete analysis of the eigenvalues and eigenvectors of an $n \times n$ matrix, it would be necessary to consider matrices with complex entries, together with the complex vector space $\mathbb{C}_n$ of ordered n-tuples of complex numbers. However, for the purposes of this text, such a complete treatment is not required.

The following theorem gives important information about the eigenvectors of a matrix.

21 THEOREM

Let A be an $n \times n$ matrix.

(i) If $r_1, r_2, \ldots, r_k$ are distinct eigenvalues of A, with $\beta_1, \beta_2, \ldots, \beta_k$ a corresponding set of eigenvectors, then the vectors $\beta_1, \beta_2, \ldots, \beta_k$ are linearly independent.

(ii) If r is an eigenvalue of A, α and β are eigenvectors associated with r, and c is a real number, then each of $\alpha + \beta \neq \emptyset$ and $c \cdot \alpha \neq \emptyset$ is an eigenvector of A associated with r.

The following example illustrates a procedure for determining an eigenvector associated with a given eigenvalue.

22 Example

Consider the 3×3 matrix B in Example 20b. Since $r_1 = -1$ is an eigenvalue, we shall determine a vector β such that $(-I - B)\beta = \emptyset$. By taking b_1, b_2, b_3 to be components of β, this vector-matrix equation can be written as

$$\begin{bmatrix} 1 & -1 & 0 \\ 5 & -3 & -1 \\ 2 & 2 & -2 \end{bmatrix} \begin{bmatrix} b_1 \\ b_2 \\ b_3 \end{bmatrix} = \begin{bmatrix} 0 \\ 0 \\ 0 \end{bmatrix}$$

which is equivalent to the system

$$b_1 - b_2 \qquad = 0$$
$$5b_1 - 3b_2 - b_3 = 0$$
$$2b_1 + 2b_2 - 2b_3 = 0$$

By eliminating variables, we get the solutions $b_1 = c$; $b_2 = c$; $b_3 = 2c$, where c is any real number. Thus the vector

$$\beta = \begin{bmatrix} 1 \\ 1 \\ 2 \end{bmatrix}$$

is an eigenvector of B associated with the eigenvalue -1. The set of all eigenvectors of B associated with -1 is merely the set of all multiples of β. The reader can verify that the vectors

$$\gamma = \begin{bmatrix} 1 \\ 3 + 2i \\ -2 + 4i \end{bmatrix} \quad \text{and} \quad \zeta = \begin{bmatrix} 1 \\ 3 - 2i \\ -2 - 4i \end{bmatrix}$$

are eigenvectors of B associated with $r_2 = 1 + 2i$ and $r_3 = 1 - 2i$, respectively.

We conclude this section with a brief description of some special types of matrices.

Let A be an $n \times n$ matrix with entries a_{ij}. The entries $a_{11}, a_{22}, \ldots, a_{nn}$ form the *main diagonal* of A. The matrix A is a *diagonal matrix* if $a_{ij} = 0$ for $i \neq j$; that is, the entries of A which are not on the main diagonal are all zero. The $n \times n$ identity matrix I is an example of a diagonal matrix. In general, a diagonal matrix has the form

$$A = \begin{bmatrix} r_1 & 0 & 0 & \cdots & 0 \\ 0 & r_2 & 0 & \cdots & 0 \\ \vdots & \vdots & \vdots & & \vdots \\ 0 & 0 & 0 & \cdots & r_n \end{bmatrix}$$

and the notation $A = \text{diag}[r_1, r_2, \ldots, r_n]$ will be used to denote this matrix.

23 THEOREM

Let $A = \text{diag}[r_1, r_2, \ldots, r_n]$. Then

(i) $\det(A) = \displaystyle\prod_{i=1}^{n} r_i = r_1 r_2 \cdots r_n$

(ii) For each i, $1 \leq i \leq n$, r_i is an eigenvalue of A, and the vector e_i, all of whose components are zero except for the ith, which equals one, is an associated eigenvector.

Suppose B is an $n \times n$ matrix with eigenvalues $r_1, r_2, \ldots, r_n$. Suppose, in addition, that B has a corresponding set of n linearly independent eigenvectors $\gamma_1, \gamma_2, \ldots, \gamma_n$. This would be the case, for example, if the eigenvalues of B are distinct. Let C be the $n \times n$ matrix whose columns are $\gamma_1, \gamma_2, \ldots, \gamma_n$. Then it follows from the eigenvalue-eigenvector equation (17) that

(24) $BC = C \cdot \text{diag}[r_1, r_2, \ldots, r_n]$

Moreover, since the columns of C are linearly independent, C^{-1} exists by Theorem 16. Thus (24) can be written as

(25) $C^{-1}BC = \text{diag}[r_1, r_2, \ldots, r_n]$

A matrix B with the property that there exists a nonsingular matrix C such that (25) holds is said to be *diagonalizable*. This is equivalent to B having a set of n linearly independent eigenvectors.

An $n \times n$ matrix A is called *upper (lower) triangular* if $a_{ij} = 0$ for $i > j$ ($i < j$). In an upper triangular matrix, each of the entries below the main diagonal is zero. Clearly, a diagonal matrix is upper triangular (as well as lower triangular). An upper triangular matrix has the form

$$A = \begin{bmatrix} a_{11} & a_{12} & a_{13} & \cdots & a_{1n} \\ 0 & a_{22} & a_{23} & \cdots & a_{2n} \\ 0 & 0 & a_{33} & \cdots & a_{3n} \\ \vdots & \vdots & \vdots & & \vdots \\ 0 & 0 & 0 & \cdots & a_{nn} \end{bmatrix}$$

26 THEOREM

Let A be an upper triangular matrix. Then

(i) $\det(A) = \displaystyle\prod_{i=1}^{n} a_{ii}$

(ii) For each i, $1 \leq i \leq n$, a_{ii} is an eigenvalue of A.

Let B be a $k \times n$ matrix with entries b_{ij}. The $n \times k$ matrix B^T whose entry in the ij position is b_{ji} is called the *transpose of B*. The columns of B are the rows of B^T. In the special case $n = 1$, B is a k-component column vector and B^T is a k-component row vector.

27 Examples

Let

$$A = \begin{bmatrix} 1 & 2 & -3 \\ 7 & -1 & 0 \end{bmatrix} \qquad B = (1, 2, 1) \qquad C = \begin{bmatrix} 2 \\ -1 \end{bmatrix}$$

Then

$$A^T = \begin{bmatrix} 1 & 7 \\ 2 & -1 \\ -3 & 0 \end{bmatrix} \qquad B^T = \begin{bmatrix} 1 \\ 2 \\ 1 \end{bmatrix} \qquad C^T = (2, -1)$$

An $n \times n$ matrix A is called *symmetric* if $A^T = A$, that is, if $a_{ij} = a_{ji}$ for all i, j. Symmetric matrices of orders 2 and 3 have the forms

$$\begin{bmatrix} a & b \\ b & c \end{bmatrix} \qquad \text{and} \qquad \begin{bmatrix} a_{11} & a_{12} & a_{13} \\ a_{12} & a_{22} & a_{23} \\ a_{13} & a_{23} & a_{33} \end{bmatrix}$$

respectively. Symmetric matrices have a number of useful properties which make them important in a variety of applications.

28 THEOREM

Let A be a symmetric matrix. Then

(i)   The eigenvalues $r_1, r_2, \ldots, r_n$ of A are real numbers.
(ii) A has n linearly independent eigenvectors $\beta_1, \beta_2, \ldots, \beta_n$ associated with these eigenvalues.
(iii) The eigenvectors $\beta_1, \beta_2, \ldots, \beta_n$ can be chosen so that if B is the matrix whose columns are $\beta_1, \beta_2, \ldots, \beta_n$, then $B^{-1} = B^T$, that is, $BB^T = B^TB = I$.

(iv) By choosing the eigenvectors as in (iii) and forming B,

$$AB = B \operatorname{diag}[r_1, r_2, \ldots, r_n]$$

or

$$B^T AB = \operatorname{diag}[r_1, r_2, \ldots, r_n]$$

Section 6 Sequences, Series, and Power Series

A (real) *sequence* is a real-valued function f whose domain is the nonnegative integers. The values of f, namely, $f(0), f(1), \ldots, f(n), \ldots$, are called the *terms* of the sequence. In addition to using the functional notation $f(n)$ to denote the terms of a sequence f, a standard convention is to use the subscript notation f_n, and to denote the sequence f by $\{f_n\}$. Both notations will be used in this text. A sequence $\{f_n\}$ is *convergent* if there exists a number L such that to each positive number ε there corresponds a positive integer N with the property $|L - f_n| < \varepsilon$ whenever $n > N$. The number L is called the *limit* of the sequence. A sequence which is not convergent is said to be *divergent*. A sequence $\{f_n\}$ is *bounded* if the function f is a bounded function, that is, if there exists a nonnegative number M such that $|f_n| \le M$ for all nonnegative integers n. A convergent sequence is bounded, but a bounded sequence is not necessarily convergent.

1 Examples

a. $\{1/(n + 1)\} = \{1, \frac{1}{2}, \frac{1}{3}, \ldots\}$.
b. $\{(-1)^n\} = \{1, -1, 1, -1, \ldots\}$.
c. $\{n^2\} = \{0, 1, 4, 9, \ldots\}$.
d. $\{2 + (-1)^n/(n + 1)\} = \{3, \frac{3}{2}, \frac{7}{3}, \frac{7}{4}, \ldots\}$.

Sequences (a) and (d) are convergent (and, hence, bounded) with limits 0 and 2, respectively. Sequences (b) and (c) are divergent; (b) is bounded and (c) is not bounded.

Let $\{a_n\}$ be a sequence, and define a new sequence $\{s_n\}$ as follows:

$$s_0 = a_0, \qquad s_1 = a_0 + a_1, \qquad \ldots, \qquad s_n = \sum_{k=0}^{n} a_k$$

By letting $n \to \infty$, we are led to consider the expression

(2) $$\sum_{k=0}^{\infty} a_k$$

This expression is called an *infinite series* whose *terms* are the terms of the given sequence $\{a_n\}$. The sequence $\{s_n\}$ is called the *sequence of partial sums* of the infinite series. An infinite series is *convergent* if and only if its sequence of partial sums is a convergent sequence. Otherwise the infinite series is said to be *divergent*. If the sequence of partial sums $\{s_n\}$ of the infinite series (2) is convergent with limit S, then S is called the *sum* of the infinite series.

Given an infinite series, the basic problem is to determine whether the series is convergent or divergent. Since it is usually very difficult to obtain an explicit expression for the sequence of partial sums of the series, a variety of tests have been developed to determine convergence or divergence. An important test is provided by the following theorem.

3 THEOREM

If the infinite series $\sum_{n=0}^{\infty} a_n$ is convergent, then $\lim_{n \to \infty} a_n = 0$.

Let $\{a_n\}$ be a sequence such that $a_n > 0$ for all nonnegative integers n. Then the infinite series $\sum_{n=0}^{\infty} a_n$ is called a *positive series*. The infinite series $\sum_{n=0}^{\infty} b_n$, where $b_n = (-1)^n a_n$, is called an *alternating series*. Most of the series studied in calculus, and most of the series which will arise in this text, belong to one of these two types. Restricting our attention to these two types of infinite series, we list the tests for convergence which will be most useful in our work.

4 COMPARISON TEST

Let $\sum_{n=0}^{\infty} a_n$ and $\sum_{n=0}^{\infty} b_n$ be positive series. If $\sum_{n=0}^{\infty} b_n$ is convergent and $a_n \le b_n$ for all n, then $\sum_{n=0}^{\infty} a_n$ is convergent. Let $\sum_{n=0}^{\infty} c_n$ and $\sum_{n=0}^{\infty} d_n$ be positive series. If $\sum_{n=0}^{\infty} c_n$ is divergent and $c_n \le d_n$ for all n, then $\sum_{n=0}^{\infty} d_n$ is divergent.

5 RATIO TEST

Let $\sum_{n=0}^{\infty} a_n$ be a positive series and assume that

$$\lim_{n \to \infty} \frac{a_{n+1}}{a_n} = L$$

exists. Then

(i) $\sum_{n=0}^{\infty} a_n$ is convergent if $L < 1$

(ii) $\sum_{n=0}^{\infty} a_n$ is divergent if $L > 1$

(iii) either convergence or divergence is possible if $L = 1$.

6 ALTERNATING SERIES TEST

Let $\sum_{n=0}^{\infty} b_n$ be an alternating series. If $|b_{n+1}| \le |b_n|$ and $\lim_{n \to \infty} b_n = 0$, then the series is convergent.

An infinite series $\sum_{n=0}^{\infty} c_n$ is *absolutely convergent* if the positive series $\sum_{n=0}^{\infty} |c_n|$ is convergent. It can be shown that an absolutely convergent series is convergent. If $\sum_{n=0}^{\infty} c_n$ is convergent, but $\sum_{n=0}^{\infty} |c_n|$ is divergent, then $\sum_{n=0}^{\infty} c_n$ is said to be *conditionally convergent*.

7 Example

The series $\sum_{n=0}^{\infty} 1/(n + 1)$ is divergent. The alternating series $\sum_{n=0}^{\infty} (-1)^n/(n + 1)$ is convergent. Thus $\sum_{n=0}^{\infty} (-1)^n/(n + 1)$ is conditionally convergent.

Let a be a real number. A *power series in powers of* $(x - a)$ is an expression of the form

$$(8) \qquad \sum_{n=0}^{\infty} a_n(x - a)^n$$

The terms of the sequence $\{a_n\}$ are called the *coefficients* of the power series (8). For the special case $a = 0$, the power series (8) has the form

$$(9) \qquad \sum_{n=0}^{\infty} a_n x^n$$

Consider the power series (8). If we replace the variable x by a real number r, then we get the infinite series

$$\sum_{n=0}^{\infty} a_n(r - a)^n$$

and so the question arises as to the convergence or divergence of this infinite series. There is always at least one value of x at which the power series (8) converges, namely, $x = a$. Thus, we are most interested in determining whether there are any other values of x at which (8) converges. The principal result in this direction is the following theorem.

10 THEOREM

Given the power series (8). If this series converges at $x = r$, $r \neq a$, then the series is absolutely convergent for all x such that $|x - a| < r$. If the series diverges at $x = s$, then the series is divergent for all x such that $|x - a| > s$.

___

The ratio test is a useful tool in studying power series. In particular, for any real number x, we have from (5)

$$\lim_{n \to \infty} \frac{|a_{n+1}(x - a)^{n+1}|}{|a_n(x - a)^n|} = \lim_{n \to \infty} \frac{|a_{n+1}|}{|a_n|} |x - a|$$

Thus, suppose

$$\lim_{n \to \infty} \frac{|a_{n+1}|}{|a_n|} = L$$

exists. Then

(i) the series is absolutely convergent for all x such that $|x - a| < 1/L$ if $L \neq 0$, and

(ii) the series is absolutely convergent for all x if $L = 0$.
     The quantity

$$R = \begin{cases} \dfrac{1}{L} & \text{if } L \neq 0 \\ \infty & \text{if } L = 0 \end{cases}$$

is called the *radius of convergence* of the power series, and the interval $(a - R, a + R)$ is called the *interval of convergence*. The power series may or may not converge at the endpoints $a - R$, $a + R$.

11 Examples

a. For the power series

$$\sum_{n=0}^{\infty} \frac{x^n}{n!}$$

the radius of convergence is $R = \infty$ since

$$L = \lim_{n \to \infty} \left| \frac{a_{n+1}}{a_n} \right| = \lim_{n \to \infty} \frac{n!}{(n+1)!} = \lim_{n \to \infty} \left(\frac{1}{n+1} \right) = 0$$

b. For the power series

$$\sum_{k=1}^{\infty} \frac{(-1)^{k+1}(x-1)^k}{k} = \sum_{n=0}^{\infty} \frac{(-1)^n (x-1)^{n+1}}{n+1}$$

we have

$$L = \lim_{n \to \infty} \frac{\left| (-1)^{n+1} \dfrac{1}{n+1} \right|}{\left| (-1)^n \dfrac{1}{n} \right|} = \lim_{n \to \infty} \frac{n}{n+1} = 1$$

Thus $R = 1$, and the series converges for all $x \in J = (0, 2)$. It is easy to show that the power series converges at $x = 2$ and diverges at $x = 0$.

If the power series

$$\sum_{n=0}^{\infty} a_n (x - a)^n$$

has radius of convergence R, then for each $c \in J = (a - R, a + R)$,

$$\sum_{n=0}^{\infty} a_n (c - a)^n$$

is a real number. Thus the power series defines a real-valued function f on J. We shall write

(12) $$f(x) = \sum_{n=0}^{\infty} a_n(x - a)^n, \qquad x \in J$$

Note that $f(a) = a_0$.

We now consider a very important property of power series which will be used extensively in Chapter 7.

13 THEOREM

If R is the radius of convergence of the power series (12), then the function f is differentiable on $J = (a - R, a + R)$,

(14) $$Df(x) = \sum_{n=1}^{\infty} na_n(x - a)^{n-1}, \qquad x \in J$$

and (14) also has R as its radius of convergence.

From (14), note that $Df(a) = a_1$. Using induction, we obtain the following general result.

15 COROLLARY

If R is the radius of convergence of the power series (12), then the function f has derivatives of all orders on J, and for each positive integer k,

(16) $$D^k f(x) = \sum_{n=k}^{\infty} n(n - 1) \cdots (n - k + 1)a_n(x - a)^{n-k}$$

The series in (16) has R as its radius of convergence, and $D^k f(a) = k!a_k$.

By use of this corollary, we can calculate the coefficients in the power series (12) in terms of the derivatives of f. Thus another representation of (12) is:

(17) $$f(x) = \sum_{n=0}^{\infty} \frac{D^n f(a)}{n!} (x - a)^n$$

and this expression should be compared with Theorem 2, Section 2.

We now consider a converse question, namely, suppose we are given a real-valued function g whose domain contains an interval of the form $J = (a - R, a + R)$. Is it possible to represent g using a power series of the form (12)? First of all, in order for g to have a power series representation in powers of $(x - a)$, it is clear that g will have to be infinitely differentiable on J. As a generalization of Theorem 2, Section 2, we have the following theorem.

18 THEOREM

Let $g \in \mathscr{C}^{\infty}(J), J = (a - R, a + R)$. Then g has the power series representation

(19) $$g(x) = \sum_{n=0}^{\infty} \frac{D^n g(a)}{n!} (x - a)^n, \qquad x \in J$$

if and only if

$$\lim_{n \to \infty} \frac{D^n g(\xi_n)}{n!} (x - a)^{n+1} = 0$$

where, for each n, ξ_n is between a and x.

The power series representation (19) is called the *Taylor series expansion of g at a*. In the special case $a = 0$, (19) becomes

(20) $$g(x) = \sum_{n=0}^{\infty} \frac{D^n g(0)}{n!} x^n$$

and this is called the *Maclaurin series expansion of g*.

21 Examples

    a. The Maclaurin series expansions of three elementary transcendental functions are:

$$e^x = \sum_{n=0}^{\infty} \frac{x^n}{n!}$$

$$\sin x = \sum_{n=0}^{\infty} \frac{(-1)^n x^{2n+1}}{(2n + 1)!}$$

$$\cos x = \sum_{n=0}^{\infty} \frac{(-1)^n x^{2n}}{(2n)!}$$

    The radius of convergence for each of these series is $R = \infty$.

    b. The Taylor series expansion for $\ln(x)$ in powers of $(x - 1)$ is:

$$\ln(x) = \sum_{n=0}^{\infty} \frac{(-1)^n (x - 1)^{n+1}}{n + 1}$$

    The radius of convergence is $R = 1$. This series converges at $x = 2$, but not at $x = 0$.

22 DEFINITION

A function f defined on an interval containing the point $x = a$ is said to be *analytic at x = a* if f has a Taylor series expansion in powers of $(x - a)$ with a positive radius of convergence.

We conclude this section by stating the basic algebraic properties of power series. These properties will be used throughout Chapter 7. Suppose

$$f(x) = \sum_{n=0}^{\infty} a_n(x - a)^n \quad \text{and} \quad g(x) = \sum_{n=0}^{\infty} b_n(x - a)^n$$

and suppose each series converges on an open interval J.

23 THEOREM

Uniqueness: If $f(x) = g(x)$ for all $x \in J$, that is, if

$$\sum_{n=0}^{\infty} a_n(x - a)^n = \sum_{n=0}^{\infty} b_n(x - a)^n$$

then $a_n = b_n$ for all n.

This result follows from the fact that the powers of $(x - a)$ are linearly independent functions.

24 THEOREM

Addition: The sum of the series f and g is the power series

$$f(x) + g(x) = \sum_{n=0}^{\infty} (a_n + b_n)(x - a)^n$$

and this series converges for all $x \in J$.

This theorem implies that the sum of two power series in powers of $(x - a)$ is calculated by adding the corresponding terms of the two series.

25 THEOREM

Multiplication: The product of the series f and g is the power series

$$f(x) \cdot g(x) = \sum_{n=0}^{\infty} c_n(x - a)^n$$

where $c_n = \sum_{k=0}^{n} a_k b_{n-k}$, and this series converges for all $x \in J$.

This theorem is the generalization to power series of the familiar rule for multiplying two polynomials.

LINEAR DIFFERENTIAL EQUATIONS

Section 7 Introduction

The purpose of this text is to present the basic theory of differential and difference equations, together with a variety of applications of this theory. Since we shall be concerned first with the theory of certain types of differential equations, we begin our discussion with an intuitive description of the concept of a differential equation and related concepts. A corresponding discussion for difference equations, together with appropriate definitions, is given in Chapter 5.

By a *differential equation* we mean, simply, an equation which contains an unknown function together with one or more of its derivatives.

1 Examples

a. A differential equation that occurs in numerous applications is:

(A) $\quad Dy = ky$

where k is a constant.

It has been observed that the rate of growth of certain biological organisms is proportional to the number of organisms present. This observation can be formulated as equation (A), where $y = y(t)$ is the number of organisms present at time t and the constant k is the growth rate.

Also, a similar situation exists with the rate of decay of a radioactive material, that is, the rate of change of the amount of material is proportional to the amount present.

This equation even occurs in commercial applications. Many savings banks advertise that they pay interest compounded continuously. Evidently this means that the rate at which your savings grow is proportional to the amount on deposit at time t.

b. Newton's law of motion, namely, that force equals mass times acceleration, is the source of many differential equations. The equation

(B) $mg = m(D^2 y)$

where m and g are constants, is a particular form of this physical law that governs the motion of a freely falling body. Here $y = y(t)$ is the position of the body at time t, m is the mass of the body, and g is a gravitational constant. In a more refined model of a falling body, we might assume that there is a retarding force that is proportional to the velocity of the body and this force is caused by air resistance. This assumption results in the equation

(C) $mg - k(Dy) = m(D^2 y)$

where k is a constant.

c. A somewhat more involved application of Newton's law yields the differential equation

(D) $$\frac{\partial^2 y}{\partial t^2} = k^2 \frac{\partial^2 y}{\partial x^2}$$

where k is a constant.

If a vibrating string is stretched between two supports on the x-axis, and $y(x, t)$ denotes for each x the y-coordinate of the string at time t, then an idealized string satisfies equation (D).

Detailed analyses of additional applications involving differential equations are contained in Chapter 11.

As suggested by Example 1, differential equations can be classified into two broad categories which are determined by the type of unknown function appearing in the equation. In particular, if the unknown function depends on a single variable (Examples 1a and 1b), then the equation is said to be an *ordinary differential equation*, since only ordinary derivatives of the unknown function will appear in the equation. On the other hand, if the unknown function depends on more than one independent variable, as in Example 1c, so that any derivatives of the unknown function which appear in the equation will be partial derivatives, then the equation is a *partial differential equation*.

Differential equations, both ordinary and partial, are also classified according to the highest-order derivative of the unknown function appearing in the equation. Specifically, the *order* of a differential equation is defined to be the order of the highest derivative of the unknown function.

2 Examples

Referring to Example 1 with respect to order and type, we find that equation (A) is a first-order ordinary differential equation, equations (B) and (C) are second-order ordinary differential equations, and equation (D) is a second-order partial differential equation.

The obvious question which we wish to investigate is that of "solving" a given differential equation, and therefore we must specify what we shall mean by the term "solution." We assume that the reader has had some experience in "solving" certain types of equations. For example, in algebra, one usually devotes some time to studying quadratic equations:

$$ax^2 + bx + c = 0$$

The objective here is to determine a number r such that when the unknown x is replaced by the number r, the left-hand side of the equation reduces to 0. As a specific example, for the equation

$$x^2 - x - 6 = 0$$

either the number 3 or the number -2 has the required property, and we say that the equation has two solutions, -2 and 3. On the other hand, the quadratic equation

$$x^2 + 1 = 0$$

does not have any real solutions.

In contrast to algebraic equations, such as those above where the unknown is replaced by a number, the objective in differential equations is to determine a function which will satisfy the equation. Accordingly, then, we say that a *solution of a differential equation* is a function defined on some interval (in the case of an ordinary differential equation) or on some region in $\mathscr{V}_n$, $n > 1$ (in the case of a partial differential equation), with the property that the equation reduces to an identity when the unknown is replaced by this function. A word of caution to the reader is appropriate here. Although the objective is to determine a function which satisfies a given differential equation, this can actually be done in only very special cases. We present the most common of these special cases in this text (first-order linear equations, Section 8; linear equations with constant coefficients, Chapter 3; special first-order equations, Chapter 10). In practice, when given a differential equation, one usually tries to analyze the equation in order to determine properties which solutions must have, and then one attempts to approximate the desired solution to within some prescribed accuracy. A basic understanding of the theory of differential equations is necessary in establishing properties of solutions, and a familiarity with the theory of difference equations is essential in developing numerical methods for approximating solutions.

3 Examples

a. Consider the differential equation

$$D^2 y = -32$$

The function $y(x) = -16x^2 + 2x + 1$ defined on $(-\infty, \infty)$ is a solution since $D^2 y(x) = -32$. In fact, it can be verified that $y(x) = -16x^2 + C_1 x + C_2$ defined on $(-\infty, \infty)$ is a solution for any choice of constants C_1 and C_2.

b. Consider the differential equation

$$\frac{\partial^2 u}{\partial y^2} = k^2 \frac{\partial^2 u}{\partial x^2}$$

The function $u(x, y) = \cos \lambda[x - ky]$, where λ is a positive constant is a solution since $\partial^2 u/\partial x^2 = -\lambda^2 \cos \lambda[x - ky]$ and $\partial^2 u/\partial y^2 = -\lambda^2 k^2 \cos \lambda[x - ky]$, so that the equation reduces to an identity when u is replaced by $\cos \lambda[x - ky]$.

c. Consider the differential equation

$$xD^2 y - Dy - 4x^3 y = 0$$

The function $y_1(x) = e^{x^2}$ is a solution since $Dy_1(x) = 2xe^{x^2}$, $D^2 y_1(x) = 2e^{x^2} + 4x^2 e^{x^2}$, and

$$x[2e^{x^2} + 4x^2 e^{x^2}] - 2xe^{x^2} - 4x^3 e^{x^2} = 0$$

In the same manner, it is easy to verify that

$$y(x) = C_1 e^{x^2} + C_2 e^{-x^2}$$

is a solution for any choice of constants C_1 and C_2.

4 EXERCISES

1. Classify the following differential equations with respect to type (that is, ordinary or partial) and order.
   a. $(Dy)^2 + xy\,Dy = \sin x$
   b. $D^2 y + e^{xy} = \tan x$
   c. $(\partial^2 u/\partial x^2) + 2(\partial^2 u/\partial x \partial y) + (\partial^2 u/\partial y^2) = 0$
   d. $(D^2 y)^3 + y = x$
   e. $D^3 y - 5x\,Dy + y = e^x - 1$
   f. $\partial u/\partial x = k(\partial u/\partial y)$

2. In each of the following, determine whether or not the given function (or functions) is a solution of the associated differential equation.
   a. $D^2 y + y = 0$; $y_1(x) = \sin x$, $y_2(x) = \cos(2x)$
   b. $D^3 y + Dy = e^x$; $y_1(x) = 1 + \sin x + \frac{1}{2}e^x$, $y_2(x) = \cos x$
   c. $x\,D^2 y + Dy = 0$; $y_1(x) = \ln(1/x)$, $y_2(x) = x^2$
   d. $(x + 1)\,D^2 y + x\,Dy - y = (x + 1)^2$; $y_1(x) = e^{-x} + x^2 + 1$
   e. $D^3 y - 5\,D^2 y + 6\,Dy = 0$; $y(x) = c_1 e^{2x} + c_2 e^{3x}$ (c_1, c_2 constants)
   f. $(\partial^2 u/\partial x^2) - (\partial^2 u/\partial y^2) = 4x + 3\cos(2y)$; $u(x, y) = (2x^3/3) - (3\cos(2y)/4) + (x - y)^2$
   g. $D^2 y - 2\,Dy + y = 0$; $y(x) = e^x(C_1 + C_2 x)$ (C_1, C_2 constants)
   h. $D^2 y - 2\,Dy + y = 0$; $y_1(x) = e^x(1 + x)$

    i. $D^2y - y = 2 - x$; $y_1(x) = e^x + x - 4$, $y_2(x) = e^{2x} + x - 4$

    j. $y\,Dy = e^{2x}$; $y_1(x) = \sqrt{e^{2x} + 1}$, $y_2(x) = e^x$

3. Determine by inspection a solution of each of the following differential equations.

    a. $Dy = 2y$ b. $Dy = 2x + \cos x$ c. $x\,Dy - y = 0$

4. In each of the following, determine, if possible, values of r for which the given differential equation has a solution of the form $y(x) = e^{rx}$.

    a. $D^2y - y = 0$ b. $Dy - y = 0$ c. $D^3y - 3\,D^2y + 2\,Dy = 0$

5. In each of the following, determine, if possible, values of r for which the given differential equation has a solution of the form $y(x) = x^r$.

    a. $x\,D^2y + Dy = 0$ b. $x^2\,D^2y + x\,Dy - y = 0$

    c. $4x^2\,D^2y - 4x\,Dy + 3y = 0$

        Readers will recall from their experiences in calculus that the study of functions of several variables (graphing, calculating derivatives, finding extrema, integration, etc.) is generally more difficult and complicated than the corresponding study of functions of a single variable. As a result, one might expect that partial differential equations would be more complicated than ordinary differential equations. This is indeed the case, and since the intent of this text is to introduce the reader to the basic theory of differential equations, together with a corresponding development of the basic theory of difference equations, we shall confine our attention to ordinary differential equations in the work which follows. Hereafter, when the term "differential equation" is used, it shall be interpreted as meaning ordinary differential equation.

        As indicated by Example 3a, a given differential equation may have many solutions. As another simple example, and one which is familiar from calculus, consider the first-order differential equation

$$Dy = f(x)$$

where $f(x)$ is a given continuous function. Finding a solution of this differential equation is equivalent to finding a function whose derivative is f, that is, finding an antiderivative of f. Also recall from calculus that if we can find one antiderivative F of f, then the family of functions $y(x) = F(x) + c$, where c is an arbitrary constant, represents all antiderivatives of f. These observations were made in Section 2.

5 Example

        Consider the differential equation

$$Dy = 2x$$

        Clearly the function $y(x) = x^2$ is a solution, and as indicated by our discussion above, each member of the family of functions $y(x) = x^2 + C$, where C is an arbitrary constant, is a solution. This family of functions is called a *one-parameter family*, the constant C being the "parameter" which distinguishes members of the family.

6 Example

Consider the second-order differential equation

$$D^2 y + y = 0$$

It is easy to verify that each of the functions $y_1(x) = \sin x$ and $y_2(x) = \cos x$ is a solution. In fact, for any real number C_1 and any real number C_2, the function $y(x) = C_1 \sin x + C_2 \cos x$ is a solution. Equivalently, each member of the *two-parameter family* of functions

$$y(x) = C_1 \sin x + C_2 \cos x$$

where C_1 and C_2 are arbitrary constants, is a solution of the differential equation.

Consider the result in Example 5, and suppose we ask this question: In the one-parameter family of solutions, $y(x) = x^2 + C$, is there at least one member which passes through the point $(1, 2)$? Replacing x by 1 and y by 2, we find $C = 1$. Thus $y = x^2 + 1$ is a solution of the differential equation which passes through the point $(1, 2)$. It is easily verified that this function is the *only* member of the one-parameter family $y(x) = x^2 + C$ which passes through the point $(1, 2)$. In general, given a point (x_0, y_0) in the plane, the function $y(x) = x^2 + (y_0 - x_0^2)$ is a solution of the differential equation which passes through (x_0, y_0), and it is the only member of the one-parameter family of solutions $y(x) = x^2 + C$ which has this property. Thus prescribing a point in the plane through which we want a solution to pass determines uniquely a member of the one-parameter family of solutions.

We found in Example 6 that the given differential equation had a two-parameter family of solutions. The fact that the first-order equation had a *one-parameter* family of solutions and the second-order equation had a *two-parameter* family of solutions is no coincidence, and it forms the basis for our work in the remainder of this chapter. Continuing with the discussion in the preceding paragraph, what conditions can we prescribe in order to determine (perhaps uniquely) a certain member of the two-parameter family $y(x) = C_1 \sin x + C_2 \cos x$? Proceeding as above, we can ask if there is a member of the family which passes through a given point in the plane. For example, the reader can verify that $y(x) = -6 \sin x$ is a solution of the second-order equation in Example 6 which passes through the point $(\pi/6, -3)$. In fact, each member of the one-parameter family $z(x) = (-6 - C_2\sqrt{3}) \sin x + C_2 \cos x$, a subfamily of the two-parameter family $y(x) = C_1 \sin x + C_2 \cos x$, is a solution of the differential equation which passes through $(\pi/6, -3)$. Thus prescribing a point in the plane is not enough to determine a unique member of the two-parameter family. This condition determined a one-parameter subfamily. Suppose, instead, we ask if there is a member of the two-parameter family $y(x) = C_1 \sin x + C_2 \cos x$ which passes through $(\pi/6, -3)$ with slope equal to 1. That is, we want a solution $y(x)$ satisfying the conditions $y(\pi/6) = -3$ and $Dy(\pi/6) = 1$. For each member of the two-parameter family, we have

$$y(x) = C_1 \sin x + C_2 \cos x$$

and

$$Dy(x) = C_1 \cos x - C_2 \sin x$$

Thus the conditions $y(\pi/6) = -3, Dy(\pi/6) = 1$, lead to the pair of equations

$$-3 = C_1 \cdot \frac{1}{2} + C_2 \cdot \frac{\sqrt{3}}{2}$$

$$1 = C_1 \cdot \frac{\sqrt{3}}{2} - C_2 \cdot \frac{1}{2}$$

in the "unknowns" C_1 and C_2. Solving for C_1 and C_2, we find

$$C_1 = \frac{\sqrt{3} - 3}{2} \quad \text{and} \quad C_2 = \frac{(1 + 3\sqrt{3})}{2}$$

Since we had a two-parameter family of solutions, we needed to specify two conditions (leading to two equations in the two "unknowns" C_1 and C_2) in order to determine a particular member of the two-parameter family.

Roughly speaking, an nth-order differential equation requires n integrations, each one of which introduces a constant of integration. Thus we expect, intuitively, to find an n-parameter family of solutions of an nth-order equation, the parameters being the constants of integration; and in order to determine a particular member of the n-parameter family, we would anticipate having to specify n conditions.

7 Example

Perhaps the simplest example of an nth-order equation which can be solved by elementary means (i.e., using calculus) is:

$$D^n y = 0$$

An n-parameter family of solutions is:

$$y(x) = C_1 x^{n-1} + C_2 x^{n-2} + \cdots + C_{n-1} x + C_n$$

each member of which is a polynomial of degree $n - 1$ or less. The function

$$y(x) = \frac{x^{n-1}}{(n-1)!}$$

is the only member of this family which satisfies the conditions

$$0 = y(0) = Dy(0) = \cdots = D^{n-2} y(0); \qquad D^{n-1} y(0) = 1$$

8 DEFINITION

An *nth-order initial-value problem* consists of an nth-order differential equation together with n conditions of the form

$$y(x_0) = a_0, \quad Dy(x_0) = a_1, \ldots, D^{n-1}y(x_0) = a_{n-1}$$

on a solution.

9 Examples

a. The differential equation $Dy = 2x$, together with the condition $y(1) = 2$, is a first-order initial-value problem. The function $y(x) = x^2 + 1$ is a solution. In fact, this function is the only solution of the initial-value problem, that is, $y(x) = x^2 + 1$ is the unique solution.

b. The problem

$$D^3y + [Dy]^2 = x + 1; \quad y(0) = 1, \quad D^2y(0) = 6$$

is *not* an initial-value problem since $Dy(0)$ is not specified.

c. Consider the initial-value problem

$$(1 - x^2)D^2y + 2x\,Dy - 2y = 0; \quad y(2) = 1, \quad Dy(2) = -1$$

A two-parameter family of solutions of the differential equation is:

$$y(x) = C_1^- x + C_2(x^2 + 1)$$

Thus

$$Dy(x) = C_1 + 2xC_2$$

By applying the initial conditions, we obtain the pair of equations

$$y(2) = 1 = 2C_1 + 5C_2$$
$$Dy(2) = -1 = C_1 + 4C_2$$

This pair of equations has the solution $C_1 = 3$, $C_2 = -1$. Therefore,

$$y(x) = 3x - (x^2 + 1)$$

is a solution of the initial-value problem.

The next examples point out two basic complexities of initial-value problems, and they introduce the two questions which are of fundamental importance in the study of differential equations. First, if we are given an initial-value problem, is there a solution? This is the question of *existence of solutions*. Second, if we know that a given initial-value problem has a solution, is this the only solution, or is there more than one solution? This is the question of *uniqueness of solutions*.

10 Example

Consider the second-order initial-value problem

$$x^2 D^2 y - 2x Dy + 2y = 0; \qquad y(0) = 1, \quad Dy(0) = -3$$

A two-parameter family of solutions of the differential equation is:

$$y(x) = C_1 x + C_2 x^2$$

Note that every member of this two-parameter family passes through $(0, 0)$; that is, if we replace x by 0, then $y = 0$, regardless of any assigned values of the constants C_1 and C_2. Thus, *no* member of this two-parameter family passes through the point $(0, 1)$, and we cannot find a solution of the initial-value problem in the two-parameter family $y = C_1 x + C_2 x^2$.

11 Example

Given the first-order initial-value problem

$$Dy = 2\sqrt{y}; \qquad y(0) = 0$$

on $[0, \infty)$. The function $y(x) = x^2$ is a solution of the differential equation since $Dy(x) = 2x = 2\sqrt{y(x)}$. Since $y(0) = 0$, $y(x) = x^2$ is a solution of this initial-value problem. It is also easy to verify that the function $y \equiv 0$ is a solution of the initial-value problem. Thus we have an example of an initial-value problem which has more than one solution. In fact, for each nonnegative number r, if we define the function $y_r(x)$ (see the graph shown here) by

$$y_r(x) = \begin{cases} 0 & 0 \le x \le r \\ (x - r)^2 & x > r \end{cases}$$

then $y_r(x)$ is a solution of the initial-value problem.

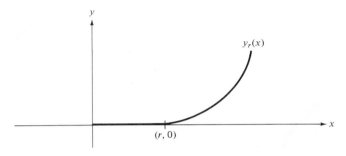

In the next section, we will consider a special class of initial-value problems, and we will establish that each such initial-value problem has a unique solution. Beyond that, we can only assure the reader that for the general class of initial-value problems under consideration in this chapter, the appropriate existence and uniqueness theorems can be established. These theorems will be stated, but their proofs will be omitted because they require more advanced mathematical techniques.

12 EXERCISES

1. Show that each member of the one-parameter family of functions

$$y(x) = Ce^{5x}$$

is a solution of the differential equation $Dy - 5y = 0$. Find a solution of the differential equation satisfying the initial condition $y(0) = 2$.

2. Show that each member of the two-parameter family of functions

$$y(x) = C_1 e^{2x} + C_2 e^{-x}$$

is a solution of the differential equation $D^2 y - Dy - 2y = 0$. Find a member of this family satisfying the initial conditions $y(0) = 2, Dy(0) = 1$.

3. Show that each member of the one-parameter family of functions

$$y(x) = \frac{1}{Ce^x + 1}$$

is a solution of the differential equation $Dy + y = y^2$. Find a solution of the differential equation satisfying the initial condition $y(1) = -1$.

4. Show that each member of the three-parameter family of functions

$$y(x) = C_1 x^2 + C_2 x + C_3$$

is a solution of the differential equation $D^3 y = 0$. Find a member of this family satisfying the initial conditions $y(2) = Dy(2) = D^2 y(2) = 0$.

5. Show that each member of the two-parameter family of functions

$$y(x) = C_1 \sin 3x + C_2 \cos 3x$$

is a solution of the differential equation $D^2 y + 9y = 0$. Find a solution of the differential equation satisfying the initial conditions $y(\pi/2) = Dy(\pi/2) = 1$.

6. Show that each member of the two-parameter family

$$y(x) = C_1 x^2 + C_2 x^2 \ln x$$

is a solution of the differential equation $x^2 D^2 y - 3x Dy + 4y = 0$. Find a member of this family satisfying the initial conditions $y(1) = 0, Dy(1) = 1$. Is there a member of the family satisfying the initial conditions $y(0) = Dy(0) = 0$? Is there a member of the family satisfying $y(0) = 0, Dy(0) = 1$? If not, why not?

7. Show that each member of the two-parameter family

$$y(x) = C_1 x + C_2 x^{1/2}$$

is a solution of the differential equation $2x^2 D^2 y - x Dy + y = 0$. Find a solution of the differential equation satisfying the initial conditions $y(4) = 1, Dy(4) = -2$. Is there a member of the family which satisfies the initial conditions $y(0) = 1, Dy(0) = 2$? If not, why not?

8. Each member of the two-parameter family of functions

$$y(x) = C_1 \sin x + C_2 \cos x$$

is a solution of the differential equation $D^2 y + y = 0$.

a. Determine whether there is a member of this family which satisfies the conditions

$$y(0) = 0; \qquad y(\pi) = 0$$

b. Show that the zero function is the only member of this family which satisfies the conditions

$$y(0) = 0; \qquad y\left(\frac{\pi}{2}\right) = 0$$

9. An nth-order differential equation together with conditions which are specified at two, or more, points on an interval J is called a *boundary-value problem*. In particular, the problems

$$D^2y + y = 0; \qquad y(0) = y(\pi) = 0$$

$$D^2y + y = 0; \qquad y(0) = y\left(\frac{\pi}{2}\right) = 0$$

are called *two-point boundary-value problems*. Boundary-value problems generally require a more advanced analysis, and a detailed study of them is beyond the scope of this text. The reader should note that the term "initial-value problem" signals that the conditions which are placed on a solution are all specified at one point, in contrast to boundary-value problems where conditions are specified at more than one point.

For each real number r, each member of the two-parameter family

$$y(x) = C_1 \sin(rx) + C_2 \cos(rx)$$

is a solution of the differential equation $D^2y + r^2y = 0$. Determine the numbers r such that the two-point boundary-value problem

$$D^2y + r^2y = 0; \qquad y(0) = y(\pi) = 0$$

has a nontrivial solution.

10. A radioactive material decays at a rate r proportional to the amount $y(t)$ present at time t. Thus $Dy = -ry$, where the minus sign signifies that the amount of material is decreasing.

a. Show that $y(t) = y_0 e^{-rt}$ is a solution of the differential equation, where y_0 is the initial mass of the material.

b. What is $\lim\limits_{t \to \infty} y(t)$?

c. The time required for the mass to decrease to one-half the initial mass is called the *half-life* of the substance. If the half-life of a certain substance is 100 years, then how much of the material is left after 10 years?

11. Suppose that a certain savings bank pays interest at a rate of 6% compounded continuously. Then the rate of growth of an investment is given by $Dy = 0.06y$.

a. Show that $y(t) = y_0 e^{0.06t}$ is a solution of the differential equation, where y_0 is the amount of the initial deposit.

b. How much money would be on deposit at the bank at the end of one year if $1000.00 was the initial deposit?

c. Which is preferable, 6% interest compounded continuously, or 7% simple interest (that is, compounded once per year)?

12. The differential equation

(A) $$D^2y + \frac{k}{m} Dy = g$$

where k, m, and g are positive constants, was given in Example 1b as the description of the position of a freely falling body subject to a retarding force proportional to its velocity.

a. Let $v(t) = Dy(t)$ be the velocity of the body. Show that the differential equation (A) can be written as

(B) $$Dv + \frac{k}{m}v = g$$

b. Show that $v(t) = (mg/k) + [v_0 - (mg/k)]e^{-kt/m}$, where $Dy(0) = v_0$ is the initial velocity, is a solution of (B).

c. Show that $\lim_{t \to \infty} v(t) = mg/k$. This is called the *terminal velocity* of the body.

d. Show that a solution of (A) is given by

$$y(t) = y_0 + \frac{mg}{k} t + \frac{m}{k}\left(v_0 - \frac{mg}{k}\right)(1 - e^{-kt/m})$$

where $y(0) = y_0$ is the initial position of the body.

Section 8 First-Order Linear Differential Equations

What seems to be a reasonable approach to the study of differential equations is to classify them insofar as possible, and to attempt to find the general properties of each class of equations. We use this approach in this chapter, beginning here with the simplest class of differential equations for which an explicit representation of the solutions can be obtained. However, before defining this class of differential equations, we introduce a notational convention that is used throughout the remainder of the book.

1 DEFINITION

The *zero function* is denoted by $\emptyset$ and is defined to be that function which assigns the value 0 to each member of its domain, that is, $\emptyset(x) = 0$ for each x in the domain of $\emptyset$. The domain of $\emptyset$ is taken to be whatever set of real numbers is appropriate in the context in which it is used.

---

2 DEFINITION

A *first-order linear differential equation* is an equation of the form

$$q(x) Dy + r(x)y = g(x)$$

where q, r, and g are continuous functions on some interval J and $q \neq \emptyset$ on J. If $g = \emptyset$ on J, then the equation is called *homogeneous*; *otherwise*, it is called *nonhomogeneous*.

---

3 Examples

a. Consider the equation

$$Dy - y = 2e^x$$

on $J = (-\infty, \infty)$. This is a first-order linear nonhomogeneous differential equation. Here $q(x) = 1$, $r(x) = -1$, and $g(x) = 2e^x$. A one-parameter family of solutions of this equation is:

$$y(x) = ce^x + 2xe^x$$

b. Consider the equation

$$x\,Dy + y = x$$

on $J = [-1, 1]$. This is a first-order linear nonhomogeneous equation. In this example, $q(x) = x$, $r(x) = 1$, and $g(x) = x$. A one-parameter family of solutions is:

$$y(x) = c\left(\frac{1}{x}\right) + \frac{x}{2}$$

c. Consider the equation

$$(x - 1)\,Dy - 2y = \varnothing$$

on $J = (-\infty, \infty)$. This is a first-order linear homogeneous equation in which $q(x) = x - 1$, $r(x) = -2$, and $g = \varnothing$. A one-parameter family of solutions is:

$$y(x) = c(x - 1)^2$$

An obvious question concerns the significance of the term "linear" in the description of the differential equation in (2). Referring to the discussion of linear transformations and linear operators in Section 4, define the operator $L_1 : \mathscr{C}^1(J) \to \mathscr{C}(J)$ by

$$L_1[F(x)] = q(x)\,D[F(x)] + r(x)F(x)$$

for any $F \in \mathscr{C}^1(J)$. Then it is easy to verify that L_1 is a linear operator. This explains the use of "linear" in describing the equation in (2). As a specific example, consider the operator

$$L_1 = (x - 1)D - 2I$$

(I is the identity operator) associated with the equation in Example 3c. The functions $F_1(x) = e^{2x}$ and $F_2(x) = x$ are members of $\mathscr{C}^1(J)$, $J = (-\infty, \infty)$, and

$$\begin{aligned}
L_1[e^{2x} + x] &= (x - 1)D[e^{2x} + x] - 2[e^{2x} + x] \\
&= (x - 1)[2e^{2x} + 1] - 2e^{2x} - 2x \\
&= 2xe^{2x} - 4e^{2x} - x - 1 = L_1[e^{2x}] + L_1[x]
\end{aligned}$$

The objective of this section is to characterize the set of all solutions of first-order linear differential equations. Examples 3b and 3c can be used to illustrate that some difficulties arise when zero is in the range of the function q, the coefficient of Dy. The numbers in J which are solutions of $q(x) = 0$ are called *singular points* of the differential equation defined in (2). The equation in Example 3a has no singular points, while the equation in Example 3b has a singular point at $x = 0$, and the equation in Example 3c has a singular point at $x = 1$. Furthermore, we see that in Example 3b the only member of the one-parameter family of solutions that is defined at the singular point $x = 0$ is $y(x) = x/2$; whereas in Example 3c all members of the one-parameter family of solutions are defined at the singular point $x = 1$. Note, however, that each solution indicated in Example 3c has the value 0 at $x = 1$, independent of the choice of the constant c. If the one-parameter family of functions characterizes all solutions of Example 3c, then we can evidently conclude that, for example, there is no solution of the initial-value problem

$$(x - 1)\, Dy - 2y = \varnothing; \qquad y(1) = 1$$

At the very least these examples imply that a fundamental question such as the existence of solutions of initial-value problems will need to be highly qualified depending on the presence and the nature of singular points on the interval J. Because of the complexities involved in analyzing differential equations with singular points, this topic is deferred to Chapter 7.

If the first-order linear equation defined in (2) has no singular points on J, then $q(x) \neq 0$ for all $x \in J$. In fact, we can conclude that q is either a positive-valued function on J or a negative-valued function defined on J. In either case, dividing by q, an equivalent equation can be obtained that has the form

(4) $Dy + p(x)y = f(x)$

where $p = r/q$, $f = g/q$, and, of course, p and f are both continuous on J. Clearly, any first-order linear differential equation without singular points can be put in the form of (4) and, conversely, the equation in (4) has no singular points. Thus, insofar as considering first-order linear equations without singular points, nothing is lost in assuming that they are in the form of (4), and this is the assumption in the remainder of this section.

5 Example

The equation in Example 3b has no singular points on the interval $J = (1, \infty)$, and on this interval the equivalent equation is:

$$Dy + \left(\frac{1}{x}\right) y = 1$$

Of course, other choices of an interval J are also possible.

6 EXERCISES

1. In each of the following, determine whether the given linear equation is homogeneous or nonhomogeneous, indicate the singular points (if any), and, when appropriate, specify a new interval J on which the equation has no singular points; put each equation in the form of (4).

   a. $(x^2 - 1)\,Dy - xy = e^x$ on $(-\infty, \infty)$ b. $x\,Dy + x^2 y = \varnothing$ on $[1, \infty)$

   c. $Dy - xy - x^2 = \varnothing$ on $(-1, 1)$ d. $x\,Dy - 2y = \varnothing$ on $[-2, 2]$

   e. $(\sin x)\,Dy + (\cos x)y = e^x$ on $(-\infty, \infty)$

2. In each of the following, show that the given one-parameter family of functions are solutions of the associated equation.

   a. $x\,Dy + y = x$ on $(1, \infty)$; $y(x) = (C/x) + (x/2)$

   b. $Dy - 2y = e^x$ on $(-\infty, \infty)$; $y(x) = Ce^{2x} - e^x$

   c. $x\,Dy - 2y = \varnothing$ on $(\tfrac{1}{2}, \infty)$; $y(x) = Cx^2$

   d. $Dy + (1/x)y = \sin x$ on $[1, \infty)$; $y(x) = (C/x) + (\sin x/x) - \cos x$

3. Determine a solution of each of the following initial-value problems. (Hint: See Problem 2.)

   a. $x\,Dy + y = x$; $y(2) = 4$ b. $Dy - 2y = e^x$; $y(0) = 2$

   c. $x\,Dy - 2y = \varnothing$; $y(1) = 0$ d. $Dy + (1/x)y = \sin x$; $y(\pi) = 0$

4. Given the linear operator $L = D + [2x/(x^2 + 1)]I$.

   a. Calculate $L[c/(x^2 + 1)]$ and $L[x^2 + 1]$.

   b. Determine a one-parameter family of solutions of $L[y] = 4x$.

   c. Find a solution of the initial-value problem

$$Dy + \left[\frac{2x}{1 + x^2}\right]y = 4x; \qquad y(1) = 3$$

Our first result solves the problem of existence and uniqueness for initial-value problems involving first-order linear homogeneous equations. Notice that the method of proof is constructive in nature, and therefore it provides a solution method for this class of equations.

7 THEOREM

Let $a \in J$ and let b be any real number. The initial-value problem

$$Dy + p(x)y = \varnothing; \qquad y(a) = b$$

has a unique solution, namely,

$$y(x) = b \exp\left[-\int_a^x p(t)\,dt\right]$$

Proof: Define the function h on J by $h(x) = \int_a^x p(t)\,dt$ and multiply the differential equation by $e^{h(x)}$. Since $e^{h(x)} > 0$ for all $x \in J$, the given differential equation and

(A) $e^{h(x)}\,Dy + p(x)e^{h(x)}y = \varnothing$

are equivalent; that is, they have the same solutions. Notice that $h \in \mathscr{C}^1(J)$ and $Dh = p$. By applying the rule for differentiating a product,

$$
\begin{aligned}
D(e^{h(x)}y) &= e^{h(x)} \, Dy + y(De^{h(x)}) \\
&= e^{h(x)} \, Dy + p(x)e^{h(x)}y
\end{aligned}
$$

This observation means that (A) can be expressed as

(B) $D(e^{h(x)}y) = \varnothing$

which is the motivation for the definition of h.

From the mean value theorem (Theorem 2, Section 2), it follows that the set of all solutions of (B) satisfies $e^{h(x)}y = c$, where c is a constant. Since (B) and the given differential equation are equivalent, it follows that the one-parameter family of functions $y(x) = ce^{-h(x)}$ represents all solutions of the homogeneous equation. Finally, to solve the initial-value problem, we note that $h(a) = 0$. Thus, $b = y(a) = ce^0 = c$, and c is uniquely determined.

Theorem 7 and its proof yield some important and useful corollaries. First, note that the method of proof provides a characterization of the set of all solutions of a first-order linear homogeneous equation.

8 COROLLARY

The set of all solutions of the first-order linear homogeneous equation

$$Dy + p(x)y = \varnothing$$

on J is given by the one-parameter family

$$y(x) = c \exp\left[-\int_a^x p(t) \, dt \right], \qquad a \in J$$

Notice that the function

$$y_1(x) = \exp\left[-\int_a^x p(t) \, dt \right]$$

is the solution of the initial-value problem

$$Dy + p(x)y = \varnothing; \qquad y(a) = 1$$

and that the one-parameter family of solutions of the differential equation is merely the set of all constant multiples of y_1.

We emphasize that in determining the one-parameter family of solutions, the choice of $a \in J$ is actually immaterial. If $d \in J$, then by the basic properties of the definite integral,

$$\int_a^x p(t) \, dt = \int_a^d p(t) \, dt + \int_d^x p(t) \, dt$$

By substituting this into the one-parameter family of Corollary 8, we obtain

$$y(x) = c \exp\left[-\int_a^d p(t)\,dt - \int_d^x p(t)\,dt\right]$$

$$= c \exp\left[-\int_a^d p(t)\,dt\right] \exp\left[-\int_d^x p(t)\,dt\right]$$

$$= k \exp\left[-\int_d^x p(t)\,dt\right]$$

where

$$k = c \exp\left[-\int_a^d p(t)\,dt\right]$$

Thus the one-parameter family in Corollary 8 and the one-parameter family

$$y(x) = k \exp\left[-\int_d^x p(t)\,dt\right], \qquad d \in J$$

are merely two representations of the same family of functions. It should be clear that we used $a \in J$ as the lower limit of integration in the proof of Theorem 7 because of its convenience in solving the given initial-value problem.

A second consequence of Theorem 7 which will be useful is the following corollary.

9 COROLLARY

If $y(x)$ is a solution of $Dy + p(x)y = \varnothing$ and $y(a) = 0$ for some $a \in J$, then $y = \varnothing$ on J.

It is clear that the zero function is always a solution of a homogeneous equation; this is the so-called *trivial solution*. All other solutions are called *nontrivial*. Another interpretation of Corollary 9 is that a solution $y(x)$ of the homogeneous equation is either the zero function on J or else $y(x)$ has no zeros on J, that is, either $y(x) < 0$ on J or $y(x) > 0$ on J.

Before proceeding to some examples, we note that in many cases an explicit specification of an interval J is not critical. Therefore, if in discussing a differential equation, or an initial-value problem, an interval is not specified, then the understanding is that J can be any interval that is appropriate. This conforms with the usual convention when dealing with functions whose domains are not stated explicitly.

10 Examples

a. Consider the first-order initial-value problem

$$Dy - \frac{2}{x-1}y = \varnothing; \qquad y(2) = 3$$

By applying Theorem 7, we get the solution

$$y(x) = 3 \exp\left[-\int_2^x -\frac{2}{t-1}\, dt\right] = 3 \exp\left[2 \int_2^x \frac{dt}{t-1}\right]$$

$$= 3 \exp[2 \cdot \ln(x-1)] = 3(x-1)^2$$

However, to emphasize the constructive nature of the proof of Theorem 7, and to illustrate the solution method which it provides, we follow the steps of the proof. Here $p(x) = -2/(x-1)$, so that

$$h(x) = \int_2^x \frac{-2}{t-1}\, dt = \ln(x-1)^{-2}$$

Thus, $e^{h(x)} = (x-1)^{-2}$, and multiplying the equation by this function gives

$$(x-1)^{-2}\, Dy - 2(x-1)^{-3}y = \varnothing$$

Since the left side of this equation is $D[(x-1)^{-2}y]$, we have

$$D[(x-1)^{-2}y] = \varnothing$$

Therefore, $(x-1)^{-2}y = c$, or $y = c(x-1)^2$ is the one-parameter family of solutions. Using the initial condition, we get

$$3 = y(2) = c(2-1)^2 = c$$

and the unique solution of the initial-value problem is $y(x) = 3(x-1)^2$.

b. By applying Theorem 7 to the initial-value problem,

$$Dy - (\sin x^2)y = \varnothing; \qquad y(0) = 1$$

we get the solution

$$y(x) = \exp\left[\int_0^x \sin t^2\, dt\right]$$

Can you find a better representation of this solution, and if not, why not?

11 EXERCISES

1. Find the one-parameter family of solutions of each of the following linear homogeneous equations.

a. $Dy - 3y = \varnothing$ b. $(x^2 + 1)\, Dy - xy = \varnothing$

c. $Dy + 2xy = \varnothing$ d. $(1 - x^2)\, Dy - y = \varnothing$

e. $x\, Dy + y = \varnothing$ f. $Dy + (\tan x)y = \varnothing$

g. $(1 - x^2)\, Dy - 2xy = \varnothing$

2. Solve each of the following initial-value problems.

 a. $Dy - 2xy = \varnothing$; $y(0) = 1$ b. $Dy - [1/(x + 1)]y = \varnothing$; $y(2) = 2$

 c. $e^x Dy - (\tan x)y = \varnothing$; $y(0) = 0$ d. $Dy + (2/x)y = \varnothing$; $y(1) = 1$

 e. $Dy + 4xy = \varnothing$; $y(3) = 0$ f. $(x^2 + 1) Dy + 2xy = \varnothing$; $y(0) = -1$

3. Consider the linear homogeneous equation $Dy + p(x)y = \varnothing$ on an interval J.

 a. Show that if $y_1(x)$ and $y_2(x)$ are each solutions of the equation, then $y(x) = y_1(x) + y_2(x)$ is also a solution of the equation.

 b. Show that if $y(x)$ is a solution of the equation, then $cy(x)$ is also a solution of the equation for any real number c.

 c. Since the set of solutions of the homogeneous equation is a subset of $\mathscr{C}^1(J)$, what can you conclude about the set of solutions of the given equation? (Hint: See Theorem 5, Section 4.)

4. Consider the linear homogeneous equation $Dy + p(x)y = \varnothing$ on an interval J.

 a. Prove that if $y(x)$ is a solution of the equation and $y(a) = 0$ for some $a \in J$, then $y = \varnothing$ on J.

 b. Show that if $y(x)$ and $z(x)$ are each solutions of the equation and $y(b) = z(b)$ for some $b \in J$, then $y(x) = z(x)$ for all $x \in J$.

 c. Show that if $y(x)$ is a nontrivial solution of the equation and $z(x)$ is any solution, then there exists a constant c such that $z(x) = c \cdot y(x)$ for all $x \in J$.

5. Consider the linear homogeneous equation $Dy + p(x)y = \varnothing$ on an interval J. Show that if $p(x) \in \mathscr{C}^1(J)$, then each solution y of the equation is a member of $\mathscr{C}^2(J)$.

6. Consider the linear homogeneous equation $Dy + ry = \varnothing$ on $[0, \infty)$, where r is a real number.

 a. Show that if $r > 0$, then any nontrivial solution y of the equation has the property $\lim_{x \to \infty} y(x) = 0$.

 b. Show that if $r < 0$, then all nontrivial solutions of the equation are unbounded on $[0, \infty)$.

 c. Describe all solutions of the equation in the case $r = 0$.

7. Consider the linear homogeneous equation $Dy + q(x)y = \varnothing$, where q is continuous on $[0, \infty)$ and $q(x) > 0$ for all x.

 a. Show that all solutions of the differential equation are bounded on $[0, \infty)$.

 b. Under what conditions will every solution y of the equation have the property $\lim_{x \to \infty} y(x) = 0$?

The remainder of this section is devoted to first-order linear non-homogeneous equations of the form (4). The first result is quite straightforward, and it illustrates that it is possible to get some information about the solutions of an initial-value problem without explicit knowledge of a solution.

12 THEOREM

The initial-value problem

$$Dy + p(x)y = f(x); \qquad y(a) = b, \quad a \in J$$

has at most one solution.

Proof: Suppose the initial-value problem has two solutions, say y_1 and y_2. Let the function u be defined by $u(x) = y_1(x) - y_2(x)$. Then

$$Du(x) + p(x)u(x) = D[y_1(x) - y_2(x)] + p(x)[y_1(x) - y_2(x)]$$
$$= Dy_1(x) - Dy_2(x) + p(x)y_1(x) - p(x)y_2(x)$$
$$= Dy_1(x) + p(x)y_1(x) - [Dy_2(x) + p(x)y_2(x)]$$
$$= f(x) - f(x) = \varnothing$$

so that u is a solution of the first-order linear homogeneous differential equation $Dy + p(x)y = \varnothing$. Also, since

$$u(a) = y_1(a) - y_2(a) = b - b = 0$$

it follows from Corollary 9 that $u = \varnothing$, which implies that $y_1 = y_2$ on J. Thus the initial-value problem cannot have two (or more) solutions.

Among other things, Theorem 12 provides the assurance that if one solution of the initial-value problem

(13) $Dy + p(x)y = f(x);$ $y(a) = b, \quad a \in J$

is found, by whatever means, then it is the unique solution of this problem.

14 THEOREM

The initial-value problem (13) has a unique solution, namely,

$$y(x) = e^{-h(x)} \left[b + \int_a^x f(t)e^{h(t)}\, dt \right]$$

where

$$h(x) = \int_a^x p(s)\, ds$$

Proof: In view of Theorem 12, it would be sufficient to show directly that $y(x)$ is a solution of (13). However, a few of the steps used in deriving $y(x)$ are also worth considering. As in the proof of Theorem 7, the idea is to multiply the differential equation by $e^{h(x)}$, thereby putting the equation into the form

(A) $D(e^{h(x)}y) = f(x)e^{h(x)}$

From equation (A), we have that the two functions, $e^{h(x)}y$ and $\int_a^x f(t)e^{h(t)}\, dt$, have the same derivative on J, and thus, by Theorem 2 in Section 2, they must differ by a constant function, that is,

$$e^{h(x)}y = \int_a^x f(t)e^{h(t)}\, dt + c$$

Solving this last equation for y gives the one-parameter family of solutions of the nonhomogeneous equation, and by applying the initial condition, we get the unique choice of the constant that specifies the solution of the initial-value problem.

___

As in the case of Theorem 7, the proof of Theorem 14 provides a characterization of the set of all solutions of a first-order nonhomogeneous equation.

15 COROLLARY

The set of all solutions of the first-order linear nonhomogeneous equation

$$Dy + p(x)y = f(x)$$

on J is given by the one-parameter family

$$y(x) = ce^{-h(x)} + e^{-h(x)} \int_a^x f(t)e^{h(t)} \, dt, \qquad a \in J$$

where $h(x) = \int_a^x p(t) \, dt$.

___

By examining the one-parameter family of solutions specified in Corollary 15, notice that the family of functions $ce^{-h(x)}$ is the one-parameter family of solutions of the homogeneous equation $Dy + p(x)y = \varnothing$. We leave it as an exercise to show that the function

$$z(x) = e^{-h(x)} \int_a^x f(t)e^{h(t)} \, dt$$

is the solution of the initial-value problem

$$Dy + p(x)y = f(x); \qquad y(a) = 0$$

The significance of these observations will become apparent in the sections which follow.

16 Examples

a. Consider the initial-value problem

$$Dy + \frac{1}{x}y = 1; \qquad y(2) = 1$$

on $J = [1, \infty)$. To emphasize the solution technique for first-order linear equations, we follow the steps outlined in the proof of Theorem 14. Here, $p(x) = 1/x$, so that

$$h(x) = \int_2^x \frac{1}{t} \, dt = \ln x - \ln 2 = \ln\left(\frac{x}{2}\right)$$

Thus

$$e^{h(x)} = e^{\ln(x/2)} = \frac{x}{2}$$

Multiplying the equation by this function yields

$$\frac{x}{2} Dy + \frac{1}{2} y = \frac{x}{2}$$

or

$$D\left[\frac{x}{2} y\right] = \frac{x}{2}$$

Thus

$$\frac{x}{2} y = \frac{x^2}{4} + c$$

and

$$y(x) = \frac{x}{2} + c\left(\frac{2}{x}\right)$$

is the one-parameter family of solutions. Applying the initial condition gives

$$1 = y(2) = \tfrac{2}{2} + c(\tfrac{2}{2})$$

or $c = 0$, and this yields the unique solution

$$y(x) = \frac{x}{2}$$

Had we used 1 as the lower limit of integration in defining h (which is more convenient than 2 in this case, since $\ln 1 = 0$), we would have found

$$h(x) = \int_1^x \frac{1}{t} \, dt = \ln x$$

and $e^{h(x)} = x$. Multiplying the equation by this function yields

$$x \, Dy + y = x$$

or

$$D[xy] = x$$

Thus $xy = (x^2/2) + k$, and

$$y(x) = \frac{x}{2} + k\left(\frac{1}{x}\right)$$

We emphasize that this is actually the same one-parameter family of functions as found above, as shown by letting the parameter $k = 2c$. Applying the initial condition yields

$$1 = y(2) = \frac{2}{2} + k\left(\frac{1}{2}\right)$$

or $k = 0$, and $y(x) = x/2$ is the unique solution as determined above.

b. To find the one-parameter family of solutions of

$$(x^2 + 1) \, Dy + xy = e^x$$

note that we can assume that $J = (-\infty, \infty)$ since the coefficient functions are each continuous on this interval and there are no singular points. Note, also, that we must first divide the equation by $x^2 + 1$ in order to obtain the form (4). Thus the equivalent equation is:

$$Dy + \frac{x}{x^2 + 1} \, y = \frac{e^x}{x^2 + 1}$$

Now

$$p(x) = \frac{x}{x^2 + 1}$$

and so

$$h(x) = \int_0^x \frac{t}{t^2 + 1} \, dt = \tfrac{1}{2} \ln(t^2 + 1) \Big|_0^x$$

$$= \tfrac{1}{2} \ln(x^2 + 1) = \ln(x^2 + 1)^{1/2}$$

Multiplying by $e^{h(x)} = (x^2 + 1)^{1/2}$ gives

$$(x^2 + 1)^{1/2} \, Dy + \frac{x}{(x^2 + 1)^{1/2}} \, y = \frac{e^x}{(x^2 + 1)^{1/2}}$$

or

$$D[(x^2 + 1)^{1/2}y] = \frac{e^x}{(x^2 + 1)^{1/2}}$$

Thus

$$(x^2 + 1)^{1/2}y = \int_0^x \frac{e^t}{(t^2 + 1)^{1/2}} \, dt + c$$

and

$$y(x) = c(x^2 + 1)^{-1/2} + (x^2 + 1)^{-1/2} \int_0^x \frac{e^t}{(t^2 + 1)^{1/2}} \, dt$$

is a representation of the one-parameter family of solutions.

17 EXERCISES

1. Find the one-parameter family of solutions of each of the following equations.
   a. $Dy - xy = x$
   b. $x \, Dy - y = x^2 e^x$ on $[1, \infty)$
   c. $Dy - y = 2e^x$
   d. $x \, Dy + y = \sec x$ on $(-\pi/2, \pi/2)$

   e. $Dy + (\tan x)y = \sec x$ on $(-\pi/2, \pi/2)$

   f. $Dy + 3x^2y = -1$

   g. $Dy + 2xy = 2x^3$

   h. $x\,Dy + [(2x + 1)/(x + 1)]y = x - 1$

   i. $Dy + (\cot x)y = \sin x$

   j. $Dy - my = e^{mx}$, where m is a constant

2. Find the solution of each of the following initial-value problems.

   a. $Dy + 2y = 4x;\ y(0) = 1$

   b. $(3x^2 + 1)\,Dy - 2xy = 6x;\ y(0) = 2$

   c. $Dy + 2xy = x^3;\ y(1) = 1$

   d. $Dy - (\tan x)y = e^{\sin x};\ y(\pi/4) = 0$

   e. $x\,Dy - (1 - x\tan x)y = x^2 \cos x;\ y(\pi/4) = 0$

3. Given the first-order linear differential equation

(A) $Dy + p(x)y = f(x) + g(x)$

   a. Show that if $u = u(x)$ is a solution of $Dy + p(x)y = f(x)$, and $v = v(x)$ is a solution of $Dy + p(x)y = g(x)$, then $y = u(x) + v(x)$ is a solution of (A). (This fact is often called the superposition principle.)

   b. Use the result in (a) to find the one-parameter family of solutions of

$$Dy + \left(\frac{1}{x}\right)y = x + e^x$$

4. Consider the linear nonhomogeneous differential equation $Dy + p(x)y = f(x)$ on an interval J. Let $a \in J$ and define h by $h(x) = \int_a^x p(t)\,dt$. Show that the function

$$z(x) = e^{-h(x)} \int_a^x f(t)e^{h(t)}\,dt$$

is the solution of the initial-value problem

$$Dy + p(x)y = f(x); \qquad y(a) = 0$$

5. a. Prove that if z_1 and z_2 are each solutions of the nonhomogeneous equation

$$Dy + p(x)y = f(x)$$

   then $y = z_1 - z_2$ is a solution of the homogeneous equation

$$Dy + p(x)y = \varnothing$$

   b. Let z_1 and z_2 be solutions of $Dy + p(x)y = f(x)$ such that $z_1(x) \neq z_2(x)$ for all x. Show that the one-parameter family of solutions of $Dy + p(x)y = f(x)$ is given by

$$y(x) = C[z_1(x) - z_2(x)] + z_1(x)$$

6. Given the first-order differential equation

(B) $Dy + p(x)y = f(x)y^r$

   where r is a real number. Equation (B) is known as the Bernoulli equation. Note that (B) is a linear equation if $r = 0$ or $r = 1$; otherwise (B) is not a linear equation.

   a. Show that (B) may be transformed into a linear first-order equation by means of the change in variable defined by $v = y^{1-r}$.

  b. Solve each of the following equations by the method suggested in (a).

    i. $Dy + (1/x)y = x^3 y^2$

    ii. $Dy + xy = x/y$

7. a. Show that the equation $Dy + ay = x + 1$, where a is a nonzero constant, has a solution of the form $y(x) = bx + c$. Obtain the values for b and c.

  b. Show that $Dy + ay = r(x)$, where a is a nonzero constant and $r(x)$ is a polynomial of degree n, has a solution which is a polynomial of degree n. What can you say if $a = 0$?

8. a. Show that the equation $Dy + by = \sin(ax)$, where a and b are nonzero constants, has a solution of the form

$$y(x) = c_1 \sin(ax) + c_2 \cos(ax)$$

    Find c_1 and c_2.

  b. Assume that $b > 0$, and let $z(x)$ be any solution of the equation in (a). What can you say about $\lim_{x \to \infty} z(x)$?

9. A certain linear nonhomogeneous equation $Dy + p(x)y = f(x)$ has the function $z(x) = x^5$ as a solution. The associated homogeneous equation $Dy + p(x)y = \varnothing$ has the function $y(x) = 1/x^4$ as a solution.

  a. Determine the solution of the initial-value problem

$$Dy + p(x)y = f(x); \qquad y(1) = 4$$

  b. What initial condition should be specified at $x = 1$ so that the corresponding solution y has the property $y(0) = 0$?

10. Consider the linear nonhomogeneous equation $Dy + ry = g(x)$ on the interval $[0, \infty)$, where r is a real number and g is a bounded function on $[0, \infty)$. Show that if $r > 0$, then every solution of the equation is bounded on $[0, \infty)$.

11. Show that if $u(x)$ and $v(x)$ are solutions of $Dy + p(x)y = f(x)$ on the interval J satisfying $u(c) = a$ and $v(c) = b$, respectively, for some $c \in J$, and if $a \neq b$, then $u(x) \neq v(x)$ for all $x \in J$.

12. Let $u(x)$ be the solution of $Dy + p(x)y = f(x)$, $y(0) = a$, on $(-\infty, \infty)$, and let $v(x)$ be the solution of $Dy + p(x)y = g(x)$, $y(0) = b$, on $(-\infty, \infty)$. Prove that if $a > b$ and $f(x) > g(x)$ for all x, then $u(x) > v(x)$ for all x.

13. Prove that if $u(x)$ is a function satisfying the inequality

$$u(x) \leq a + \int_0^x f(t)u(t)\, dt$$

on $[0, \infty)$, where a is a nonnegative constant and f is a nonnegative function, then

$$u(x) \leq a \cdot \exp \int_0^x f(t)\, dt$$

on $[0, \infty)$.

Section 9 Higher-Order Linear Equations

    The higher-order linear equations introduced in this section include the first-order linear differential equations as a special case. The basic assumptions and terminology which are presented here will be used throughout the remainder of this chapter.

1 DEFINITION

A differential equation of the form

$$q_0(x)\, D^n y + q_1(x)\, D^{n-1}y + \cdots + q_{n-1}(x)\, Dy + q_n(x)y = g(x)$$

where $n \geq 1, q_0, q_1, \ldots, q_n$ and g are continuous functions on an interval J, and q_0 is not the zero function ($q_0 \neq \varnothing$) on J, is called an *nth-order linear differential equation*. The functions $q_0, q_1, \ldots, q_n$ are called *coefficient functions*. If $g = \varnothing$ on J, the differential equation is called *homogeneous*; otherwise it is called *nonhomogeneous*.

    An nth-order differential equation which does not have this form, that is, an nth-order differential equation that is not linear, is called *nonlinear*.

---

2 Examples

  a. Taking $n = 1$ in the definition above gives the first-order linear differential equation

$$q_0(x)\, Dy + q_1(x)y = g(x)$$

    on J, which has the same form as in Definition 2, Section 8.

  b. The equation $D^3 y + 4\, Dy = 16e^{2x}$ on $J = (-\infty, \infty)$ is a third-order linear nonhomogeneous differential equation, where $q_1 = q_3 = \varnothing$, $q_0(x) = 1$, $q_2(x) = 4$, and $g(x) = 16e^{2x}$. In this equation, the coefficient functions are all constant functions.

  c. The equation $x^2\, D^2 y - x\, Dy + y = \varnothing$ on $J = (-1, 1)$ is a second-order linear homogeneous differential equation with $q_0(x) = x^2$, $q_1(x) = -x$, and $q_2(x) = 1$.

  d. The equation $D^2 y + 4y^2 = \sin(xy)$ is a second-order nonlinear differential equation.

    A separate treatment of first-order linear differential equations was given in Section 8 because explicit representations of their solutions are possible. This is not the case with higher-order linear equations. In fact, there is no general technique for obtaining solutions of higher-order linear equations, although certain special higher-order linear equations can be solved explicitly. This situation is somewhat analogous to the problem of finding zeros of polynomial functions.

    The notion of singular points, as illustrated in Examples 3b and 3c in Section 8, extends to higher-order equations. The numbers $x \in J$ such that $q_0(x) = 0$ are called the *singular points* of the linear equation defined in (1). A treatment of linear differential equations with singular points is deferred to Chapter 7. However, the next example illustrates some of the problems created by the presence of singular points.

3 Example

The only singular point of the second-order equation in Example 2c is $x = 0$. It can be verified that each of the functions $y_1(x) = x$ and $y_2(x) = x \ln(x)$ is a solution. Notice that although y_1 is defined at the singular point, the solution y_2 is not defined there. In general, linear equations with a singular point have solutions which fail to exist at the singular point.

If an nth-order linear differential equation has no singular points on an interval J, then it can be expressed in the form

(4) $D^n y + p_1(x) D^{n-1} y + \cdots + p_{n-1}(x) Dy + p_n(x) y = f(x)$

on J, where the coefficient functions $p_1, p_2, \ldots, p_n$ and the function f are continuous on the interval J. The coefficient functions in (1) and (4) are related by $p_i = q_i/q_0$, for $i = 1, 2, \ldots, n$, and $f = g/q_0$.

This chapter, including the previous section, is devoted to the study of those differential equations which are, or can be, expressed in the form (4). As before, if an interval J is not specified with an equation of the form (4), then J is assumed to be any interval in which the coefficient functions and f are continuous.

5 DEFINITION

The nth-order operator $L:\mathscr{C}^n(J) \to \mathscr{C}(J)$ associated with equation (4) is defined by

$$L = D^n + p_1(x) D^{n-1} + \cdots + p_{n-1}(x) D + p_n(x) I$$

(I is the identity operator), where for each $y \in \mathscr{C}^n(J)$,

$$L[y(x)] = D^n y(x) + p_1(x) D^{n-1} y(x) + \cdots + p_{n-1}(x) Dy(x)$$
$$+ p_n(x) y(x)$$

An immediate advantage of introducing the nth-order operator L is that the differential equation (4) can be expressed in the more compact form

$$L(y) = f(x)$$

Other advantages derived from the study of L will become apparent.

6 Examples

a. Consider the third-order operator

$$L = D^3 + 4 D$$

Notice that L is the operator associated with the equation in Example 2b, and that

$$L(e^{2x}) = D^3(e^{2x}) + 4 D(e^{2x}) = 8e^{2x} + 8e^{2x} = 16e^{2x}$$

Also,

$$L[\cos(2x)] = D^3 \cos(2x) + 4 D \cos(2x)$$
$$= 8 \sin(2x) - 8 \sin(2x) = \varnothing$$

Thus $y(x) = e^{2x}$ is a solution of the nonhomogeneous equation $L(y) = 16e^{2x}$, while $y(x) = \cos(2x)$ is a solution of the homogeneous equation $L(y) = \varnothing$. It can also be verified that $y(x) = e^{2x} + \cos(2x)$ is a solution of $L(y) = 16e^{2x}$.

b. The second-order operator associated with Example 2c is:

$$L = D^2 - \frac{1}{x} D + \frac{1}{x^2} I$$

on $J = [1, \infty)$. We have that $L[x] = \varnothing$ and that $L[x(\ln x)^2] = 2/x$. Thus, $y(x) = x$ is a solution of $L[y] = \varnothing$ on J, and $y(x) = x[\ln x]^2$ is a solution of $L[y] = 2/x$ on J.

As suggested in Example 25, Section 4, L is a linear operator from the vector space $\mathscr{C}^n(J)$ into the vector space $\mathscr{C}(J)$. The proof that L is linear depends only on the definition of L and the well-known rules of differentiation given in (3) and (4) in Section 2.

7 THEOREM

The nth-order operator L, as defined in (5), is a linear operator.

Proof: If b is a number and $y(x) \in \mathscr{C}^n(J)$, then $b \cdot y(x) \in \mathscr{C}^n(J)$ and

$$L[b \cdot y(x)] = D^n[b \cdot y(x)] + p_1(x) D^{n-1}[b \cdot y(x)] + \cdots$$
$$+ p_n(x) \cdot b \cdot y(x)$$
$$= b D^n y(x) + bp_1(x) D^{n-1} y(x) + \cdots + bp_n(x)y(x)$$
$$= b\{D^n y(x) + p_1(x) D^{n-1} y(x) + \cdots + p_n(x)y(x)\}$$
$$= bL[y(x)]$$

If $y(x)$ and $z(x)$ are members of $\mathscr{C}^n(J)$, then $[y(x) + z(x)] \in \mathscr{C}^n(J)$ and

$$L[y(x) + z(x)] = D^n[y(x) + z(x)] + p_1(x) D^{n-1}[y(x) + z(x)]$$
$$+ \cdots + p_n(x)[y(x) + z(x)]$$
$$= D^n y(x) + D^n z(x) + p_1(x)[D^{n-1}y(x) + D^{n-1}z(x)]$$
$$+ \cdots + p_n(x)[y(x) + z(x)]$$

Removing the parentheses and rearranging the items yields the equation

$$L[y(x) + z(x)] = D^n y(x) + p_1(x) D^{n-1} y(x) + \cdots + p_n(x)y(x)$$
$$+ D^n z(x) + p_1(x) D^{n-1} z(x) + \cdots + p_n(x)z(x)$$
$$= L[y(x)] + L[z(x)]$$

As indicated in Example 25, Section 4, the linear operator L defined in (5) is often called a *linear differential operator*, or more precisely, an *nth-order linear differential operator*. In Section 8, it was mentioned that the term "linear" in the description of the first-order differential equation was motivated by the fact that the associated operator L_1 is a linear operator. Clearly, this same remark applies to the differential equations specified in Definition 1 and equation (4).

8 EXERCISES

1. In each of the following, determine whether the given differential equation is homogeneous or nonhomogeneous, indicate the singular points (if any), and, when appropriate, specify a new interval J on which the equation has no singular points; put each equation into the form (4).
   a. $x^2(x + 1) D^2y - 2y = e^{-x} + 1$ on $(-\infty, -\frac{1}{2})$
   b. $x(x - 1) D^2y - x Dy + y = \varnothing$ on $(\frac{1}{3}, \infty)$
   c. $x D^3y - D^2y - x Dy + y = \varnothing$ on $(-\infty, \infty)$
   d. $x^2(2x - 1) D^3y + 4x^2 Dy + x^2y = e^{3x} - 2x^3$ on $(-3, 3)$
   e. $(x^2 - 3x + 2) D^2y - x Dy + e^xy = \varnothing$ on $(4, \infty)$
2. Given the linear operator $L = (1 - 2x^2) D^2 + 2 D + 4I$.
   a. Calculate $L[e^x]$.
   b. Calculate $L[2x^3 + 3x]$.
   c. Find a quadratic polynomial $p(x)$ such that $L[p(x)] = \varnothing$.
3. Given the linear operator $L = D^2 + 4x D + (4x^2 + 2)I$.
   a. Calculate $L[x^2 - x]$.
   b. Calculate $L[xe^{-x^2}]$.
   c. Show that there is a solution of the homogeneous equation $L[y] = \varnothing$ of the form $y(x) = e^{ax^2}$.
4. Given the linear operator $L = (1 - x^2) D^2 - x D + I$.
   a. Calculate $L[(x^2 - 1)^{1/2}]$.
   b. Calculate $L[x^2]$.
   c. Show that there is (or is not) a polynomial of degree 3, $p(x)$, such that $L[p(x)] = x^3$.

While the concept of an initial-value problem was defined in general in Section 7, it is worthwhile to give a precise statement of this idea for the equations under consideration in the remainder of the chapter.

9 DEFINITION

An *nth-order linear initial-value problem* consists of a linear differential equation

$$L(y) = f(x)$$

on J, where L is an nth-order linear operator, together with a set of n *initial conditions*

$$y(a) = b_0, Dy(a) = b_1, \ldots, D^{n-1}y(a) = b_{n-1}$$

where $a \in J$ and $b_0, b_1, \ldots, b_{n-1}$ are real numbers. A *solution* of this initial-value problem is a function $y = y(x) \in \mathscr{C}^n(J)$ that is a solution of the differential equation and that satisfies the n initial conditions.

In Section 8, a direct proof that each first-order initial-value problem has a unique solution was given. The next theorem is an extension of this result to higher-order initial-value problems. Although this result is of fundamental importance, its proof is beyond the scope of an elementary presentation.

10 THEOREM

Each nth-order linear initial-value problem has a unique solution.

A reasonable proof of Theorem 10 is of roughly the same order of complexity as a proof that each polynomial p of degree $n \geq 1$ has a root.

11 Example

The differential equation

$$D^3 y + 4\, Dy = \varnothing$$

together with the initial conditions

$$y(0) = 0; \qquad Dy(0) = 0; \qquad D^2 y(0) = 1$$

is a third-order linear initial-value problem. A three-parameter family of solutions of the differential equation is:

$$y(x) = c_1 + c_2 \sin(2x) + c_3 \cos(2x)$$

By differentiating twice, we obtain the two additional equations

$$Dy(x) = 2c_2 \cos(2x) - 2c_3 \sin(2x)$$

$$D^2 y(x) = -4c_2 \sin(2x) - 4c_3 \cos(2x)$$

Applying the three initial conditions, we get the three equations

$$c_1 + c_3 = 0 = y(0)$$
$$2c_2 = 0 = Dy(0)$$
$$-4c_3 = 1 = D^2 y(0)$$

It is easy to verify that $c_1 = \frac{1}{4}$, $c_2 = 0$, and $c_3 = -\frac{1}{4}$ is the unique solution of this system of equations, and therefore,

$$y(x) = \tfrac{1}{4} - \tfrac{1}{4} \cos(2x)$$

is the unique solution of the initial-value problem.

The study of higher-order linear equations is organized in the same way that first-order equations were treated; that is, homogeneous equations are considered first, and then nonhomogeneous equations. The final item in this section is to introduce some useful symbolism for the solutions of the two kinds of higher-order equations, and to notice how the two sets of solutions are related.

12 DEFINITION

Given an nth-order linear operator L on the interval J and $f \in \mathscr{C}(J)$. The set of all solutions of

$$L[y] = f(x)$$

is denoted by $\mathscr{S}$; the set of all solutions of

$$L[y] = \varnothing$$

is denoted by $\mathscr{H}$.

Evidently, both $\mathscr{S}$ and $\mathscr{H}$ are subsets of $\mathscr{C}^n(J)$. Notice that $\mathscr{H}$ depends only on the linear operator L while $\mathscr{S}$ depends upon both L and f.

13 THEOREM

If $L[y] = f(x)$ is an nth-order linear differential equation, and $y_1 \in \mathscr{S}$ and $y_2 \in \mathscr{S}$, then $(y_1 - y_2) \in \mathscr{H}$.

Proof: It follows from Theorem 7 that

$$L[y_1(x) - y_2(x)] = L[y_1(x)] - L[y_2(x)]$$

and since y_1 and y_2 are members of $\mathscr{S}$, we have that $L[y_1(x)] = L[y_2(x)] = f(x)$. Thus $L[y_1(x) - y_2(x)] = \varnothing$, or $(y_1 - y_2) \in \mathscr{H}$.

14 COROLLARY

If $u \in \mathscr{S}$, then for each $v \in \mathscr{S}$ there exists $y \in \mathscr{H}$ such that $v = u + y$.

This corollary is merely a restatement of Theorem 13 but in a form that indicates a principal application. In essence, it means that if one solution of $L(y) = f(x)$ is known, together with all solutions of $L(y) = \varnothing$, then all solutions of $L(y) = f(x)$ are known. In the specific case of first-order nonhomogeneous equations, this observation has already been established. In particular, see the representation of the set of solutions as given in Corollary 15, Section 8.

15 EXERCISES

1. Given the linear operator $L = D^n + p_1(x) D^{n-1} + \cdots + p_n(x)I$. Show that if the coefficient functions $p_k(x) \in \mathscr{C}^1(J), k = 1, 2, \ldots, n$, and $y(x)$ is a solution of $L[y] = \varnothing$, then $y(x) \in \mathscr{C}^{n+1}(J)$. See also Exercises 11, Section 8, Problem 5.

2. Given the linear operator $L = D^2 - 2x D + 2I$.
   a. Find $L[x^n]$ where n is a positive integer.
   b. Determine a solution of $L[y] = \varnothing$.
   c. Show that there is (or is not) a polynomial of degree 3, $p(x)$, such that $L[p(x)] = x^2$.
3. Given the second-order linear equation $x^2 D^2 y - x Dy + y = 8x^3$ on $(\frac{1}{2}, \infty)$.
   a. Show that each of $y_1(x) = x + 2x^3$ and $y_2(x) = x \cdot \ln(x) + 2x^3$ is a solution of the equation.
   b. Show that $y(x) = x$ is a solution of the associated homogeneous equation $x^2 D^2 y - x Dy + y = \varnothing$.
   c. Determine a second solution of the associated homogeneous equation [other than a constant multiple of $y(x) = x$].
4. Given the linear operator $L = D^2 - (6/x^2)I$.
   a. Show that $y(x) = cx^3 + x^3 \ln(x) - 2x^2$ is a solution of $L[y] = 5x + 8$ for all values of the constant c.
   b. Show that each of $y_1(x) = x^{-2} + x^3 \ln(x) - 2x^2$ and $y_2(x) = x^3 \ln(x) - 2x^2$ is a solution of $L[y] = 5x + 8$.
   c. Find the solution of the initial-value problem $L[y] = 5x + 8; y(1) = 0, Dy(1) = 1$. (Hint: Find a two-parameter family of solutions of the equation.)

Section 10 Homogeneous Equations

As suggested by the organization of Section 8 and by the concluding remarks in Section 9, a systematic development of the basic theory of linear differential equations begins with an examination of homogeneous equations, that is, with equations of the form

(1) $D^n y + p_1(x) D^{n-1} y + \cdots + p_n(x)y = \varnothing$

where $p_1, p_2, \ldots, p_n$ are members of $\mathscr{C}(J)$ for some interval J. The objective of this section is to give a characterization of the set $\mathscr{H}$ of solutions of equation (1). In the case where $n = 1$, this characterization of $\mathscr{H}$ will reduce to Corollary 8, Section 8. The results of this section have a natural interpretation in the context of linear algebra. In this context, our objective is to show that $\mathscr{H}$ is a vector space, in fact, a subspace of $\mathscr{C}^n(J)$, of dimension n. Since the standard terminology used in studying linear differential equations is taken from linear algebra, and since many of the results depend upon facts from linear algebra, the reader is urged to consult Sections 4 and 5 for reference and review.

The nth-order linear operator L associated with (1), namely,

$$L = D^n + p_1(x) D^{n-1} + \cdots + p_n(x)I$$

is a linear transformation from the vector space $\mathscr{C}^n(J)$ into the vector space $\mathscr{C}(J)$. Thus, in the language of linear algebra, the set $\mathscr{H}$ of solutions of $L[y] = \varnothing$ is the null space of L, and hence it is a subspace of $\mathscr{C}^n(J)$ (see Theorem 27, Section 4). While this application of a general fact from linear algebra is a proof that $\mathscr{H}$ is a subspace of $\mathscr{C}^n(J)$, we state and prove this result in the form in which it will be used in this chapter.

2 THEOREM

If y_1 and y_2 are members of $\mathscr{H}$ and c is any real number, then each of the functions $y_1 + y_2$ and cy_1 are members of $\mathscr{H}$.

Proof: Since y_1 and y_2 are members of $\mathscr{H}$, $L[y_1(x)] = L[y_2(x)] = \varnothing$. Since L is a linear operator,

$$L[y_1(x) + y_2(x)] = L[y_1(x)] + L[y_2(x)] = \varnothing + \varnothing = \varnothing$$

and

$$L[cy_1(x)] = cL[y_1(x)] = c\varnothing = \varnothing$$

This theorem together with Theorem 5, Section 4, gives, as a corollary, the result mentioned above.

3 COROLLARY

The set $\mathscr{H}$ of solutions of equation (1) is a subspace of $\mathscr{C}^n(J)$.

An immediate and useful consequence of the result that $\mathscr{H}$ is a vector space is the fact that any linear combination of members of $\mathscr{H}$ is also a member of $\mathscr{H}$. Stated explicitly in our context, we have the following.

4 COROLLARY

If k is any positive integer, $y_1(x)$, $y_2(x)$, ..., $y_k(x)$ are members of $\mathscr{H}$ and $c_1, c_2, \ldots, c_k$ are real numbers, then the function y defined by

$$y(x) = c_1 y_1(x) + c_2 y_2(x) + \cdots + c_k y_k(x) = \sum_{i=1}^{k} c_i y_i(x)$$

is a member of $\mathscr{H}$.

It is clear that the zero function $\varnothing$ is a solution of any homogeneous equation, and, in particular, $\varnothing$ is a solution of equation (1). For this reason, the zero function is called the *trivial solution*. All other members of $\mathscr{H}$ are called *nontrivial* solutions. This is the terminology introduced in Section 8 in the special case of the first-order linear differential equations.

Since $\mathscr{H}$, as a subspace of $\mathscr{C}^n(J)$, is a vector space, information about the dimension of $\mathscr{H}$ is essential information about the nature of solutions of equation (1). Recalling that $\mathscr{C}^n(J)$ is not a finite dimensional vector space (see Examples 16 and 18 in Section 4), the remainder of this section is devoted to showing that the dimension of $\mathscr{H}$ is n, the order of (1). This fact will lead to a characterization of all solutions of equation (1).

To establish this result, it is necessary to have a clear understanding and working knowledge of the fundamental concepts of linear independence and dependence. Although these concepts are discussed in some detail in Section 4, it will be advantageous to reconsider them here in the special context of the vector space $\mathscr{C}(J)$ and its subspaces $\mathscr{C}^k(J)$.

5 DEFINITION

If $f_1(x), f_2(x), \ldots, f_k(x)$ are members of $\mathscr{C}(J)$, for some interval J, and if the only linear combination of these functions that is equal to the zero function on J is the trivial linear combination, that is, if

$$c_1 f_1(x) + c_2 f_2(x) + \cdots + c_k f_k(x) = \varnothing$$

implies $c_1 = c_2 = \cdots = c_k = 0$, then this set of functions is said to be *linearly independent* on J. If, on the other hand, some nontrivial linear combination of these functions is equal to the zero function on J, that is, if

$$c_1 f_1(x) + c_2 f_2(x) + \cdots + c_k f_k(x) = \varnothing$$

and not all of the c_i's are 0, then the set of functions is said to be *linearly dependent* on J.

At this point, it would be worthwhile to review Section 4, where a number of examples of these two concepts are given in a variety of settings.

6 Examples

a. Consider the functions $f_1(x) = 1$, $f_2(x) = x$, and $f_3(x) = x^2$ on $J = (-\infty, \infty)$, and the equation

$$c_1 \cdot 1 + c_2 \cdot x + c_3 \cdot x^2 = \varnothing$$

on J. Since each of the functions f_1, f_2, f_3, and $\varnothing$ are members of $\mathscr{C}^\infty(J)$, it follows by differentiation that

$$c_1 \cdot \varnothing + c_2 \cdot 1 + c_3 \cdot 2x = \varnothing$$

and

$$c_1 \cdot \varnothing + c_2 \cdot \varnothing + c_3 \cdot 2 = \varnothing$$

on J. The last equation implies $c_3 = 0$, which in turn implies $c_2 = 0$ from the second equation. Finally, $c_2 = c_3 = 0$ implies, from the first equation, that $c_1 = 0$, and thus the functions are linearly independent. An alternate proof is suggested in Example 12, Section 4.

Note that the three equations above can be represented in the vector-matrix form

$$\begin{bmatrix} 1 & x & x^2 \\ \varnothing & 1 & 2x \\ \varnothing & \varnothing & 2 \end{bmatrix} \cdot \begin{bmatrix} c_1 \\ c_2 \\ c_3 \end{bmatrix} = \begin{bmatrix} \varnothing \\ \varnothing \\ \varnothing \end{bmatrix}$$

and that the unique solution vector is $(0, 0, 0)^T$ (T denotes transpose).

b. The functions $f_1(x) = 1, f_2(x) = x, f_3(x) = x^2$, and $f_4(x) = 2x^2 - 1$ are linearly dependent on $J = (-\infty, \infty)$ since the nontrivial linear combination

$$1 \cdot f_1(x) + 0 \cdot f_2(x) - 2 \cdot f_3(x) + 1 \cdot f_4(x) = \varnothing$$

on J.

c. The functions $f_1(x) = x^3$ and $f_2(x) = |x^3|$ are linearly independent on $J = (-\infty, \infty)$ since if

$$c_1 x^3 + c_2 |x|^3 = \varnothing$$

then

$$c_1 + c_2 = 0$$

and

$$-c_1 + c_2 = 0$$

by taking $x = 1$ and $x = -1$, respectively. However, the only solution of this pair of equations is $c_1 = c_2 = 0$, and we conclude that f_1 and f_2 are linearly independent on J.

7 EXERCISES

1. In each of the following, determine whether the given set of functions is linearly dependent or linearly independent.
   a. $\sin ax, \cos ax$ on $(-\infty, \infty)$
   b. $1, x, |x|$ on $(-1, 1)$
   c. $1, \cos x, \cos 2x$ on $(-2\pi, 2\pi)$
   d. $1, \ln x, \ln x^2$ on $(0, \infty)$
   e. e^{ax}, xe^{ax} on $(-\infty, \infty)$
   f. $x^2 - x + 3, 2x^2 + x, 2x - 4$ on $(-\infty, \infty)$
   g. $\sin x, \cos x, \sin[x + (\pi/4)]$ on $(-\infty, \infty)$
   h. $e^{ax}, e^{bx}, a \neq b$, on $(-\infty, \infty)$

2. Let $f_1, f_2, \ldots, f_k$ be continuous functions on an interval J.
   a. Prove that if one of the functions is the zero function, that is, if $f_i(x) = \varnothing$ for some i, $1 \leq i \leq k$, then the set of functions is linearly dependent.
   b. Prove that if there exist numbers c and d such that $f_1(x) = cf_2(x) + df_3(x)$ for all $x \in J$, then the functions are linearly dependent.

   The problem of establishing the linear independence or dependence of a given set of vectors in some vector space $\mathscr{V}$ is usually quite difficult. However, in the case of testing n solutions of the nth-order homogeneous equation (1) for independence or dependence, we have an effective procedure.

8 DEFINITION

Let $y_1, y_2, \ldots, y_n$ be n members of $\mathscr{H}$. The $n \times n$ matrix-valued function

$$W(x) = \begin{bmatrix} y_1(x) & y_2(x) & \cdots & y_n(x) \\ Dy_1(x) & Dy_2(x) & \cdots & Dy_n(x) \\ \vdots & \vdots & & \vdots \\ D^{n-1}y_1(x) & D^{n-1}y_2(x) & \cdots & D^{n-1}y_n(x) \end{bmatrix}$$

on J is called the *Wronski matrix* of $y_1, y_2, \ldots, y_n$. The determinant of this matrix, $\det[W(x)]$, is called the *Wronskian* of $y_1, y_2, \ldots, y_n$.

We emphasize that in forming the Wronski matrix, the definition requires n solutions of the nth-order homogeneous equation $L[y] = \varnothing$. The number of solutions, the order of the Wronski matrix, and the order of the differential equation are all equal. Since the solutions of $L[y] = \varnothing$ are members of $\mathscr{C}^n(J)$, it follows from this definition, together with the definition of the determinant function, that the Wronskian of n members of $\mathscr{H}$ is continuously differentiable on J, that is, $\det[W(x)] \in \mathscr{C}^1(J)$.

9 Examples

a. Consider the second-order homogeneous equation

$$D^2y - \frac{1}{x} Dy + \frac{1}{x^2} y = \varnothing$$

on $J = [1, \infty)$. It can be verified that $y_1(x) = x$ and $y_2(x) = x \ln(x)$ are solutions of the equation (see Example 6b, Section 9). The Wronski matrix of this pair of solutions is:

$$W(x) = \begin{bmatrix} x & x \ln(x) \\ 1 & \ln x + 1 \end{bmatrix}$$

and $\det[W(x)] = x$ on J.

Notice that if c_1 and c_2 are real numbers and $z = c_1 y_1 + c_2 y_2$, then the pair of equations

$$z(x) = c_1 x + c_2 x \ln x$$

$$Dz(x) = c_1 + c_2(1 + \ln x)$$

can be represented in the vector-matrix form

$$W(x)\gamma = \begin{bmatrix} x & x \ln x \\ 1 & 1 + \ln x \end{bmatrix} \begin{bmatrix} c_1 \\ c_2 \end{bmatrix} = \begin{bmatrix} z(x) \\ Dz(x) \end{bmatrix}$$

on J, where $\gamma = (c_1, c_2)^T$.

b. Consider the third-order homogeneous equation

$$D^3y - 4 D^2y + 4 Dy = \varnothing$$

on $J = (-\infty, \infty)$. Each of the functions $y_1(x) = 1$, $y_2(x) = e^{2x}$, and $y_3(x) = 2 - e^{2x}$ is a solution of the equation. The Wronski matrix corresponding to these three solutions is:

$$W(x) = \begin{bmatrix} 1 & e^{2x} & 2 - e^{2x} \\ \varnothing & 2e^{2x} & -2e^{2x} \\ \varnothing & 4e^{2x} & -4e^{2x} \end{bmatrix}$$

and $\det[W(x)] = \varnothing$ on J.

If c_1, c_2, and c_3 are real numbers and $z = c_1 y_1 + c_2 y_2 + c_3 y_3$, then the system of equations

$$z(x) = c_1 + c_2 e^{2x} + c_3(2 - e^{2x})$$

$$Dz(x) = 2c_2 e^{2x} - 2c_3 e^{2x}$$

$$D^2 z(x) = 4c_2 e^{2x} - 4c_3 e^{2x}$$

can be written in the vector-matrix form

$$[W(x)]\gamma = \begin{bmatrix} 1 & e^{2x} & 2 - e^{2x} \\ \varnothing & 2e^{2x} & -2e^{2x} \\ \varnothing & 4e^{2x} & -4e^{2x} \end{bmatrix} \begin{bmatrix} c_1 \\ c_2 \\ c_3 \end{bmatrix} = \begin{bmatrix} z(x) \\ Dz(x) \\ D^2 z(x) \end{bmatrix}$$

on J, where $\gamma = (c_1, c_2, c_3)^T$. In particular, note that

$$W(x)\beta = \begin{bmatrix} \varnothing \\ \varnothing \\ \varnothing \end{bmatrix}$$

where $\beta = (-2, 1, 1)^T$.

The application of the Wronski matrix to the problem of establishing the linear independence or dependence of n members of $\mathscr{H}$ is given by the following theorem.

10 THEOREM

Let $y_1, y_2, \ldots, y_n$ be n members of $\mathscr{H}$, and let W be their Wronski matrix. These functions are linearly dependent on J if and only if their Wronskian is the zero function on J, that is, if and only if $\det[W(x)] = \varnothing$ on J.

Proof: Assume that the solutions $y_1, y_2, \ldots, y_n$ are linearly dependent on J. Then there is a linear combination

(A) $c_1 y_1(x) + c_2 y_2(x) + \cdots + c_n y_n(x) = \varnothing$

on J, where not all of the coefficients $c_1, c_2, \ldots, c_n$ are zero.

Since equation (A) involves functions which are n times differentiable, we obtain, by differentiating $n - 1$ times, a system of n equations of the form

(B) $c_1 D^p y_1(x) + c_2 D^p y_2(x) + \cdots + c_n D^p y_n(x) = \varnothing$

on J, $p = 0, 1, 2, \ldots, n - 1$. Note that $p = 0$ yields (A). Now choose any point $a \in J$. Evaluating each of the equations in (B) at $x = a$ yields the homogeneous system of linear algebraic equations:

(C)

$$
\begin{array}{llll}
c_1 y_1(a) & + c_2 y_2(a) & + \cdots + c_n y_n(a) & = 0 \\
c_1 Dy_1(a) & + c_2 Dy_2(a) & + \cdots + c_n Dy_n(a) & = 0 \\
\quad\vdots & \quad\vdots & \quad\vdots & \\
c_1 D^{n-1} y_1(a) & + c_2 D^{n-1} y_2(a) & + \cdots + c_n D^{n-1} y_n(a) & = 0
\end{array}
$$

Next, let $\gamma = (c_1, c_2, \ldots, c_n)^T$ be the vector of coefficients in the nontrivial linear combination (A) and notice that the system (C) can be expressed as

$$[W(a)]\gamma = \varnothing$$

where $W(a)$ is the Wronski matrix of $y_1, y_2, \ldots, y_n$, evaluated at a, and $\varnothing$ is the n-component zero vector. Since $\gamma \neq \varnothing$, it follows from Theorem 16, Section 5, that $\det[W(a)] = 0$. Finally, since $a \in J$ was an arbitrary point, it follows that

$$\det[W(x)] = \varnothing$$

on J.

Now assume that $\det[W(x)] = \varnothing$ on J. Choose any $b \in J$. Then $\det[W(b)] = 0$, and it follows from Theorem 16, Section 5, that there is a nonzero constant vector $\gamma = (c_1, c_2, \ldots, c_n)^T$ such that

(D) $[W(b)]\gamma = \varnothing$

We use the components of γ to form a nontrivial linear combination z of $y_1, y_2, \ldots, y_n$, that is,

(E) $z(x) = c_1 y_1(x) + c_2 y_2(x) + \cdots + c_n y_n(x)$

According to Corollary 4, $z \in \mathscr{H}$.

By differentiating (E) $n - 1$ times, we obtain, with (E), the system of n equations

$$D^p z(x) = c_1 \, D^p y_1(x) + c_2 \, D^p y_2(x) + \cdots + c_n \, D^p y_n(x)$$

$p = 0, 1, \ldots, n - 1$, which can be represented in the vector-matrix form

(F) $[W(x)]\gamma = \begin{bmatrix} z(x) \\ Dz(x) \\ \vdots \\ D^{n-1}z(x) \end{bmatrix}$

on J. Evaluating (F) at $x = b$ and using (D), we have

(G) $z(b) = Dz(b) = \cdots = D^{n-1}z(b) = 0$

Thus z is a solution of $L[y] = \varnothing$ satisfying the initial conditions (G). However, the trivial solution $y(x) = \varnothing$ is also a solution of $L[y] = \varnothing$ satisfying (G) and so, by the uniqueness theorem, Theorem 10 in Section 9, $z(x) = \varnothing$ on J. Finally, from the definition of the solution z, it follows that there is a nontrivial linear combination of the functions $y_1, y_2, \ldots, y_n$ which is equal to the zero function on J. Thus, this set of functions is linearly dependent.

An examination of the second part of the proof of Theorem 10 shows that we have actually established the following result.

11 COROLLARY

Let $y_1, y_2, \ldots, y_n$ be n members of $\mathscr{H}$, and let W be their Wronski matrix. If for some $b \in J$, $\det[W(b)] = 0$, then the n solutions are linearly dependent on J.

The significance of this observation is that having $\det[W(x)] = 0$ at *one* point of J is enough to insure that the n solutions are linearly dependent on J. This, in turn, implies that $\det[W(x)] = 0$ for *all* $x \in J$ by the first part of Theorem 10.

12 COROLLARY

Let $y_1, y_2, \ldots, y_n$ be n members of $\mathscr{H}$, and let W be their Wronski matrix. These functions are linearly independent on J if and only if their Wronskian is nonzero on J, that is, if and only if $\det[W(x)] \neq 0$ for all $x \in J$.

The practical conclusion which can be drawn from Theorem 10 and its corollaries is the following test for linear independence or dependence: Let $y_1, y_2, \ldots, y_n$ be n members of $\mathscr{H}$, form their Wronski matrix W, calculate the Wronskian $\det[W(x)]$, and evaluate $\det[W]$ at a convenient point $a \in J$. If $\det W[a] = 0$, the solutions are linearly dependent; if $\det[W(a)] \neq 0$, the solutions are linearly independent.

13 Examples

a. Refer to Example 9a. The functions $y_1(x) = x$ and $y_2(x) = x \ln(x)$ are solutions of

$$D^2 y - \frac{1}{x} Dy + \frac{1}{x^2} y = \varnothing$$

on $J = [1, \infty)$. Let W denote the Wronski matrix of y_1 and y_2. Since $\det[W(x)] = x \neq 0$ on J, y_1 and y_2 are linearly independent.

b. Refer to Example 9b. The functions $y_1(x) = 1$, $y_2(x) = e^{2x}$, and $y_3(x) = 2 - e^{2x}$ are solutions of

$$D^3 y - 4 D^2 y + 4 Dy = \varnothing$$

on $J = (-\infty, \infty)$. Let W denote the Wronski matrix of y_1, y_2, and y_3. Since $\det[W(x)] = \varnothing$ on J, these functions are linearly dependent.

c. The functions $y_1(x) = 1$, $y_2(x) = \sin 2x$, and $y_3(x) = \cos 2x$ are solutions of

$$D^3 y + 4 Dy = \varnothing$$

on $J = (-\infty, \infty)$. Their Wronski matrix is:

$$W(x) = \begin{bmatrix} 1 & \sin 2x & \cos 2x \\ \emptyset & 2\cos 2x & -2\sin 2x \\ \emptyset & -4\sin 2x & -4\cos 2x \end{bmatrix}$$

and $\det[W(x)] = 1(-8\cos^2 2x - 8\sin^2 2x) = -8$. Thus, these solutions are linearly independent on J.

14 EXERCISES

1. Consider the differential equation

$$D^2 y - 2\, Dy + 8y = \emptyset$$

on $J = (-\infty, \infty)$.
   a. Show that the functions $y_1(x) = e^{4x}$ and $y_2(x) = e^{-2x}$ are linearly independent solutions on J. Thus, $y(x) = c_1 e^{4x} + c_2 e^{-2x}$ is a two-parameter family of solutions.
   b. Determine a solution of the equation satisfying the initial conditions $y(0) = 2$, $Dy(0) = 2$.

2. Consider the differential equation

$$x^2\, D^2 y - x\, Dy - 3y = \emptyset$$

on $J = (0, \infty)$.
   a. Show that $y_1(x) = x^3$ and $y_2(x) = x^{-1}$ are linearly independent solutions of the equation on J.
   b. Determine a solution of the equation that satisfies the initial conditions $y(1) = 3$, $Dy(1) = -1$.

3. Refer to Example 9b. Show that the function $y(x) = xe^{2x}$ is a solution of the differential equation

$$D^3 y - 4\, D^2 y + 4\, Dy = \emptyset$$

on $(-\infty, \infty)$, and show that the solutions $y_1(x) = 1$, $y_2(x) = e^{2x}$, and $y(x) = xe^{2x}$ are linearly independent.

4. The functions $y_1(x) = 1$, $y_2(x) = x$, and $y_3(x) = x^2$ are solutions of the differential equation $D^3 y = \emptyset$ on $J = (-\infty, \infty)$. By using the Wronskian, show that these functions are linearly independent.

5. Which of the following pairs of solutions of the differential equation

$$x^2\, D^2 y - 2x\, Dy + 2y = \emptyset$$

on $(0, \infty)$ are linearly independent, and which are linearly dependent?
   a. $\{x, x^2\}$
   b. $\{x - 2x^2, -2x + 4x^2\}$
   c. $\{x - x^2, 2x\}$

6. The concept of the Wronski matrix and the Wronskian can be extended as follows: let $f_1(x), f_2(x), \ldots, f_k(x)$ be k functions such that $f_i(x) \in \mathscr{C}^{k-1}(J)$, $i = 1, 2, \ldots, k$.

Then the *Wronski matrix* of this set is the $k \times k$ matrix

$$W(x) = \begin{bmatrix} f_1(x) & f_2(x) & \cdots & f_k(x) \\ Df_1(x) & Df_2(x) & \cdots & Df_k(x) \\ \vdots & \vdots & & \vdots \\ D^{k-1}f_1(x) & D^{k-1}f_2(x) & \cdots & D^{k-1}f_k(x) \end{bmatrix}$$

and the *Wronskian* is $\det[W(x)]$. Note that $\det[W(x)] \in \mathscr{C}(J)$. Prove that if the functions $f_1, f_2, \ldots, f_k$ are linearly dependent on the interval J, then $\det[W(x)] = \varnothing$ on J. (Hint: See the proof of Theorem 10.)

7. Let $f_1, f_2, \ldots, f_k$ be members of $\mathscr{C}^{k-1}(J)$ and let W be their Wronski matrix. Prove that if $\det[W(x)] \neq 0$ for at least one $x \in J$, then the functions are linearly independent on J.

8. Note that Problems 6 and 7 do not say anything about the independence or dependence of the set $f_1, f_2, \ldots, f_k$ in the case $\det[W(x)] = \varnothing$ on J. In particular, the functions $f_1(x) = x^3$ and $f_2(x) = |x|^3$ are members of $\mathscr{C}^1(J)$, where $J = (-\infty, \infty)$, and they are linearly independent (see Example 7c). Show that $\det[W(x)] = \varnothing$ on J for this pair of functions. Does this contradict Theorem 10 (or Corollary 12)? Explain.

9. Use the Wronskian to test the following sets of functions for linear independence.
   a. $\{\sin ax, \cos ax\}$ b. $\{e^{ax}, e^{bx}\}, a \neq b$
   c. $\{x, x^2, 2x^2 - x\}$ d. $\{x, x - 1, (x - 1)/x\}$ on $(0, \infty)$
   e. $\{x, x + 1, x - 1\}$ f. $\{e^{ax} \sin bx, e^{ax} \cos bx\}$

   If the Wronskian test is inconclusive, then establish independence or dependence by some other method.

10. Prove that the functions $f_1(x) = x^r e^{ax}$ and $f_2(x) = x^s e^{bx}$ are linearly independent on $(0, \infty)$ if and only if $r \neq s$, or $a \neq b$, or both.

11. Let f and g be in $\mathscr{C}^1(J)$ and assume that $g(x) \neq 0$ for all $x \in J$. Prove that if W is the Wronski matrix of f and g and $\det[W] = \varnothing$ on J, then f and g are linearly dependent.

12. Suppose $f \in \mathscr{C}^1(J)$ and $f \neq \varnothing$. Prove that the functions f and xf are linearly independent on J.

Recall that the objective of this section is to give a characterization of the set $\mathscr{H}$ of solutions of $L[y] = \varnothing$. Stated in the context of linear algebra, the objective is to show that $\mathscr{H}$ is a finite dimensional vector space [a subspace of $\mathscr{C}^n(J)$] having dimension n equal to the order of the equation. This can be accomplished by showing that any solution of $L[y] = \varnothing$ can be represented as a linear combination of n linearly independent solutions of the equation.

15 THEOREM

There exists a set $\{y_1, y_2, \ldots, y_n\}$ of n linearly independent members of $\mathscr{H}$ such that if y is any member of $\mathscr{H}$, then

$$y(x) = c_1 y_1(x) + c_2 y_2(x) + \cdots + c_n y_n(x)$$

on J, where the constants $c_1, c_2, \ldots, c_n$ are uniquely determined.

Proof: Fix any point $b \in J$ and consider the n sets of initial conditions:

(I_1) $y(b) = 1;$ $Dy(b) = D^2 y(b) = \cdots = D^{n-1} y(b) = 0$

(I_2) $y(b) = 0;$ $Dy(b) = 1;$ $D^2 y(b) = \cdots = D^{n-1} y(b) = 0$

$\vdots$

(I_n) $y(b) = Dy(b) = \cdots = D^{n-2} y(b) = 0;$ $D^{n-1} y(b) = 1$

By Theorem 10, Section 9, each of the initial-value problems $L[y] = \varnothing$, with initial conditions (I_k) has a unique solution $y_k(x)$, $k = 1$, $2, \ldots, n$. It should be clear that the Wronski matrix $W(x)$ of the solutions $y_1(x), y_2(x), \ldots, y_n(x)$ at $x = b$ is:

$$
W(b) = \begin{bmatrix} 1 & 0 & 0 & \cdots & 0 \\ 0 & 1 & 0 & \cdots & 0 \\ \vdots & \vdots & \vdots & & \vdots \\ 0 & 0 & 0 & \cdots & 1 \end{bmatrix} = I
$$

the $n \times n$ identity, and $\det[W(b)] = 1$. Thus, by Theorem 10, these functions are linearly independent on J.

Now suppose $y(x)$ is any member of $\mathscr{H}$.

The values of y and its first $n - 1$ derivatives at b can be used to define a linear combination $z(x)$ of $y_1(x), y_2(x), \ldots, y_n(x)$, namely,

$$z(x) = y(b) \cdot y_1(x) + Dy(b) \cdot y_2(x) + \cdots + D^{n-1} y(b) \cdot y_n(x)$$

By evaluating z and its derivatives at $x = b$, and using the initial conditions of $y_1(x), y_2(x), \ldots, y_n(x)$, that is, $(I_1), (I_2), \ldots, (I_n)$, it follows that $z(b) = y(b), Dz(b) = Dy(b), \ldots, D^{n-1} z(b) = D^{n-1} y(b)$.

Since $z(x) \in \mathscr{H}$ (Corollary 5), and $y(x)$ and $z(x)$ have the same initial values at $x = b$, we conclude (Theorem 10, Section 9) that $y(x) = z(x)$ on J. Thus, any member of $\mathscr{H}$ can be uniquely represented as a linear combination of the solutions $y_1(x), y_2(x), \ldots, y_n(x)$.

Some observations concerning Theorem 15 are appropriate here. First, the particular choices of the initial conditions (I_k), $k = 1, 2, \ldots, n$, were made for convenience in calculating the Wronskian at $x = b$ and, even more importantly, for the convenience in representing an arbitrary member of $\mathscr{H}$ as a linear combination of the solutions $y_1, y_2, \ldots, y_n$. We emphasize that *any* n sets of initial conditions such that $\det[W(b)] \neq 0$ would have served just as well.

Stated in the language of linear algebra, Theorem 15 is as follows.

16 COROLLARY

The vector space $\mathscr{H}$ of solutions of $L[y] = \varnothing$ has dimension n.

Since $\mathscr{H}$ is a vector space of dimension n, it follows from Theorem 17, Section 4, that any set of n linearly independent solutions of $L[y] = \varnothing$ forms a basis for $\mathscr{H}$. This fact suggests the following terminology.

17 DEFINITION

Any set of n linearly independent members of $\mathcal{H}$, that is, any set of n linearly independent solutions of $L[y] = \emptyset$, is called a *solution basis* for $L[y] = \emptyset$.

We know from Theorem 15 that $L[y] = \emptyset$ has a solution basis. The next result gives a characterization of the set $\mathcal{H}$ in terms of any solution basis.

18 THEOREM

Let $y_1, y_2, \ldots, y_n$ be a solution basis for $L[y] = \emptyset$. Then $y \in \mathcal{H}$ if and only if y has a unique representation as a linear combination of $y_1, y_2, \ldots, y_n$, that is, if and only if

$$y(x) = c_1 y_1(x) + c_2 y_2(x) + \cdots + c_n y_n(x)$$

on J, for a unique choice of constants $c_1, c_2, \ldots, c_n$.

Proof: From Corollory 4, every linear combination of $y_1, y_2, \ldots, y_n$ is a member of $\mathcal{H}$.

Suppose $y \in \mathcal{H}$. Choose any point $b \in J$. Let W be the Wronski matrix of the solutions $y_1, y_2, \ldots, y_n$. Since these solutions are linearly independent, their Wronskian, $\det[W]$, is nonzero on J, and, in particular, $\det[W(b)] \neq 0$. Therefore, by Theorem 16, Section 5, the system of equations

$$\textbf{(A)} \qquad [W(b)]\xi = \begin{bmatrix} y(b) \\ Dy(b) \\ \vdots \\ D^{n-1}y(b) \end{bmatrix}$$

has a unique solution $\gamma = (c_1, c_2, \ldots, c_n)^T$. Let z be the member of $\mathcal{H}$ defined by

$$z(x) = c_1 y_1(x) + c_2 y_2(x) + \cdots + c_n y_n(x)$$

By evaluating z and its derivatives, $Dz, D^2z, \ldots, D^{n-1}z$ at $x = b$, and using (A), it follows that

$$z(b) = y(b), Dz(b) = Dy(b), \ldots, D^{n-1}z(b) = D^{n-1}y(b)$$

Thus, by the uniqueness theorem, Theorem 10, Section 9,

$$z(x) = y(x) = c_1 y_1(x) + c_2 y_2(x) + \cdots + c_n y_n(x)$$

on J.

The characterization of the set $\mathcal{H}$ given in Theorem 18 is of fundamental importance. The following is a restatement of this theorem using the terminology introduced in Section 7.

19 COROLLARY

Let $y_1, y_2, \ldots, y_n$ be a solution basis for $L[y] = \emptyset$. Then the set $\mathscr{H}$ of all solutions of $L[y] = \emptyset$ is represented by the n-parameter family

$$y(x) = c_1 y_1(x) + c_2 y_2(x) + \cdots + c_n y_n(x)$$

To coordinate the results in this section with the treatment of first-order homogeneous equations in Section 8, note that in the case $n = 1$, the solution $y_1(x) = e^{h(x)}$, where $h(x) = -\int_a^x p_1(t)dt$, is a solution basis for the equation $L_1[y] = Dy + p_1(x)y = \emptyset$. Thus by Corollary 19, the one-parameter family $y(x) = c_1 y_1(x)$ represents the set of all solutions of $L_1[y] = \emptyset$, and this is precisely the result stated in Corollary 8, Section 8.

20 Examples

a. Consider the third-order homogeneous equation

$$D^3 y + 4 Dy = \emptyset$$

on $J = (-\infty, \infty)$. The functions $y_1(x) = 1$, $y_2(x) = \sin 2x$, and $y_3(x) = \cos 2x$ are solutions. As seen in Example 13c, these solutions are linearly independent on J. Thus the three-parameter family

$$y(x) = c_1 \cdot 1 + c_2 \sin 2x + c_3 \cos 2x$$

characterizes all solutions of the equation. See, also, Example 11, Section 9.

b. Given the second-order initial-value problem

$$D^2 y - \frac{1}{x} Dy + \frac{1}{x^2} y = \emptyset; \qquad y(1) = 2, Dy(1) = 0$$

on $J = [1, \infty)$. The functions $y_1(x) = x$ and $y_2(x) = x \ln(x)$ are linearly independent solutions of the differential equation, as seen in Example 13a. Thus $y_1(x)$ and $y_2(x)$ form a solution basis for the equation and the two-parameter family of functions

$$y(x) = c_1 x + c_2 x \ln(x)$$

represents all solutions. Differentiating, we obtain the equation

$$Dy(x) = c_1 + c_2[1 + \ln(x)]$$

Finally, applying the initial conditions yields the pair of equations

$$2 = c_1 + c_2 \ln(1) = c_1$$
$$0 = c_1 + c_2[1 + \ln 1] = c_1 + c_2$$

having the unique solution $c_1 = 2$, $c_2 = -2$. Therefore, $y(x) = 2x - 2x \ln(x)$ is the unique solution of the initial-value problem.
    Notice that the pair of equations $y(x) = c_1 x + c_2 x \ln(x)$ and $Dy(x) = c_1 + c_2[1 + \ln(x)]$ can be written in the vector-matrix form

$$[W(x)]\gamma = \begin{bmatrix} x & x \ln(x) \\ 1 & 1 + \ln(x) \end{bmatrix} \begin{bmatrix} c_1 \\ c_2 \end{bmatrix} = \begin{bmatrix} y(x) \\ Dy(x) \end{bmatrix}$$

Applying the initial conditions, we have

$$[W(1)]\gamma = \begin{bmatrix} 1 & 0 \\ 1 & 1 \end{bmatrix} \begin{bmatrix} c_1 \\ c_2 \end{bmatrix} = \begin{bmatrix} 2 \\ 0 \end{bmatrix} = \begin{bmatrix} y(1) \\ Dy(1) \end{bmatrix}$$

or

$$\begin{bmatrix} c_1 \\ c_2 \end{bmatrix} = \begin{bmatrix} 1 & 0 \\ -1 & 1 \end{bmatrix} \begin{bmatrix} 2 \\ 0 \end{bmatrix} = \begin{bmatrix} 2 \\ -2 \end{bmatrix}$$

yielding the unique solution found above.

21 EXERCISES

1. Consider the differential equation

(A) $D^2 y - 3\, Dy - 4y = \varnothing$

on $(-\infty, \infty)$.
   a. Find two values of r such that the function $y(x) = e^{rx}$ is a solution of (A).
   b. Determine a solution basis for (A) and give the two-parameter family of solutions.
   c. Find the solution of (A) satisfying the initial conditions $y(0) = 1$ and $Dy(0) = 0$.
2. Consider the differential equation

(B) $D^2 y - \left(\dfrac{2}{x}\right) Dy - \left(\dfrac{4}{x^2}\right) y = \varnothing$

on $(1, \infty)$.
   a. Show that there are two solutions of (B) of the form $y(x) = x^r$.
   b. Determine a solution basis for (B) and give the two-parameter family of solutions.
   c. Find the solution of (B) satisfying the initial conditions $y(2) = Dy(2) = 0$.
3. Consider the differential equation

(C) $D^3 y - 2\, D^2 y - Dy + 2y = \varnothing$

   a. Show that (C) has solutions of the form $y(x) = e^{rx}$.
   b. Determine a solution basis for (C) and give the three-parameter family of solutions.
   c. Find the solution of (C) satisfying the initial conditions $y(0) = Dy(0) = 0$ and $D^2 y(0) = 1$.

4. Consider the differential equation

(D) $(x^2 + 2x - 1) D^2 y - 2(x + 1) Dy + 2y = \varnothing$

on $J = (1, \infty)$.

a. Show that (D) has a linear polynomial and a quadratic polynomial as solutions.

b. Determine a solution basis for (D) and give the two-parameter family of solutions.

5. Given the differential equation $D^2 y + p_1(x) Dy + p_2(x)y = \varnothing$, where p_1 and p_2 are continuous functions on an interval J. Show that if u and v are linearly independent solutions of the equation, then there does not exist a point $x_0 \in J$ such that $u(x_0) = v(x_0) = 0$.

6. Let u and v be nontrivial solutions of the differential equation in Problem 5. Prove that u and v are linearly dependent if and only if there exists a constant k such that $u(x) = kv(x)$ for all $x \in J$.

7. Let u and v be linearly independent solutions of the equation in Problem 5, and let W be their Wronski matrix.

a. Show that $\det[W]$ is a solution of the first-order equation $Dy + p_1(x)y = \varnothing$ and obtain an explicit form for $\det[W]$.

b. What can you say about $\det[W]$ in the case $p_1 = \varnothing$?

8. Let f_1 and f_2 be functions in $\mathscr{C}^2(J)$, for some interval J, such that $\det[W(x)] \neq 0$ for all $x \in J$, where W is the Wronski matrix of f_1 and f_2. Calculate the third-order determinant $\det[U]$, where

$$U = \begin{bmatrix} y & f_1 & f_2 \\ Dy & Df_1 & Df_2 \\ D^2 y & D^2 f_1 & D^2 f_2 \end{bmatrix}$$

and show that f_1 and f_2 form a solution basis for the second-order homogeneous equation $\det[U] = \varnothing$ on J. Explain the restriction $\det[W(x)] \neq 0$ for all $x \in J$. Note that this gives a procedure for determining a linear homogeneous differential equation having a given set of functions as a solution basis. This procedure can be extended to higher-order equations.

9. In each of the following, determine a second-order homogeneous equation having the given pair of functions as a solution basis.

a. $\{1, x\}$ b. $\{e^{2x}, e^{-2x}\}$

c. $\{x, x \ln x\}$ (See Example 9a.)   d. $\{\sin 2x, \cos 2x\}$

e. $\{x, x^2\}$ (See Example 10, Section 7.)

10. Let u and v be solutions of the linear homogeneous equation $D^2 y + p_1(x) Dy + p_2(x)y = \varnothing$ on J. Let $a \in J$, and suppose that

$$u(a) = b_1, \qquad Du(a) = b_2$$

and

$$v(a) = c_1, \qquad Dv(a) = c_2$$

Under what conditions on b_1, b_2, c_1, and c_2 will the functions u and v be linearly independent on J?

11. Let u and v be linearly independent solutions of the second-order homogeneous equation $D^2 y + p_1(x) Dy + p_2(x)y = \varnothing$, and assume that $u(a) = u(b) = 0$ with

$u(x) \neq 0$ on (a, b). Prove that there exists exactly one point $c \in (a, b)$ such that $v(c) = 0$. (Hint: Show first that $v(a) \neq 0$, $v(b) \neq 0$, then consider u/v and use Rolle's theorem.)

Section 11 Nonhomogeneous Equations

We complete the development of the basic theory of linear differential equations by considering the nth-order nonhomogeneous equation, that is, equations of the form

(1) $L(y) = f(x)$

where f is a member of $\mathscr{C}(J)$ and $f \neq \varnothing$. Here, as before, L denotes an nth-order linear operator of the form

(2) $L = D^n + p_1(x) D^{n-1} + \cdots + p_n(x) I$

on the interval J. The objective of this section is to characterize the set $\mathscr{S}$ of solutions of (1).

As suggested in Section 9, the set $\mathscr{H}$ of solutions of the associated homogeneous equation

(3) $L(y) = \varnothing$

plays a fundamental role in the characterization of the set $\mathscr{S}$. The homogeneous equation (3) is called the *reduced equation* of (1).

Whereas both $\mathscr{S}$ and $\mathscr{H}$ are subsets of the vector space $\mathscr{C}^n(J)$, $\mathscr{H}$ is a subspace and $\mathscr{S}$ is not. To confirm that $\mathscr{S}$ is not a subspace of $\mathscr{C}^n(J)$, it is sufficient to observe that the zero function is not a member of $\mathscr{S}$ since $L[\varnothing] = \varnothing \neq f$. Also, the difference of two members of $\mathscr{S}$ is a member of $\mathscr{H}$, as noted in Theorem 13, Section 9. It is this fact that leads to a characterization of $\mathscr{S}$.

4 THEOREM

If $z(x)$ is a solution of $L[y] = f(x)$ and $y_1(x), y_2(x), \ldots, y_n(x)$ is a solution basis for $L[y] = \varnothing$, then $y(x) \in \mathscr{S}$ if and only if

(A) $y(x) = \sum_{i=1}^{n} c_i y_i(x) + z(x)$

for unique constants $c_1, c_2, \ldots, c_n$.

Proof: If $y(x) \in \mathscr{S}$, then by Theorem 13, Section 9, $[y(x) - z(x)] \in \mathscr{H}$. But, by Theorem 18, Section 10, each member of $\mathscr{H}$ is a unique linear combination of members of a solution basis for $L[y] = \varnothing$.

Thus

$$y(x) - z(x) = \sum_{i=1}^{n} c_i y_i(x)$$

for unique constants $c_1, c_2, \ldots, c_n$, and $y(x)$ is as specified in (A).

Conversely, if $y(x)$ is as specified in (A), then

$$L[y(x)] = L\left(\sum_{i=1}^{n} c_i y_i(x) + z(x)\right) = \sum_{i=1}^{n} c_i L[y_i(x)] + L[z(x)] = f(x)$$

since L is a linear operator, and $L[y_i(x)] = \varnothing, i = 1, 2, \ldots, n$, and $L[z(x)] = f(x)$.

Before proceeding to an example, there are several useful interpretations of this theorem which are worth considering. First, equation (A) in the theorem is a representation of all members of $\mathscr{S}$ in the form of an n-parameter family of functions. Thus, to know all members of $\mathscr{S}$, it is enough to know one member of $\mathscr{S}$ together with all members of $\mathscr{H}$. However, all members of $\mathscr{H}$ can be represented in terms of a solution basis for $L[y] = \varnothing$. Therefore, any solution basis for $L[y] = \varnothing$ together with one solution of $L[y] = f(x)$ is enough to characterize all solutions of $L[y] = f(x)$.

Notice that if $n = 1$, then the one-parameter family (A) in Theorem 4 has the form

(A₁) $\qquad y(x) = c_1 y_1(x) + z(x)$

where y_1 is a nonzero (that is, linearly independent) solution of $L_1[y] = Dy + p_1(x)y = \varnothing$, and z is a solution of $L_1[y] = f(x)$. In particular, by letting

$$y_1(x) = e^{-h(x)}$$

and

$$z(x) = e^{-h(x)} \int_a^x f(t) e^{h(t)} \, dt$$

where $h(x) = \int_a^x p_1(t) \, dt$, the one-parameter family (A₁) is the same as the one-parameter family specified in Corollary 15, Section 8.

A third observation concerning this theorem is of a geometric nature. Notice that Theorem 4 implies that $\mathscr{S}$ can be formed by taking a single member $z(x) \in \mathscr{S}$ and adding it to each member of $\mathscr{H}$. Roughly speaking, this amounts to translating the vector space $\mathscr{H}$ by an amount $z(x)$. Subsets formed in this manner are usually called *affine subspaces*. As an example of an affine subspace that has a simple geometric interpretation, consider the vector space $\mathscr{V}_2$ consisting of all pairs of real numbers (x_1, x_2). If **H** is the subset of $\mathscr{V}_2$ that satisfies the equation $x_2 = mx_1$, and **S** is the subset that satisfies the equation $x_2 = mx_1 + b$, where $b \neq 0$, then **H** is a one-dimensional subspace of $\mathscr{V}_2$ and **S** is an affine subspace of $\mathscr{V}_2$. The figure gives the usual geometric interpretation. Notice that **S** can be formed by adding $(0, b)$ to each member of **H**.

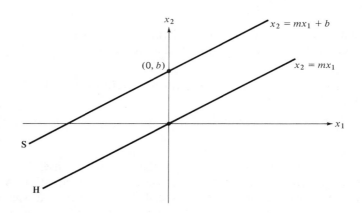

5 Example

The third-order linear operator

$$L = D^3 + 4D$$

was used in Examples 2b, 6a, and 11 of Section 9 and in Example 20a of Section 10. An initial-value problem

$$L[y] = 16e^{2x}; \qquad y(0) = b_0, \quad Dy(0) = b_1, \quad D^2y(0) = b_2$$

can be solved by first observing that a solution basis for the reduced equation $L[y] = \varnothing$ is $y_1(x) = 1$, $y_2(x) = \sin(2x)$, and $y_3(x) = \cos(2x)$. Next, since $L[e^{2x}] = 16e^{2x}$, $z(x) = e^{2x}$ is a solution of the nonhomogeneous equation. Thus, the family of solutions of $L[y] = 16e^{2x}$ is given by

(B) $$y(x) = c_1 + c_2 \sin(2x) + c_3 \cos(2x) + e^{2x}$$

In order to apply the initial conditions, it is necessary to differentiate (B) twice to obtain

$$Dy(x) = 2c_2 \cos(2x) - 2c_3 \sin(2x) + 2e^{2x}$$

$$D^2y(x) = -4c_2 \sin(2x) - 4c_3 \cos(2x) + 4e^{2x}$$

These two equations together with (B) can be put into matrix form as

(C) $$\begin{bmatrix} y(x) \\ Dy(x) \\ D^2y(x) \end{bmatrix} = \begin{bmatrix} 1 & \sin(2x) & \cos(2x) \\ \varnothing & 2\cos(2x) & -2\sin(2x) \\ \varnothing & -4\sin(2x) & -4\cos(2x) \end{bmatrix} \begin{bmatrix} c_1 \\ c_2 \\ c_3 \end{bmatrix} + \begin{bmatrix} e^{2x} \\ 2e^{2x} \\ 4e^{2x} \end{bmatrix}$$

Notice that the matrix in (C) is the Wronski matrix of y_1, y_2, and y_3.

By evaluating (C) at $x = 0$, we have

$$\begin{bmatrix} b_0 \\ b_1 \\ b_2 \end{bmatrix} = \begin{bmatrix} 1 & 0 & 1 \\ 0 & 2 & 0 \\ 0 & 0 & -4 \end{bmatrix} \begin{bmatrix} c_1 \\ c_2 \\ c_3 \end{bmatrix} + \begin{bmatrix} 1 \\ 2 \\ 4 \end{bmatrix}$$

or

$$\begin{bmatrix} 1 & 0 & 1 \\ 0 & 2 & 0 \\ 0 & 0 & -4 \end{bmatrix} \begin{bmatrix} c_1 \\ c_2 \\ c_3 \end{bmatrix} = \begin{bmatrix} b_0 - 1 \\ b_1 - 2 \\ b_2 - 4 \end{bmatrix}$$

Therefore

$$\begin{bmatrix} c_1 \\ c_2 \\ c_3 \end{bmatrix} = \begin{bmatrix} 1 & 0 & \frac{1}{4} \\ 0 & \frac{1}{2} & 0 \\ 0 & 0 & -\frac{1}{4} \end{bmatrix} \begin{bmatrix} b_0 - 1 \\ b_1 - 2 \\ b_2 - 4 \end{bmatrix}$$

and

$$c_1 = b_0 + \frac{b_2}{4} - 2$$

$$c_2 = \frac{b_1}{2} - 1$$

$$c_3 = \frac{-b_2}{4} + 1$$

is the unique choice of constants in (B) that solves the initial-value problem.

The importance of obtaining just one solution of the nonhomogeneous equation (1) is indicated in Theorem 4. While there is no general method to find $z(x)$ such that $L[z(x)] = f(x)$, it is sometimes possible to use the specific form of L, or f, to solve this problem. The next result provides a useful fact concerning equation (1) in case f is, or can be, expressed as a sum of two functions. This result is another consequence of the fact that L is a linear operator, and it is usually called the *superposition principle*.

6 THEOREM

If $z_1(x)$ is a solution of $L[y] = f_1(x)$ and $z_2(x)$ is a solution of $L[y] = f_2(x)$, then $z(x) = z_1(x) + z_2(x)$ is a solution of $L[y] = f_1(x) + f_2(x)$.

Proof: Since L is a linear operator,

$$L[z(x)] = L[z_1(x)] + L[z_2(x)] = f_1(x) + f_2(x)$$

and this proves the theorem.

7 Example

Consider the third-order nonhomogeneous equation

$$L[y] = D^3y + 4\,Dy = 16e^{2x} + x$$

It can be verified that $z_2(x) = (x^2/8)$ is a solution of $L[y] = x$. As seen in Example 5, $z_1(x) = e^{2x}$ is a solution of $L[y] = 16e^{2x}$ and $y_1(x) = 1$, $y_2(x) = \sin 2x$, $y_3(x) = \cos 2x$ is a solution basis of $L[y] = \varnothing$. We can now conclude that $z(x) = e^{2x} + (x^2/8)$ is a solution of $L[y] = 16e^{2x} + x$, and that

$$y(x) = c_1 + c_2 \sin 2x + c_3 \cos 2x + e^{2x} + \frac{x^2}{8}$$

represents all solutions of the equation.

Finally, we present an obvious generalization of Theorem 6 which is easy to verify.

8 THEOREM

If $z_i(x)$ is a solution of $L[y] = f_i(x)$ for $i = 1, 2, \ldots, k$, then $z(x) = \sum_{i=1}^{k} z_i(x)$ is a solution of $L(y) = \sum_{i=1}^{k} f_i(x)$.

9 EXERCISES

1. Consider the linear operator $L = D^2 + I$.
   a. Show that $L[\cos x] = L[\sin x] = \varnothing$.
   b. Show that $L[3x^2 - 2] = 3x^2 + 4$.
   c. Solve the initial-value problem $L[y] = 3x^2 + 4$, $y(0) = 1$, $Dy(0) = -1$.
2. Consider the linear operator $L = x\,D^2 + D$.
   a. Show that $L[(x^2/4) + x] = x + 1$.
   b. Calculate $L[\ln(x) + (x^2/4) + x]$.
   c. Determine the two-parameter family of solutions of $L[y] = x + 1$.
3. Consider the linear operator $L = x^2\,D^2 - 2x\,D + 2I$.
   a. Show that $L[x] = \varnothing$.
   b. Show that each of the functions $z_1(x) = x + x^3[\frac{1}{2}\ln(x) - \frac{3}{4}]$ and $z_2(x) = x^2 + x^3[\frac{1}{2}\ln(x) - \frac{3}{4}]$ is a solution of $L[y] = x^3 \ln(x)$.
   c. Determine the two-parameter family of solutions of $L[y] = x^3 \ln(x)$.
4. Consider the linear operator $L = D^3 - D^2 - D + I$.
   a. Show that each of the functions $y_1 = e^x$, $y_2 = xe^x$, and $y_3 = e^{-x}$ is a solution of the homogeneous equation $L[y] = \varnothing$.
   b. Show that $z(x) = 2 + 2x + xe^{-x}$ is a solution of the nonhomogeneous equation $L[y] = 2x + 4e^{-x}$.
   c. Find the solution of the initial-value problem $L[y] = 2x + 4e^{-x}$, $y(0) = Dy(0) = D^2y(0) = 0$.

5. Consider the equation $L[y] = D^2y + p_1(x) Dy + p_2(x)y = f(x)$. Suppose that each of $z_1(x) = x^2 - 1 + x \ln(x)$ and $z_2(x) = x \ln(x)$ is a solution of $L[y] = f(x)$, and that the function $y_1(x) = x$ is a solution of $L[y] = \varnothing$.
   a. Find the two-parameter family of solutions of $L[y] = \varnothing$.
   b. Find the two-parameter family of solutions of $L[y] = f(x)$.
   c. Find the solution of $L[y] = f(x)$ satisfying $y(1) = 0, Dy(1) = 1$.

6. Prove Theorem 8.

7. Consider the linear operator $L = D^2 - 4D + 3I$.
   a. Show that the equation $L[y] = \sin x$ has a solution of the form $z_1(x) = A \sin x + B \cos x$.
   b. Show that the equation $L[y] = e^x$ has a solution of the form $z_2(x) = Cxe^x$.
   c. Show that $y_1(x) = e^x$ and $y_2(x) = e^{3x}$ are linearly independent solutions of $L[y] = \varnothing$.
   d. Determine the two-parameter family of solutions of $L[y] = \sin x + e^x$.

8. Given the second-order equation $L_2[y] = D^2y + p_1(x) Dy + p_2(x)y = f(x)$ on the interval J. Let y_1, y_2, and y_3 be solutions of this equation, and let

$$U = \begin{bmatrix} y_1 & y_2 & y_3 \\ Dy_1 & Dy_2 & Dy_3 \\ 1 & 1 & 1 \end{bmatrix}$$

Show that if $\det[U(x)] \neq 0$ for all x on J, then $v_1 = y_2 - y_1$ and $v_2 = y_3 - y_1$ are linearly independent solutions of the reduced equation, and

$$y(x) = c_1v_1(x) + c_2v_2(x) + y_1(x)$$

is the two-parameter family of solutions of $L_2[y] = f(x)$.

Section 12 Second-Order
Linear Differential Equations

The majority of the results presented in this chapter concern the nature of solutions of linear differential equations, with very little specific information on obtaining solutions. The exception to this was Section 8 in which all solutions of first-order linear equations were characterized. This section contains some specific methods for obtaining solutions of higher-order equations. While the methods are somewhat cumbersome to describe, they are, when applicable, very effective. This is especially so in the case of second-order linear equations.

To begin, we consider a special class of second-order operators, namely,

(1) $L_1 = D^2 + q(x) D$

on the interval J.

2 THEOREM

A solution basis for $L_1[y] = \varnothing$ is the pair of functions $y_1(x) = 1$ and $y_2(x) = \int_a^x e^{h(t)} dt$, where $a \in J$ and $h(t) = -\int_a^t q(s) ds$.

Proof: Evidently, any constant function is a solution of $L_1[y] = \varnothing$, and so $y_1(x) = 1$ is merely a convenient choice. To obtain the second solution, notice that

$$L_1[y_1] = D(Dy) + q(x)(Dy)$$

That is, this second-order equation can be interpreted as a first-order equation in the unknown function Dy. By Theorem 7, Section 8, a solution of $Du + q(x)u = \varnothing$ is $u(x) = e^{h(x)}$, where this solution satisfies the initial condition $u(a) = 1$. Therefore, if $Dy = e^{h(x)}$, then $y(x)$ is a solution of $L_1[y] = \varnothing$. Thus

$$y_2(x) = \int_a^x e^{h(t)}\, dt$$

is a solution of $L_1[y] = \varnothing$. It is easy to verify that the Wronskian of the pair $y_1(x)$, $y_2(x)$ is $\det[W(x)] = e^{h(x)} \neq \varnothing$ and, consequently, $y_1(x)$, $y_2(x)$ is a solution basis for $L_1[y] = \varnothing$.

---

3 Example

The initial-value problem

$$D^2 y - 2x\, Dy = \varnothing; \qquad y(0) = 0, \quad Dy(0) = 1$$

can be solved by first applying Theorem 2 to obtain a solution basis for this second-order equation. In this example, $0 \in J = (-\infty, +\infty)$, $h(t) = -\int_0^t -2s\, ds = t^2$, and $y_2(x) = \int_0^x e^{t^2}\, dt$. Thus the solutions of this equation are of the form $y(x) = c_1 \cdot 1 + c_2 \int_0^x e^{t^2}\, dt$. Differentiating this equation, we obtain

$$Dy(x) = c_2 e^{x^2}$$

and applying the initial conditions gives

$$y(0) = 0 = c_1$$

$$Dy(0) = 1 = c_2$$

Therefore, $y(x) = \int_0^x e^{t^2}\, dt$ is the solution of the initial-value problem. Since $\int_0^x e^{t^2}\, dt$ is not an elementary function, and since techniques for calculating this integral involve infinite series, we shall leave the solution in this integral form as a convenient representation.

Notice that the reasons that it is possible to find all solutions of $L_1[y] = \varnothing$ are as follows: (i) one solution of this equation is obvious, and (ii) the equation can be considered as a lower-order equation (in this case, first-order) in Dy. Observations of this kind are generally beneficial in analyzing linear differential equations, and the technique of finding solutions by solving a lower-order equation is often referred to as a *deflation rule*, or an *order-reducing method*.

Theorem 6, Section 3, provides a familiar illustration of a deflation rule for finding the zeros of a polynomial. In particular, if p is a polynomial of degree n, and if a root r of p is known, then $p(x) = (x - r) \cdot q(x)$, where $q(x)$ has degree $n - 1$. Now the remaining roots of p are the roots of q, and the problem of finding all n roots of p has been "reduced" to finding the roots of the lower-degree polynomial q.

An analogous situation exists in the case of nth-order, linear homogeneous differential equations. Specifically, for the nth-order linear equation $L[y] = \varnothing$, a solution basis consisting of n functions is required. If $y_1 \neq \varnothing$ is a solution of the equation, then a deflation rule is a method to use y_1 to derive a homogeneous equation of order $n - 1$ whose solutions can be used to construct a solution basis for $L[y] = \varnothing$. The next result is a deflation rule for second-order homogeneous equations. It can be stated in terms of the general second-order operator

(4) $L_2 = D^2 + p_1(x) D + p_2(x)I$

on an interval J.

5 THEOREM

Let $y_1(x)$ be a solution of $L_2[y] = \varnothing$ such that $y_1(x) \neq 0$ for all $x \in J$. Let $a \in J$ and define y_2 by

$$y_2(x) = y_1(x) \int_a^x \frac{\exp[-\int_a^t p_1(s)\, ds]}{y_1{}^2(t)}\, dt$$

Then y_2 is a solution of $L_2[y] = \varnothing$ and y_1, y_2 form a solution basis.

Proof: For convenience in the proof, let $v(t) = \exp[-\int_a^t p_1(s)\, ds]$. Rather than merely verifying that $y_2(x)$ is a solution, we present a derivation of this result which exposes the mechanics of the deflation rule. To do this, let

(A) $y_2(x) = y_1(x)u(x)$

where $u(x)$ is a function to be determined. Then

$$Dy_2(x) = u(x) Dy_1(x) + y_1(x) Du(x)$$

and

$$D^2 y_2(x) = u(x) D^2 y_1(x) + 2 Du(x) Dy_1(x) + y_1(x) D^2 u(x)$$

Now $L_2[y_2(x)] = D^2 y_2(x) + p_1(x) Dy_2(x) + p_2(x)y_2(x)$, so that by using the expressions for y_2, Dy_2, and $D^2 y_2$, we get (after rearranging terms)

$$\begin{aligned} L_2[y_2(x)] &= u(x)[D^2 y_1(x) + p_1(x) Dy_1(x) + p_2(x)y_1(x)] \\ &\quad + y_1(x) D^2 u(x) + [2 Dy_1(x) + p_1(x)y_1(x)] Du(x) \\ &= u(x)L_2[y_1(x)] + y_1(x) D^2 u(x) \\ &\quad + [2 Dy_1(x) + p_1(x)y_1(x)] Du(x) \end{aligned}$$

Since $L_2[y_1(x)] = \varnothing$, it follows that $L_2[y_2(x)] = \varnothing$ if and only if

(B) $y_1(x) D^2 u(x) + [2 Dy_1(x) + p_1(x)y_1(x)] Du(x) = \varnothing$

Thus any function $u(x)$ which is a solution of (B) can be used in (A) to define a solution of $L_2[y] = \emptyset$.

Equation (B) is a second-order linear homogeneous equation and the assumption on $y_1(x)$ implies that (B) has no singular points on J. Therefore, (B) is equivalent to

(C) $$D^2u + \left[\frac{2\,Dy_1(x)}{y_1(x)} + p_1(x)\right]Du = \emptyset$$

By Theorem 2, a solution of (C) is $u_2(x) = \int_a^x e^{h(t)}\,dt$, where

$$h(t) = -\int_a^t \left[\frac{2\,Dy_1(s)}{y_1(s)} + p_1(s)\right]ds$$

Integrating this expression, we have

$$h(t) = -2\int_a^t \frac{Dy_1(s)}{y_1(s)}\,ds - \int_a^t p_1(s)\,ds$$

$$= -2\ln[y_1(t)] + 2\ln[y_1(a)] - \int_a^t p_1(s)\,ds$$

Thus $e^{h(t)} = [y_1(a)]^2[y_1(t)]^{-2}v(t)$, and

$$u_2(x) = [y_1(a)]^2 \int_a^x v(t)[y_1(t)]^{-2}\,dt$$

is a solution of (C). Since the constant $y_1(a) \neq 0$,

$$u(x) = [y_1(a)]^{-2}u_2(x)$$

is also a solution of (C), and this choice gives the specified form for $y_2(x)$ in (A). Finally, the pair $y_1(x)$, $y_2(x)$ is a solution basis for $L_2[y] = \emptyset$ since the Wronskian is $\det[W(x)] = v(x) \neq \emptyset$. This is left as an exercise.

___

6 Example

The second-order equation

$$D^2y - \left(\frac{1}{x}\right)Dy + \left(\frac{1}{x^2}\right)y = \emptyset$$

on $[1, \infty)$ was presented in Example 19b, Section 10. Notice that $y_1(x) = x$ is a solution of this equation that satisfies the hypothesis of Theorem 5. Here,

$$v(t) = \exp\left[-\int_1^t \left(\frac{-1}{s}\right)ds\right] = \exp[\ln(t)] = t$$

so that

$$y_2(x) = x\int_1^x \frac{t}{t^2}\,dt = x\ln(x).$$

A more general version of Theorem 5 is possible in that the requirement $y_1(x) \neq 0$ for all $x \in J$ can be relaxed. As an illustration, consider

$$D^2 y + y = \varnothing$$

on $J = (-\infty, \infty)$, and $y_1(x) = \cos x$. Here, y_1 has an infinite number of zeros on J. However, Theorem 5 is applicable on $J_1 = (-\pi/2, \pi/2)$ since $y_1(x) \neq 0$ for all $x \in J_1$. Thus

$$y_2(x) = \cos x \int_0^x \frac{dt}{\cos^2 t}$$

$$= \cos(x) \tan(x)$$

$$= \sin(x)$$

for all $x \in J_1$. Of course, $y_2(x) = \sin x$ is a solution of the second-order equation on J. The situation illustrated here holds in general, and some comments on the general result are contained in the exercises.

7 EXERCISES

In each of the following, find a solution basis for the given equation by starting with the given solution, and specify an interval J.

1. $D^2 y + 4 Dy + 4y = \varnothing$, $y_1(x) = e^{-2x}$
2. $D^2 y - [(2x - 1)/x] Dy + [(x - 1)/x]y = \varnothing$, $y_1(x) = e^x$
3. $(1 - x^2) D^2 y + 2x Dy - 2y = \varnothing$, $y_1(x) = x$
4. $x^2 D^2 y - 2x Dy + 2y = \varnothing$, $y_1(x) = x$
5. $x D^2 y - Dy - 4x^3 y = \varnothing$, $y_1(x) = e^{x^2}$
6. $x D^2 y - (x + 1) Dy + y = \varnothing$, $y_1(x) = e^x$
7. $(e^x + 1) D^2 y - 2 Dy - e^x y = \varnothing$, $y_1(x) = e^x - 1$
8. The function $y_1(x) = x$ is a solution of the differential equation

$$(x - 1) D^2 y - x Dy + y = \varnothing$$

Find the two-parameter family of solutions of the equation and determine the solution satisfying the initial conditions $y(2) = 2 + e^2$, $Dy(2) = 1 + e^2$.

9. Verify that the function $y_1(x) = x^{-1/2} \sin x$ is a solution of the differential equation

$$x^2 D^2 y + x Dy + (x^2 - \tfrac{1}{4})y = \varnothing$$

and determine the two-parameter family of solutions.

10. The function $y_1(x) = x$ is a solution of the differential equation

$$(1 - x^2) D^2 y - 2x Dy + 2y = \varnothing$$

Find the two-parameter family of solutions of the equation and determine the solution of the initial-value problem $y(0) = 0$, $Dy(0) = 1$.

11. Let $y_1(x)$ be a solution of $D^2 y + p_1(x) Dy + p_2(x)y = \varnothing$ such that $y_1(x) \neq 0$ for all x on an interval J. Let y_2 be the solution of the equation as specified in Theorem 5. Show that y_1 and y_2 are linearly independent on J.

12. In the proof of Theorem 5, it was assumed that the solution y_1 of $D^2y + p_1(x)Dy + p_2(x)y = \emptyset$ was nonzero on J. Prove that if $y_1(b) = 0$ for some $b \in J$, and if y_2 is the solution specified by Theorem 5, then $\lim_{x \to b} y_2(x)$ exists and is nonzero. (Thus y_2 has a removable discontinuity at $x = b$.)

13. Let $y_1(x)$ and $y_2(x)$ be nontrivial solutions of the homogeneous equation $D^2y + p_1(x)Dy + p_2(x)y = \emptyset$ on the interval J. Show that if $y_1(b) = y_2(b) = 0$ for some $b \in J$, then $y_2(x) = ky_1(x)$ for all $x \in J$, where k is the constant $Dy_2(b)/Dy_1(b)$.

The next result provides a method for producing a solution of a second-order linear nonhomogeneous differential equation. This method, which can be generalized to include higher-order equations, is called *variation of parameters*.

8 THEOREM

If $y_1(x)$, $y_2(x)$ is a solution basis for $L_2[y] = \emptyset$, $\det[W(x)]$ is the Wronskian of this pair of functions, and $a \in J$, then

$$z(x) = y_1(x) \int_a^x \frac{-y_2(t)f(t)\,dt}{\det[W(t)]} + y_2(x) \int_a^x \frac{y_1(t)f(t)\,dt}{\det[W(t)]}$$

is a solution of $L_2[y] = f(x)$.

Proof: Again, rather than a verification that z is a solution of $L_2[y] = f(x)$, we present a derivation of the result that illustrates the techniques involved.

Let $z(x) = y_1(x)u(x) + y_2(x)v(x)$, where u and v are functions to be determined so that $L_2[z(x)] = f(x)$. The fact that a choice of u and v exists, and the method used here to determine them, are results of much experimentation. After careful examination of this proof, the reader should be able to supply justification for some of the seemingly arbitrary steps.

Since $Dz(x) = y_1(x)\,Du(x) + u(x)\,Dy_1(x) + y_2(x)\,Dv(x) + v(x)\,Dy_2(x)$,

(A) $y_1(x)\,Du(x) + y_2(x)\,Dv(x) = \emptyset$

has the effect of removing the first derivative term in the unknown functions from this expression for $Dz(x)$. Thus if condition (A) is imposed on u and v, then

$$Dz(x) = u(x)\,Dy_1(x) + v(x)\,Dy_2(x)$$

and

$$D^2z(x) = u(x)\,D^2y_1(x) + Dy_1(x)\,Du(x) + v(x)\,D^2y_2(x) + Dy_2(x)\,Dv(x)$$

Now, using the expressions for $z(x)$, $Dz(x)$, and $D^2z(x)$, we have (after rearranging terms)

$$L_2[z(x)] = u(x)L_2[y_1(x)] + v(x)L_2[y_2(x)] + Dy_1(x)\,Du(x)$$
$$+ Dy_2(x)\,Dv(x)$$
$$= Dy_1(x)\,Du(x) + Dy_2(x)\,Dv(x)$$

since $L_2[y_1(x)] = L_2[y_2(x)] = \varnothing$. From this we can conclude that if u and v satisfy the two equations (A) and

(B) $\qquad Dy_1(x)\,Du(x) + Dy_2(x)\,Dv(x) = f(x)$

then $L_2[z(x)] = f(x)$. Equations (A) and (B) can be written compactly as

(C) $\qquad W(x) \begin{bmatrix} Du(x) \\ Dv(x) \end{bmatrix} = \begin{bmatrix} \varnothing \\ f(x) \end{bmatrix}$

where $W(x)$ is the Wronski matrix for $y_1(x)$, $y_2(x)$; and, since $\det[W(x)] \neq 0$, equation (C) has the unique solution

$$Du(x) = \frac{-y_2(x)f(x)}{\det[W(x)]}$$

$$Dv(x) = \frac{y_1(x)f(x)}{\det[W(x)]}$$

Solutions of these two first-order equations are

$$u(x) = \int_a^x \frac{-y_2(t)f(t)\,dt}{\det[W(t)]}$$

$$v(x) = \int_a^x \frac{y_1(t)f(t)\,dt}{\det[W(t)]}$$

and this provides the specified form of $z(x)$.

As indicated in Section 11, a solution basis for $L_2[y] = \varnothing$, along with one solution of $L_2[y] = f(x)$, is all that is needed to characterize all solutions of $L_2[y] = f(x)$. Thus an immediate consequence of Theorem 8 is the following corollary.

9 COROLLARY

If $y_1(x)$, $y_2(x)$ is a solution basis for $L_2[y] = \varnothing$, then each solution of $L_2[y] = f(x)$ is of the form

$$y(x) = c_1 y_1(x) + c_2 y_2(x) + z(x)$$

where $z(x)$ is as specified in Theorem 8.

---

10 Example

We continue with Example 6 in which the operator is

$$L_2 = D^2 - \left(\frac{1}{x}\right)D + \left(\frac{1}{x^2}\right)I$$

on $J = [1, +\infty)$, and a solution basis for $L_2[y] = \varnothing$ is the pair of functions $y_1(x) = x$ and $y_2(x) = x\ln(x)$. The Wronski matrix for this pair is:

$$W(x) = \begin{bmatrix} x & x\ln(x) \\ 1 & 1+\ln(x) \end{bmatrix}$$

and $\det[W(x)] = x$. Thus a solution of $L_2[y] = f(x)$ is:

$$z(x) = x \int_1^x \frac{-t\cdot\ln(t)f(t)\,dt}{t} + x\ln(x)\int_1^x \frac{tf(t)\,dt}{t}$$

or

$$z(x) = -x\int_1^x \ln(t)f(t)\,dt + x\ln(x)\int_1^x f(t)\,dt$$

In particular, if $f(x) = 2/x$, then $z(x) = x[\ln(x)]^2$, and this was noted in Example 6b, Section 9. To solve the initial-value problem

$$D^2y - \left(\frac{1}{x}\right)Dy + \left(\frac{1}{x^2}\right)y = \frac{2}{x}; \qquad y(1) = 0, \quad Dy(1) = 1$$

note that each solution of the nonhomogeneous equation is of the form

$$y(x) = c_1x + c_2x\ln(x) + x[\ln(x)]^2$$

and

$$Dy(x) = c_1 + c_2[1 + \ln(x)] + [\ln(x)]^2 + 2[\ln(x)]$$

Thus

$$y(1) = 0 = c_1 \qquad \text{and} \qquad Dy(1) = 1 = c_1 + c_2$$

so that

$$y(x) = x\ln(x) + x[\ln(x)]^2$$

is the unique solution.

Examples 6 and 10 illustrate the combined use of Theorem 5 and Theorem 8. In this case, having just one solution of $L_2[y] = \varnothing$ is enough information to solve every initial-value problem of the form

$$L_2[y] = f(x); \qquad y(a) = b_1, \quad Dy(a) = b_2$$

11 EXERCISES

In each of the following equations, use the given solution basis of the corresponding reduced equation to find a solution of the given equation.

1. $D^2y + y = \sec^3 x$; $y_1(x) = \sin x$, $y_2(x) = \cos x$
2. $D^2y + y = \cot x$
3. $(x+1)D^2y + x\,Dy - y = (x+1)^2$; $y_1(x) = x$, $y_2(x) = e^{-x}$
4. $D^2y + 4y = x$; $y_1(x) = \sin 2x$, $y_2(x) = \cos 2x$
5. $D^2y + y = \sec x$

6. $D^2y - 2\,Dy + y = xe^x$; $y_1(x) = e^x$, $y_2(x) = xe^x$

7. $D^2y - 5\,Dy + 6y = e^x \cos x$; $y_1(x) = e^{2x}$, $y_2(x) = e^{3x}$

8. $D^2y - y = \sinh x$; $y_1(x) = e^x$, $y_2(x) = e^{-x}$

9. $(x - 1)\,D^2y - x\,Dy + y = (x - 1)^2$; $y_1(x) = x$, $y_2(x) = e^x$

10. $x^2\,D^2y - 2x\,Dy + 2y = x^3 \ln x$; $y_1(x) = x$, $y_2(x) = x^2$

11. $(x^2 - 1)\,D^2y - 2x\,Dy + 2y = (x^2 - 1)^2$; $y_1(x) = x$, $y_2(x) = x^2 + 1$

12. Consider the differential equation $D^2y + Dy - 2y = e^x - x^2$. A solution of the reduced equation is $y_1(x) = e^x$. Determine the the two-parameter family of solutions of the given equation.

13. Consider the differential equation $x^2\,D^2y + x\,Dy - y = 2x$. Show that the reduced equation has a solution of the form $y(x) = x^r$. Determine the two-parameter family of solutions of the given equation. (Hint: Put the equation in standard form.)

14. Consider the differential equation $x^4\,D^2y + 2x^3\,Dy + y = \cos(1/x)$. A solution of the reduced equation is $y_1(x) = \sin(1/x)$. Find the two-parameter family of solutions of the given equation.

15. Find the solution of the initial-value problem

$$(x^2 + 1)\,D^2y - 2x\,Dy + 2y = (x^2 + 1)^2; \qquad y(0) = 0, \quad Dy(0) = 1$$

given that $y_1(x) = x$ is a solution of the reduced equation.

16. Show that the solution z specified in Theorem 8 is the solution of the initial-value problem

$$L_2[y] = D^2y + p_1(x)\,Dy + p_2(x)y = f(x); \qquad y(a) = Dy(a) = 0$$

17. Using Problem 16, show that if y is the solution of the initial-value problem

$$L_2[y] = \varnothing; \qquad y(a) = b_0, \quad Dy(a) = b_1$$

then $y + z$ is the solution of

$$L_2[y] = f(x); \qquad y(a) = b_0, \quad Dy(a) = b_1$$

The remainder of this section can be omitted without loss of continuity.

Theorems 2, 5, and 8 can be generalized in various ways. Three such generalizations are presented without proof for the nth-order linear operator

$$\textbf{(12)} \qquad L = D^n + p_1(x)\,D^{n-1} + \cdots + p_{n-1}(x)\,D + p_n(x)I$$

on J.

13 THEOREM

If the coefficient functions in L satisfy

$$p_n = p_{n-1} = \cdots = p_{n-k} = \varnothing$$

on J, for some integer k, $1 \le k \le n$, then the k linearly independent functions $y_1(x) = 1$, $y_2(x) = x$, $\ldots$, $y_k(x) = x^{k-1}$ are solutions of

$$L[y] = \varnothing$$

The deflation rule for the nth-order operator L that is an extension of Theorem 5 will necessarily involve considerable symbolism. However, the main idea behind the rule is to assume that a function $y_1(x)$ is known such that $L[y_1(x)] = \varnothing$, and then let $y(x) = y_1(x)u(x)$, where u is a function to be determined such that $y_1(x)u(x)$ is a solution of $L[y] = \varnothing$. This is exactly what was done in the proof of Theorem 5.

Since the determination of $L[y_1(x)u(x)]$ will involve the differentiation of a product of the two functions, Leibnitz's rule, that is, Theorem 8, Section 2, is all that is needed for the derivation of the deflation rule. The details are omitted but they are not difficult to verify. To state the deflation rule, we need to specify an operator of order $n - 1$ defined in terms of the nth-order operator L.

If $p_1(x)$, $p_2(x)$, ..., $p_n(x)$ are the coefficient functions of L in (12), $p_0(x) = 1$, and $y_1(x) \in \mathscr{C}^n(J)$, then define the operator L_3 by

$$L_3 = q_0(x) D^{n-1} + q_1(x) D^{n-2} + \cdots + q_{n-2}(x) D + q_{n-1}(x)I$$

on J, where

$$q_j(x) = \sum_{i=0}^{j} \binom{n-i}{j-i} p_i(x) D^{j-i} y_1(x)$$

14 THEOREM

Let $y_1(x)$ be a solution of $L[y] = \varnothing$ such that $y_1(x) \neq 0$ for all $x \in J$. If $a \in J$ and $v_1(x)$, $v_2(x)$, ..., $v_{n-1}(x)$ is a solution basis for $L_3[v] = \varnothing$, then the set of $n - 1$ functions

$$y_{i+1}(x) = y_1(x) \int_a^x v_i(t)\, dt \qquad i = 1, 2, \ldots, n - 1$$

together with $y_1(x)$, is a solution basis for $L[y] = \varnothing$.

Notice that the coefficient function q_0 in the definition of L_3 is $q_0(x) = y_1(x)$, so that the assumption $y_1(x) \neq 0$ for all $x \in J$ assures that $L_3[v] = \varnothing$ has no singular points on J. If $y_1(x)$ has zeros on J, then, as discussed in the second-order case, it is possible to apply Theorem 14 on a subinterval J_1 of J on which y_1 is nonzero.

The extension of the method of variation of parameters to the nth-order case also calls for a few special definitions in order to make the formula reasonably compact.

If $W(x)$ is the Wronski matrix of a solution basis $y_1(x)$, $y_2(x)$, ..., $y_n(x)$ for $L[y] = \varnothing$ and $f(x) \in \mathscr{C}(J)$, then we define $W_i(x)$ to be the matrix formed by replacing the ith column of $W(x)$ with $[\varnothing, \ldots, \varnothing, f(x)]^T$. For example, if $n = 2$, then

$$W_1(x) = \begin{bmatrix} \varnothing & y_2(x) \\ f(x) & Dy_2(x) \end{bmatrix}, \qquad W_2(x) = \begin{bmatrix} y_1(x) & \varnothing \\ Dy_1(x) & f(x) \end{bmatrix}$$

where

$$W(x) = \begin{bmatrix} y_1(x) & y_2(x) \\ Dy_1(x) & Dy_2(x) \end{bmatrix}$$

15 THEOREM

Let $y_1(x)$, $y_2(x)$, ..., $y_n(x)$ be a solution basis for $L[y] = \varnothing$, and let $W(x)$ and $W_i(x)$ be as defined above. A solution of $L[y] = f(x)$ is:

$$z(x) = \sum_{i=1}^{n} y_i(x) \int_a^x \frac{\det[W_i(t)] \, dt}{\det[W(t)]}$$

where $a \in J$.

Once again, a proof of Theorem 15 is omitted. However, the idea is to find functions $u_1, u_2, \ldots, u_n$ so that

$$z(x) = y_1(x)u_1(x) + y_2(x)u_2(x) + \cdots + y_n(x)u_n(x)$$

is a solution of $L[y] = f(x)$. The particular form of the matrices $W_i(x)$ that were discussed above come from applying Cramer's rule, a topic in many linear algebra texts, to the problem.

16 EXERCISES

1. Determine the three-parameter family of solutions of $D^3y = \cos x$ by:
   a. doing the obvious; and
   b. using the results of this section.
2. One solution of $x^3 \, D^3y - 3x^2 \, D^2y + 6x \, Dy - 6y = \varnothing$ is $y_1(x) = x$. Determine the three-parameter family of solutions of this equation.
3. Consider the third-order nonhomogeneous equation

   $$D^3y + p_1(x) \, D^2y + p_2(x) \, Dy + p_3(x)y = f(x)$$

   on J. Let y_1, y_2, and y_3 be linearly independent solutions of the reduced equation, and let W be their Wronski matrix. Verify Theorem 15 in this case where $n = 3$.
4. Determine the three-parameter family of solutions of the third-order equation

   $$D^3y - 2 \, D^2y - Dy + 2y = 2e^x$$

   given that $y_1(x) = e^x$, $y_2(x) = e^{2x}$, $y_3(x) = e^{-x}$ is a solution basis of the reduced equation.
5. Consider the linear operator $L = x^2 \, D^3 + 2x \, D^2 - 6 \, D$.
   a. Show that the homogeneous equation $L[y] = \varnothing$ has three linearly independent solutions of the form $y(x) = x^r$.
   b. Determine the three-parameter family of solutions of $L[y] = 24x^3$.
6. Consider the third-order homogeneous equation

   $$L[y] = D^3y + p_1(x) \, D^2y + p_2(x) \, Dy + p_3(x)y = \varnothing$$

   Let y_1 and y_2 be linearly independent solutions of $L[y] = \varnothing$.

  a. Put $z = u \cdot y_1$ and calculate the equation of order 2 which is satisfied by Du in order that $L[z(x)] = \varnothing$. [Hint: See the proof of Theorem 5.]

  b. Let $M[y] = \varnothing$ denote the second-order equation found in (a). Show that $v = D(y_2/y_1)$ is a solution of $M[y] = \varnothing$ and use this fact to reduce the order of this equation to an equation of order 1.

7. Two solutions of $x\,D^3y - D^2y - x\,Dy + y = \varnothing$ are $y_1(x) = x$ and $y_2(x) = e^x$. Determine a solution basis for the equation by using the procedure outlined in Problem 6.

LINEAR DIFFERENTIAL EQUATIONS WITH CONSTANT COEFFICIENTS*

Section 13 Introduction

The main topic in this chapter is the analysis of linear differential equations whose coefficients are constant functions. In terms of nth-order linear operators, this amounts to assuming that

(1) $L = D^n + p_1(x) D^{n-1} + \cdots + p_{n-1}(x) Dy + p_n(x)I$

has a more specialized form, namely, each coefficient $p_i(x)$, $i = 1, 2, \ldots, n$, is a constant function.

As usual, no symbolic distinction is made between constants and constant functions. Our only exception to this convention has been the use of $\varnothing$ to represent the zero function. Thus, in this chapter, the linear operators under consideration are of the form

(2) $L = D^n + a_1 D^{n-1} + \cdots + a_{n-1} D + a_n I$

where each coefficient is a constant (function). Since the coefficients in (2) are continuous on the interval $(-\infty, \infty)$ there is no restriction on the interval J associated with (2).

In order to emphasize and take advantage of the special nature of the linear operators having the form (2), a more suggestive notation will be used.

* This chapter depends upon the material in Chapter 2.

3 DEFINITION

Let $P(t) = a_0 t^n + a_1 t^{n-1} + \cdots + a_{n-1} t + a_n$ be a polynomial of degree n. The *nth-order operator polynomial, P(D), associated with P(t)* is defined by

$$P(D) = a_0 D^n + a_1 D^{n-1} + \cdots + a_{n-1} D + a_n I$$

Since the polynomial $P(t)$ in Definition 3 has degree n, the leading coefficient $a_0 \neq 0$. The $P(D)$ representation of nth-order linear operators with constant coefficients is used because of the close relationship between the linear operator $P(D)$ and the polynomial $P(t)$. This relationship will become apparent as the properties of $P(D)$ are derived. Note that, in contrast to the general linear operators studied in the preceding chapter, we are not requiring the leading coefficient to be the constant (function) 1. The primary reason for that requirement in the previous chapter was notational convenience, and a resulting simplification in formulas and representations of solutions. The special nature of the linear operators $P(D)$ allows us to relax this requirement in this chapter.

The general theory of linear equations, as developed in the preceding chapter, suggests that we concentrate first on homogeneous equations. An nth-order linear homogeneous differential equation with constant coefficients is expressed as

(4) $P(D)[y] = a_0 D^n y + a_1 D^{n-1} y + \cdots + a_{n-1} Dy + a_n y = \varnothing$

Notice that not only are all of the coefficient functions in (4) continuous on $(-\infty, \infty)$, each of them is actually infinitely differentiable on this interval. This fact has an effect on the differentiability of solutions of the equations (see Exercises 15, Section 9, Problem 1).

5 THEOREM

If $y(x)$ is a solution of $P(D)[y] = \varnothing$, then $y(x) \in \mathscr{C}^\infty(-\infty, \infty)$.

Proof: Since $y(x)$ is a solution of $P(D)[y] = \varnothing$, $y(x) \in \mathscr{C}^n(-\infty, \infty)$. Substituting $y(x)$ into the equation and solving for $a_0 D^n y(x)$, we have

(A) $a_0 D^n y(x) = -a_1 D^{n-1} y(x) - a_2 D^{n-2} y(x) - \cdots - a_{n-1} Dy(x)$
$$- a_n y(x)$$

Now, $y(x) \in \mathscr{C}^n(-\infty, \infty)$ implies $D^k y(x) \in \mathscr{C}^{n-k}(-\infty, \infty), k = 1, 2, \ldots, n-1$. Thus each term on the right-hand side of (A) has a continuous first derivative at least. This implies that $D^n y(x)$ has a continuous first derivative. Therefore, $y(x) \in \mathscr{C}^{n+1}(-\infty, \infty)$. This is the first step in an inductive argument. The inductive hypothesis is that $y(x) \in \mathscr{C}^{n+m}(-\infty, \infty)$ for some positive integer m. Then, differentiating (A) m times, we obtain the equation

(B) $a_0 D^{n+m} y(x) = -a_1 D^{n+m-1} y(x) - \cdots - a_{n-1} D^{m+1} y(x)$
$$- a_n D^m y(x)$$

Here, by the inductive hypothesis, each term on the right-hand side of (B) is in $\mathscr{C}^1(-\infty, \infty)$, so that by differentiating (B) we have $y(x) \in \mathscr{C}^{n+m+1}(-\infty, \infty)$. This proves the theorem.

In addition to the fact that the solutions of $P(D)[y] = \varnothing$ are infinitely differentiable, we have a related result concerning the nature of these solutions.

6 THEOREM

If $y(x)$ is a solution of $P(D)[y] = \varnothing$, then $D^m y(x)$ is a solution of $P(D)[y] = \varnothing$ for each positive integer m.

Proof: Let $y(x)$ be a solution of $P(D)[y] = \varnothing$ and let m be a positive integer. Since, by Theorem 5, $y(x) \in \mathscr{C}^\infty(-\infty, \infty)$, it follows that both sides of

$$a_0 D^n y(x) + a_1 D^{n-1} y(x) + \cdots + a_{n-1} Dy(x) + a_n y(x) = \varnothing$$

are infinitely differentiable. Thus

$$D^m[a_0 D^n y(x) + a_1 D^{n-1} y(x) + \cdots + a_{n-1} Dy(x) + a_n y(x)]$$
$$= D^m \varnothing = \varnothing$$

or

$$a_0 D^n[D^m y(x)] + a_1 D^{n-1}[D^m y(x)] + \cdots + a_{n-1} D[D^m y(x)]$$
$$+ a_0 D^m y(x) = \varnothing$$

Therefore, $P(D)[D^m y(x)] = \varnothing$.

---

7 Example

One solution of

$$D^2 y + y = \varnothing$$

is $y_1(x) = \sin x$. By Theorem 6, $D[\sin x] = \cos x = y_2(x)$ is also a solution of this equation. Since y_1 and y_2 are linearly independent and the equation has order 2, this pair of functions is a solution basis for the equation. Thus no "new" solutions (that is, no solutions independent of $\sin x$ and $\cos x$) are to be found by taking higher derivatives of $y_1(x)$.

A first-order linear differential operator with constant coefficients is defined in terms of a first-degree polynomial

(8) $\qquad P_1(t) = a_0 t + a_1$

The corresponding operator polynomial is $P_1(D) = a_0 D + a_1 I$. The first-order homogeneous equation

(9) $\qquad P_1(D)[y] = \varnothing$

has a solution of the form

$$y(x) = b \exp\left(\frac{-a_1 x}{a_0}\right)$$

where this solution satisfies the initial condition $y(0) = b$. This solution of (9) is obtained by applying Theorem 7, Section 8. Also, as determined in the proof of Theorem 7, Section 8, every solution of (9) is a multiple of the exponential function $y = \exp(-a_1 x/a_0)$. A natural extension of this result is to find conditions for exponential solutions of the nth-order equation (4).

10 THEOREM

If $P(D)$ is the nth-order operator polynomial associated with

$$P(t) = a_0 t^n + a_1 t^{n-1} + \cdots + a_{n-1} t + a_n$$

then $P(D)[e^{bx}] = P(b)e^{bx}$.

Proof: Since $D^k e^{bx} = b^k e^{bx}$ for each positive integer k, it follows that

$$
\begin{aligned}
P(D)[e^{bx}] &= a_0 b^n e^{bx} + a_1 b^{n-1} e^{bx} + \cdots + a_{n-1} b e^{bx} + a_n e^{bx} \\
&= [a_0 b^n + a_1 b^{n-1} + \cdots + a_{n-1} b + a_n] e^{bx} \\
&= P(b) e^{bx}
\end{aligned}
$$

——————————

    An interesting interpretation concerning this result relating nth-order operator polynomials and exponential functions is its analogy with the eigenvalue-eigenvector problem for matrices. In particular, if A is a square matrix and $y \neq \varnothing$ is a vector such that $Ay = \lambda y$ for some number λ, then λ is called an eigenvalue of A and y is a corresponding eigenvector. Equivalently, the matrix A transforms y into a multiple of itself. Theorem 10 shows that the linear operator $P(D)$ transforms any exponential function into a multiple of itself, and the multiplier (eigenvalue) is $P(b)$.

    With respect to solving the homogeneous equation $P(D)[y] = \varnothing$, Theorem 10 clearly shows that if $P(b) = 0$, that is, if b is a root of the polynomial $P(t)$, then $y(x) = e^{bx}$ is a solution of $P(D)[y] = \varnothing$. Conversely, if $y(x) = e^{bx}$ is a solution of $P(D)[y] = \varnothing$, then $P(b)e^{bx} = \varnothing$, from which we conclude $P(b) = 0$, or b is a root of $P(t)$. Thus we have the following result.

11 COROLLARY

Let $P(D)$ be the nth-order operator polynomial associated with

$$P(t) = a_0 t^n + a_1 t^{n-1} + \cdots + a_{n-1} t + a_n$$

The exponential function $y(x) = e^{bx}$ is a solution of $P(D)[y] = \varnothing$ if and only if b is a root of the polynomial $P(t)$.

——————————

12 Examples

a. Consider the first-degree polynomial $P_1(t) = a_0 t + a_1$. The corresponding first-order homogeneous equation is:

$$P_1(D)[y] = a_0\, Dy + a_1 y = \varnothing$$

Using Corollary 8, Section 8, the solutions have the form

(A) $$y(x) = c \exp\left(\frac{-a_1 x}{a_0}\right)$$

Note that $b = -a_1/a_0$ is the root of $P_1(t)$ so that we could have arrived at the solutions (A) by using Corollary 11.

b. The roots of $P(t) = 2t^2 + 3t - 2$ are $r_1 = \frac{1}{2}$ and $r_2 = -2$. Thus, $y_1(x) = e^{x/2}$ and $y_2(x) = e^{-2x}$ are solutions of

$$2\, D^2 y + 3\, Dy - 2y = \varnothing$$

13 EXERCISES

1. Let $P(t) = t^2 - 3t - 4$ and consider $P(D)[y] = \varnothing$.
   a. Find the two roots of $P(t)$ and determine two solutions of $P(D)[y] = \varnothing$.
   b. Show that the two solutions in (a) are linearly independent and hence constitute a solution basis for $P(D)[y] = \varnothing$.
   c. Determine the two-parameter family of solutions of $P(D)[y] = \varnothing$.
2. Let $P(t) = t^2 + 2t + 2$ and consider $P(D)[y] = \varnothing$.
   a. Show that $y_1(x) = e^{-x} \sin x$ is a solution of $P(D)[y] = \varnothing$.
   b. Determine a solution $y_2(x)$ of $P(D)[y] = \varnothing$ which is independent of y. (Hint: See Theorem 6.)
   c. Determine the two-parameter family of solutions of $P(D)[y] = \varnothing$.
3. Let $P(t) = t^2 - 6t + 9$ and consider $P(D)[y] = \varnothing$.
   a. Show that $y_1(x) = e^{3x}$ is a solution of $P(D)[y] = \varnothing$.
   b. Can Theorem 6 be used to obtain a second solution of $P(D)[y] = \varnothing$? If Theorem 6 is used, will the resulting solution be independent of y_1? Explain.
   c. Use Theorem 5, Section 12, to obtain a second solution y_2 of $P(D)[y] = \varnothing$ which is independent of y_1.
4. Consider the differential equation $D^{n+1} y = \varnothing$. Obtain a solution basis for this equation by:
   a. doing the obvious; and
   b. showing that $y_1(x) = x^n$ is a solution and by applying Theorem 6.
5. Let $P(t) = (t - 1)^3 = t^3 - 3t^2 + 3t - 1$. One solution of $P(D)[y] = \varnothing$ should be obvious from Corollary 11. Can you determine a solution basis for this equation?
6. In each of the following, determine a polynomial $P(t)$ such that the specified set of functions is a solution basis for $P(D)[y] = \varnothing$.
   a. $\{e^{2x}, e^{-x}\}$ b. $\{e^{4x}, e^{-2x}, e^{x/2}\}$ c. $\{1, e^x, e^{-x}\}$

Section 14 Second-Order Homogeneous Equations

We continue the study of linear equations with constant coefficients by considering in detail second-order operator polynomials, that is, by considering operator polynomials associated with the quadratic polynomials

(1) $P_2(t) = a_0 t^2 + a_1 t + a_2$

There are several reasons for concentrating on second-order operator polynomials at this stage in the development. As a result of Corollary 11, Section 13, we know that $y(x) = e^{bx}$ is a solution of $P_2(D)[y] = \varnothing$ if and only if b is a root of (1), and the roots r_1, r_2 of (1) are:

(2)
$$r_1 = \frac{-a_1 + \sqrt{a_1{}^2 - 4a_0 a_2}}{2a_0}$$

$$r_2 = \frac{-a_1 - \sqrt{a_1{}^2 - 4a_0 a_2}}{2a_0}$$

Also, quadratic polynomials are the lowest-degree polynomials which exhibit the various possibilities for roots, namely, distinct roots, multiple roots, and complex roots. Finally, the results established here motivate and illustrate the formal procedures which will be developed for analyzing nth-order operator polynomials.

Consider the second-order homogeneous equation with constant coefficients

(3) $P_2(D)[y] = a_0 D^2 y + a_1 Dy + a_2 y = \varnothing$

The object is to find a solution basis for (3). From Corollary 11, Section 13, the functions

$$y_1(x) = e^{r_1 x}, \qquad y_2(x) = e^{r_2 x}$$

are each solutions of (3), and there are three possible cases determined by the nature of the numbers r_1 and r_2.

CASE I. r_1, r_2 real and unequal.

In this case, it is easily verified that $y_1(x)$ and $y_2(x)$ are linearly independent. In particular, the Wronskian of y_1 and y_2, $\det[W(x)]$, is given by

$$\det[W(x)] = (r_2 - r_1)e^{(r_1 + r_2)x} \neq 0$$

ⱶor all x, since $r_1 \neq r_2$. Thus $y_1(x) = e^{r_1 x}$, $y_2(x) = e^{r_2 x}$ is a solution basis for (3) in this case, and

$$y(x) = c_1 e^{r_1 x} + c_2 e^{r_2 x}$$

is the two-parameter family of solutions.

4 Example

Consider the second-order homogeneous equation

$$D^2y + Dy - 6y = \varnothing$$

The corresponding quadratic polynomial is:

$$P(t) = t^2 + t - 6 = (t + 3)(t - 2)$$

whose roots are $r_1 = -3$ and $r_2 = 2$. Thus $y_1(x) = e^{-3x}$, $y_2(x) = e^{2x}$ is a solution basis for the equation, and

$$y(x) = c_1 e^{-3x} + c_2 e^{2x}$$

is the two-parameter family of solutions.

CASE II. r_1, r_2 real and equal.

In this case, the functions $y_1(x) = e^{r_1 x} = y_2(x)$ are linearly dependent. Thus the pair $y_1(x)$, $y_2(x)$ does not form a solution basis for (3). We must find a second solution of (3) which is independent of $y_1(x) = e^{r_1 x}$. An application of Theorem 6, Section 13, does not provide a solution basis since $Dy_1(x) = r_1 e^{r_1 x} = r_1 y_1(x)$, and so $y_1(x)$ and $Dy_1(x)$ are linearly dependent. Since $y_1(x) \neq 0$ for all x, Theorem 5, Section 12, can be applied. In particular, after dividing (3) by a_0 and using $a = 0$ as the lower limit of integration in the statement of Theorem 5, Section 12, we have

$$y_2(x) = e^{r_1 x} \int_0^x \frac{\exp\left[-\int_0^t \left(\frac{a_1}{a_0}\right) ds\right]}{[e^{r_1 t}]^2} dt = e^{r_1 x} \int_0^x \frac{\exp[-a_1 t/a_0]}{e^{2r_1 t}} dt$$

Now, $r_1 = r_2$, and so it follows from equations (2) that $r_1 + r_2 = 2r_1 = -a_1/a_0$. Therefore,

$$y_2(x) = e^{r_1 x} \int_0^x \frac{e^{-2r_1 t}}{e^{-2r_1 t}} dt = e^{r_1 x} \int_0^x dt = xe^{r_1 x}$$

is a solution of (3) which is independent of $y_1(x) = e^{r_1 x}$. Thus $y_1(x)$, $y_2(x)$ is a solution basis for (3) in this case, and

$$y(x) = c_1 e^{r_1 x} + c_2 x e^{r_1 x}$$

is the two-parameter family of solutions. Note that if we had known the solution $y_2(x) = xe^{r_1 x}$, then, since $Dy_2(x) = e^{r_1 x} + r_1 xe^{r_1 x}$, it follows that $y_1(x) = e^{r_1 x}$ is a second solution which is independent of $y_2(x)$.

5 Example

Consider the second-order homogeneous equation

$$4 D^2y + 4 Dy + y = \varnothing$$

The corresponding quadratic polynomial is $P(t) = 4t^2 + 4t + 1 = 4(t + \frac{1}{2})^2$, and $-\frac{1}{2}$ is a root of multiplicity 2. Thus, by our general result above, $y_1(x) = e^{-x/2}$, $y_2(x) = xe^{-x/2}$ is a solution basis for the equation and

$$y(x) = c_1 e^{-x/2} + c_2 x e^{-x/2}$$

is the two-parameter family of solutions.

CASE III. r_1, r_2 complex conjugates.

Assume that $r_1 = a + ib$ and $r_2 = a - ib, b \neq 0$. The two solutions corresponding to these roots are $y_1(x) = e^{(a + ib)x}$, $y_2(x) = e^{(a - ib)x}$. Since $r_1 \neq r_2$, these functions are linearly independent and hence form a solution basis for (3). However, each of the functions $y_1(x)$ and $y_2(x)$ has the disadvantage of being a complex-valued function of the real variable x. Our interest up to this point has been in obtaining real-valued solutions, and so we want to convert, if possible, the complex-valued solutions $y_1(x)$ and $y_2(x)$ into a linearly independent pair of real-valued solutions. Using Euler's formula, (3) of Section 3, we can write

$$y_1(x) = e^{(a + ib)x} = e^{ax}e^{ibx} = e^{ax}(\cos bx + i \sin bx)$$

$$y_2(x) = e^{(a - ib)x} = e^{ax}e^{-ibx} = e^{ax}(\cos bx - i \sin bx)$$

Now, since $P_2(D)$ is a linear operator, $u(x) = \frac{1}{2}[y_1(x) + y_2(x)] = e^{ax} \cos bx$ is a solution of the equation, and it is a real-valued function. From Theorem 6, Section 13, $Du = ae^{ax} \cos bx - be^{ax} \sin bx$ is also a solution of the equation. Thus

$$P_2(D)[ae^{ax} \cos bx - be^{ax} \sin bx] = \varnothing$$

$$aP_2(D)[e^{ax} \cos bx] - bP_2(D)[e^{ax} \sin bx] = \varnothing$$

$$a \cdot \varnothing - bP_2(D)[e^{ax} \sin bx] = \varnothing$$

Thus it follows that $v(x) = e^{ax} \sin bx$ is a solution of the equation. Note, also, that the difference $y_1 - y_2$ of the two solutions above yields the solution $2ie^{ax} \sin bx$, and this observation provides an alternative derivation of the solution $v(x) = e^{ax} \sin bx$. It is readily verified that the Wronskian of the pair of real-valued solutions $u(x) = e^{ax} \cos bx$, $v(x) = e^{ax} \sin bx$ is given by

$$\det[W(x)] = be^{ax} \neq 0$$

for all x. Thus $u(x), v(x)$ is a solution basis for (3) in this case, and

$$y(x) = e^{ax}(c_1 \cos bx + c_2 \sin bx)$$

is the two-parameter family of solutions.

6 Examples

a. Consider the second-order homogeneous equation

$$D^2 y - 2 Dy + 5y = \varnothing$$

The corresponding quadratic polynomial is $P(t) = t^2 - 2t + 5$, and its roots are $r_1 = 1 + 2i$ and $r_2 = 1 - 2i$. Thus $y_1(x) = e^{(1+2i)x}$ and $y_2(x) = e^{(1-2i)x}$ are linearly independent (complex-valued) solutions. Using the device indicated above for converting this pair into linearly independent, real-valued solutions, we find $u(x) = e^x \cos 2x$ and $v(x) = e^x \sin 2x$ as real-valued solutions forming a solution basis for the equation. The two-parameter family of solutions is given by

$$y(x) = e^x(c_1 \cos 2x + c_2 \sin 2x)$$

b. The second-order homogeneous equation

$$D^2 y + y = \emptyset$$

has been used in a variety of examples (see Example 6, Section 7, and Example 7, Section 13). The corresponding quadratic polynomial is $P(t) = t^2 + 1$, and its roots are $r_1 = i$ and $r_2 = -i$. Thus $y_1(x) = e^{ix}$ and $y_2(x) = e^{-ix}$ are linearly independent, complex-valued solutions which can be converted into the real-valued solutions $u(x) = \cos x$ and $v(x) = \sin x$, as seen in the examples cited above.

We summarize the discussion of second-order homogeneous equations with constant coefficients by the following theorem.

7 THEOREM

Given the second-order homogeneous equation

$$P_2(D)[y] = a_0 D^2 y + a_1 Dy + a_2 y = \emptyset$$

Let r_1 and r_2 be the roots of the corresponding quadratic polynomial $P_2(t) = a_0 t^2 + a_1 t + a_2$. Then:

(i) $y_1(x) = e^{r_1 x}$, $y_2(x) = e^{r_2 x}$ is a solution basis for the equation when r_1 and r_2 are *real* and *unequal*,

(ii) $y_1(x) = e^{rx}$, $y_2(x) = xe^{rx}$ is a solution basis for the equation when r_1 and r_2 are *real* and *equal*, $r_1 = r_2 = r$,

(iii) $y_1(x) = e^{ax} \cos bx$, $y_2(x) = e^{ax} \sin bx$ is a solution basis for the equation when r_1 and r_2 are *complex conjugates*, $r_1 = a + ib$, $r_2 = a - ib$, $b \neq 0$.

8 EXERCISES

Determine a solution basis and the two-parameter family of solutions in each of the following.

1. $D^2 y + 4 Dy + 3y = \emptyset$
2. $D^2 y + 2 Dy + y = \emptyset$
3. $D^2 y - 2 Dy = \emptyset$
4. $D^2 y - 4y = \emptyset$

5. $2 D^2y - 5 Dy + 2y = \emptyset$ 6. $D^2y + 4y = \emptyset$

7. $D^2y + 4 Dy + 4y = \emptyset$ 8. $D^2y + 2 Dy + 3y = \emptyset$

9. $D^2y - 2 Dy + y = \emptyset$ 10. $D^2y - Dy - 30y = \emptyset$

Determine the two-parameter family of solutions in Problems 11–14.

11. $D^2y - y = 2e^x - x^2$ 12. $D^2y - 2 Dy - 3y = e^{4x}$

13. $D^2y + y = 4xe^x$ 14. $D^2y + y = 4 \sin(x)$

15. Let $P(t) = t^2 - 2at + a^2 + b^2, b \neq 0$. Show that

$$P(D)[e^{ax} \cos(bx)] = P(D)[e^{ax} \sin(bx)] = \emptyset$$

16. Let $P(t) = t^2 - 2at + a^2 + b^2, b \neq 0$.
   a. Show that there exists a unique real number c such that $P(D)[e^{cx}] = b^2 e^{cx}$.
   b. Show that if $\gamma < b^2$, then there does not exist a real number r such that $P(D)[e^{rx}] = \gamma e^{rx}$.
   c. Show that if $\gamma > b^2$, then there exist two real numbers r and s such that $P(D)[e^{rx}] = \gamma e^{rx}$ and $P(D)[e^{sx}] = \gamma e^{sx}$.

17. Let $P(t) = at^2 + bt + c$ and let $r + is$ be a complex number. Show that

$$P(D)[e^{rx} \cos(sx)] = e^{rx}\{\cos(sx) Re[P(r + is)] - \sin(sx) Im[P(r + is)]\}$$

Obtain a corresponding expression for $P(D)[e^{rx} \sin(sx)]$.

18. Let $P(t) = at^2 + bt + c$. Use Problem 17 to show that the functions

$$y_1(x) = e^{rx} \cos(sx), \qquad y_2(x) = e^{rx} \sin(sx)$$

are solutions of $P(D)[y] = \emptyset$ if and only if $r + is$ is a root of $P(t)$.

19. Let $P(t) = at^2 + bt + c$, and consider the nonhomogeneous equation

(A) $P(D)[y] = ke^{rx}$

k and r real numbers.
   a. Show that if $P(r) \neq 0$, then $z(x) = ke^{rx}/P(r)$ is a solution of (A).
   b. Show that if $P(r) = 0$ and $DP(r) \neq 0$, then $z(x) = kxe^{rx}/DP(r)$ is a solution of (A).
   c. Show that if $P(r) = DP(r) = 0$, then $z(x) = kx^2 e^{rx}/D^2 P(r)$ is a solution of (A).

Section 15 Homogeneous Equations

In this section, we show that a solution basis for the homogeneous equation $P(D)[y] = \emptyset$, where $P(D)$ is an nth-order operator polynomial, can be specified in terms of the roots of the nth-degree polynomial P. This fact is a generalization of the results that were derived case-by-case for second-order operator polynomials in the previous section.

Some applicable results on polynomials were presented in Section 3. Of particular use in this section is Theorem 9, Section 3, in which the multiplicity of a root of a polynomial is characterized. This leads to a convenient classification of polynomials. A polynomial of degree n is called *simple* if it has n distinct roots. Thus $P(t) = a_0 t^n + a_1 t^{n-1} + \cdots + a_n$ is simple if and only if $P(t) = a_0(t - b_1)(t - b_2) \cdots (t - b_n)$, where $b_i \neq b_j$ for $i \neq j$. In this case, $P(b_1) = P(b_2) = \cdots = P(b_n) = 0$. Therefore, if P is a simple polynomial, then Corollary 11, Section 13, implies that $y_j(x) = e^{b_j x}, j = 1, 2, \ldots, n$, are solutions of $P(D)[y] = \varnothing$. If this set of functions is linearly independent, then they form a solution basis for $P(D)[y] = \varnothing$.

1 THEOREM

If $b_1, b_2, \ldots, b_n$ are distinct (real or complex) numbers, then the set of n functions $y_j(x) = e^{b_j x}, j = 1, 2, \ldots, n$, is linearly independent.

Proof: This proof uses mathematical induction. The theorem is true if $n = 1$ since $c_1 e^{b_1 x} = \varnothing$ implies that $c_1 = 0$. We assume that the theorem is true for $n - 1 \geq 1$ and consider

(A) $c_1 e^{b_1 x} + c_2 e^{b_2 x} + \cdots + c_n e^{b_n x} = \varnothing$

However, (A) is equivalent to

$$c_n = -\sum_{j=1}^{n-1} c_j \exp[(b_j - b_n)x]$$

and differentiating both sides of this equation gives

(B) $$\varnothing = -\sum_{j=1}^{n-1} c_j(b_j - b_n) \exp[(b_j - b_n)x]$$

Notice in (B) that $b_j - b_n \neq 0$ and $(b_j - b_n) \neq (b_k - b_n)$ for $k \neq j$ since the b_j's are distinct. Thus by the inductive assumption, the $n - 1$ functions in (B) are linearly independent and so $c_1(b_1 - b_n) = c_2(b_2 - b_n) = \cdots = c_{n-1}(b_{n-1} - b_n) = 0$, which implies that $c_1 = c_2 = \cdots = c_{n-1} = 0$. Using this information in (A) yields $c_n = 0$, and we can conclude that the n functions are linearly independent.

2 COROLLARY

If P is a simple polynomial of degree n and $b_1, b_2, \ldots, b_n$ are the (distinct) roots of P, then the set of n functions $y_1(x) = e^{b_1 x}, y_2(x) = e^{b_2 x}, \ldots, y_n(x) = e^{b_n x}$ is a solution basis for $P(D)[y] = \varnothing$.

3 Examples

a. If $P(t) = t^3 + 3t^2 - t - 3$, then $P(1) = P(-1) = P(-3) = 0$, or $P(t) = (t - 1)(t + 1)(t + 3)$. Thus a solution basis for

$$P(D)[y] = D^3 y + 3 D^2 y - Dy - 3y = \varnothing$$

is $y_1(x) = e^x$, $y_2(x) = e^{-x}$, $y_3(x) = e^{-3x}$, and each solution of $P(D)[y] = \varnothing$ is of the form

$$y(x) = c_1 e^x + c_2 e^{-x} + c_3 e^{-3x}$$

b. If $P(t) = t^3 - 3t^2 + 9t + 13$, then $P(-1) = P(2 + 3i) = P(2 - 3i) = 0$, or $P(t) = (t + 1)(t^2 - 4t + 13)$. Thus a solution basis for $P(D)[y] = \varnothing$ is

$$y_1(x) = e^{-x}, \quad y_2(x) = e^{(2+3i)x}, \quad y_3(x) = e^{(2-3i)x}$$

Now the functions $y_2(x)$ and $y_3(x)$ are complex-valued. But, as we saw in the preceding section, Case III, this pair of functions can be replaced by real-valued functions involving sines and cosines. In particular, replacing the pair $e^{(2+3i)x}$, $e^{(2-3i)x}$ by $e^{2x} \cos 3x$, $e^{2x} \sin 3x$, we have $y_1(x) = e^{-x}$, $y_2(x) = e^{2x} \cos 3x$, $y_3(x) = e^{2x} \sin 3x$ as a solution basis for the equation. Finally, each solution of $P(D)[y] = \varnothing$ is of the form

$$y(x) = c_1 e^{-x} + e^{2x}(c_2 \cos 3x + c_3 \sin 3x)$$

In the analysis of the homogeneous equation $P(D)[y] = \varnothing$, where $P(t)$ is a simple polynomial, we made no distinction between real and complex roots. Notice that we treated the complex-valued function e^{cx}, where c is a complex number, as if it were a real-valued function. Specifically, this was done in forming linear combinations, establishing linear independence, and calculating derivatives. To put this treatment on a more firm foundation would involve the calculus of complex-valued functions and the algebra of complex vector spaces. However, since the complex roots of a real polynomial occur in conjugate pairs, $a + ib$ and $a - ib$, the associated solutions $e^{(a+ib)x}$ and $e^{(a-ib)x}$ can be replaced by the linearly independent pair of real-valued solutions $e^{ax} \cos(bx)$, $e^{ax} \sin(bx)$, as was indicated in Section 14. In order to make the remainder of our presentation concise, we continue to develop the theory in terms of the roots of the polynomial, without making a distinction between the real and complex roots. Since we are dealing with real differential equations, we naturally want to characterize the set of real-valued solutions. Thus, in actual practice, we assume that the reader will replace complex-valued solution pairs by the associated pair of real-valued functions. An alternative approach which avoids a discussion of complex-valued functions is outlined at the end of this section.

In order to motivate the formal results and procedures of this section, we consider two preliminary examples.

4 Examples

a. Consider the homogeneous equation

$$D^3 y - 3 D^2 y + 4y = \varnothing$$

Here $P(t) = t^3 - 3t^2 + 4 = (t + 1)(t - 2)(t - 2) = (t + 1)(t - 2)^2$. The three roots of $P(t)$ are $-1, 2, 2$ and, according to Corollary 11, Section 13, the functions $y_1(x) = e^{-x}, y_2(x) = e^{2x}$, and $y_3(x) = e^{2x}$ are solutions of $P(D)[y] = \varnothing$. Clearly, these functions are not linearly independent, and, consequently, they do not form a solution basis for the equation. Referring to a similar situation in Section 14, Case II, consider the function $u(x) = xe^{2x}$. Calculating the derivatives of u, we find

$$Du(x) = 2xe^{2x} + e^{2x}$$

$$D^2u(x) = 4xe^{2x} + 4e^{2x}$$

$$D^3u(x) = 8xe^{2x} + 12e^{2x}$$

It is easy to show that $P(D)[u(x)] = \varnothing$, and that the functions $y_1(x)$, $y_2(x)$, and $u(x)$ form a solution basis for $P(D)[y] = \varnothing$.

b. Consider the homogeneous equation

$$D^3y - 6\,D^2y + 12Dy - 8y = \varnothing$$

with $P(t) = t^3 - 6t^2 + 12t - 8 = (t - 2)^3$. Thus 2 is a root of $P(t)$ of multiplicity 3. Of course, $y_1(x) = e^{2x}$ is a solution of $P(D)[y] = \varnothing$, and it is to be expected that $y_2(x) = xe^{2x}$ is also a solution of the equation. Finally, it would now seem reasonable to consider $y_3(x) = x^2e^{2x}$ as a possible solution. By direct calculation, it is easy to verify that $P(D)[y_3(x)] = \varnothing$.

It is easy to show that the functions $y_1(x)$, $y_2(x)$, and $y_3(x)$ are linearly independent, so that they form a solution basis for $P(D)[y] = \varnothing$.

The remainder of this section is concerned with the characterization of a solution basis for $P(D)[y] = \varnothing$, where P is an arbitrary polynomial. In this regard, the main result of this section is Theorem 14. As indicated by the examples above, this characterization will involve computations of the form $P(D)[x^je^{bx}]$, and we shall consider these computations in a sequence of intermediate steps. For the most part, these intermediate steps, or lemmas, are easy to verify, and their proofs are left as exercises. Throughout the development which follows, the reader should keep in mind that the basic objective is to determine n linearly independent solutions of $P(D)[y] = \varnothing$, where n is the degree of the polynomial P. The significance of this chapter, in contrast to Chapter 2, is that when the roots of the polynomial P are known, then an explicit representation of a solution basis, and, consequently, of the n-parameter family of solutions, can be given.

While the characterization stated in Theorem 14 is plausible based upon Examples 4, there is much to be learned in studying the steps which lead to this main result.

5 LEMMA

If j and k are positive integers, then

$$D^j x^k = \binom{k}{j} j! \, x^{k-j}$$

where

$$\binom{k}{j} = \frac{k!}{j! \, (k - j)!}$$

if $k \geq j$ and

$$\binom{k}{j} = 0$$

if $j > k$.

Proof: Exercise.

---

6 LEMMA

If k and m are nonnegative integers and $b \neq 0$, then

$$D^m(x^k e^{bx}) = e^{bx} \sum_{j=0}^{k} \binom{k}{j}\binom{m}{j} j! \, b^{m-j} x^{k-j}$$

Proof: By applying Leibnitz's rule (Theorem 8, Section 2), we have that

$$D^m(x^k e^{bx}) = \sum_{j=0}^{m} \binom{m}{j} D^j x^k \, D^{m-j} e^{bx}$$

Clearly, $D^{m-j} e^{bx} = b^{m-j} e^{bx}$. Thus applying Lemma 5 gives the final result.

Notice that Lemma 6 implies that if $b \neq 0$, then $D^m(x^k e^{bx}) = e^{bx} Q(x)$, where Q is a polynomial of degree k. In fact, the explicit coefficients of Q are given in Lemma 6. This observation can be generalized in two ways. The proofs are left as exercises.

7 LEMMA

If $P(t) = a_0 t^n + a_1 t^{n-1} + \cdots + a_{n-1} t + a_n$ is a polynomial of degree n, and k is a positive integer, then

(a) $P(D)[x^k e^{bx}] = e^{bx} Q(x)$, where Q is a polynomial of degree k or less; and
(b) $D^k[P(x)e^{bx}] = e^{bx} R(x)$, where R is a polynomial of degree n, provided $b \neq 0$.

It was shown in one of the three cases examined in Section 14 that $P_2(D)[xe^{bx}] = \varnothing$ when $P_2(t) = a_0(t - b)^2$. Thus according to Lemma

7(a), $y(x) = x^k e^{bx}$ is a solution of $P(D)[y] = \varnothing$ if and only if $Q(x) = \varnothing$, that is, all of the coefficients of the polynomial Q are zero. It is for this reason that we investigate the explicit form of the polynomial Q in Lemma 7(a).

Before considering the general case, we show all details for the third-degree polynomial

$$P_3(t) = a_0 t^3 + a_1 t^2 + a_2 t + a_3$$

According to Lemma 7(a),

$$P_3(D)[x^k e^{bx}] = e^{bx} Q(x) = e^{bx}(c_0 x^k + c_1 x^{k-1} + \cdots + c_{k-1} x + c_k)$$

Now,

$$P_3(D)[x^k e^{bx}] = a_0 D^3[x^k e^{bx}] + a_1 D^2[x^k e^{bx}] \\ + a_2 D[x^k e^{bx}] + a_3 x^k e^{bx}$$

and by Lemma 6,

$$a_0 D^3[x^k e^{bx}] = a_0 e^{bx}\left[\binom{k}{0}\binom{3}{0} 0! \, b^3 x^k + \binom{k}{1}\binom{3}{1} 1! \, b^2 x^{k-1} \right. \\ \left. + \binom{k}{2}\binom{3}{2} 2! \, b x^{k-2} + \binom{k}{3}\binom{3}{3} 3! \, x^{k-3}\right]$$

$$a_1 D^2[x^k e^{bx}] = a_1 e^{bx}\left[\binom{k}{0}\binom{2}{0} 0! \, b^2 x^k + \binom{k}{1}\binom{2}{1} 1! \, b x^{k-1} \right. \\ \left. + \binom{k}{2}\binom{2}{2} 2! \, x^{k-2}\right]$$

$$a_2 D[x^k e^{bx}] = a_2 e^{bx}\left[\binom{k}{0}\binom{1}{0} 0! \, b x^k + \binom{k}{1}\binom{1}{1} 1! \, x^{k-1}\right]$$

$$a_3 x^k e^{bx} = a_3 e^{bx}\binom{k}{0}\binom{0}{0} 0! \, x^k$$

From these four equations we see that the coefficients $c_0, c_1, \ldots, c_k$ are given by

$$c_0 = \binom{k}{0}\left[a_0 \binom{3}{0} 0! \, b^3 + a_1 \binom{2}{0} 0! \, b^2 + a_2 \binom{1}{0} 0! \, b + a_3 \binom{0}{0} 0!\right]$$

$$c_1 = \binom{k}{1}\left[a_0 \binom{3}{1} 1! \, b^2 + a_1 \binom{2}{1} 1! \, b + a_2 \binom{1}{1} 1!\right]$$

$$c_2 = \binom{k}{2}\left[a_0 \binom{3}{2} 2! \, b + a_1 \binom{2}{2} 2!\right]$$

$$c_3 = \binom{k}{3}\left[a_0 \binom{3}{3} 3!\right]$$

and $c_4 = c_5 = \cdots = c_k = 0$. The coefficients c_0, c_1, c_2, c_3 can be simplified to

$$c_0 = \binom{k}{0} [a_0 b^3 + a_1 b^2 + a_2 b + a_3] = \binom{k}{0} P_3(b)$$

$$c_1 = \binom{k}{1} [3a_0 b^2 + 2a_1 b + a_2] = \binom{k}{1} DP_3(b)$$

$$c_2 = \binom{k}{2} [6a_0 b + 2a_1] = \binom{k}{2} D^2 P_3(b)$$

$$c_3 = \binom{k}{3} 6a_0 = \binom{k}{3} D^3 P_3(b)$$

In summary,

$$P_3(D)[x^k e^{bx}] = e^{bx} \left[\binom{k}{0} P_3(b) x^k + \binom{k}{1} DP_3(b) x^{k-1} \right.$$

$$\left. + \binom{k}{2} D^2 P_3(b) x^{k-2} + \binom{k}{3} D^3 P_3(b) x^{k-3} \right]$$

$$= e^{bx} \sum_{j=0}^{k} \binom{k}{j} D^j P_3(b) x^{k-j}$$

Since $P_3(t)$ has degree 3, $a_0 \neq 0$. Thus

$$c_3 = \binom{k}{3} 6a_0 = 0$$

if and only if

$$\binom{k}{3} = 0$$

that is, $k \leq 2$. From this we can conclude that $y(x) = x^k e^{bx}$ cannot be a solution of $P_3(D)y = \varnothing$ if $k \geq 3$. Another consequence of this expression is, for example, $P_3(D)[x^2 e^{bx}] = \varnothing$ if and only if $P_3(b) = DP_3(b) = D^2 P_3(b) = 0$, that is, b is a root of P_3 of multiplicity 3. Additional conclusions of this type can be derived from the following statement of the general result.

8 THEOREM

If P is a polynomial of degree n and k is a nonnegative integer, then

$$P(D)[x^k e^{bx}] = e^{bx} \sum_{j=0}^{k} \binom{k}{j} D^j P(b) x^{k-j}$$

Proof: Let $P(t) = a_0 t^n + a_1 t^{n-1} + \cdots + a_{n-1} t + a_n$. Then

(A) $$P(D)[x^k e^{bx}] = \sum_{m=0}^{n} a_k D^{n-m}[x^k e^{bx}]$$

Notice that each of the $n + 1$ terms on the right side of (A) is, by Lemma 6, a polynomial of degree k times e^{bx}. We use Lemma 6 to determine the coefficient of x^{k-j} in each of these $n + 1$ polynomials. Thus

$$c_j = a_0 \binom{k}{j}\binom{n}{j} j!\, b^{n-j} + a_1 \binom{k}{j}\binom{n-1}{j} j!\, b^{n-1-j} + \cdots$$

$$+ \, a_{n-j} \binom{k}{j}\binom{j}{j} j!\, b^0$$

where

$$P(D)[x^k e^{bx}] = e^{bx}[c_0 x^k + c_1 x^{k-1} + \cdots + c_i x^{k-i} + \cdots + c_k x^0]$$

Therefore,

$$c_j = \binom{k}{j}\left[a_0 \binom{n}{j} j!\, b^{n-j} + a_1 \binom{n-1}{j} j!\, b^{n-1-j} + \cdots \right.$$

$$\left. + \, a_{n-j} \binom{j}{j} j!\, b^0 \right]$$

However,

$$D^j P(t) = a_0 D^j t^n + a_1 D^j t^{n-1} + \cdots + a_{n-j} D^j t^j$$

so that Lemma 5 gives

$$D^j P(t) = a_0 \binom{n}{j} j!\, t^{n-j} + a_1 \binom{n-1}{j} j!\, t^{n-1-j} + \cdots$$

$$+ \, a_{n-j} \binom{j}{j} j!\, t^0$$

Thus

$$c_j = \binom{k}{j} D^j P(b)$$

and this proves the theorem.

---

9 EXERCISES

Each of the following polynomials P is simple. In each case, enough information is given to determine all of the roots of P. Determine the n-parameter family of solutions for each of the homogeneous equations $P(D)[y] = \varnothing$.

1. $P(t) = t^3 - 6t^2 + 11t - 6$, $r_1 = 1$ is a root
2. $P(t) = t^3 + t + 10$, $r_1 = -2$ is a root
3. $P(t) = t^4 - 2t^3 + t^2 + 8t - 20$, $r_1 = 1 + 2i$ is a root
4. $P(t) = t^4 - 3t^2 - 4$
5. $P(t) = t^4 - 4t^3 + 14t^2 - 4t + 13$, $r_1 = i$ is a root
6. $P(t) = t^3 + t^2 - 4t - 4$, $r_1 = -1$ is a root

7. Prove Corollary 2.

8. Calculate each of the following derivatives and compare the results with the formula in Lemma 6.
   a. $D^3[xe^{2x}]$ b. $D^3[x^3e^{2x}]$ c. $D^3[x^5e^{2x}]$

9. Let $P(D) = D^3 - 4D^2 + 5D - 2I$. Calculate each of the following.
   a. $P(D)[x^2e^x]$ b. $P(D)[x^2e^{2x}]$ c. $P(D)[x^2e^{-x}]$
   Can you explain these results? (Hint: Consider the roots of $P(t)$.)

10. Let $P(D) = D^2 - 2D + I$. Calculate each of the following.
    a. $P(D)[xe^x]$ b. $P(D)[x^4e^x]$ c. $P(D)[x^3e^{2x}]$

11. Prove Lemma 5.

12. Prove Lemma 7.

13. Show that $P(D)[\sinh x] = \emptyset$ if and only if $P(1) = P(-1) = 0$.

Theorem 8 includes Theorem 10, Section 13, as a special case, since if $k = 0$, then

$$P(D)[e^{bx}] = e^{bx}\binom{0}{0}D^0P(b)x^0 = e^{bx}P(b)$$

which is the formula in Theorem 10, Section 13. Notice also that $P(D)[x^ke^{bx}] = \emptyset$ if and only if

$$\sum_{j=0}^{k}\binom{k}{j}D^jP(b)x^{k-j} = \emptyset$$

and, since

$$\binom{k}{j} \neq 0$$

when $k \geq j$ and the powers of x are linearly independent, this implies that $P(b) = DP(b) = \cdots = D^kP(b) = 0$. Thus Theorem 8 can be used to characterize solutions of $P(D)[y] = \emptyset$.

10 THEOREM

If $P(t) = (t - b)^kR(t)$, where k is a positive integer and R is a polynomial, then $y_1(x) = e^{bx}, y_2(x) = xe^{bx}, \ldots, y_k(x) = x^{k-1}e^{bx}$ are linearly independent solutions of $P(D)[y] = \emptyset$.

Proof: It follows from Theorem 9, Section 3, that $P(b) = DP(b) = \cdots = D^{k-1}P(b) = 0$, so that Theorem 8 implies $y_j(x) = x^{j-1}e^{bx}$ is a solution, where $j = 1, 2, \ldots, k$. These k solutions are linearly independent since if

$$c_1e^{bx} + c_2xe^{bx} + \cdots + c_kx^{k-1}e^{bx} = \emptyset$$

then

$$c_1 + c_2x + \cdots + c_kx^{k-1} = \emptyset$$

However, as was shown in several examples (see Example 12, Section 4, and Example 6, Section 10), this implies that $c_1 = c_2 = \cdots = c_k = 0$.

11 Examples

a. If $P(t) = -2t^3 + 6t^2 - 6t + 2$, then

$$DP(t) = -6t^2 + 12t - 6$$

$$D^2 P(t) = -12t + 12$$

Thus

$$P(1) = DP(1) = D^2 P(1) = 0$$

or

$$P(t) = -2(t - 1)^3$$

A solution basis for $P(D)[y] = \varnothing$ is:

$$y_1(x) = e^x, \qquad y_2(x) = xe^x, \qquad y_3(x) = x^2 e^x$$

b. If $P(t) = (t - 1)^2(t + 2)^3(t - 7)$, then P is a sixth-degree polynomial and $P(D)[y] = \varnothing$ is a homogeneous equation of order six. Theorem 10 gives that $y_1(x) = e^x$, $y_2(x) = xe^x$, $y_3(x) = e^{-2x}$, $y_4(x) = xe^{-2x}$, $y_5(x) = x^2 e^{-2x}$, and $y_6(x) = e^{7x}$ are six solutions of this equation. The question now is this: Is this set of six solutions linearly dependent or linearly independent? This question has not been answered by Theorem 10. However, a linear combination of these six functions is:

$$c_1 e^x + c_2 xe^x + c_3 e^{-2x} + c_4 xe^{-2x} + c_5 x^2 e^{-2x} + c_6 e^{7x}$$

or

$$(c_1 + c_2 x)e^x + (c_3 + c_4 x + c_5 x^2)e^{-2x} + c_6 e^{7x}$$

Notice that $Q_1(x) = c_1 + c_2 x$ is a polynomial of degree one or less, $Q_2(x) = c_3 + c_4 x + c_5 x^2$ is a polynomial of degree two or less, and either $Q_3(x) = c_6$ is a polynomial of degree zero or $Q_3(x) = \varnothing$. Thus each linear combination of $y_1, y_2, \ldots, y_6$ is of the form

$$Q_1(x)e^x + Q_2(x)e^{-2x} + Q_3(x)e^{7x}$$

where $Q_1, Q_2,$ and Q_3 are polynomials as described above.

As remarked in the above example, Theorem 10 shows how to find n solutions of $P(D)[y] = \varnothing$, where P is a polynomial of degree n whose roots together with their multiplicities are known. If P is a simple polynomial, then the n solutions are a solution basis, as was shown in Corollary 2. A generalization of Corollary 2 is what is needed to completely characterize all solutions of $P(D)[y] = \varnothing$ for an arbitrary polynomial P.

To begin, we know from Theorem 10 that there are solutions of $P(D)[y] = \varnothing$ of the form

(12) $S_j = \{e^{b_j x}, xe^{b_j x}, \ldots, x^{k_j - 1}e^{b_j x}\}$

where k_j is a positive integer.

A proof that any set of functions of the form of S_j is linearly independent is given in Theorem 10. However, a set of solutions of $P(D)[y] = \varnothing$ can, according to Theorem 10, be of the form of a union of sets of functions as specified in (12), that is $S_1 \cup S_2 \cup \cdots \cup S_m$, for some positive integer m. The question now is whether such a union of m sets, where $b_1, b_2, \ldots, b_m$ are distinct numbers, is a linearly independent set.

13 THEOREM

Let $b_1, b_2, \ldots, b_m$ be distinct (real or complex) numbers and let $S_j, j = 1, 2, \ldots, m$, be sets of functions as specified in (12). The set of functions $S_1 \cup S_2 \cup \cdots \cup S_m$ is linearly independent.

Proof: It was shown in the proof of Theorem 10 that this statement is true for $m = 1$. We assume that it is true for $m - 1 \geq 1$ and use an inductive argument in a manner similar to that in the proof of Theorem 1. If a linear combination of the functions in $S_1 \cup S_2 \cup \cdots \cup S_m$ is equal to the zero function, then we have that

(A) $Q_1(x)e^{b_1 x} + Q_2(x)e^{b_2 x} + \cdots + Q_m(x)e^{b_m x} = \varnothing$

where each Q_j is a polynomial of degree at most $k_j - 1$. Equation (A) can be put in the form

(B) $Q_m(x) = -Q_1(x)e^{r_1 x} - Q_2(x)e^{r_2 x} - \cdots - Q_{m-1}(x)e^{r_{m-1} x}$

where $r_j = b_j - b_m$ and $r_1, r_2, \ldots, r_{m-1}$ are distinct numbers. There is an integer k such that $D^k Q_m(x) = \varnothing$, so that by differentiating both sides of (B), we have

(C) $\varnothing = R_1(x)e^{r_1 x} + R_2(x)e^{r_2 x} + \cdots + R_{m-1}(x)e^{r_{m-1} x}$

where $D^k Q_j(x)e^{r_j x} = R_j(x)e^{r_j x}$. From Lemma 7, the degree of $R_j(x)$ is the degree of $Q_j(x)$. From the inductive assumption, we have that the right side of (C) is a linear combination of linearly independent functions. Thus $R_1(x) = R_2(x) = \cdots = R_{m-1}(x) = \varnothing$, and therefore, $Q_1(x) = Q_2(x) = \cdots = Q_{m-1}(x) = \varnothing$. This, in turn, means from equation (B) that $Q_m(x) = \varnothing$. Therefore, the only solution of equation (A) is $Q_j(x) = \varnothing, j = 1, 2, \ldots, m$, and so $S_1 \cup S_2 \cup \cdots \cup S_m$ is a linearly independent set of functions.

14 THEOREM

If $P(t) = a_0(t - b_1)^{k_1}(t - b_2)^{k_2} \cdots (t - b_m)^{k_m}$, where $b_1, b_2, \ldots, b_m$ are distinct (real or complex) numbers, then a solution basis for $P(D)[y] = \emptyset$ is $S_1 \cup S_2 \cup \cdots \cup S_m$, where

$$S_j = \{e^{b_j x}, xe^{b_j x}, \ldots, x^{k_j - 1} e^{b_j x}\}$$

$j = 1, 2, \ldots, m$.

Proof: Theorem 10 implies that each member of S_j is a solution of $P(D)[y] = \emptyset$, and Theorem 13 implies that $S_1 \cup S_2 \cup \cdots \cup S_m$ is a linearly independent set.

15 Example

If $P(t) = 6(t - 3)^4 (t^2 - 4t + 53)^3 (t + 2)$, then a solution basis for the eleventh-order equation $P(D)[y] = \emptyset$ is:

$$y_1(x) = e^{3x} \qquad y_2(x) = xe^{3x}$$
$$y_3(x) = x^2 e^{3x} \qquad y_4(x) = x^3 e^{3x}$$
$$y_5(x) = e^{2x} \cos(7x) \qquad y_6(x) = e^{2x} \sin(7x)$$
$$y_7(x) = xe^{2x} \cos(7x) \qquad y_8(x) = xe^{2x} \sin(7x)$$
$$y_9(x) = x^2 e^{2x} \cos(7x) \qquad y_{10}(x) = x^2 e^{2x} \sin(7x)$$
$$y_{11}(x) = e^{-2x}$$

The factor $(t^2 - 4t + 53)^3 = (t - 2 - 7i)^3 (t - 2 + 7i)^3$ gives the solutions of the form $x^j e^{(2 + 7i)x}$ and $x^j e^{(2 - 7i)x}$, $j = 0, 1, 2$. However, for exactly the same reasons as presented in Section 14, the real solutions $x^j e^{2x} \cos(7x)$ and $x^j e^{2x} \sin(7x)$ are used here.

To summarize, suppose $P(D)[y] = \emptyset$ is an nth-order homogeneous equation, where $P(D)$ is the operator polynomial associated with the nth-degree polynomial

$$P(t) = a_0 t^n + a_1 t^{n-1} + \cdots + a_{n-1} t + a_n$$

In order to characterize the set of solutions of the equation as an n-parameter family of functions, we must determine a solution basis for the equation. Using the results of this section, specifically, Theorem 14, we can obtain a solution basis if we can express the polynomial $P(t)$ in the factored form

$$P(t) = a_0 (t - b_1)^{k_1} (t - b_2)^{k_2} \cdots (t - b_m)^{k_m}$$

where $k_1, k_2, \ldots, k_m$, are positive integers, $\sum_{j=1}^{m} k_j = n$, and $b_1, b_2, \ldots, b_m$ are distinct numbers, real or complex. Writing $P(t)$ in this factored form is equivalent to finding the m distinct roots of $P(t)$ together with their multiplicities.

16 EXERCISES

1. Find the n-parameter family of solutions for each of the following differential equations.
   a. $D^3y - y = \emptyset$
   b. $D^3y - 5\,Dy - 2y = \emptyset$
   c. $D^3y + 6\,D^2y + 13\,Dy = \emptyset$
   d. $D^3y + 2\,D^2y - Dy - 2y = \emptyset$
   e. $D^4y + 3\,D^2y + 2y = \emptyset$
   f. $D^4y + D^3y + D^2y = \emptyset$
   g. $D^6y + 4\,D^5y + 6\,D^4y + 2\,D^3y - 5\,D^2y - 6\,Dy - 2y = \emptyset$
   h. $D^5y + 4\,D^3y + 4\,Dy = \emptyset$
   i. $D^3y - 2\,D^2y + 3\,Dy - 2y = \emptyset$
   j. $D^4y - 4\,D^2y = \emptyset$
   k. $D^4y - 4\,D^3y + 5\,D^2y - 4\,Dy + 4y = \emptyset$

2. Find the solution of each of the following initial-value problems.
   a. $D^4y - y = \emptyset$; $y(0) = Dy(0) = 0$, $D^2y(0) = D^3y(0) = 1$
   b. $D^3y - 6\,D^2y + 11\,Dy - 6y = \emptyset$; $y(0) = Dy(0) = 0$, $D^2y(0) = 4$
   c. $D^4y + 4\,D^2y = \emptyset$; $y(0) = 1$, $Dy(0) = 0$, $D^2y(0) = 1$, $D^3y(0) = -1$
   d. $D^3y - D^2y + 9\,Dy - 9y = \emptyset$; $y(0) = Dy(0) = 0$, $D^2y(0) = 2$

3. Determine a second-order linear homogeneous equation with constant coefficients having the given pair of functions as a solution basis.
   a. $y_1(x) = e^{-2x}$, $y_2(x) = e^{3x}$ b. $y_1(x) = e^{5x}$, $y_2(x) = xe^{5x}$
   c. $y_1(x) = e^{-x}\sin 2x$, $y_2(x) = e^{-x}\cos 2x$

4. In each of the following, determine a linear homogeneous equation with constant coefficients, and of lowest order, having the given set of functions as solutions.
   a. $y_1(x) = e^x$, $y_2(x) = \sin x$ b. $y_1(x) = xe^{-x}$, $y_2(x) = e^{2x}$
   c. $y_1(x) = \sin x$, $y_2(x) = \sin 2x$ d. $y_1(x) = 1$, $y_2(x) = x$, $y_3(x) = e^x$

5. Give a direct proof that the set of solutions in Example 11b is linearly independent.

6. Consider the homogeneous equation $D^4y - r^4y = \emptyset$, where r is a positive number.
   a. Show that $y_1(x) = \sin rx$, $y_2(x) = \cos rx$, $y_3(x) = \sinh rx$, $y_4(x) = \cosh rx$ is a solution basis for the equation.
   b. Show that the two-point boundary problem

$$D^4y - r^4y = \emptyset; \qquad y(0) = Dy(0) = y(1) = Dy(1) = 0$$

   has a nontrivial solution if and only if $\cos r \cosh r = 1$.

7. Consider the homogeneous equation $D^3y - D^2y - Dy + y = \emptyset$.
   a. Suppose y is a solution of the equation such that $y(0) = Dy(0) = 0$. What value should be assigned to $D^2y(0)$ so that $y(1) = e^2 - 1$?
   b. What initial conditions should be assigned so that the resulting solution y has the property $\lim_{x \to \infty} y(x) = 0$?

8. Let $P(t) = a_0t^n + a_1t^{n-1} + \cdots + a_{n-1}t + a_n$.
   a. Show that if P has the property that its real roots are negative and its complex roots have negative real parts, then every solution y of $P(D)[y] = \emptyset$ has limit zero as $x \to \infty$.

b. Give a necessary and sufficient condition that all solutions of $P(D)[y] = \emptyset$ are bounded on $[0, \infty)$.

9. Consider the second-order homogeneous equation

$$C(D)[y] = ax^2 D^2y + bx\, Dy + cy = \emptyset$$

on $(0, \infty)$, where a, b, and c are constants. Equations of this form are called *Cauchy-Euler equations*.

a. Show that the change of independent variable determined by $t = \ln x$ transforms $C(D)[y] = \emptyset$ into a second-order equation with constant coefficients.

b. Using the change of variable suggested in (a), find the two-parameter family of solutions of each of the following equations.

(i) $x^2 D^2y - x\, Dy + 2y = \emptyset$

(ii) $x^2 D^2y - x\, Dy + 5y = \emptyset$

The remainder of this section can be omitted without loss of continuity.

At the beginning of this section, it was indicated that there is an alternative approach to the characterization of the solutions of the homogeneous equation $P(D)[y] = \emptyset$, and that this approach does not involve the algebra and calculus of complex-valued functions. Of course, since real polynomials of degree $n \geq 2$ will, in general, have complex roots, it is not possible to avoid the arithmetic of complex numbers.

A re-examination of the results in this section will show that the characterization of the solutions of $P(D)[y] = \emptyset$ is essentially contained in the identity

$$(17) \qquad P(D)[x^k e^{bx}] = e^{bx} \sum_{j=0}^{k} \binom{k}{j} D^j P(b) x^{k-j}$$

in Theorem 8. The alternative approach to the characterization of the solutions of $P(D)[y] = \emptyset$ requires an extension of this identity.

As before, let $P(t) = a_0 t^n + a_1 t^{n-1} + \cdots + a_n$ be a real polynomial of degree n. It can be shown that for each nonnegative integer k and each pair of real numbers a and b,

$$(18) \qquad \begin{aligned} P(D)[x^k e^{ax} \cos(bx)] &= e^{ax} \cos(bx) \sum_{j=0}^{k} \binom{k}{j} Re[D^j P(a + ib)] x^{k-j} \\ &\quad - e^{ax} \sin(bx) \sum_{j=0}^{k} \binom{k}{j} Im[D^j P(a + ib)] x^{k-j} \end{aligned}$$

A derivation of this identity can be based upon suitably extended versions of Lemmas 6 and 7, so that (18) is really another application of Leibnitz's rule. The details are omitted, but some observations about identity (18) are worth considering.

Recall from Section 3 that if $c = a + ib$ is a complex number, then $Re(c) = a$ is called the real part of c, and $Im(c) = b$ is called the imaginary part of c. While $P(t), DP(t), \ldots, D^k P(t)$ are real polynomials, their values for $t = a + ib$, namely, $P(a + ib), DP(a + ib), \ldots, D^k P(a + ib)$, are, in general, complex numbers. Thus $Re[D^jP(a + ib)]$ and $Im[D^jP(a + ib)]$ are symbols that denote certain real numbers, so that the right side of (18) is a real-valued function of a real variable. In fact, if $b = 0$ then, since $\cos(0x) = 1$ and $\sin(0x) = 0$, (18) simplifies to (17) in this special case.

Also, the complex number $a + ib$ is a root of the polynomial $P(t)$ if and only if $P(a + ib) = 0$; and $P(a + ib) = 0$ if and only if $Re[P(a + ib)] = Im[P(a + ib)] = 0$. With these observations, it is not too difficult to prove the following theorem.

19 THEOREM

If P is a real polynomial, $a + ib$ is a complex number, and k is a nonnegative integer, then the following three conditions are equivalent:

(a) $P(D)[x^k e^{ax} \cos(bx)] = \varnothing$
(b) $P(a + ib) = DP(a + ib) = \cdots = D^kP(a + ib) = 0$
(c) $(t^2 - 2at + a^2 + b^2)^{k+1}$ is a factor of $P(t)$.

Proof: Exercise.

___

It follows from identity (18) that if $x^k e^{ax} \cos(bx)$ is a solution of $P(D)[y] = \varnothing$, then the set of functions of the form $y_j(x) = x^j e^{ax} \cos(bx)$, where $j = 0, 1, \ldots, k$, is a solution of $P(D)[y] = \varnothing$. This observation together with an application of Theorem 6, Section 13, can be used to prove the following theorem.

20 THEOREM

If $P(D)[x^k e^{ax} \cos(bx)] = \varnothing$, then

$$P(D)[x^j e^{ax} \sin(bx)] = \varnothing$$

for $j = 0, 1, \ldots, k$.

Proof: Exercise.

___

With attention to a few additional details, some of which are mentioned in the exercises, this approach to an analysis of $P(D)[y] = \varnothing$ can be completed.

21 EXERCISES

1. Prove Theorem 19.
2. Prove Theorem 20.
3. Prove that set

$$\{e^{ax} \cos bx, xe^{ax} \cos bx, \ldots, x^k e^{ax} \cos bx, e^{ax} \sin bx, xe^{ax} \sin bx, \ldots, x^k e^{ax} \sin bx\}$$

is linearly independent.

4. Prove that the set $\{e^{ax}\cos bx, e^{ax}\sin bx, e^{cx}\cos dx, e^{cx}\sin dx\}$ is linearly independent if and only if either $a \neq c$ or $b \neq d$, or both.

5. Prove that the union of sets of the form

$$\{e^{a_jx}\cos b_jx, xe^{a_jx}\cos b_jx, \ldots, x^{k_j}e^{a_jx}\cos b_jx, e^{a_jx}\sin b_jx, \ldots, x^{k_j}e^{a_jx}\sin b_jx\}$$

$j = 1, 2, \ldots, m$, where $a_j + ib_j \neq a_p + ib_p$ when $j \neq p$, is a linearly independent set.

6. Let $P(t) = a_0t^n + a_1t^{n-1} + \cdots + a_{n-1}t + a_n$ have the distinct real roots $r_1, r_2, \ldots, r_k$ with multiplicities $m_1, m_2, \ldots, m_k$, respectively, and the distinct complex roots $a_1 + ib_1, a_2 + ib_2, \ldots, a_p + ib_p$ with multiplicities $q_1, q_2, \ldots, q_p$, respectively. Determine a solution basis for $P(D)[y] = \varnothing$.

Section 16 Operator Polynomials and Nonhomogeneous Equations

To conclude this chapter, we consider nonhomogeneous linear differential equations with constant coefficients. These equations can be expressed in the form

(1) $P(D)[y] = f(x) \neq \varnothing$

where, as before,

$$P(t) = a_0t^n + a_1t^{n-1} + \cdots + a_n$$

Since the solutions of the homogeneous equation

(2) $P(D)[y] = \varnothing$

were characterized in Section 15, it follows from the results in Chapter 2 that only one solution of (1) is required to characterize all solutions of (1).

A particularly effective method to determine a solution of the nonhomogeneous equation (1), in an important special case, can be explained in terms of operator polynomials. To do this, however, additional properties of operator polynomials need to be established, and this is the first consideration of this section.

It was shown in Section 15 that members of a solution basis for the homogeneous equation (2) can be determined from the roots, together with their multiplicities, of the polynomial $P(t)$. Of course, a number r is a root of P of multiplicity $k + 1$ if and only if $(t - r)^{k+1}$ is a factor of P. For convenience, these facts, together with the explicit form of the solution of (2), are summarized in the following table.

3 TABLE   If $P(t)$ is a polynomial and k is a nonnegative integer, then

A solution of $P(D)[y] = \varnothing$ is:	If and only if a factor of $P(t)$ is:
(a) x^ke^{rx}	$(t - r)^{k+1}$
(b) $x^ke^{ax}\cos bx, b \neq 0$	$(t^2 - 2at + a^2 + b^2)^{k+1}$
(c) $x^ke^{ax}\sin bx, b \neq 0$	$(t^2 - 2at + a^2 + b^2)^{k+1}$

Table 3 is merely a restatement of the result in Theorem 14, Section 15. In fact, the information in this table was used in Section 15 to specify members of a solution basis of $P(D)[y] = \emptyset$ in terms of the factors of $P(t)$.

Another application of Table 3 is this: Given a collection of functions of the type that appear in the table, we can determine the factors of a polynomial P so that the functions are solutions of $P(D)[y] = \emptyset$.

4 Examples

a. Let $y_1(x) = e^{-x}$, $y_2(x) = e^x \sin(2x)$, and $y_3(x) = xe^{3x}$. To find a polynomial P so that $P(D)[y_1(x)] = \emptyset$, notice that $y_1(x)$ is a solution if and only if $[t - (-1)]^1 = (t + 1)$ is a factor of $P(t)$; $y_2(x)$ is a solution if and only if $[t^2 - 2t + 1^2 + 2^2]^1 = (t^2 - 2t + 5)$ is a factor of $P(t)$; and $y_3(x)$ is a solution if and only if $(t - 3)^{1+1} = (t - 3)^2$ is a factor of $P(t)$. Thus a choice of P is $P(t) = (t + 1)(t^2 - 2t + 5)(t - 3)^2$, and with this choice of P, $P(D)[y] = \emptyset$ is a fifth-order equation with solution basis $y_1(x)$, $y_2(x)$, $y_3(x)$, $y_4(x)$, $y_5(x)$, where $y_4(x) = e^x \cos(2x)$ and $y_5(x) = e^{3x}$. Therefore, every function of the form

$$y(x) = c_1 e^{-x} + c_2 e^x \sin(2x) + c_3 x e^{3x} + c_4 e^x \cos(2x) + c_5 e^{3x}$$

is a solution of $P(D)[y] = \emptyset$. Notice that P is not unique. For example, if

$$Q(t) = 8(t + 1)(t^2 - 2t + 5)(t - 3)^4(t + 6)$$

then y_1, y_2, and y_3 are solutions of $Q(D)[y] = \emptyset$. In fact, for any polynomial $R(t)$,

$$Q(t) = (t + 1)(t^2 - 2t + 5)(t - 3)^2 \cdot R(t)$$

has the property that $y_1(x)$, $y_2(x)$, and $y_3(x)$ are solutions of $Q(D)[y] = \emptyset$.

b. Let $y_1(x) = x^2$, $y_2(x) = e^{-7x} \cos(4x)$, and $y_3(x) = \sin(x)$. Using the table, the factors of $P(t)$ are:

$$(t - 0)^{2+1} = t^3$$

$$[t^2 + 14t + (-7)^2 + (4)^2]^1 = (t^2 + 14t + 65)$$

$$(t^2 + 0t + 0^2 + 1^2) = (t^2 + 1)$$

respectively. Thus $P(t) = t^3(t^2 + 14t + 65)(t^2 + 1)$ and all solutions of $P(D)[y] = \emptyset$ are of the form

$$y(x) = c_1 + c_2 x + c_3 x^2 + c_4 e^{-7x} \cos(4x) + c_5 e^{-7x} \sin(4x)$$
$$+ c_6 \cos(x) + c_7 \sin(x)$$

Note that y_1, y_2, and y_3 are included in this seven-parameter family of functions. Evidently, $P(t)$ is a polynomial of lowest degree such that $P(D)[y_1] = P(D)[y_2] = P(D)[y_3] = \emptyset$.

c. Let $y_1(x) = e^{3x}$, $y_2(x) = x^2 e^{3x}$, and $y_3(x) = xe^{-x}\cos(2x)$. In this example there is no need to consider $y_1(x)$ since $y_2(x)$ is a solution of $P(D)[y] = \varnothing$ if and only if $(t - 3)^{2+1} = (t - 3)^3$ is a factor of $P(t)$; and if $(t - 3)^3$ is a factor of $P(t)$, then $P(D)[e^{3x}] = P(D)[xe^{3x}] = P(D)[x^2 e^{3x}] = \varnothing$, so that both $y_1(x)$ and $y_2(x)$ are solutions of $P(D)[y] = \varnothing$. The factor associated with $y_3(x)$ is:

$$(t^2 + 2t + (-1)^2 + (2)^2)^{1+1} = (t^2 + 2t + 5)^2$$

so that $P(t) = (t - 3)^3(t^2 + 2t + 5)^2$ is a lowest-degree polynomial for which $y_1(x)$, $y_2(x)$, and $y_3(x)$ are solutions of $P(D)[y] = \varnothing$. All solutions of this equation are of the form

$$y(x) = (c_1 + c_2 x + c_3 x^2)e^{3x} + (c_4 + c_5 x)e^{-x}\cos(2x)$$
$$+ (c_6 + c_7 x)e^{-x}\sin(2x)$$

The observations made in these examples, namely, ways to produce an operator polynomial with prescribed properties, lead to the next concept.

5 DEFINITION

If f is a function and $P(t)$ is a polynomial such that $P(D)[f(x)] = \varnothing$, then $P(t)$ is called a *nullifying polynomial for f*.

Not every function has a nullifying polynomial. First of all, f would have to be in $\mathscr{C}^n(J)$, where n is the degree of P. Also, it follows from Section 15 that f has a nullifying polynomial if and only if f can be represented as a linear combination of functions of the form found in Table 3. For some examples of nullifying polynomials, we can use the functions that appeared in Examples 4.

6 Examples

a. If $f(x) = -3e^{-x} + 4e^x \sin(2x) - 7xe^{3x}$, then, from Example 4a,

$$P(t) = (t + 1)(t^2 - 2t + 5)(t - 3)^2$$

is a nullifying polynomial for f. Thus $P(D)[f(x)] = \varnothing$. Notice that $P(t)$ is not only a nullifying polynomial for f, but it is a nullifying polynomial for every function of the form

$$y(x) = c_1 e^{-x} + c_2 e^x \sin(2x) + c_3 xe^{3x} + c_4 e^x \cos(2x) + c_5 e^{3x}$$

Also, it is clear from Example 4a that $P(t)$ is not unique, but it is a nullifying polynomial for f of lowest degree.

b. Following Example 4b, a nullifying polynomial for

$$f(x) = 6x^2 - 4e^{-7x}\cos(4x) + 3\sin(x)$$

is $P(t) = t^3(t^2 + 14t - 65)(t^2 + 1)$.

c. From Example 4c, we have that a nullifying polynomial for

$$f(x) = (9 - 3x + 4x^2)e^{3x} - 2xe^{-x}\cos(2x)$$

is $P(t) = (t - 3)^3(t^2 + 2t + 5)^2$.

d. There is no nullifying polynomial for $f(x) = e^{-x} + \cos(x^2)$ since $\cos(x^2)$ cannot be represented as a linear combination of functions in Table 3. That is, there is no nonzero polynomial $P(t)$ such that $P(D)[\cos(x^2)] = \varnothing$.

7 EXERCISES

1. Consider the set of functions $S = \{x, e^{-2x}, \sin(x)\}$.
   a. Is there a polynomial $P(t)$ of degree 4 such that S is a subset of a solution basis for $P(D)[y] = \varnothing$? If so, find $P(t)$. If not, why not?
   b. Determine a polynomial $Q(t)$ of degree 7 and leading coefficient $b_0 = 3$ such that S is a subset of a solution basis for $Q(D)[y] = \varnothing$.
   c. Determine a homogeneous differential equation with constant coefficients and of least order such that S is a subset of a solution basis for the equation.

2. In each of the following, determine, if possible, a nonzero polynomial $P(t)$ such that the given set of functions is contained in a solution basis for $P(D)[y] = \varnothing$.
   a. $\{e^x, 1\}$ b. $\{\sin(2x), e^{-x}, x^2\}$
   c. $\{e^x \cos(3x), \sin(3x), xe^x\}$ d. $\{xe^{-x}, \cos(x), \ln(x), x^{-2}\}$

3. In each of the following, determine a nullifying polynomial of least degree for the given function.
   a. $f(x) = xe^x + \cos(x) + \sin(x)$
   b. $f(x) = x^2(1 + e^{-x} - \cos 3x)$
   c. $f(x) = 7xe^{-x} + 2x^2 - x + 3 + e^{-2x}\cos(x)$
   d. $f(x) = e^{-2x} + xe^x \sin(2x)$
   e. $f(x) = \cos(2x) + e^x \cos(2x) + 1$
   f. $f(x) = x^4 + x\sin(x) + 3e^{-5x} + \sinh(x)$

According to the definition of the derivative operators $D^m, m = 0, 1, 2, \ldots$, given in Section 2, if h is any function in $\mathscr{C}^{k+m}(J)$, for some interval J, then each of $D^k(D^m[h])$ and $D^m(D^k[h])$ exists and $D^k(D^m[h]) = D^m(D^k[h]) = D^{k+m}[h]$. This identity can be expressed in terms of operator polynomials as follows: D^m and D^k are the operator polynomials associated with the polynomials $P(t) = t^m$ and $Q(t) = t^k$, respectively. The operator polynomials $P(D)$ and $Q(D)$ can be composed in either order, and the two possible compositions are equal. Finally, the composition of $P(D)$ and $Q(D)$ (in either order) is the operator polynomial $R(D)$ associated with the product

$$R(t) = P(t)Q(t) = Q(t)P(t) = t^{k+m}$$

Now suppose that $Q(D)$ is the operator polynomial associated with

$$Q(t) = b_0 t^k + b_1 t^{k-1} + \cdots + b_{k-1}t + b_k$$

and let h be any function in $\mathscr{C}^{k+m}(J)$. Then

$$
\begin{aligned}
Q(D)(D^m[h]) &= b_0\, D^k(D^m[h]) + b_1\, D^{k-1}(D^m[h]) + \cdots \\
&\quad + b_{k-1}\, D(D^m[h]) + b_k\, D^m[h] \\
&= b_0\, D^{k+m}[h] + b_1\, D^{k+m-1}[h] + \cdots \\
&\quad + b_{k-1}\, D^{m+1}[h] + b_k\, D^m[h]
\end{aligned}
$$

or $Q(D)[D^m[h]] = R(D)[h]$, where

$$
R(t) = b_0 t^{k+m} + b_1 t^{k+m-1} + \cdots + b_{k-1} t^{m+1} + b_k t^m = Q(t)\cdot t^m
$$

Composing $Q(D)$ and D^m in the other order, and using the linearity of D^m, we have

$$
\begin{aligned}
D^m(Q(D)[h]) &= D^m(b_0\, D^k[h] + b_1\, D^{k-1}[h] + \cdots \\
&\quad + b_{k-1}\, D[h] + b_k h) \\
&= b_0\, D^{k+m}[h] + b_1\, D^{k+m-1}[h] + \cdots \\
&\quad + b_{k-1}\, D^{m+1}[h] + b_k\, D^m[h] \\
&= Q(D)(D^m[h]) = R(D)[h]
\end{aligned}
$$

Again, the composition of the operator polynomials D^m and $Q(D)$, in either order, is the operator polynomial $R(D)$ associated with the product $R(t) = t^m Q(t) = Q(t)t^m$.

The general result is given by the following theorem.

8 THEOREM

Let $P(D)$ and $Q(D)$ be the operator polynomials associated with the polynomials

$$
P(t) = a_0 t^n + a_1 t^{n-1} + \cdots + a_{n-1}t + a_n
$$

and

$$
Q(t) = b_0 t^k + b_1 t^{k-1} + \cdots + b_{k-1}t + b_k
$$

respectively. Then, for any $h \in \mathscr{C}^{k+n}(J)$,

$$
Q(D)(P(D)[h]) = P(D)(Q(D)[h])
$$

and $Q(D)[P(D)] = P(D)[Q(D)] = R(D)$ is the operator polynomial associated with the product $R(t) = Q(t)P(t) = P(t)Q(t)$.

Proof: Let $h \in \mathscr{C}^{k+n}(J)$. Since $Q(D)$ is a linear operator,

$$
\begin{aligned}
Q(D)(P(D)[h]) &= Q(D)[a_0\, D^n[h] + a_1\, D^{n-1}[h] + \cdots \\
&\quad + a_{n-1}\, D[h] + a_n h] \\
&= a_0 Q(D)(D^n[h]) + a_1 Q(D)(D^{n-1}[h]) + \cdots \\
&\quad + a_{n-1} Q(D)(D[h]) + a_n Q(D)[h]
\end{aligned}
$$

The proof can be completed by expanding the terms $Q(D)(D^m[h])$, $m = 0$, $1, \ldots, n$, and combining the coefficients of the powers of D. The completion of the proof is left as an exercise.

9 Example

This example illustrates that Theorem 8 holds in general only for linear differential operators with constant coefficients. Consider the linear operators

$$L_1 = D \quad \text{and} \quad L_2 = D^2 - x D + x^2 I$$

Let $h(x) = x^3$. Then

$$\begin{aligned}
L_1(L_2[h(x)]) &= D(D^2[x^3] - x D[x^3] + x^2 \cdot x^3) \\
&= D(6x - 3x^3 + x^5) \\
&= 6 - 9x^2 + 5x^4
\end{aligned}$$

and

$$\begin{aligned}
L_2(L_1[h[x]]) &= (D^2 - x D + x^2 I)[D[x^3]] \\
&= D^2(3x^2) - x D(3x^2) + x^2(3x^2) \\
&= 6 - 6x^2 + 3x^4
\end{aligned}$$

Thus $L_1(L_2[h(x)]) \neq L_2(L_1[h(x)])$. Is there any function h in $\mathscr{C}^3(-\infty, \infty)$ such that $L_1(L_2[h(x)]) = L_2(L_1[h(x)])$?

Suppose that $P(t)$ and $Q(t)$ are polynomials and $R(t) = P(t)Q(t)$. It is easy to see that if r is a root of either $P(t)$ or $Q(t)$, then r is a root of $R(t)$. The following corollary of Theorem (8) provides an analogous result for homogeneous equations with constant coefficients.

10 COROLLARY

Let $P(D)$ and $Q(D)$ be the operator polynomials associated with

$$P(t) = a_0 t^n + a_1 t^{n-1} + \cdots + a_{n-1} t + a_n$$
$$Q(t) = b_0 t^k + b_1 t^{k-1} + \cdots + b_{k-1} t + b_k$$

respectively, and let $R(t) = P(t)Q(t) = Q(t)P(t)$. If $y(x)$ is a solution of either $P(D)[y] = \varnothing$ or $Q(D)[y] = \varnothing$, then $y(x)$ is a solution of $R(D)[y] = \varnothing$.

Proof: Suppose $y(x)$ is a solution of $P(D)[y] = \varnothing$. Then

$$R(D)[y(x)] = Q(D)(P(D)[y(x)]) = Q(D)[\varnothing] = \varnothing$$

11 Examples

    a. Corollary 10 can be used to obtain another interpretation of the homogeneous equation $P(D)[y] = \varnothing$. If

$$P(t) = a_0(t - b_1)^{k_1}(t - b_2)^{k_2} \cdots (t - b_m)^{k_m}$$

    where the numbers b_i are distinct, and if $Q_i(t) = (t - b_i)^{k_i}$, $i = 1, 2, \ldots, m$, then the solutions $e^{b_i x}$, $xe^{b_i x}, \ldots, x^{k_i - 1}e^{b_i x}$ of $Q_i(D)[y] = \varnothing$ are solutions of $P(D)[y] = \varnothing$. This observation is also a consequence of Theorem 14, Section 15.

    b. Another interpretation of Corollary 10 amounts to a deflation theorem for $P(D)[y] = \varnothing$. If, for example, $P(b) = DP(b) = \cdots = D^k P(b) = 0$ and $D^{k+1}P(b) \neq 0$, then $(t - b)^{k+1}$ is a factor of $P(t)$. Therefore, $P(t) = (t - b)^{k+1}Q(t)$ for some polynomial $Q(t)$. This implies that $y_1(x) = e^{bx}$, $y_2(x) = xe^{bx}, \ldots, y_{k+1}(x) = x^k e^{bx}$ are members of a solution basis for $P(D)[y] = \varnothing$, and the remaining members of this solution basis are solutions of the lower-order equation $Q(D)[y] = \varnothing$.

Let $P(D)$ be an nth-order operator polynomial. From the general theory of linear differential equations developed in Chapter 2, the solutions of the nonhomogeneous equation $P(D)[y] = f(x)$ consist of all functions of the form

$$y(x) = c_1 y_1(x) + c_2 y_2(x) + \cdots + c_n y_n(x) + z(x)$$

where $y_1(x), y_2(x), \ldots, y_n(x)$ is a solution basis for $P(D)[y] = \varnothing$, c_1, $c_2, \ldots, c_n$ are real numbers, and $z(x)$ is a solution of $P(D)[y] = f(x)$. The functions $y_1, y_2, \ldots, y_n$ can be characterized using the methods presented in Section 15. Our next result provides the basis for a technique to obtain a solution of the nonhomogeneous equation in the special case that the function $f(x)$ is a linear combination of functions from Table 3.

12 THEOREM

Let $P(t)$ be a polynomial, $Q(t)$ be a nullifying polynomial for f, and $R(t) = P(t)Q(t)$. If $z(x)$ is a solution of $P(D)[y] = f(x)$, then $z(x)$ is a solution of $R(D)[y] = \varnothing$.

Proof: If $Q(D)$, $R(D)$, and $z(x)$ are as described in the hypothesis, we have

$$R(D)[z(x)] = Q(D)(P(D)[z(x)]) = Q(D)[f(x)] = \varnothing$$

We now describe a technique for obtaining a solution of $P(D)[y] = f(x)$ in the case where f has a nullifying polynomial $Q(t)$. By Theorem 12, the set of solutions of $P(D)[y] = f(x)$ is a subset of the set of solutions of

$R(D)[y] = \emptyset$, where $R(t) = Q(t)P(t)$. The method consists of determining all solutions of $R(D)[y] = \emptyset$ and choosing from this set a solution of $P(D)[y] = f(x)$. Before presenting a precise description of this method, we consider some examples in detail.

13 Examples

a. Consider the nonhomogeneous equation

(A) $D^2y - 2\,Dy - 3y = 9e^{4x} + 7e^{3x}$

Here, $P(t) = t^2 - 2t - 3 = (t - 3)(t + 1)$ and $f(x) = 9e^{4x} + 7e^{3x}$. A solution basis for the reduced equation

(B) $P(D)[y] = \emptyset$

is $y_1(x) = e^{3x}$, $y_2(x) = e^{-x}$, and a nullifying polynomial for f is:

$Q(t) = (t - 4)(t - 3) = t^2 - 7t + 12$

Now,

$$R(t) = Q(t)P(t) = [(t - 4)(t - 3)][(t - 3)(t + 1)]$$
$$= (t - 4)(t - 3)^2(t + 1)$$

By Theorem 12, the solutions of (A) are solutions of

(C) $R(D)[y] = \emptyset$

Examining the factors of $R(t)$ and $P(t)$, a solution basis for (C) consists of

$$z_1(x) = xe^{3x}, \qquad z_2(x) = e^{4x}, \qquad y_1(x), \qquad y_2(x)$$

where $y_1(x)$ and $y_2(x)$ are the solutions of (B) indicated above. All solutions of (C) are of the form

$$z(x) = c_1z_1(x) + c_2z_2(x) + c_3y_1(x) + c_4y_2(x)$$

and so we look for a solution of (A) in this four-parameter family; that is, we consider the equation

(D) $P(D)[z(x)] = f(x)$

By the linearity of $P(D)$ and the form of $z(x)$, equation (D) can be written

$$c_1P(D)[z_1(x)] + c_2P(D)[z_2(x)]$$
$$+ c_3P(D)[y_1(x)] + c_4P(D)[y_2] = f(x)$$

Since $y_1(x)$ and $y_2(x)$ are solutions of (B), equation (D) can be simplified to

$$c_1P(D)[xe^{3x}] + c_2P(D)[e^{4x}] = 9e^{4x} + 7e^{3x}$$

A straightforward calculation gives

$$P(D)[xe^{3x}] = 4e^{3x} \qquad \text{and} \qquad P(D)[e^{4x}] = 5e^{4x}$$

Therefore,

$$c_1 4e^{3x} + c_2 5e^{4x} = 9e^{4x} + 7e^{3x}$$

or

$$(4c_1 - 7)e^{3x} + (5c_2 - 9)e^{4x} = \varnothing$$

Since e^{3x} and e^{4x} are linearly independent, it follows that $c_1 = \frac{7}{4}$ and $c_2 = \frac{9}{5}$, and these values are uniquely determined.

Now $z(x) = \frac{7}{4}xe^{3x} + \frac{9}{5}e^{4x}$ is a solution of (A), and all solutions of (A) are represented by the two-parameter family

$$y(x) = c_1 e^{3x} + c_2 e^{-x} + \tfrac{7}{4}xe^{3x} + \tfrac{9}{5}e^{4x}$$

b. Consider the equation

$$P(D)[y] = D^2 y + y = 3x - 2e^x \cos(4x) = f(x)$$

A solution basis for the reduced equation is $y_1(x) = \cos x$, $y_2(x) = \sin x$, and a nullifying polynomial for f is $Q(t) = t^2(t^2 - 2t + 17)$. Thus $R(t)$ in factored form is

$$R(t) = t^2(t^2 - 2t + 17)(t^2 + 1)$$

and a solution basis for $R(D)[y] = \varnothing$ is:

$$z_1(x) = 1, \qquad z_2(x) = x, \qquad z_3(x) = e^x \cos(4x)$$

$$z_4(x) = e^x \sin(4x), \qquad y_1(x), \qquad y_2(x)$$

Now the solutions of $P(D)[y] = f(x)$ are contained in the six-parameter family

$$z(x) = c_1 z_1(x) + c_2 z_2(x) + c_3 z_3(x) + c_4 z_4(x) + c_5 y_1(x) + c_6 y_2(x)$$

and so we solve the equation $P(D)[z(x)] = f(x)$. But, $P(D)[y_1(x)] = P(D)[y_2(x)] = \varnothing$, and thus $P(D)[z(x)] = f(x)$ can be written as

$$c_1 P(D)[1] + c_2 P(D)[x] + c_3 P(D)[e^x \cos(4x)]$$
$$+ c_4 P(D)[e^x \sin(4x)] = f(x)$$

In this case,

$$P(D)[1] = 1, \qquad P(D)[x] = x$$

$$P(D)[e^x \cos(4x)] = -14e^x \cos(4x) - 8e^x \sin(4x)$$

$$P(D)[e^x \sin(4x)] = 8e^x \cos(4x) - 14e^x \sin(4x)$$

Therefore,

$$c_1 + c_2 x + c_3[-14e^x \cos(4x) - 8e^x \sin(4x)]$$
$$+ c_4[8e^x \cos(4x) - 14e^x \sin(4x)] = 3x - 2e^x \cos(4x)$$

yielding the four equations

$$c_1 = 0$$
$$c_2 = 3$$
$$-14c_3 + 8c_4 = -2$$
$$-8c_3 - 14c_4 = 0$$

This implies $c_3 = \frac{7}{65}$, and $c_4 = -\frac{4}{65}$. With these values for $c_1, c_2,$ $c_3,$ and $c_4,$ a solution of $P(D)[y] = f(x)$ is:

$$z(x) = 3x + (\tfrac{1}{65})e^x[7\cos(4x) - 4\sin(4x)]$$

The two-parameter family of solutions of $P(D)[y] = f(x)$ is:

$$y(x) = c_1 \cos x + c_2 \sin x + 3x + (\tfrac{1}{65})e^x[7\cos(4x) - 4\sin(4x)]$$

Having the two-parameter family enables us to solve any initial-value problem associated with $P(D)[y] = f(x)$. For example, if $y(0) = 1$ and $Dy(0) = \frac{9}{65}$ are the initial conditions, then it is easy to show that $c_1 = \frac{58}{65}$ and $c_2 = -3$.

The procedure followed in these examples is known as the *method of undetermined coefficients*. The theoretical basis for this method is contained in Theorem 14, Section 15, Table 3, Corollary 10, and Theorem 12. In summary form, the steps which comprise this method are as follows.

14 THE METHOD OF UNDETERMINED COEFFICIENTS

Let $P(D)[y] = f(x)$, where P is a polynomial of degree n and $f(x)$ has a nullifying polynomial. To find $z(x)$ such that $P(D)[z(x)] = f(x)$, do the following.

a. Find a solution basis $y_1(x), y_2(x), \ldots, y_n(x)$ for $P(D)[y] = \varnothing$.
b. Find a nullifying polynomial $Q(t)$ for f.
c. Form $R(t) = Q(t)P(t)$ and determine a solution basis

$$\{z_1(x), z_2(x), \ldots, z_k(x), y_1(x), y_2(x), \ldots, y_n(x)\}$$

for $R(D)[y] = \varnothing$.
d. Let $z(x) = c_1 z_1(x) + c_2 z_2(x) + \cdots + c_k z_k(x)$ and find the coefficients $c_i,$ $i = 1, 2, \ldots, k$, such that

$$c_1 P(D)[z_1(x)] + c_2 P(D)[z_2(x)] + \cdots + c_k P(D)[z_k(x)] = f(x)$$

Step (a) requires factoring $P(t)$ and applying Theorem 14, Section 15. In step (b), we use Table 3. Note the advantage of using a nullifying polynomial for f of least degree. The higher-order homogeneous equation formed in step (c) has, by Corollary 10, $y_1(x), y_2(x), \ldots, y_n(x)$ as members of its solution basis. The remaining functions $z_1(x), z_2(x), \ldots, z_k(x)$ are found by applying Theorem 14, Section 15. By Theorem 12, there is a function $z(x)$ of the form

$$z(x) = c_1 z_1(x) + c_2 z_2(x) + \cdots + c_k z_k(x)$$
$$+ c_{k+1} y_1(x) + \cdots + c_{k+n} y_n(x)$$

which satisfies the equation $P(D)[y] = f(x)$. In other words, there must exist values for the constants c_i such that $P(D)[z(x)] = f(x)$. Using the linearity of $P(D)$, we have

$$c_1 P(D)[z_1(x)] + \cdots + c_k P(D)[z_k(x)] + c_{k+1} P(D)[y_1(x)] + \cdots$$
$$+ c_{k+n} P(D)[y_n(x)] = f(x)$$

Observe that from step (a), $P(D)[y_i(x)] = \varnothing, i = 1, 2, \ldots, n$, and thus values for $c_{k+1}, c_{k+2}, \ldots, c_{k+n}$ can be assigned arbitrarily. The choice $c_{k+1} = c_{k+2} = \cdots = c_{k+n} = 0$ simplifies the calculations and yields step (d).

15 EXERCISES   *odd*

Find the n-parameter family of solutions for each of the following differential equations.

1. $D^2 y - 3 Dy + 2y = \sin(x)$
2. $D^2 y + 2 Dy + y = x + e^x$
3. $D^3 y + 6 D^2 y + 9 Dy + 4y = e^{-x}$
4. $D^2 y - 4 Dy + 5y = 3e^{-x} + x^2 - 2x$
5. $D^2 y + Dy - 2y = 3xe^x$
6. $D^2 y + y = 2 \cos(x)$
7. $D^2 y - 2 Dy + y = x \cos(x)$
8. $D^2 y + 3 Dy - 4y = e^{-4x} + xe^{-x}$
9. $D^2 y + 4 Dy + 4y = xe^{-2x}$
10. $D^3 y + Dy = 6x^2 + \sin x$
11. $D^2 y + Dy - 2y = 2e^{-x} + e^{2x}$
12. $D^2 y + y = x \sin x$
13. $D^2 y - 2 Dy + 2y = e^x \cos x$
14. $D^2 y + 2 Dy + 2y = 8e^{-x} \sin x$

Find the solution of each of the following initial-value problems.

15. $D^2 y - y = 1; y(0) = 0, Dy(0) = 1$
16. $D^3 y - 8y = e^{2x}; y(0) = Dy(0) = D^2 y(0) = 0$
17. $D^2 y + Dy - 6y = \varnothing; y(0) = Dy(0) = 0$
18. $D^2 y + y = \cot(x); y(\pi/2) = Dy(\pi/2) = 0$
19. $D^2 y - Dy - 2y = \sin(2x); y(0) = 1, Dy(0) = -1$
20. Complete the proof of Theorem 8.
21. Consider the linear operator $P(D) = D^2 + r^2 I$ on $[0, \infty)$, where r is a positive number.
    a. Determine the two-parameter family of solutions of $P(D)[y] = \sin ax, a \neq r$.
    b. Determine the two-parameter family of solutions $P(D)[y] = \sin rx$.

c. For what values (if any) of b are all solutions of $P(D)[y] = \sin bx$ bounded? For what values of b are all solutions unbounded? Explain.

22. Consider the differential equation $D^2y - Dy - 2y = 4e^{-x}$ on $J = [0, \infty)$.

   a. Show that the equation has both bounded and unbounded solutions on J.

   b. Determine what initial conditions should be specified at $x = 0$ so that the resulting solutions will be bounded.

The remainder of this section can be omitted without loss of continuity.

The method of variation of parameters presented in Section 12 provides a means for constructing a solution of a nonhomogeneous equation once a solution basis for its reduced equation is known. Thus for nonhomogeneous equations of the form $P(D)[y] = f(x)$, where $P(D)$ is an operator polynomial, we have two techniques available for finding a solution of the equation. Each method has its advantages and disadvantages. The method of undetermined coefficients is applicable only when $f(x)$ has a nullifying polynomial. When this method is applicable, however, it is usually preferable to variation of parameters because the calculations are simpler and no integrations are required. Variation of parameters, on the other hand, has the advantage of being applicable whenever a solution basis for the reduced equation $P(D)[y] = \emptyset$ is known, regardless of the form of function $f(x)$.

16 Examples

a. Consider the nonhomogeneous equation

$$P(D)[y] = \frac{1}{x}$$

where $P(t) = t^2 + 1$. The method of undetermined coefficients is not applicable since $f(x) = 1/x$ does not have a nullifying polynomial. However, the method of variation of parameters, that is, Theorem 8, Section 12, can be used to obtain a solution of this equation. A solution basis for $P(D)[y] = \emptyset$ was found in Example 13b and is $y_1(x) = \cos x$, $y_2(x) = \sin x$, and the Wronskian of this pair of functions is $\det[W(x)] = 1$. Thus

$$z(x) = -\cos x \int_1^x \frac{\sin t}{t}\, dt + \sin x \int_1^x \frac{\cos t}{t}\, dt$$

is a solution of

$$P(D)[y] = \frac{1}{x}$$

Note that the function z can be written

$$z(x) = \int_1^x \sin(x - t)t^{-1}\, dt$$

b. Consider

$$P(D)[y] = \frac{1}{x} + 3x - 2e^x \cos(4x)$$

where $P(t) = t^2 + 1$. Using the results of (a) above, Example 13b, and the superposition principle, that is, Theorem 6, Section 11, a solution of this equation is:

$$z(x) = -\cos x \int_1^x \frac{\sin t}{t} dt + \sin x \int_1^x \frac{\cos t}{t} dt + 3x$$

$$+ \tfrac{1}{65} e^x [7 \cos(4x) - 4 \sin(4x)]$$

We conclude this section with some further observations concerning the method of undetermined coefficients. Notice that the procedure outlined in (14) leads to a characterization of all solutions of the nonhomogeneous equation $P(D)[y] = f(x)$. In most applications, this is precisely what is needed. However, using the results of this section, it is possible to modify the method of undetermined coefficients so that step (a) in (14) is not required. Of course, by deleting step (a) in (14), we lose the characterization of all solutions of $P(D)[y] = f(x)$. In particular, the modified method of undetermined coefficients is a method for finding one solution of $P(D)[y] = f(x)$ without knowledge of a solution basis for $P(D)[y] = \varnothing$.

The important step in the method of undetermined coefficients is the determination of the functions $z_1(x), z_2(x), \ldots, z_k(x)$ in step (c) of (14). This set, which we denote by $Z = \{z_1(x), z_2(x), \ldots, z_k(x)\}$, consists of those members of a solution basis for $R(D)[y] = \varnothing$ which are *not* solutions of $P(D)[y] = \varnothing$. The modification gives an alternative method for determining Z.

17 MODIFIED UNDETERMINED COEFFICIENTS

Let $P(D)[y] = f(x)$, where P is a polynomial of degree n and $f(x)$ has a nullifying polynomial. To find $z(x)$ such that $P(D)[z(x)] = f(x)$, do the following.

a. Find a nullifying polynomial $Q(t)$ for f, where

$$Q(t) = (t - r_1)^{s_1}(t - r_2)^{s_2} \cdots (t - r_j)^{s_j}$$

and the r's are distinct.

b. Calculate $P(r_i)$, $i = 1, 2, \ldots, j$.

c. If $P(r_i) \neq 0$, then $\{e^{r_i x}, x e^{r_i x}, \ldots, x^{s_i - 1} e^{r_i x}\} \subset Z$.

d. If $P(r_i) = DP(r_i) = \cdots = D^{m_i} P(r_i) = 0$; $D^{m_i + 1} P(r_i) \neq 0$, then

$$\{x^{m_i + 1} e^{r_i x}, \ldots, x^{s_i + m_i} e^{r_i x}\} \subset Z$$

e. Let $z(x)$ be a linear combination of members of Z and find the coefficients by solving $P(D)[z(x)] = f(x)$.

The verification of the steps outlined in this procedure is left as an exercise. We illustrate this technique with some examples.

18 Examples

a. Consider the nonhomogeneous equation

$$D^3 y + 4 D^2 y + Dy + y = xe^{-x} + \cos x$$

where

$$P(t) = t^3 + 4t^2 + t + 1$$

A nullifying polynomial for f is:

$$Q(t) = (t + 1)^2(t^2 + 1) = (t + 1)^2(t + i)(t - i)$$

Following step (b) in (17), $P(-1) = 3$ and $P(i) = P(-i) = -3$. Thus case (c) of (17) holds for the roots of $Q(t)$ and $Z = \{e^{-x}, xe^{-x}, \cos x, \sin x\}$. Let

$$z(x) = c_1 e^{-x} + c_2 xe^{-x} + c_3 \cos x + c_4 \sin x$$

and determine values for c_1, c_2, c_3, and c_4 as in the method of undetermined coefficients; that is, solve

$$c_1 P(D)[e^{-x}] + c_2 P(D)[xe^{-x}] + c_3 P(D)[\cos(x)] + c_4 P(D)[\sin(x)]$$
$$= xe^{-x} + \cos(x)$$

b. Consider the nonhomogeneous equation

$$D^4 y - D^3 y - Dy + y = x^2 e^x + e^{-x}$$

where $P(t) = t^4 - t^3 - t + 1$. A nullifying polynomial for f is:

$$Q(t) = (t - 1)^3(t + 1)$$

Following step (b) in (17), $P(1) = 0$ and $P(-1) = 4$. From step (c) in (17), $e^{-x} \in Z$. Calculating the derivatives of P and evaluating them at 1, as indicated by step (d) in (17), $P(1) = DP(1) = 0$ and $D^2 P(1) = 6$. Thus $\{x^2 e^x, x^3 e^x, x^4 e^x\} \subset Z$, and $Z = \{e^{-x}, x^2 e^x, x^3 e^x, x^4 e^x\}$. Finally, let

$$z(x) = c_1 e^{-x} + c_2 x^2 e^x + c_3 x^3 e^x + c_4 x^4 e^x$$

and follow the method of undetermined coefficients.

19 EXERCISES

1. Determine the two-parameter family of solutions of

$$D^2 y - 2 Dy + y = 4e^x \ln x$$

2. Determine the two-parameter family of solutions of

$$D^2 y + 4 Dy + 4y = e^{-2x} + \frac{e^{-2x}}{x^2}$$

(Hint: Use the superposition principle.)

Let $P(t) = t^4 + t^3 - t - 1$. Determine a nullifying polynomial $Q(t)$ for each of the following functions, and determine a solution of $P(D)[y] = f(x)$ by using (17).

3. $f(x) = -2e^{-x}$
4. $f(x) = 2 \sin(x) - 2e^{-x}$
5. $f(x) = 3 \sin(2x) - 2 \cos(2x)$
6. $f(x) = 3e^x + e^{-x} + 2xe^x$
7. Verify the steps of the modified method of undetermined coefficients.

The following two problems provide a basis for an alternate approach to the study of linear equations with constant coefficients.

8. Let h be a function in $\mathscr{C}^\infty(-\infty, \infty)$, that is, h is infinitely differentiable, and let r be a real number. Prove that for every positive integer k,

$$D^k[e^{rx}h(x)] = e^{rx}(D + r)^k[h(x)]$$

where $(D + r)^k$ is the operator polynomial corresponding to the polynomial $P(t) = (t + r)^k$. (Hint: Use induction.) Show also that

$$D^k[e^{-rx}h(x)] = e^{-rx}(D - r)^k[h(x)]$$

9. Let $P(t) = a_0 t^n + a_1 t^{n-1} + \cdots + a_{n-1}t + a_n$. Let $h \in \mathscr{C}^n(-\infty, \infty)$ and let r be any real number. Prove that

$$P(D)[e^{rx}h(x)] = e^{rx}P(D + r)[h(x)]$$

THE FINITE CALCULUS*

Section 17 Introduction

This chapter contains a brief account of those topics from the finite calculus which are useful in analyzing functional equations. The use of the term "calculus" generally suggests a study of functions which have properties defined in terms of limits. Thus there is the calculus of differentiable and integrable functions. Essentially, the finite calculus is the study of functions for which the concept of limits is either not defined or of no importance. For the most part, the functions considered in this chapter will have as their domain either an unbounded interval or a set of consecutive integers.

1 Example

Let N denote the set of nonnegative integers and consider the function defined by

$$f(k) = k!$$

where $k \in N$. Thus $f(0) = 0! = 1$, $f(1) = 1! = 1$, $f(2) = 2! = 2$, $f(3) = 3! = 6$, and so on. For this function, the symbols $Df(3)$ and $\lim_{k \to \pi} f(k)$ are both undefined. Nonetheless, it is easy to verify that f is a solution of the functional equation

$$f(k + 1) = (k + 1)f(k)$$

on N.

* Most of the material in this chapter is independent of the previous chapters. The exception is Section 20 which depends upon Chapters 2 and 3.

In that portion of the finite calculus that is presented here, the emphasis is on the study of two operators whose properties are similar to those of the derivative operator D.

2 DEFINITION

If f is a function and x and $x + 1$ are in the domain of f, then

$$\Delta f(x) = f(x + 1) - f(x)$$

is called the *difference of f at x*, and Δ is called the *difference operator*;

$$Ef(x) = f(x + 1)$$

is called the *displacement of f at x*, and E is called the *displacement operator*. Finally, it is convenient to continue to use I to denote the *identity operator*, that is,

$$If(x) = f(x)$$

for all x in the domain of f.

Notice that just as in the case of derivative, the concepts of difference and displacement are point concepts; that is, these operators are defined at each appropriate number x in the domain of a given function f. Notice also that each of Δf and Ef is a function whose domain is a subset of the domain of f.

3 Examples

a. The equation in Example 1, that is,

$$f(k + 1) = (k + 1)f(k)$$

can be written as

$$f(k + 1) - f(k) = kf(k)$$

or

$$\Delta f(k) = kf(k)$$

and $f(k) = k!$ is a solution of this equation.

b. Let $f(x) = \sin(x)$ on the interval $J = (-\infty, \infty)$. Then

$$E \sin(x) = \sin(x + 1) = \sin(x) \cdot \cos(1) + \cos(x) \cdot \sin(1)$$

and the domain of $E \sin(x)$ is J.

c. Let $g(x) = x^2$ on $J = [0, \infty)$. Then

$$\Delta g(x) = (x + 1)^2 - x^2 = x^2 + 2x + 1 - x^2 = 2x + 1$$

and the domain of Δg is J.

d. The following table defines a function f along with Ef and Δf.

k	$f(k)$	$Ef(k)$	$\Delta f(k)$
3	0	8	8
4	8	-3	-11
5	-3	0	3
6	0	-7	-7
7	-7	9	16
8	9	1	-8
9	1	*	*

* Not defined.

As an example of how one of the entries is computed, observe that

$$\Delta f(7) = f(7 + 1) - f(7) = f(8) - f(7) = 9 - (-7) = 16$$

Of course, $Ef(9)$ and $\Delta f(9)$ are undefined since $9 + 1 = 10$ is not in the domain of f.

For the remainder of this chapter we shall adopt the convention of writing Δf and Ef without explicit reference to the domain of f, the assumption being that these operators will be applied to f only at appropriate points in the domain of f. This is the same convention which is standard for the derivative operator D.

In order to see how Δ is related to D, suppose that g is a differentiable function. The definition of $Dg(b)$ is:

$$Dg(b) = \lim_{h \to 0} \frac{g(b + h) - g(b)}{h}$$

Let f be defined by

$$f(x) = \left(\frac{1}{h}\right) g(hx)$$

Since $f(b/h) = (1/h)g(b)$ and

$$f\left[\left(\frac{b}{h}\right) + 1\right] = f\left(\frac{b + h}{h}\right) = \left(\frac{1}{h}\right) g(b + h)$$

it follows that

$$\left(\frac{1}{h}\right) g(b + h) - \left(\frac{1}{h}\right) g(b) = \Delta f\left(\frac{b}{h}\right)$$

Therefore,

$$Dg(b) = \lim_{h \to 0} \frac{g(b + h) - g(b)}{h} = \lim_{h \to 0} \Delta f\left(\frac{b}{h}\right)$$

In this sense, the operator Δ is, in the limit, the operator D.

The study of the properties of Δ and E constitute the *difference calculus* which is, in many respects, similar to the differential calculus. The *summation calculus*, which is discussed in Section 21, has many points of similarity to integral calculus.

It is clear from Definition 2 that $\Delta f(x)$ exists if and only if $Ef(x)$ exists since both require that $x + 1$ be in the domain of f. In fact,

$$\Delta f(x) = Ef(x) - If(x) = (E - I)f(x)$$

This equality holds independently of the specific choice of the function f, and writing $Ef(x) - If(x) = (E - I)f(x)$ defines the operator $E - I$. A convenient way to record this relationship between Δ and E is to write

(4) $\Delta \equiv E - I$

and to refer to (4) as an *operator identity*.

Similarly, since

$$Ef(x) = f(x + 1) - f(x) + f(x) = \Delta f(x) + If(x) = (\Delta + I)f(x)$$

we write

(5) $E \equiv \Delta + I$

and this is another operator identity.

As a third example of an operator identity, notice that

$$E[\Delta f(x)] = E[f(x + 1) - f(x)] = f(x + 2) - f(x + 1)$$

and

$$\Delta[Ef(x)] = \Delta[f(x + 1)] = f(x + 2) - f(x + 1)$$

Thus $E[\Delta f(x)] = \Delta[Ef(x)]$, or more briefly,

(6) $E \Delta \equiv \Delta E$

There is an extensive theory of the algebra of operators which generalizes the concepts that are merely illustrated in (4), (5), and (6). However, only a few operator identities are needed in this chapter so these examples are offered instead of a detailed treatment of the algebra of operators.

7 EXERCISES

1. Let $f(k) = k!$, $g(k) = \cos(k\pi)$, and $h(k) = 3(-1)^k$. Calculate the following.
   a. $\Delta f(3)$ b. $Eg(7)$ c. $Eh(5)$
   d. $h(4) + 3g(4)$ e. $\Delta h(k)$ f. $\Delta(1/f(k))$
   g. $E[h(k) - 3g(k)]$ h. $\Delta[f(k)g(k)]$

2. Refer to Example 3 and calculate the following.
    a. $\Delta(\Delta x^2)$ b. $E(E \sin x)$ c. $E(\Delta k!)/(k + 1)$
    d. $\Delta[Ef(5)]$, where f is defined in the table.
3. Show that if f is a constant function, then $\Delta f = \varnothing$ and $Ef = f$.
4. Determine Δx^3 and compare this function with Dx^3.
5. Show that if $f \in \mathscr{C}^1 (J)$, where $[0, 1] \subset J$, then there is a number $x \in (0, 1)$ such that

$$\Delta f(0) = Df(x)$$

6. Suppose $f \in \mathscr{C}(-\infty, \infty)$ and f is an increasing function. What properties do Δf and Ef have?
7. Let $f(x) = \sqrt{x}$ on $J = [0, \infty)$. Is this function a solution of the functional equation

$$f(x) + Ef(x) = [\Delta f(x)]^{-1}?$$

8. Is $\Delta D \equiv D \Delta$ an operator identity on the set of functions in $\mathscr{C}^1(-\infty, \infty)$?

Section 18 Differences and Displacements

In developing the properties of Δ and E, it will be assumed that all of the functions under consideration have appropriate domains.

1 THEOREM

Both E and Δ are linear operators.

Proof: It is sufficient to prove this for E since the result for Δ will follow from the operator identity (4), Section 17. If f and g are two functions, and b and c are real numbers, then

$$E[bf(x) + cg(x)] = bf(x + 1) + cg(x + 1) = bEf(x) + cEg(x)$$

Thus E is a linear operator.

Higher-order differences and displacements are defined inductively by

$$\Delta^n f = \Delta[\Delta^{n-1}f]$$
$$E^n f = E[E^{n-1}f]$$

for each integer $n > 1$, and it is convenient to define

$$E^0 \equiv \Delta^0 \equiv I$$

It follows from the linearity of Δ and E that Δ^n and E^n are linear operators for all positive integers n. Also, it is easy to verify using the definitions of E and E^n that

(2) $E^n f(x) = f(x + n)$

Before considering $\Delta^n f$ for arbitrary $n > 1$, notice that

$$\Delta^2 f(x) = \Delta[f(x + 1) - f(x)] = \Delta f(x + 1) - \Delta f(x)$$
$$= f(x + 2) - 2f(x + 1) + f(x)$$

or

$$\Delta^2 f(x) = E^2 f(x) - 2Ef(x) + f(x)$$

This implies that

(3) $\Delta^2 \equiv E^2 - 2E + I$

is an operator identity. Moreover, it is easy to verify that

(4) $E^2 - 2E + I \equiv (E - I)^2$

is also an operator identity, so that (3) and (4) can be combined to give

(5) $\Delta^2 \equiv (E - I)^2 \equiv E^2 - 2E + I$

Keep in mind that the superscript in $(E - I)^2$ does not mean a second power, but rather is a compact way of denoting $(E - I)(E - I)$. Nonetheless, identity (4) is of exactly the same form as

$$(b - c)^2 = b^2 - 2bc + c^2$$

where b and c are real numbers. These observations are generalized by the following theorem, the proof of which is left as an exercise.

6 THEOREM

If n is a positive integer, then

$$\Delta^n \equiv (E - I)^n \equiv \sum_{i=0}^{n} (-1)^i \binom{n}{i} E^{n-i}$$

and

$$E^n \equiv (\Delta + I)^n \equiv \sum_{i=0}^{n} \binom{n}{i} \Delta^{n-i}$$

_____

7 Examples

a. Consider the functions $f(x) = \sin(x)$ and $g(x) = x$ on $J = (-\infty, \infty)$. Then

$$E^3 f(x) = E^3 \sin(x) = \sin(x + 3)$$

and

$$\Delta^2 g(x) = \Delta^2 x = E^2 x - 2Ex + Ix = x + 2 - 2(x + 1) + x = \varnothing$$

on J.

b. Consider the exponential function $h(x) = 2^x$. Since

$$\Delta h(x) = \Delta 2^x = 2^{x+1} - 2^x = 2 \cdot 2^x - 2^x = 2^x(2 - 1) = 2^x$$

it follows that $\Delta^n 2^x = 2^x$ for all positive integers n. Compare this result with the relationship between $y = e^x$ and D.

c. Let h be the function defined on the nonnegative integers N by

$$h(k) = 2^k - 5 \cdot (3)^{-k} + k^2 + 4$$

Then

$$\begin{aligned}
\Delta h(k) &= \Delta[2^k] - 5 \cdot \Delta[3^{-k}] + \Delta[k^2] + \Delta[4] \\
&= [2^{k+1} - 2^k] - 5[3^{-(k+1)} - 3^{-k}] \\
&\quad + [(k+1)^2 - k^2] + [4 - 4] \\
&= 2^k + (\tfrac{10}{3})3^{-k} + 2k + 1
\end{aligned}$$

and

$$\begin{aligned}
Eh(k) &= E[2^k] - 5 \cdot E[3^{-k}] + E[k^2] + E[4] \\
&= 2^{k+1} - 5 \cdot 3^{-(k+1)} + (k+1)^2 + 4
\end{aligned}$$

8 EXERCISES

1. Calculate the following.
   a. $\Delta^2 k!$ b. $E^2 k!$ c. $\Delta^3 7^k$
   d. $\Delta[(k!)2^k]$ e. $\Delta[(k!)/2^k]$ f. $\Delta[2^{-k}(k!)]$
   g. $\Delta[2^k/(k!)]$ h. $\Delta^4[2^{-k}(k!)]$
2. Verify the operator identities (4) and (5).
3. Show that $E^2 \equiv (\Delta + I)^2$.
4. Let $f(k) = (-1)^k + (-2)^k + (-3)^k$, where $k \in N$. Calculate

$$\Delta f(k), \qquad \Delta^2 f(k), \qquad \Delta^3 f(k), \qquad \Delta^4 f(k)$$

5. Prove Theorem 6.
6. Solve the following equations.
   a. $\Delta x^2 = 0$ b. $\Delta^2 x^3 = 0$ c. $\Delta^4 2^x = 0$
   d. $E \sin x = E \cos x$ e. $\Delta^2 x^3 = D^2 x^3$ f. $E^4 x^2 = 1$
   g. $Ex^2 = \Delta^2 x$

As mentioned previously, some of the properties of the difference operator are similar to those of the derivative operator. This is illustrated by considering differences of products and quotients.

9 THEOREM

For any functions f and g,

$$\Delta[f \cdot g] = \Delta f \cdot Eg + f \cdot \Delta g$$

Proof: Let f and g be functions such that x and $x + 1$ are in the domains of f and g. Then

$$\Delta[f(x)g(x)] = f(x + 1)g(x + 1) - f(x)g(x)$$
$$= f(x + 1)g(x + 1) - f(x)g(x + 1)$$
$$+ f(x)g(x + 1) - f(x)g(x)$$
$$= g(x + 1)[f(x + 1) - f(x)]$$
$$+ f(x)[g(x + 1) - g(x)]$$
$$= Eg(x)[\Delta f(x)] + f(x)[\Delta g(x)]$$

and the theorem follows.

We leave as an exercise the proof of the quotient rule.

10 THEOREM

For any functions f and g,

$$\Delta\left[\frac{f}{g}\right] = \frac{g \cdot \Delta f - f \cdot \Delta g}{g \cdot Eg}$$

Computing higher-order differences of the product of two functions leads to a formula that is analogous to Leibnitz's rule for differentation, Theorem 8, Section 2. Before treating the general case, consider $\Delta^2[f \cdot g]$:

$$\Delta^2[fg] = \Delta\{\Delta[fg]\} = \Delta[(\Delta f)(Eg) + f \Delta g]$$
$$= (\Delta^2 f)(E^2 g) + (\Delta f)(\Delta Eg) + (\Delta f)(E \Delta g) + f(\Delta^2 g)$$

However, $\Delta Eg = E \Delta g$, as observed in operator identity (6), Section 17. Thus

$$\Delta^2[fg] = (\Delta^2 f)(E^2 g) + 2(\Delta f)(E \Delta g) + f(\Delta^2 g)$$

which can be written as

(11) $\quad \Delta^2[fg] = \binom{2}{0}(\Delta^2 f)(E^2 \Delta^0 g) + \binom{2}{1}(\Delta f)(E \Delta g) + \binom{2}{2}(\Delta^0 f)(E^0 \Delta^2 g)$

where $\Delta^0 = E^0 = I$ is the identity operator. Also, the binomial coefficients were inserted into (11) to illustrate their occurrence in the general formula. The proof of the general result can be established by induction, and it is left as an exercise.

12 THEOREM

Leibnitz's rule: For any functions f and g,

$$\Delta^n[f \cdot g] = \sum_{i=0}^{n} \binom{n}{i}(\Delta^{n-i} f)(E^{n-i} \Delta^i g)$$

The difference formulas presented in this section can be represented in a variety of equivalent forms by using operator identities. The following example illustrates an alternative form of the product rule, Theorem 9. Additional results along these lines are contained in the exercises.

13 Example

From Theorem 9,

$$\Delta[fg] = (\Delta f)(Eg) + f(\Delta g)$$

and using the operator identity $E = \Delta + I$, we have

$$\Delta[fg] = (\Delta f)(\Delta g + g) + f(\Delta g)$$

or

$$\Delta[fg] = f(\Delta g) + g(\Delta f) + (\Delta f)(\Delta g)$$

14 EXERCISES

1. Prove Theorem 10.
2. Show that $\Delta[fg] = g(\Delta f) + (\Delta g)(Ef)$. Does this contradict Theorem 9?
3. Use Theorem 10 and an appropriate operator identity to show that

$$\Delta\left(\frac{f}{g}\right) = \frac{g(\Delta f) - f(\Delta g)}{g^2 + g(\Delta g)}$$

4. Show that

$$\Delta g^{-1} = \frac{-\Delta g}{g^2 + g(\Delta g)}$$

5. Show that $\Delta g^2 = 2g(\Delta g) + (\Delta g)^2$.
6. Use the results in Problem 5 to obtain an expression for Δg^3 in terms of g and Δg.
7. Can you characterize the class of functions for which

$$\Delta g^2 = (\Delta g)^2 ?$$

8. Use Leibnitz's rule to prove that

$$\Delta^n x^2 2^x = 2^x[x^2 + 4nx + 4n^2 - 2n]$$

$n = 0, 1, 2, \ldots$.
9. Prove Theorem 12.
10. Let $f(k)$ be a function whose domain is N. Find a representation of

$$\Delta^n[(-1)^k f(k)]$$

in terms of the difference of f.

The final topic of this section concerns a class of functions which has especially simple differences.

15 DEFINITION

If f is a function such that $\Delta f(x) = 0$ whenever x and $x + 1$ are in the domain of f, that is, if $\Delta f = \varnothing$, then f is said to be *periodic*.

It follows from this definition that periodic functions have properties with respect to Δ which are similar to the properties of constant functions with respect to D. As a further illustration of this analogy, consider the product rule for differences. From Theorem 9,

(16) $\quad \Delta[fg] = f \, \Delta g$

whenever f is periodic, and this corresponds to the differentiation rule

$$D[ch] = c \, Dh$$

which holds whenever c is a constant and h is differentiable. Periodic functions will also be involved in the treatment of summation in Section 21.

It is clear that any constant function is periodic. However, not all periodic functions are constant functions.

17 Example

The function defined by $f(x) = \sin(2\pi x)$ on the interval $J = (-\infty, \infty)$ is periodic since

$$\Delta \sin(2\pi x) = \sin[2\pi(x + 1)] - \sin(2\pi x)$$
$$= \sin[2\pi x + 2\pi] - \sin(2\pi x) = \varnothing$$

The fact that $\sin[2\pi x + 2\pi] = \sin(2\pi x)$ is a consequence of the addition formula for the sine function.

By using the operator identity $E \equiv \Delta + I$, it is easy to verify the following theorem.

18 THEOREM

A function f is periodic if and only if $Ef(x) = f(x)$ whenever x and $x + 1$ are in the domain of f.

This theorem has a corollary which will be useful in Chapter 5.

19 COROLLARY

Let f be a function with domain N, the set of nonnegative integers. Then f is periodic if and only if f is a constant function.

Proof: If f is a constant function, that is, if

$$f(0) = f(1) = \cdots = f(k) = \cdots$$

then it is clear that f is periodic.

Now suppose f is periodic, so that $Ef = f$. It follows that

$$E^2 f = E(Ef) = Ef = f$$

and, in general, $E^n f = f$ for all nonnegative integers n. But, from (2), $E^n f(0) = f(n)$. Therefore, we have

$$f(n) = E^n f(0) = f(0)$$

for all nonnegative integers n, and f is a constant function.

Recall that if f and g are differentiable functions and $Df = Dg$, then $f = g + c$ for some constant c. The analogue of this result for the difference operator Δ is the final theorem of this section.

20 THEOREM

If f and g are functions such that $\Delta f = \Delta g$, then $f = g + p$, where p is a periodic function.

---

21 EXERCISES

1. Prove Theorem 18.
2. Prove Theorem 20.
3. Is a product of periodic functions a periodic function?
4. If $f(x)$ is periodic, is $[f(x)]^{-1}$ periodic?
5. Find all solutions of $\Delta f(x) = f(x)$. (Hint: $f(x) = 2^x$ is a solution of this equation.)
6. Find all solutions of $\Delta f(x) = 2x + 1$.
7. If g is periodic, is $f(x) = \cos[g(x)]$ periodic?
8. Find all real numbers b such that $f(x) = b^x$ is periodic.

Section 19 Differences of Elementary Functions

Since Δ is a linear operator, it follows that if

$$P(x) = a_0 x^n + a_1 x^{n-1} + \cdots + a_n$$

is a polynomial of degree n, then

(1) $\Delta P(x) = a_0 \, \Delta x^n + a_1 \, \Delta x^{n-1} + \cdots + a_{n-1} \, \Delta x$

For this reason, we begin by considering differences of the power functions. Notice that

$$\Delta x^0 = \Delta 1 = 1 - 1 = \varnothing$$

$$\Delta x^1 = x + 1 - x = 1$$

$$\Delta x^2 = (x + 1)^2 - x^2 = x^2 + 2x + 1 - x^2 = 2x + 1$$

In general,

(2) $\qquad \Delta x^n = (x + 1)^n - x^n = \left[\sum_{j=0}^{n} \binom{n}{j} x^{n-j} \right] - x^n = \sum_{j=1}^{n} \binom{n}{j} x^{n-j}$

for any positive integer n. Thus Δx^n is a polynomial of degree $n - 1$, and this observation together with (1) implies the following result.

3 THEOREM

If $P(x)$ is a polynomial of degree $n \geq 1$, then $\Delta P(x)$ is a polynomial of degree $n - 1$ and $\Delta^{n+1} P(x) = \varnothing$.

___

The statement of Theorem 3 is also correct if the operator Δ is replaced by the operator D. Note, however, that since $Dx^n = nx^{n-1}$, the calculation of $DP(x)$ is quite simple, while it is clear from equation (2) that a determination of $\Delta P(x)$ would be considerably more complicated. It is for this reason that a new class of functions is introduced.

4 DEFINITION

The functions defined by

$$x^{(n)} = x(x - 1)(x - 2) \cdots (x - n + 1) = \prod_{j=0}^{n-1} (x - j)$$

$$x^{(-n)} = \frac{1}{(x + n)^{(n)}} = \frac{1}{(x + n)(x + n - 1) \cdots (x + 1)} = \prod_{j=1}^{n} (x + j)^{-1}$$

where n is a positive integer, together with

$$x^{(0)} = 1$$

are called the *factorial power functions*.

___

In this definition, the symbol Π is used to denote a product in a compact form just as the symbol Σ is used to denote a sum.

Notice that a positive factorial power is a polynomial in the usual sense. For example,

$$x^{(2)} = x(x - 1) = x^2 - x$$

5 THEOREM

If m is an integer, then

$$\Delta x^{(m)} = m[x^{(m-1)}]$$

Proof: If $m = 0$, then the theorem is true since $x^{(0)} = 1$. Suppose $m \geq 1$.

Then

$$\Delta x^{(m)} = (x + 1)^{(m)} - x^{(m)} = \prod_{j=0}^{m-1} (x + 1 - j) - \prod_{j=0}^{m-1} (x - j)$$
$$= [x + 1 - (x - m + 1)][x(x - 1)(x - 2) \cdots (x - m + 2)]$$

or

$$\Delta x^{(m)} = m[x^{(m-1)}]$$

The case $m \leq -1$ is left as an exercise.

A main advantage in using factorial powers in the difference calculus is the analogy between

$$\Delta x^{(m)} = m[x^{(m-1)}]$$

and

$$Dx^m = mx^{m-1}$$

Factorial powers also provide an alternative way to represent some familiar expressions.

6 Examples

a. If r is a real number and j is a positive integer, then

(A) $D^j x^r = r^{(j)} x^{r-j}$

In particular,

$$Dx^r = rx^{r-1} = r^{(1)} x^{r-1}$$

and

$$D^2 x^r = r(r - 1)x^{r-2} = r^{(2)} x^{r-2}$$

Thus a proof of (A) follows by induction.

b. If n is a positive integer, then $n^{(n)} = n!$ since

$$n^{(n)} = n(n - 1)(n - 2) \cdots (n - n + 1) = n!$$

c. If m and n are integers and $m \geq n > 0$, then

$$m^{(n)} = m(m - 1) \cdots (m - n + 1) = \frac{m!}{(m - n)!}$$

Also,

$$\frac{m^{(n)}}{n!} = \frac{m!}{n! (m - n)!} = \binom{m}{n}$$

where

$$\binom{m}{n}$$

denotes the binomial number.

The connection between factorial powers and binomial numbers illustrated by Example 6c leads to an extension of the definition of binomials.

7 DEFINITION

If $n \geq 0$ is an integer, then the function defined by

$$\binom{x}{n} = \frac{x^{(n)}}{n!}, \qquad x \in (-\infty, \infty)$$

is called the nth *binomial function*.

_____

Since Δ is a linear operator, it follows from Theorem 5 and Definition 7 that

$$\Delta \binom{x}{n} = \frac{\Delta x^{(n)}}{n!} = \frac{n x^{(n-1)}}{n!} = \frac{x^{(n-1)}}{(n-1)!} = \binom{x}{n-1}$$

that is, if $n \geq 1$, then

(8) $$\Delta \binom{x}{n} = \binom{x}{n-1}$$

9 EXERCISES

1. Calculate the following factorial powers.
   a. $5^{(3)}$ b. $3^{(5)}$ c. $0^{(-2)}$
   d. $(\frac{1}{2})^{(3)}$ e. $(-2)^{(-3)}$ f. $1/(-2)^{(3)}$
   g. $6^{(5)}$

2. Prove Theorem 5 for the case $m < 0$.

3. Show that if $n > 0$, then
   $$x^{(-n)}(x+n)^{(n)} = x^{(0)} = 1$$

4. What are the solutions of the following?
   a. $x^{(n)} = 0$, where $n > 0$ b. $[x^{(n)}]^{-1} = 0$, where $n < 0$

5. Express the following functions in terms of factorial powers.
   a. $f(x) = (x+1)(x+2)(x+3)$ b. $f(x) = x^2 - 2x$
   c. $f(x) = x(x+2)(x+3)$ d. $f(x) = x^3 + 5x^2 + 6x$
   e. $f(x) = x/(x+1)$ f. $f(x) = 1/(x+4)(x+1)$

6. Calculate the following.
   a. $\Delta x^{(-5)}$ b. $\Delta \binom{x}{4}$ c. $\Delta[x^{(2)}x^{(-2)}]$

7. Calculate the following binomial numbers.
   a. $\binom{4}{7}$ b. $\binom{-2}{2}$
   c. $\binom{\frac{1}{2}}{3}$ d. $\binom{-\frac{1}{2}}{3}$
   e. $(-1)^n \binom{2n-1}{n} - \binom{-n}{n}$, where $n \geq 1$

8. Show that any set of distinct nonnegative factorial power functions is a linearly independent set in $\mathscr{C}(-\infty, \infty)$.

9. Establish the following identities.

a. $(-1)^n \binom{2n-1}{n} = \binom{-n}{n}$ for $n > 0$

b. $0^{(-n)} = 1/n!$ for $n \geq 0$

c. $\sum_{j=0}^{n} (-1)^j \binom{n}{j} (x + n - j)^n = n!$ for $n \geq 0$ and $x \in (-\infty, \infty)$

The behavior of the difference operator with respect to factorial powers motivates the following definition.

10 DEFINITION

If n is a nonnegative integer and $b_0 \neq 0$, $b_1, b_2, \ldots, b_n$ are constants, then the function q defined by

$$q(x) = b_0 x^{(n)} + b_1 x^{(n-1)} + \cdots + b_{n-1} x^{(1)} + b_n$$

is called a *factorial polynomial of degree n.*

Of course, a factorial polynomial of degree n is a polynomial of degree n in the usual sense. For example, if

(11) $f(x) = 3x^{(2)} + 7x^{(1)} - 9$

then

$$f(x) = 3[x^2 - x] + 7x - 9$$

or

(12) $f(x) = 3x^2 + 4x - 9$

Thus (11) and (12) are two different representations of the same function. Evidently, (11) would be more convenient to use in computing Δf, whereas (12) would be preferable for use in determining Df. In general, in order to emphasize the distinction between these two possible forms, a polynomial in the usual sense will be referred to as a *power polynomial*. Thus the polynomial f above is a polynomial of degree two, and (11) defines the representation of f as a factorial polynomial, while (12) is the representation of f as a power polynomial.

A connection between the two representations of a polynomial is contained in the next theorem. This result also shows another analogy between Δ and D.

13 THEOREM

If q is a polynomial of degree $n \geq 0$, then

(a) $q(x) = \dfrac{\Delta^n q(0)}{n!} x^{(n)} + \dfrac{\Delta^{n-1} q(0)}{(n-1)!} x^{(n-1)} + \cdots + \dfrac{\Delta q(0)}{1!} x^{(1)} + \dfrac{q(0)}{0!}$

and

(b) $\quad q(x) = \dfrac{D^n q(0)}{n!} x^n + \dfrac{D^{n-1} q(0)}{(n-1)!} x^{n-1} + \cdots + \dfrac{Dq(0)}{1!} x + \dfrac{q(0)}{0!}$

Proof: We prove only part (a). The proof of (b) is similar, and, moreover, is probably familiar from calculus.

Assume that q is represented as a factorial polynomial, that is,

$$q(x) = b_0 x^{(n)} + b_1 x^{(n-1)} + \cdots + b_{n-1} x^{(1)} + b_n$$

Evaluating q at $x = 0$ gives $b_n = q(0)$. The first difference of q is:

$$\Delta q(x) = n b_0 x^{(n-1)} + (n-1) b_1 x^{(n-2)} + \cdots + b_{n-1}$$

and at $x = 0$, we have $b_{n-1} = \Delta q(0)$. The second difference of q is:

$$\Delta^2 q(x) = n(n-1) b_0 x^{(n-2)} + (n-1)(n-2) b_1 x^{(n-3)} + \cdots$$
$$+ 2 \cdot 1 b_{n-2}$$

and at $x = 0$, we have

$$b_{n-2} = \frac{\Delta^2 q(0)}{2!}$$

In general,

$$\Delta^j q(x) = n^{(j)} b_0 x^{(n-j)} + (n-1)^{(j)} b_1 x^{(n-1-j)} + \cdots$$
$$+ (j+1)^{(j)} b_{n-j-1} x^{(1)} + j^{(j)} b_{n-j}$$

so that

$$b_{n-j} = \frac{\Delta^j q(0)}{j^{(j)}} = \frac{\Delta^j q(0)}{j!}$$

and (a) holds.

While Theorem 13 could be used to transform a power polynomial into a factorial polynomial, or a factorial polynomial into a power polynomial, this would entail a considerable amount of calculation. The reason for this is the difficulties which are encountered in computing $\Delta^j x^k$ and $D^j x^{(k)}$. A more effective method to transform a factorial polynomial into a power polynomial is the next topic.

14 DEFINITION

The *Sterling numbers of order n*, denoted by $s(n, j)$, are defined for each nonnegative integer n and all integers j by

$$s(n, j) = 0 \qquad \text{if} \qquad j < 0$$
$$s(n, j) = 0 \qquad \text{if} \qquad j > n$$

and

$$x^{(n)} = \sum_{j=0}^{n} s(n, j) x^j \qquad \text{if} \quad 0 \le j \le n$$

Notice that

$$x^{(1)} = x = s(1, 0) + s(1, 1)x$$

and

$$x^{(2)} = x(x - 1) = x^2 - x = s(2, 0) + s(2, 1)x + s(2, 2)x^2$$

Thus

$$s(1, 0) = 0, \qquad s(1, 1) = 1$$
$$s(2, 0) = 0, \qquad s(2, 1) = -1, \qquad s(2, 2) = 1$$

It follows from Definition 14 that the Sterling numbers of order n are the coefficients in the representation of factorial power $x^{(n)}$ as a power polynomial. The following is a brief table of Sterling numbers; a more extensive table can be found in various handbooks. See *Handbook of Mathematical Functions*, N.B.S. Applied Mathematics Series No. 55.

15 TABLE $s(n, j)$—Sterling numbers of order n.

			j			
n	0	1	2	3	4	5
0	1	0	0	0	0	0
1	0	1	0	0	0	0
2	0	-1	1	0	0	0
3	0	2	-3	1	0	0
4	0	-6	11	-6	1	0
5	0	24	-50	35	-10	1

The Sterling numbers of order n are the members of the range of a function whose domain is the set of integers. For example, $s(4, j)$ can be interpreted as the function defined by

$$s(4, j) = 0 \qquad \text{if} \qquad j \le 0$$
$$s(4, 1) = -6$$
$$s(4, 2) = 11$$
$$s(4, 3) = -6$$
$$s(4, 4) = 1$$
$$s(4, j) = 0 \qquad \text{if} \qquad j \ge 5$$

In this sense, the Sterling numbers represent a family of functions. The next theorem contains a functional equation which the Sterling numbers satisfy and which can be used to compute a table of Sterling numbers.

16 THEOREM

If n and j are integers and $n \geq 0$, then

$$s(n + 1, j) = s(n, j - 1) - n\, s(n, j).$$

Proof: Notice that $n \geq 0$ implies that

$$x^{(n+1)} = x^{(n)}(x - n) = xx^{(n)} - nx^{(n)}$$

Thus, using the definition of Sterling numbers,

(A) $$\sum_{j=0}^{n+1} s(n + 1, j)x^j = x \sum_{j=0}^{n} s(n, j)x^j - n \sum_{j=0}^{n} s(n, j)x^j$$

We leave it as an exercise to show that (A) can be rewritten as

(B) $$\sum_{j=0}^{n+1} [s(n + 1, j) - s(n, j - 1) + n\, s(n, j)]x^j = \emptyset$$

The result now follows from equation (B) and the fact that $\{1, x, x^2, \dots , x^{n+1}\}$ is a linearly independent set of functions.

---

17 Examples

a. If $q(x) = 8x^{(3)} - x^{(2)} + 7x^{(1)} - 4$, then, by using Table 15, it follows that

$$q(x) = 8(2x - 3x^2 + x^3) - (-x + x^2) + 7x - 4$$

or

$$q(x) = 8x^3 - 25x^2 + 24x - 4$$

b. For any positive integer n,

$$Dx^{(n)} = D\left[\sum_{j=0}^{n} s(n, j)x^j \right] = \sum_{j=0}^{n} js(n, j)x^{j-1}$$

Just as the transformation of a factorial polynomial of degree n into a power polynomial suggests the definition of the Sterling numbers of order n, the reverse problem of representing a power polynomial as a factorial polynomial motivates the next definition.

18 DEFINITION

The *inverse Sterling numbers of order* n, denoted $\sigma(n, j)$, are defined for each nonnegative integer n and for all integers j by

$$\sigma(n, j) = 0 \qquad \text{if} \qquad j < 0$$
$$\sigma(n, j) = 0 \qquad \text{if} \qquad j > n$$

and

$$x^n = \sum_{j=0}^{n} \sigma(n, j)x^{(j)} \qquad \text{if} \qquad 0 \le j \le n$$

The analogue of Table 15 for the inverse Sterling numbers is the following table.

19 TABLE $\sigma(n, j)$—Inverse Sterling numbers of order n

				j		
n	0	1	2	3	4	5
0	1	0	0	0	0	0
1	0	1	0	0	0	0
2	0	1	1	0	0	0
3	0	1	3	1	0	0
4	0	1	7	6	1	0
5	0	1	15	25	10	1

There is a functional equation for the inverse Sterling numbers which can be used to generate a more extensive table. This result is an analogue of Theorem 16, and the proof is left as an exercise.

20 THEOREM

If n and j are integers and $n \ge 0$, then

$$\sigma(n + 1, j) = \sigma(n, j - 1) + j\sigma(n, j)$$

21 EXERCISES

1. Complete the proof of Theorem 13.
2. Use Table 15 to transform the following factorial polynomials into power polynomials.
   a. $f(x) = x^{(4)}$ b. $f(x) = x^{(3)} - 3x^{(2)} - 5x$
   c. $f(x) = x^{(4)} + 3x^{(3)} + 7$ d. $f(x) = x^{(5)} + 10x^{(4)} - 24x^{(1)}$
3. Complete the proof of Theorem 16.
4. Use Theorem 16 and Table 15 to determine the Sterling numbers of order six and seven.
5. Use Table 19 to transform the following power polynomials into factorial polynomials.
   a. $f(x) = x^4$ b. $f(x) = x^3 - 3x^2$ c. $f(x) = x^5 - 10x^4 - x$

6. Prove Theorem 20.

7. Show that no inverse Sterling number is negative.

8. Let S denote the 6×6 matrix of Sterling numbers in Table 15 and let T denote the 6×6 matrix of inverse Sterling numbers in Table 19. Show that $S^{-1} = T$. Can you explain why this is so?

9. Show that if $f(x)$ is a polynomial of degree n, then there exist unique constants $c_0, c_1, \ldots, c_n$ such that

$$f(x) = c_0 \binom{x}{n} + c_1 \binom{x}{n-1} + \cdots + c_{n-1} \binom{x}{1} + c_n \binom{x}{0}$$

This representation is called a *binomial polynomial*.

10. In our place system of representing integers, the number $374 = 3(10^2) + 7(10^1) + 4(10^0)$. Can this number be represented uniquely as

$$a_0 10^{(2)} + a_1 10^{(1)} + a_2 10^{(0)} ?$$

What about

$$b_0 \binom{10}{2} + b_1 \binom{10}{1} + b_2 \binom{10}{0} ?$$

Can you generalize these observations?

11. If

$$f(x) = x^4 - 3x^{(2)} + \binom{x}{2} - 9$$

compute $\Delta f(x)$, $Df(x)$, and $Ef(x)$, and express each answer as a power, factorial, and binomial polynomial.

An exhaustive treatment of the other elementary functions, that is, exponential, trigonometric, and so on, would be a lengthy and repetitious undertaking. The basic properties of the elementary functions, as developed in a calculus course, are all that are needed to continue an independent study of the behavior of these functions with respect to differences and displacements. Therefore, only a few additional examples are mentioned here.

As observed in Example 7b, Section 18, and elsewhere,

(22) $\quad \Delta 2^x = 2^x$

and this is analogous to $De^x = e^x$. In general,

(23) $\quad \Delta a^x = a^{x+1} - a^x = (a-1)a^x$

while

(24) $\quad Ea^x = a(a^x)$

Notice that (24) is analogous to $De^{ax} = ae^{ax}$

If $a > 1$, then the inverse of the function a^x is $\log_a(x)$. We omit the explicit references to a, the base of the logarithm function, and note that

$$\Delta \log(x) = \log(x + 1) - \log(x) = \log\left(\frac{x + 1}{x}\right) = \log\left(1 + \frac{1}{x}\right)$$

Some other examples that are easy to verify are:

(25) $\Delta \sin(ax) = 2 \sin\left(\dfrac{a}{2}\right) \cos[a(x + \frac{1}{2})]$

(26) $\Delta \cos(ax) = -2 \sin\left(\dfrac{a}{2}\right) \sin[a(x + \frac{1}{2})]$

27 EXERCISES

1. Verify the difference formulas (25) and (26).
2. Show that if n is a positive integer, then

$$\Delta \log[x^{(n)}] = \sum_{j=0}^{n-1} \log[1 + (x - j)^{-1}]$$

3. If n is a negative integer, find a difference formula for $\log[x^{(n)}]$.
4. Suppose that $\Delta^7 f(k) = 9$ for all $k \in N$. What kind of function is f?
5. Show that $\Delta k^k \geq k^k$ for all $k \geq 1$.
6. Let f be a function and define $f^{(m)}(x)$ by $f^{(m)}(x) = f(x)f(x - 1) \cdots f(x - m + 1)$, for each positive integer m. Show that $\Delta[(ax + b)^{(m)}] = ma(ax + b)^{(m-1)}$.
7. Calculate $\Delta(ax^2)^{(m)}$ and $\Delta[ax^{(2)}]^{(m)}$.

Section 20 Equidimensional Differential Equations

The factorial polynomials, as defined in Section 19, are useful in analyzing a class of linear differential equations that is called *equidimensional equations*.

1 DEFINITION

An nth-order linear differential operator of the form

$$C = a_0 x^n D^n + a_1 x^{n-1} D^{n-1} + \cdots + a_{n-1} x D + a_n I$$

where $a_0 \neq 0$, $a_1, a_2, \ldots, a_n$ are constants, is called an nth-order *Cauchy operator*; and

$$L = b_0(cx + d)^n D^n + b_1(cx + d)^{n-1} D^{n-1} + \cdots$$
$$+ b_{n-1}(cx + d) D + b_n I$$

where $b_0 \neq 0$, $b_1, b_2, \ldots, b_n$, $c \neq 0$, and d are constants, is called an nth-order *Legendre operator*. The *Cauchy equation* (also called the *Euler equation*) is:

$$C(y) = \varnothing$$

and the *Legendre equation* is

$$L(y) = \varnothing$$

Each of these equations is called an equidimensional equation.

The two factorial polynomials of degree n that are associated with these differential equations are:

(2) $\qquad P(t) = a_0 t^{(n)} + a_1 t^{(n-1)} + \cdots + a_{n-1} t^{(1)} + a_n$

and

(3) $\qquad Q(t) = b_0 c^n t^{(n)} + b_1 c^{n-1} t^{(n-1)} + \cdots + b_{n-1} c t^{(1)} + b_n$

4 THEOREM

If C and L are the nth-order linear operators defined in Definition 1, P and Q are the factorial polynomials defined in (2) and (3), and r is a constant, then

$$C(x^r) = P(r)x^r$$

and

$$L[(cx + d)^r] = Q(r)(cx + d)^r$$

Proof: Since $D^j x^r = r^{(j)} x^{r-j}$, it follows that

$$
\begin{aligned}
C(x^r) &= a_0 x^n\, D^n x^r + a_1 x^{n-1}\, D^{n-1} x^r + \cdots + a_{n-1} x\, D x^r + a_n x^r \\
&= a_0 x^n r^{(n)} x^{r-n} + a_1 x^{n-1} r^{(n-1)} x^{r-n+1} + \cdots \\
&\quad + a_{n-1} x r^{(1)} x^{r-1} + a_n x^r \\
&= (a_0 r^{(n)} + a_1 r^{(n-1)} + \cdots + a_{n-1} r^{(1)} + a_n) x^r \\
&= P(r) x^r
\end{aligned}
$$

Similarly, since $D^j (cx + d)^r = c^j r^{(j)} (cx + d)^{r-j}$, it follows that

$$L[(cx + d)^r] = Q(r)(cx + d)^r$$

Theorem 4 is of the same general form as Theorem 10, Section 13. Thus an analysis of the equidimensional equations can be based upon the roots of the factorial polynomials P and Q. In fact, the development in Chapter 3 can be used as a model for a systematic study of the equidimensional equations. Only a few of the properties of the Cauchy equation, $C(y) = \varnothing$, and its related factorial polynomial, $P(t)$, are presented in this section. Corresponding facts about the Legendre equation, as well as the equidimensional equations in general, are contained in the exercises.

Since $C(y) = \varnothing$ has a singular point at $x = 0$, the discussion of this equation will be simplified by considering the equation on the interval $J = (0, \infty)$. The behavior of this equation on intervals that include zero is treated in Chapter 7.

Consider the second-order Cauchy operator

(5) $\qquad C_2 = a_0 x^2 D^2 + a_1 x D + a_2 I$

and its associated polynomial

(6) $P_2(t) = a_0 t^{(2)} + a_1 t^{(1)} + a_2$

It follows from Theorem 4 that

$$C_2(x^r) = \varnothing$$

if and only if

$$P_2(r) = 0$$

Thus the same three cases considered in Section 14 must be examined here. In order to solve $P_2(r) = 0$, it is more convenient to represent P_2 as a power polynomial, that is,

(7) $P_2(t) = a_0 t(t - 1) + a_1 t + a_2 = a_0 t^2 + (a_1 - a_0)t + a_2$

The roots of P_2 are:

(8)
$$r_1 = \frac{(a_0 - a_1) + \sqrt{(a_0 - a_1)^2 - 4a_0 a_2}}{2a_0}$$

$$r_2 = \frac{(a_0 - a_1) - \sqrt{(a_0 - a_1)^2 - 4a_0 a_2}}{2a_0}$$

Notice that the two roots in (8) amount to a "quadratic formula" for the roots of the second-degree factorial polynomial $P_2(t)$.

CASE I. r_1 and r_2 are real and unequal, that is,

$$(a_0 - a_1)^2 - 4a_0 a_2 > 0$$

In this case, it is a simple matter to verify that $y_1(x) = x^{r_1}$, $y_2(x) = x^{r_2}$ is a linearly independent pair of solutions of $C_2(y) = \varnothing$ on the interval $J = (0, \infty)$.

CASE II. $r_1 = r_2$, that is, $(a_0 - a_1)^2 - 4a_0 a_2 = 0$.

In this case, it follows from (8) that

(9) $r_1 = r_2 = \dfrac{a_0 - a_1}{2a_0}$ and $(a_0 - a_1)^2 = 4a_0 a_2$

Thus, $y_1(x) = x^{r_1}$ is a solution of $C_2(y) = \varnothing$. However, a pair of linearly independent solutions is what is needed. By considering the equivalent equation

(10) $\left(D^2 + \dfrac{a_1}{a_0 x} D + \dfrac{a_2}{a_0 x^2} I \right) y = \varnothing$

we can obtain a second solution of $C_2(y) = \varnothing$ using the method in Theorem 5, Section 12. Either by applying Theorem 5, Section 12, or by direct verification, it can be shown that

$$y_2(x) = x^{r_1} \ln(x)$$

is a solution of $C_2(y) = \varnothing$, and that the pair of functions $y_1(x) = x^{r_1}$, $y_2(x) = x^{r_1} \ln(x)$ is a solution basis.

CASE III. r_1 and r_2 are complex, that is,

$$(a_0 - a_1)^2 - 4a_0a_2 < 0$$

In this case, we have

$$r_1 = u + iv \qquad \text{and} \qquad r_2 = u - iv$$

where

$$u = \frac{a_0 - a_1}{2a_0}$$

and

$$v = \frac{[4a_0a_2 - (a_0 - a_1)^2]^{1/2}}{2a_0}$$

Since $x^{r_1} = \exp[r_1 \ln(x)] = \exp[(u + iv) \ln(x)]$, it follows from equation (3), Section 3, that

$$x^{r_1} = x^u\{\cos[v \ln(x)] + i \sin[v \ln(x)]\}$$

and

$$x^{r_2} = x^u\{\cos[v \ln(x)] - i \sin[v \ln(x)]\}$$

This is a similar situation to that which was discussed in Section 14. Either by following the line of reasoning in Section 14 or by direct verification, it follows that

$$y_1(x) = x^u \cos[v \ln(x)]$$

and

$$y_2(x) = x^u \sin[v \ln(x)]$$

is a solution basis for $C_2(y) = \varnothing$ in this case.

11 Examples

    a. The second-order equation

$$[-4x^2 D^2 + 12x D + 9I]y = \varnothing$$

    has the associated factorial polynomial

$$P_2(t) = -4t^{(2)} + 12t^{(1)} + 9$$

    Since $P_2(-\tfrac{1}{2}) = P_2(\tfrac{9}{2}) = 0$, it follows that $y_1(x) = x^{-1/2}$ and $y_2(x) = x^{9/2}$ are linearly independent solutions of this equation.

b. The equation

$$[3x^2 D^2 - x D + (\tfrac{4}{3})I]y = \varnothing$$

has the associated polynomial

$$P_2(t) = 3t^{(2)} - t^{(1)} + \tfrac{4}{3}$$

Since $P_2(\tfrac{2}{3}) = 0$ and $\tfrac{2}{3}$ is a root of multiplicity two, it follows that $y_1(x) = x^{2/3}$, $y_2(x) = x^{2/3} \ln(x)$ is a solution basis for the equation.

c. The equation

$$[x^2 D^2 + 5x D + (\tfrac{25}{4})I]y = \varnothing$$

has the associated polynomial

$$P_2(t) = t^{(2)} + 5t^{(1)} + (\tfrac{25}{4})$$

Since $P_2(-2 - \tfrac{3}{2}i) = P_2(-2 + \tfrac{3}{2}i) = 0$, it follows that

$$y_1(x) = x^{-2} \cos[(\tfrac{3}{2}) \ln(x)]$$

and

$$y_2(x) = x^{-2} \sin[(\tfrac{3}{2}) \ln(x)]$$

is a solution basis for the equation.

12 EXERCISES

1. Verify that each pair of functions in the three-case analysis of $C_2(y) = \varnothing$ are solutions.
2. Form the Wronski matrix for the pair of solutions of $C_2(y) = \varnothing$ in each of the three cases, and show that each pair of solutions is a solution basis.
3. Complete the analysis of Case II by applying Theorem 5, Section 12, to obtain $y_2(x)$.
4. Use the initial conditions $y(1) = 0$ and $Dy(1) = 5$ with each of the three equations in Example 11 and solve the three initial-value problems.
5. Let $C_2 = -4x^2 D^2 + 12x D + 9I$ and let $g(x) = b_0 x^2 + b_1 x + b_2$. Calculate $C_2[g(x)]$ and use this result to show that if $f(x)$ is a polynomial of degree two or less, then the nonhomogeneous equation

$$C_2(y) = f(x)$$

   has a solution that is a polynomial of degree two or less.
6. Solve the initial-value problem

$$-4x^2 D^2 y + 12x Dy + 9y = 100x^2 + 21x - 9$$
$$y(1) = 4, \quad Dy(1) = 14$$

7. Let $C_2 = a_0 x^2 D^2 + a_1 x D + a_2 I$ and $P_2(t) = a_0 t^{(2)} + a_1 t^{(1)} + a_2$. Show each of the following.
   a. If $g(x)$ is a polynomial of degree two or less, then $C_2[g(x)]$ is a polynomial of degree two or less.

   b. If $f(x)$ is a polynomial of degree two, then there exists a unique polynomial $g(x)$ of degree two such that $C_2[g(x)] = f(x)$ if and only if $P_2(0) \neq 0, P_2(1) \neq 0$, and $P_2(2) \neq 0$.

   c. If $f(x)$ is a polynomial of degree two and $P_2(2) = 0$, then $C_2[f(x)]$ is a polynomial of degree one or less.

   d. If the roots of $P_2(t)$ are complex and $f(x)$ is a polynomial of degree two or less, then

$$C_2(y) = f(x)$$

   has a polynomial solution.

8. Show that every solution of

$$x^2 D^2y - 6x\, Dy + 12y = 9x^2 - 3x + 2$$

is a polynomial of degree four or less.

9. Show that the change of variable $x = e^t$ transforms the Cauchy equation

$$a_0 x^2 D^2 y + a_1 x\, Dy + a_2 y = \varnothing$$

into a linear equation with constant coefficients. [Hint: $Dy = dy/dx = (dy/dt)(dt/dx) = (dy/dt)(1/x)$.]

10. Let $L_2 = b_0(cx + d)^2 D^2 - b_1(cx + d)D + b_2 I$ be a second-order Legendre operator. Characterize a solution basis for $L_2(y) = \varnothing$ by considering the three cases for the roots of the associated polynomial $Q(t)$ as defined in (3).

11. Find a second-order Legendre operator L_2 such that $y_1(x) = (2x - 1)$ and $y_2(x) = (2x - 1)^{-1}$ are solutions of $L_2(y) = \varnothing$.

12. Solve the initial-value problem

$$(4x^2 - 4x + 1) D^2 y + (4x - 2)\, Dy - 4y = -2$$
$$y(3) = \tfrac{3}{2}, \quad Dy(3) = -\tfrac{2}{5}$$

    The preceding discussion together with the results of Chapter 3, provides the necessary information for treating Cauchy equations of arbitrary order n. In fact, it is possible to develop a table which is similar to Table 3, Section 16, and which gives the relationship between the roots of $P(t)$ and the solutions of $C(y) = \varnothing$. The details are omitted for the sake of brevity, but they are easy to supply.

13 TABLE   If $P(t)$ is the factorial polynomial defined in (2) and k is a nonnegative integer, then

A solution of $C(y) = \varnothing$ is:	If and only if a factor of $P(t)$ is:
(a) $[\ln(x)]^k x^r$	$(t - r)^{k+1}$
(b) $[\ln(x)]^k x^a \cos[b \ln(x)]$	$(t^2 - 2at + a^2 + b^2)^{k+1}$
(c) $[\ln(x)]^k x^a \sin[b \ln(x)]$	$(t^2 - 2at + a^2 + b^2)^{k+1}$

14 Example

    Associated with the third-order Cauchy equation

$$(x^3 D^3 - 3x^2 D^2 + 7x\, D - 8I)y = \varnothing$$

is the third-degree factorial polynomial

$$P(t) = t^{(3)} - 3t^{(2)} + 7t^{(1)} - 8$$

By transforming $P(t)$ into a power polynomial, we find that

$$P(t) = (t - 2)^3$$

Thus $t - 2$ is a root of $P(t)$ of multiplicity three. It can be shown directly that

$$y_1(x) = x^2$$

$$y_2(x) = x^2 \ln(x)$$

$$y_3(x) = x^2[\ln(x)]^2$$

are three solutions of the differential equation and that they are linearly independent.

Notice also that Table 13 has the information that is needed to develop a *method of undetermined coefficients* for the nonhomogeneous Cauchy equation

$$C(y) = f(x)$$

This and much more information about the equidimensional equations can be developed by utilizing the methods of Chapter 3.

15 EXERCISES

1. Can you find constants c_1 and c_2 such that

   $$y(x) = c_1 + c_2 \ln(x)$$

   is a solution of $x^3 D^3 y - 3x^2 D^2 y + 7x Dy - 8y = 12 - 8 \ln(x)$?

2. Show that all solutions of $x^2 D^2 y + (1 - 2a)x Dy + (a^2 + b^2)y = \emptyset$ can be represented as

   $$y(x) = y(1)x^a \cos[b \ln(x)] + \frac{Dy(1) - ay(1)}{b} x^a \sin[b \ln(x)]$$

   provided $b \neq 0$.

3. Find a solution basis for

   $$x^3 D^3 y + 5x^2 D^2 y - 4x Dy + 4y = \emptyset$$

4. Find a Cauchy equation whose solutions are:

   $$y_1(x) = 1 - 2x$$

   $$y_2(x) = 7$$

   $$y_3(x) = x^2 + 8$$

5. Solve the initial-value problem

   $$x^3 D^3 y = 2; \qquad y(1) = 3, \quad Dy(1) = 4, \quad D^2 y(1) = 1$$

6. Generalize the results in Problem 7 of Exercises 12 to higher-order Cauchy operators.

7. Verify that the entries in Table 13 are correct.

8. If n is a positive integer, what are the solutions of

$$x^n D^n y = \varnothing?$$

9. Find all solutions of

$$8(\tfrac{1}{2}x + 6)^3 D^3 y + 20(\tfrac{1}{2}x + 6)^2 D^2 y - 8(\tfrac{1}{2}x + 6) Dy + 4y = \varnothing$$

10. Solve the initial-value problem

$$x^2 D^2 y - 5x Dy + 9y = -8x^{-1}(1 - 2 \cdot \ln x)$$

$$y(1) = 1, \quad Dy(1) = 7$$

Section 21 Summation

We have seen that the difference calculus is similar in many respects to the differential calculus. The purpose of this section is to develop the finite calculus analogue of the definite integral. This analogue is called the summation calculus, and there will be a number of similarities to the integral calculus. In presenting the basic properties of the summation calculus, we shall continue with the convention of Section 18, namely, that all of the functions under consideration have appropriate domains.

Recall that if f is a continuous function on an interval J, and if g is a function such that $Dg = f$ on J, that is, g is an antiderivative of f, then

(1) $$\int_a^b f(x)\, dx = g(b) - g(a), \qquad a, b \in J$$

See Theorem 6, Section 2. In addition, if the function h also has the property that $Dh = f = Dg$ on J, then $h = g + c$ for some constant c. Conversely, if $h = g + c$, where c is a constant, then it is clear that $Dh = Dg = f$. In the language of Chapter 2, this is equivalent to saying that the first-order differential equation

$$Dy = f(x)$$

has the one-parameter family of solutions $y(x) = g(x) + c$. However, from calculus, the more familiar terminology is to call this family the indefinite integral of f, and to denote this by

(2) $$\int f(x)\, dx = g(x) + c$$

We now give a corresponding development for the finite calculus.

3 THEOREM

Given a function f. If g and h are functions such that

$$\Delta g(x) = f(x) = \Delta h(x)$$

for all x in the domain of f, then $h = g + p$, where p is a periodic function. Conversely, if $\Delta g(x) = f(x)$ for all x in the domain of f, and $h = g + p$, where p is periodic, then $\Delta h(x) = f(x)$ for all x in the domain of f.

Proof: Suppose $\Delta g = f = \Delta h$. Then $\Delta[h - g] = \varnothing$, which implies that $h - g = p$ is a periodic function. Thus $h = g + p$.
   If $\Delta g = f$ and $h = g + p$, where p is periodic, then

$$\Delta h = \Delta[g + p] = \Delta g + \Delta p = f$$

4 DEFINITION

Given a function f. If g is a function such that $\Delta g(x) = f(x)$ for all x in the domain of f, then the collection of all functions of the form $g(x) + p(x)$, where p is periodic, is called the *indefinite sum of f* and is denoted by

$$\sum f(x) = g(x) + p(x)$$

   Note that the indefinite sum of a function is analogous to the indefinite integral (2), that is, each denotes a collection, or family, of functions. In the special cases where the domain of f is N, the set of nonnegative integers, the set of periodic functions on N is merely the set of constant functions as shown in Corollary 19, Section 18. Thus the indefinite sum of f is the one-parameter family

(5) $\sum f(k) = g(k) + c,$ $k \in N$

where $\Delta g = f$ and c is any constant.

6 Examples

a. Since $\Delta x^{(4)} = 4x^{(3)}$, it follows that

$$\sum 4x^{(3)} = x^{(4)} + p(x)$$

where p is any periodic function.

b. Since $\Delta x^{(n)} = nx^{(n-1)}$ for any integer n, we have the summation formula

$$\sum x^{(n)} = \frac{x^{(n+1)}}{n + 1} + p(x), \qquad n \neq -1$$

where p is any periodic function.

c. Let N denote the set of nonnegative integers and consider the exponential function $h(k) = a^k$, $a \neq 1$. According to (23) in Section 19, $\Delta a^k = (a - 1)a^k$. Thus

$$\sum a^k = \frac{a^k}{a - 1} + c$$

where c is any constant.

From the discussion and examples above, it should be clear that the periodic functions in the summation calculus play the same role as the constants of integration in the integral calculus. To prepare an extensive table of summation formulas would be of the same order of difficulty as preparing a table of integrals. However, each difference formula implies a corresponding summation formula, just as each differentiation formula implies an integration formula. In particular, the following formulas are immediate consequences of the linearity of Δ.

$$\sum [f(x) + g(x)] = \sum f(x) + \sum g(x)$$

and

$$\sum c \cdot f(x) = c \cdot \sum f(x)$$

for any constant c. In fact, from (16) in Section 18,

$$\sum p(x)f(x) = p(x) \sum f(x)$$

for any periodic function p.

By a proper interpretation of the rule for differencing a product as presented in Theorem 9, Section 18, we have a "summation by parts" formula:

(7) $$\sum f(x) \Delta g(x) = f(x)g(x) - \sum [Eg(x) \Delta f(x)]$$

The derivation of this result follows the same line of reasoning as that used in deriving the "integration by parts" rule.

8 Examples

a. Consider the indefinite sum

$$\sum [x^3 - 4x^2]$$

In order to determine a function g such that $\Delta g = x^3 - 4x^2$, we first express the power polynomial $f(x) = x^3 - 4x^2$ as a factorial polynomial. From Table 19, Section 19,

$$x^3 - 4x^2 = x^{(3)} + 3x^{(2)} + x^{(1)} - 4[x^{(2)} + x^{(1)}]$$
$$= x^{(3)} - x^{(2)} - 3x^{(1)}$$

Therefore,

$$\sum [x^3 - 4x^2] = \sum [x^{(3)} - x^{(2)} - 3x^{(1)}]$$
$$= \frac{x^{(4)}}{4} - \frac{x^{(3)}}{3} - \frac{3x^{(2)}}{2} + p(x)$$

where p is periodic.

b. To determine the indefinite sum

$$\sum x2^x$$

we apply (7) with $f(x) = x$ and $\Delta g(x) = 2^x$, or $g(x) = 2^x$. Thus

$$\sum x2^x = x2^x - \sum 2^{x+1}1 = x2^x - 2 \sum 2^x$$

and

$$\sum x2^x = (x - 2)2^x + p(x)$$

where p is periodic.

9 EXERCISES

1. Derive the "summation by parts" formula (7).
2. Determine the following indefinite sums.

   a. $\sum -5x^{(-6)}$ b. $\sum \binom{x}{3}$ c. $4 \sum x^{(1)}x^{(-3)}$

   d. $\sum (-7)^x$ e. $\sum \sin(2x)$ f. $\sum \cos(8x)$

3. Show that if $r \neq 0$ is a constant, then
   a. $\sum \sin(rx) = \{-\cos[r(x - \frac{1}{2})]/2 \sin(r/2)\} + p(x)$
   b. $\sum \cos(rx) = \{\sin[r(x - \frac{1}{2})]/2 \sin(r/2)\} + p(x)$

4. Show that
   a. $\sum x(4)^x = [(3x - 4)4^x/9] + p(x)$
   b. $\sum x(-4)^x = [-(5x - 4)(-4)^x/25] + p(x)$

5. If $r \neq 1$ is a constant, find a general summation formula for

   $$\sum x(r)^x$$

6. Show that

   $$\sum x \cos x = \frac{\cos x + 2x \sin(\frac{1}{2}) \sin(x - \frac{1}{2})}{[2 \sin(\frac{1}{2})]^2} + p(x)$$

7. Find a summation formula for

   $$\sum x \sin x$$

8. Verify the following general summation formulas.
   a. $2 \sin(b/2) \sum \sin(a + bx) = -\cos[a - (b/2) + bx] + p(x)$
   b. $2 \sin(b/2) \sum \cos(a + bx) = \sin[a - (b/2) + bx] + p(x)$

The analogue of the definite integral $\int_a^b f(x) \, dx$ in the summation calculus is the definite (or finite) sum

$$S = f(a) + f(a + 1) + f(a + 2) + \cdots + f(a + m)$$

where it is assumed that $a, a + 1, a + 2, \ldots, a + m$ are in the domain of f. Note that this sum can be written in the equivalent form

(10) $$S = \sum_{i=0}^{m} f(a + i)$$

and for notational reasons, it will be convenient to adopt this form for all definite sums.

In contrast to the definite integral $\int_a^b f(x) \, dx$, where an exact calculation of this number usually requires a function g such that $Dg = f$, definite sums can always be calculated exactly. This can be seen by writing out the terms in (10).

$$\sum_{i=0}^{m} f(a + i) = f(a) + f(a + 1) + \cdots + f(a + m)$$

Note that this is just a finite sum of real numbers. Of course, if m is large, or if f is a complicated function (or both), the computations involved could be prohibitive. For example, consider

$$\sum_{i=0}^{49} (i + 7)^{(3)} = 7 \cdot 6 \cdot 5 + 8 \cdot 7 \cdot 6 + 9 \cdot 8 \cdot 7 + \cdots + 56 \cdot 55 \cdot 54$$

Although this sum can be determined exactly, no one would seriously attempt to perform all of the indicated computations. It is in cases such as this that the summation calculus can be used to great advantage.

The next result could be called the fundamental theorem of the summation calculus. It is analogous to (1), the fundamental theorem of integral calculus.

11 THEOREM

If the domain of f includes $\{a, a + 1, a + 2, \ldots, a + n - 1\}$ and $\Delta g = f$, then

$$\sum_{i=0}^{n-1} f(a + i) = g(a + n) - g(a) = g(x) \Big|_{a}^{a+n}$$

Proof: By definition,

$$\Delta g(x) = g(x + 1) - g(x) = f(x)$$

Thus

$$\sum_{i=0}^{n-1} f(a + i) = \sum_{i=0}^{n-1} [g(a + i + 1) - g(a + i)]$$

or

$$\sum_{i=0}^{n-1} f(a + i) = \sum_{i=0}^{n-1} g(a + i + 1) - \sum_{i=0}^{n-1} g(a + i)$$

Now, all but two terms on the right side of this equation appear twice and with opposite sign. Thus, after cancellation,

$$\sum_{i=0}^{n-1} f(a + i) = g(a + n) - g(a)$$

12 Examples

a. Consider the definite sum

$$\sum_{i=0}^{49} (i + 7)^{(3)} = 7 \cdot 6 \cdot 5 + 8 \cdot 7 \cdot 6 + \cdots + 56 \cdot 55 \cdot 54$$

Here we have $f(x) = x^{(3)}$. Since $\Delta x^{(4)}/4 = x^{(3)}$, it follows from Theorem 11 that

$$\sum_{i=0}^{49} (i + 7)^{(3)} = \frac{x^{(4)}}{4}\bigg|_7^{57} = \frac{57^{(4)}}{4} - \frac{7^{(4)}}{4} = 2{,}369{,}850$$

b. Consider the definite sum

$$S = \sum_{j=-7}^{9} j2^j$$

The first step is to rewrite this sum in the form (10.) This is accomplished by setting $j = -7 + i$ to obtain

$$S = \sum_{i=0}^{16} (-7 + i)2^{-7+i}$$

We now have $f(x) = x2^x$ and $a = -7$. As noted in Example 8b, $\Delta[2^x(x - 2)] = x2^x$. Therefore, $g(x) = 2^x(x - 2)$, and

$$S = g(-7 + 17) - g(-7) = 2^{10}(8) - 2^{-7}(-9) = 2^{13} + 9 \cdot 2^{-7}$$

Theorem 11 can also be useful in evaluating infinite series, as illustrated in the next example. In the example, note the similarity to the procedure for treating improper integrals of the form $\int_a^\infty f(x)\, dx$.

13 Example

In order to compute

$$S = \sum_{i=0}^{\infty} \frac{1}{(i + 1)(i + 2)(i + 3)} = \frac{1}{1 \cdot 2 \cdot 3} + \frac{1}{2 \cdot 3 \cdot 4} + \frac{1}{4 \cdot 5 \cdot 6} + \cdots$$

recall (see Section 6) that if S exists, then

$$S = \lim_{n \to \infty} S_n$$

where

$$S_{n-1} = \sum_{i=0}^{n-1} \frac{1}{(i + 1)(i + 2)(i + 3)} = \sum_{i=0}^{n-1} (i)^{(-3)}$$

Determination of the partial sum S_n can, in this case, be done by applying Theorem 11 with $f(x) = x^{(-3)}$ and $a = 0$. Since

$$\Delta\left[\frac{x^{(-2)}}{-2}\right] = x^{(-3)}$$

we have $g(x) = x^{(-2)}/-2$ and

$$S_{n-1} = g(n) - g(0) = \frac{n^{(-2)}}{-2} - \frac{0^{(-2)}}{-2}$$

From the definition of factorial powers we have that

$$n^{(-2)} = \frac{1}{(n+2)(n+1)}$$

and

$$0^{(-2)} = \frac{1}{(0+2)(0+1)} = \tfrac{1}{2}$$

Therefore,

$$S_{n-1} = \tfrac{1}{4} - \frac{1}{2(n+2)(n+1)}$$

and

$$S = \lim_{n\to\infty} S_n = \tfrac{1}{4}$$

Although there is much more to the theory of summation, as well as all of the topics of the finite calculus, this brief introduction is an adequate background for the main purposes of this book, that is, studying differential and difference equations.

(14) EXERCISES

1. Calculate the following definite sums.
   a. $\sum_{i=8}^{61} i$
   b. $\sum_{i=-9}^{9} [i^2 - i]$
   c. $\sum_{i=4}^{9} \binom{i}{3}$
   d. $\sum_{i=0}^{6} \cos(8i)$
   e. $\sum_{i=0}^{n-1} (-1)^i$
   f. $\sum_{i=0}^{n-1} i(-1)^i$

2. Find a general formula for the following sums, where n is a positive integer.
   a. $\sum_{i=1}^{n} i$
   b. $\sum_{i=1}^{n} i^2$
   c. $\sum_{i=1}^{n} i^3$
   d. $\sum_{i=1}^{n} i^4$

3. Calculate the following definite sums and determine if they are negative or positive.
   a. $\sum_{i=0}^{72} i^2(-1)^i$
   b. $\sum_{j=4}^{13} [(-1)^j \cos(j + \tfrac{1}{2})]$
   c. $-2\sum_{j=3}^{60} [(-1)^j/(j+1)(j+3)]$
   d. $\sum_{j=-17}^{12} (j^2 - j)$

4. Find an expression for the partial sums of each of the following infinite series. In each case, either show that the series is divergent or find the limit of the sequence of partial sums.
   a. $\sum_{i=0}^{\infty} [1/(\sqrt{i+2} + \sqrt{i+1})]$ (Hint: Compute $\Delta x^{1/2}$.)
   b. $\sum_{j=0}^{\infty} (-1)^j$
   c. $\sum_{i=0}^{\infty} [1/(i+3)(i+1)]$ (Hint: $1/(x+3)(x+1) = (x+2)x^{(-3)}$)
   d. $\tfrac{1}{2} + \tfrac{2}{4} + \tfrac{3}{8} + \tfrac{4}{16} + \tfrac{5}{32} + \cdots$
   e. $\sum_{i=0}^{\infty} (-1)^i \binom{50}{i}$
   f. $\sum_{i=0}^{\infty} i[\tfrac{3}{4}]^i$
   g. $\sum_{i=1}^{\infty} \arctan[1/(i^2 + i + 1)]$ (Hint: Calculate $\Delta \arctan x$.)

5

*LINEAR DIFFERENCE EQUATIONS**

Section 22 Introduction

In Chapter 2, the vector space $\mathscr{C}^n(J)$ played a central role in the study of linear differential equations. This vector space is defined in terms of concepts from the calculus, namely, continuity and differentiability of functions. The first topic in this chapter is to consider a real vector space which has a similar role in the finite calculus.

Throughout this chapter, N is used to denote the set of all nonnegative integers, that is, $N = \{0, 1, 2, 3, \ldots\}$. This notational convention will be restated several times. All functions in this chapter will have N as their domain.

1 DEFINITION

Let N denote the set of nonnegative integers. The set of all real-valued functions whose domain is N, along with the usual definition of addition of functions and multiplication of a function by a real number, is a vector space and is denoted by V_∞.

It is easy to verify that V_∞ is a real vector space. This observation was made in Example 2, Section 4. In fact, V_∞ is the set of all real sequences. While sequences are usually associated with a subscripted symbolism, we continue to use the same functional notation of the previous chapters. If the claim that a sequence is a special case of the general concept of function is confusing, then perhaps a few examples will clarify matters.

* This chapter depends upon the material in Chapters 2, 3, and 4.

2 Examples

a. Consider the function defined on N by

$$f(k) = 2^k, \qquad k \in N$$

Clearly $f \in V_\infty$, and whether f is specified as above, or as $f_k = 2^k$, is unimportant. Likewise, specific members of the range of f can be denoted by $f(0) = 2^0 = 1$, $f(1) = 2^1 = 2$, $f(2) = 2^2 = 4$, $f(3) = 2^3 = 8$, and so on, or by $f = \{2^0, 2^1, 2^2, 2^3, \ldots\}$. It is often useful to notice apparent properties of functions. In this case, f is an increasing function on N, and f is unbounded. As a sequence, note that $\lim\limits_{k \to \infty} f(k)$ does not exist.

b. Consider the function g defined on N by $g(0) = 0$, $g(1) = 1$, and $g(k + 2) = g(k + 1) + g(k)$ for $k \in N$. In this case,

$$g(2) = g(0 + 2) = g(0 + 1) + g(0) = 1 + 0 = 1$$

$$g(3) = g(2) + g(1) = 1 + 1 = 2$$

and so on. In sequential notation, we have that

$$g = \{0, 1, 1, 2, 3, 5, 8, 13, 21, 34, 55, \ldots\}$$

where each new term in the sequence is found by adding its two predecessors. The numbers in the range of g are known as the Fibonacci numbers. Evidently, $g \in V_\infty$. Note that g is an increasing, unbounded function.

c. Consider the function h defined on N by

$$h(k) = \left[2 + \frac{(-1)^k}{k + 1} \right], \qquad k \in N$$

The first several terms in this sequence are:

$$h(0) = 3, \qquad h(1) = \tfrac{3}{2}, \qquad h(2) = \tfrac{7}{3}, \qquad h(3) = \tfrac{7}{4}$$

This member of V_∞ is neither increasing nor decreasing. However, h is a bounded function, and it is easy to show that $\lim\limits_{k \to \infty} h(k) = 2$.

If $f \in V_\infty$, then both Ef and Δf are also members of V_∞. In fact, it is easy to verify that for each positive integer n, $E^n f$ and $\Delta^n f$ belong to V_∞. Since E^n and Δ^n are linear operators on V_∞, as observed in Section 18, the next definition parallels Definition 5, Section 9.

3 DEFINITION

The *nth-order operator* $\Lambda : V_\infty \to V_\infty$ is defined by

$$\Lambda = E^n + p_1(k)E^{n-1} + \cdots + p_n(k)I$$

where $p_i \in V_\infty$ for $i = 1, 2, \ldots, n$, and for each $f \in V_\infty$,

$$\Lambda[f(k)] = E^n f(k) + p_1(k)E^{n-1}f(k) + \cdots + p_n(k)f(k)$$

In the study of linear difference equations, Λ has the same role that the nth-order operator L has in the theory of linear differential equations. It is easy to show that Λ is a linear operator on V_∞; a slight modification of the proof of Theorem 7, Section 9, is all that is needed.

4 Example

The third-order operator

$$\Lambda = E^3 + k^{(2)}E^2 + 7E + kI$$

when applied to $f(k) = 5^k$ gives

$$\begin{aligned}
\Lambda[5^k] &= 5^{k+3} + k^{(2)}5^{k+2} + 7(5^{k+1}) + k5^k \\
&= 5^k[125 + 25k^{(2)} + 35 + k] \\
&= 5^k[25k^{(2)} + k + 160]
\end{aligned}$$

Notice that both $f(k)$ and $\Lambda[f(k)]$ are in V_∞.

5 EXERCISES

1. Show that V_∞ together with the usual definition of addition and scalar multiplication is a vector space.
2. Let B_∞ be the set of all bounded infinite sequences and let C_∞ be the set of all convergent infinite sequences. Are these two subsets of V_∞ also subspaces?
3. Represent the operator Λ in Example 4 in terms of Δ.
4. Let $\Lambda = E^3 - 5E^2 + 3E + 9I$ and calculate $\Lambda[f(k)]$ for each of the following.
   a. $f(k) = 1$ b. $f(k) = (-1)^k$ c. $f(k) = (3)^k$
   d. $f(k) = 2^k$ e. $f(k) = k(3)^k$ f. $f(k) = (7k - 16)(3)^k$
   g. $f(k) = k^{(2)}$
5. Show that Λ, as defined in Definition 3, is a linear operator on V_∞.

The concept of a linear difference equation is defined in terms of the linear operator Λ.

6 DEFINITION

Let $p_1(k), p_2(k), \ldots, p_n(k)$ and $f(k)$ be members of V_∞. If $p_n(k) \neq 0$ for all $k \in N$, then

$$[E^n + p_1(k)E^{n-1} + \cdots + p_n(k)I]y = f(k)$$

is an nth-order linear difference equation. If $f(k) = \emptyset$, then this difference equation is called homogeneous; otherwise it is nonhomogeneous. If

$$\Lambda = E^n + p_1(k)E^{n-1} + \cdots + p_n(k)I$$

then the difference equation can be written more compactly as

$$\Lambda(y) = f(k)$$

_____

Definition 6 is a convenient standard form for nth-order linear difference equations. Many equations can be put into this form by elementary methods. An example of this is given below.

7 Example

The functional equation

(A) $z(j) = \cos[(j + 2)\pi]z(j - 1) + (j + 4)^{(2)}z(j - 2)$

$j \in \{1, 2, \dots\}$, can be put into standard form by first defining

$j = k + 1, \quad k \in N$

Then

$z(k + 1) = \cos[(k + 3)\pi]z(k) + (k + 5)^{(2)}z(k - 1)$

Next, let

$y(k) = z(k - 1)$

for $k \in N$, so that

$y(k + 2) = \cos[(k + 3)\pi]y(k + 1) + (k + 5)^{(2)}y(k)$

or

(B) $[E^2 - \cos[(k + 3)\pi]E - (k + 5)^{(2)}I]y(k) = \varnothing$

which is the form specified in Definition 6. Equations (A) and (B) are related by the change of variables

$j = k + 1 \quad$ and $\quad y(k) = z(k - 1)$

and it is clear that (B) is a second-order linear homogeneous difference equation.

The format for the remainder of this chapter is to present a theory of linear difference equations in a manner that closely parallels the treatment of differential equations in Chapters 2 and 3. The many analogies between differential equations and difference equations are used to condense the presentation in this chapter.

8 EXERCISES

1. Transform the difference equation

$$(\Delta^2 + k\,\Delta - 2I)y = 2(k + 1)$$

into the form of Definition 6 and check to see which, if any, of the following are solutions.

a. $y(k) = 1 - 2^k$ b. $y(k) = k^{(3)}$ c. $y(k) = 1 - k^2$

d. $y(k) = k^{(2)}$ e. $y(k) = k^{(-1)}$

2. Transform the following difference equations into the form specified in Definition 6.

a. $(j^2 + 1)z(j - 3) + j^{(2)}z(j - 2) + 2^j z(j - 1) = (-1)^j + z(j); j = 3, 4, 5, \ldots)$

b. $V(7 + j) + V(8 + j) = 3^j V(j); j = 0, 1, 2, \ldots$

c. $\Delta u(j) + u(j + 4) = E^3 u(j) + \cos(3j); j = 0, 1, 2, \ldots$

Section 23   Existence and Uniqueness

In Section 9, the uniqueness theorem, that is, Theorem 10, for linear differential equations was stated without proof. The analogous result for linear difference equations is somewhat simpler, and so the details are presented in this section.

1 DEFINITION

An *nth-order linear initial-value problem* consists of a linear difference equation

$$\Lambda(y) = f(k)$$

where Λ is an *n*th-order linear operator, together with a set of *n initial conditions*

$$y(0) = a_0; Ey(0) = a_1; \ldots; E^{n-1}y(0) = a_{n-1}$$

where $a_0, a_1, \ldots, a_{n-1}$ are real numbers. A *solution* of the initial-value problem is a function that is a solution of the difference equation and that satisfies the initial conditions.

Notice that the *n* initial conditions can also be expressed as

$$y(0) = a_0; y(1) = a_1; y(2) = a_2; \ldots; y(n - 1) = a_{n-1}$$

since $E^i y(0) = y(i)$.

2 THEOREM

An *n*th-order linear initial-value problem has a unique solution.

Proof: The *n*th-order equation

(A) $E^n y(k) + p_1(k)E^{n-1}y(k) + \cdots + p_n(k)y(k) = f(k)$

can be written in the form

(B) $E^n y(k) = f(k) - \{p_1(k)E^{n-1}y(k) + p_2(k)E^{n-2}y(k) + \cdots + p_n(k)y(k)\}$

In particular, for $k = 0$, equation (B) is:

$$y(n) = f(0) - \{p_1(0)y(n - 1) + p_2(0)y(n - 2) + \cdots + p_n(0)y(0)\}$$

Thus $y(n)$ is uniquely determined by the initial conditions $y(0)$, $y(1)$, ..., $y(n - 1)$. Proceeding inductively, suppose that for some integer $m > 0$, y satisfies (B). Then $y(m)$, $y(m + 1)$, ..., $y(m + n)$ are known, and by taking $k = m + 1$ in (B), $y(m + n + 1)$ is uniquely determined. In this way, a solution of (A) can be constructed, and it is the only solution that satisfies the specified initial conditions.

The remainder of this chapter is devoted to various approaches for characterizing the solutions of linear difference equations. As a first step, we consider the first-order homogeneous equation

(3) $Ey(k) + p(k)y(k) = \varnothing$

This equation is equivalent to

$$y(k + 1) = -p(k)y(k)$$

and thus

$$y(1) = -p(0)y(0)$$
$$y(2) = -p(1)y(1) = -p(1)[-p(0)y(0)] = (-1)^2 p(1)p(0)y(0)$$
$$y(3) = -p(2)y(2) = (-1)^3 p(2)p(1)p(0)y(0)$$
$$\vdots$$
$$y(k) = (-1)^k p(k - 1)p(k - 2) \cdots p(1)p(0)y(0)$$

Let

(4) $$t(k) = (-1)^k p(0)p(1) \cdots p(k - 1) = (-1)^k \prod_{i=0}^{k-1} p(i)$$

for $k = 1, 2, \ldots$, and let $t(0) = 1$. Then the solutions of (3) can be represented by

(5) $y(k) = t(k)y(0)$

These observations yield the following theorem.

6 THEOREM

The initial-value problem

$$Ey(k) + p(k)y(k) = \varnothing; \qquad y(0) = b$$

has the unique solution

$$y(k) = b \cdot t(k)$$

where $t(k)$ is given by (4).

7 Examples

a. If

$$Ey(k) - \left(\frac{k+2}{k+1}\right) y(k) = \emptyset$$

then $p(k) = -[(k+2)/(k+1)]$. Therefore,

$$t(k) = (-1)^k \prod_{i=0}^{k-1} \left[-\left(\frac{i+2}{i+1}\right)\right]$$

$$= (-1)^{2k} \prod_{i=0}^{k-1} \left(\frac{i+2}{i+1}\right)$$

$$= \tfrac{2}{1} \cdot \tfrac{3}{2} \cdots \cdots \left(\frac{k+1}{k}\right) = k+1$$

or $t(k) = k + 1$. Thus

$$y(k) = t(k)y(0) = (k+1)y(0)$$

represents all solutions of this first-order equation.

b. Consider the initial-value problem

$$Ey(k) + \frac{2^{k+1}}{(k+1)^2} y(k) = \emptyset; \qquad y(0) = 3$$

Here,

$$p(k) = \frac{2^{k+1}}{(k+1)^2}$$

so that

$$t(k) = (-1)^k \prod_{i=0}^{k-1} \frac{2^{i+1}}{(i+1)^2} = (-1)^k \left[\frac{2}{1^2} \cdot \frac{2^2}{2^2} \cdot \frac{2^3}{3^2} \cdots \cdots \frac{2^k}{k^2}\right]$$

Since

$$\sum_{j=1}^{k} j = \frac{k(k+1)}{2}$$

$t(k)$ can also be written

$$t(k) = (-1)^k \frac{[2^{k(k+1)}]^{1/2}}{(k!)^2}$$

Thus

$$y(k) = \frac{(-1)^k 3 [2^{k(k+1)}]^{1/2}}{(k!)^2}$$

is the solution of the initial-value problem.

Before considering the first-order nonhomogeneous equation, it would be well to review one of the restrictions in the definition of a linear difference equation given in Definition 6, Section 22. In order for equation (3) to satisfy this definition, it is required that $p(k) \neq 0$ for all $k \in N$. The necessity for this assumption will become apparent in the derivation of the representation for the set of solutions of the nonhomogeneous equation. From (4), notice that $t(k) \neq 0$ for all $k \in N$ if and only if $p(k) \neq 0$ for all $k \in N$. In addition, note that if $p(j) = 0$ for some $j \in N$, then $t(j + 1) = 0$, and, in fact, $t(k) = 0$ for all $k \geq j + 1$.

The first-order nonhomogeneous equation has the form

(8) $Ey(k) + p(k)y(k) = f(k)$

9 THEOREM

The one-parameter family of functions

$$y(k) = t(k) \sum_{i=0}^{k-1} \frac{f(i)}{t(i + 1)} + ct(k)$$

where c is any constant, represents the set of all solutions of equation (8).

Proof: By definition, $p(k) \neq 0$ for all $k \in N$. Therefore, as observed above, $t(k)$, given by (4), is nonzero for all $k \in N$. Thus equation (8) is equivalent to

(A) $\dfrac{y(k + 1)}{t(k + 1)} + \dfrac{p(k)y(k)}{t(k + 1)} = \dfrac{f(k)}{t(k + 1)}$

Notice that

$$\frac{p(k)}{t(k + 1)} = \frac{p(k)}{-p(k)t(k)} = -\frac{1}{t(k)}$$

so that (A) is equivalent to

$$\frac{y(k + 1)}{t(k + 1)} - \frac{y(k)}{t(k)} = \frac{f(k)}{t(k + 1)}$$

which can be written

(B) $\Delta\left[\dfrac{y(k)}{t(k)}\right] = \dfrac{f(k)}{t(k + 1)}$

Hence, from Theorem 11, Section 21,

$$\frac{y(k)}{t(k)} = \sum_{i=0}^{k-1} \frac{f(i)}{t(i + 1)} + c$$

is a solution of (B). Therefore,

$$y(k) = t(k) \sum_{i=0}^{k-1} \frac{f(i)}{t(i + 1)}$$

is a solution of (8), and thus the one-parameter family of solutions of (8) is as specified in the statement of the theorem.

The reader should compare the one-parameter family of solutions of equation (8) with the one-parameter family of solutions of the first-order nonhomogeneous differential equation specified in Corollary 15, Section 8. Also, the one-parameter family of solutions of (8) can be represented as

$$y(k) = t(k) \sum \frac{f(k)}{t(k + 1)}$$

since a "constant of summation" is associated with the indefinite sum. This form of the solution of equation (8) is used in the next example.

10 Example

Consider the first-order nonhomogeneous difference equation

$$Ey(k) - \left[\frac{k + 2}{k + 1}\right] y(k) = k(k + 2)2^k$$

From Example 7a, $t(k) = k + 1$, so that $t(k + 1) = k + 2$. Thus, using Theorem 9, the solutions are:

$$y(k) = (k + 1) \sum \frac{k(k + 2)2^k}{(k + 2)}$$

or

$$y(k) = (k + 1) \sum k2^k$$

We have, from Example 8b, Section 21, that

$$\sum k2^k = 2^k(k - 2) + c$$

Therefore, the one-parameter family of solutions of the equation is:

$$y(k) = (k + 1)(k - 2)2^k + c(k + 1)$$

To solve the initial-value problem

$$Ey(k) - \left[\frac{k + 2}{k + 1}\right] y(k) = k(k + 2)2^k; \qquad y(0) = 3$$

we have that

$$3 = y(0) = (0 + 1)(0 - 2){\cdot}2^0 + c(0 + 1) = -2 + c$$

or $c = 5$. Thus the solution of the initial-value problem is:

$$y(k) = (k + 1)(k - 2)2^k + 5(k + 1)$$

11 EXERCISES

1. Solve the following initial-value problems.
   a. $Ey(k) - [(k + 1)/(k + 3)]y(k) = \varnothing; \; y(0) = \frac{1}{2}$
   b. $Ey(k) + 3y(k) = \varnothing; \; y(0) = 1$
   c. $Ey(k) + [(k + 2)/(k + 1)]y(k) = \varnothing; \; y(0) = 1$
   d. $Ey(k) = [(k^2 + 2k + 2)/(k^2 + 1)]y(k); \; y(0) = 3$
   e. $Ey(k) - [(2k + 3)/(2k + 1)]y(k) = \varnothing; \; y(0) = 1$
   f. $Ey(k) = [(k^2 + 4k + 3)/(k^2 + 4k + 4)]y(k); \; y(0) = 2$
   g. $Ey(k) = [(2k + 4)/(k + 1)]y(k); \; y(0) = 1$
   h. $Ey(k) + [3(k + 4)/(k + 3)]y(k) = \varnothing; \; y(0) = 3$

2. Solve the following initial-value problems.
   a. $Ey(k) - [(k + 1)/(k + 3)]y(k) = 3; \; y(0) = 3$
   b. $Ey(k) - [(k + 1)/(k + 3)]y(k) = 3; \; y(0) = 5$
   c. $Ey(k) - [(2k + 3)/(2k + 1)]y(k) = 8/(2k + 1); \; y(0) = 0$
   d. $Ey(k) - [2(k + 2)^{(2)}/(k + 1)^2]y(k) = -(k + 3)/(k + 1); \; y(0) = 0$
   e. $Ey(k) = [(k + 1)(k + 3)/(k + 2)^2]y(k) + (k + 3); \; y(0) = 0$
   f. $Ey(k) + (k + 1)y(k) = (k + 2)^{(2)} + (k + 2)^{(1)} + 1; \; y(0) = 2$
   g. $Ey(k) - y(k) = \ln[(k + 2)/(k + 1)]; \; y(0) = -4$
   h. $Ey(k) = (k + 2)y(k) + k(2)^k; \; y(0) = 2$

We conclude this section with an observation which is motivated by the proof of the uniqueness theorem, Theorem 2. In particular, this proof indicates that a step-by-step calculation of the solution of an initial-value problem involving a linear difference equation is always possible. This is in contrast to initial-value problems for differential equations where, in general, it is not possible to obtain the exact solution. This situation is similar to the comparison of a definite sum with a definite integral, as made in Section 21. The following example illustrates the step-by-step method of solution of an initial-value problem.

12 Example

Consider the initial-value problem

$$E^2y(k) - \left[\frac{k + 1}{k + 2}\right]Ey(k) - \frac{2}{(k + 1)(k + 2)}y(k) = \varnothing$$

$$y(0) = 2, \quad y(1) = 4$$

The difference equation can be written in the form

(A) $\quad y(k + 2) = \left[\frac{k + 1}{k + 2}\right]y(k + 1) + \frac{2}{(k + 1)(k + 2)}y(k)$

Thus, by letting $k = 0$ in (A), and using the initial conditions, we get

$$y(2) = (\tfrac{1}{2})4 + (\tfrac{2}{2})2 = 4$$

For $k = 1, 2, 3, \ldots$, we have

$$y(3) = \tfrac{2}{3}y(2) + \tfrac{1}{3}y(1) = (\tfrac{2}{3})4 + (\tfrac{1}{3})4 = 4$$

$$y(4) = \tfrac{3}{4}y(3) + \tfrac{1}{6}y(2) = (\tfrac{3}{4})4 + (\tfrac{1}{6})4 = \tfrac{11}{3}$$

$$y(5) = \tfrac{4}{5}y(4) + \tfrac{1}{10}y(3) = (\tfrac{4}{5})(\tfrac{11}{3}) + (\tfrac{1}{10})4 = \tfrac{10}{3}$$

and so on. Of course, this procedure can be continued, and as many terms of the solution $y(k)$ as desired can be calculated.

13 EXERCISES

1. In each of the following, determine $y(5)$ by a step-by-step calculation.
   a. $E^2 y(k) + k^2 y(k) = 2y(k) + (-2)^k$; $y(0) = 0, y(1) = 0$
   b. $\Delta^2 y(k) + 2Ey(k) = \varnothing$; $y(0) = 1, y(1) = 0$
   c. $3E^2 y(k) + (k^{(2)} - 1)y(k) = 1$; $y(0) = 0, y(1) = 0$
   d. $\Delta^2 y(k) = 2$; $y(0) = 0, y(1) = 0$
   e. $E^2 y(k) = 2^k y(k) + (-2)^k$; $y(0) = 2, y(1) = 0$

Section 24 General Theory

As remarked in Section 22, the nth-order operator

(1) $\Lambda = E^n + p_1(k)E^{n-1} + \cdots + p_n(k)I$

is a linear operator. Let H denote the set of all solutions of the homogeneous equation

(2) $\Lambda(y) = \varnothing$

that is, $y(k) \in H$ if and only if $\Lambda[y(k)] = \varnothing$. Thus if

$$\Lambda(y_1) = \Lambda(y_2) = \cdots = \Lambda(y_j) = \varnothing$$

for some integer j, then it follows that all linear combinations of $y_1, y_2, \ldots, y_j$ are solutions of (2), since

$$\Lambda[c_1 y_1(k) + c_2 y_2(k) + \cdots + c_j y_j(k)]$$
$$= c_1 \Lambda(y_1) + c_2 \Lambda(y_2) + \cdots + c_j(y_j) = \varnothing$$

In the language of linear algebra, this observation can be stated as follows.

3 THEOREM

The set H of all solutions of $\Lambda[y] = \varnothing$ is a subspace of the vector space V_∞.

As was the case in Chapter 2, what is needed at this point is a way to represent all members of H. This leads to the concept of linear independence, which is reviewed again here.

4 DEFINITION

Let $y_1(k), y_2(k), \ldots, y_j(k)$ be members of V_∞. If the only linear combination of these functions that is equal to the zero function is the trivial linear combination, that is, if

$$c_1 y_1(k) + c_2 y_2(k) + \cdots + c_j y_j(k) = \varnothing$$

implies that $c_1 = c_2 = \cdots = c_j = 0$, then this set of functions is called *linearly independent*. If some nontrivial linear combination of these functions is equal to the zero function, then the set of functions is called *linearly dependent*.

---

5 Examples

      a. If $y_1(k) = 1$, $y_2(k) = k$, and $y_3(k) = k^{(2)}$, then

(A) $c_1 + c_2 k + c_3 k^{(2)} = \varnothing$

      implies that

$$c_2 + 2c_3 k = \varnothing$$

      and

$$2c_3 = \varnothing$$

      These last two equations are obtained by applying Δ and Δ^2 to both sides of (A). Thus $c_1 = c_2 = c_3 = 0$, and the functions are linearly independent.

      b. If $y_1(k) = 1$, $y_2(k) = k$, $y_3(k) = k^{(2)}$, and $y_4(k) = 2k^{(2)} - 1$, then, since

$$y_1(k) - 2y_3(k) + y_4(k) = \varnothing$$

      this set of functions is linearly dependent.

      Associated with any finite set of members of V_∞ is a matrix which is, in form, very similar to the Wronski matrix.

---

6 DEFINITION

Let $f_1(k), f_2(k), \ldots, f_m(k)$ be members of V_∞. The $m \times m$ matrix

$$C(k) = \begin{bmatrix} f_1(k) & f_2(k) & \cdots & f_m(k) \\ Ef_1(k) & Ef_2(k) & \cdots & Ef_m(k) \\ \vdots & \vdots & & \vdots \\ E^{m-1}f_1(k) & E^{m-1}f_2(k) & \cdots & E^{m-1}f_m(k) \end{bmatrix}$$

is called the *Casorati matrix* of $f_1, f_2, \ldots, f_m$.

For a given set of functions $f_1, f_2, \ldots, f_m$, $C(k)$ is a matrix-valued function of an integer variable, while $\det[C(k)]$ is a real-valued function of an integer variable.

The use of the Casorati matrix in establishing linear independence or dependence is given by the next theorem. This theorem is the analogue of the result stated in Exercises 14, Section 10, Problem 6, in terms of the Wronski matrix.

7 THEOREM

If $f_1, f_2, \ldots, f_m$ are members of V_∞ and these m functions are linearly dependent on N, then

$$\det[C(k)] = \varnothing$$

on N.

Proof: If the functions are linearly dependent on N, then there exist constants $c_1, c_2, \ldots, c_m$, not all zero, such that

$$c_1 f_1(k) + c_2 f_2(k) + \cdots + c_m f_m(k) = \varnothing$$

on N. Hence for each nonnegative integer i, $0 \le i \le m - 1$,

(A) $\qquad c_1 E^i f_1(k) + c_2 E^i f_2(k) + \cdots + c_m E^i f_m(k) = \varnothing$

Let $\gamma = (c_1, c_2, \ldots, c_m)^T$ be the vector of coefficients in (A). Then, using matrix notation, the system of equations (A) can be written

$$C(k)\gamma = \varnothing$$

for each $k \in N$, where $\varnothing$ denotes the vector of zero functions. For each $k \in N$, this is a linear homogeneous system of algebraic equations, and since γ is not the zero vector, it follows from Theorem 16, Section 5, that the matrix is singular, that is,

$$\det[C(k)] = \varnothing$$

8 COROLLARY

If $\det[C(k)] \ne 0$ for some $k \in N$, then $f_1, f_2, \ldots, f_m$ are linearly independent on N.

Of course, this result does not exclude the possibility of having a linearly independent set of functions whose Casorati matrix C has the property $\det[C(k)] = 0$ for all $k \in N$. For a simple example, consider the following functions: $f_1(k) = 0$ for $k \ne 4$, $f_1(4) = 1$, and $f_2(k) = 0$ for $k \ge 1$, $f_2(0) = 1$. Clearly, f_1 and f_2 are linearly independent while

$$C(k) = \begin{bmatrix} f_1(k) & f_2(k) \\ Ef_1(k) & Ef_2(k) \end{bmatrix}$$

has both entries in at least one column equal to zero for all $k \in N$. Hence,

$$\det[C(k)] = \varnothing.$$

At this point, it is convenient to digress from the study of the Casorati matrix to consider the following related result.

9 THEOREM

If $z(k)$ is a solution of the homogeneous equation (2), and if for some $m \in N$, $z(m) = Ez(m) = \cdots = E^{n-1}z(m) = 0$, then $z(k)$ is the zero function.

Proof: If $m = 0$, then $z(k)$ is the unique solution of (2) with initial values $z(0) = Ez(0) = \cdots = E^{n-1}z(0) = 0$. Since the zero function is an obvious solution of (2), it follows that in this case $z(k)$ is the zero function. If $m > 0$, then

(A) $\qquad z(k) = \left[\dfrac{-1}{p_n(k)}\right] [E^n + p_1(k)E^{n-1} + \cdots + p_{n-1}(k)E]z(k)$

Notice that (A) is well-defined since, in Definition 6, Section 22, it is specified that $p_n(k) \neq 0$ for all $k \in N$. By taking $k = m - 1$ in (A), we get that $z(m - 1) = 0$, and if $m - 1 \neq 0$, then $k = m - 2$ implies that $z(m - 2) = 0$. Continuing in this way, it follows that

$$z(0) = Ez(0) = \cdots = E^{n-1}z(0) = 0$$

and so in this case $z(k)$ is the zero function.

The following theorem is the difference equation analogue of Corollary 11, Section 10. The reader should note that the result of Theorem 9 is an essential step in the proof.

10 THEOREM

If $y_1(k), y_2(k), \ldots, y_n(k)$ are n solutions of the homogeneous equation (2), $C(k)$ is the Casorati matrix of these functions, and if for some $m \in N$, $\det[C(m)] = 0$, then these solutions are linearly dependent on N.

Proof: If $\det[C(m)] = 0$, then there is a vector $\gamma = (c_1, c_2, \ldots, c_n)^T$ such that γ is not the zero vector and

(A) $\qquad C(m)\gamma = \varnothing$

Let $z(k) = c_1 y_1(k) + c_2 y_2(k) + \cdots + c_n y_n(k)$. It follows from Theorem 3 that $z(k)$ is a solution of (2). Also, an immediate consequence of (A) is that

$$z(m) = Ez(m) = \cdots = E^{n-1}z(m) = 0$$

Thus from Theorem 9, $z(k)$ is the zero function, and hence

$$c_1 y_1(k) + c_2 y_2(k) + \cdots + c_n y_n(k) = \varnothing$$

Since not all c_i are zero, the solutions $y_1(k)$, $y_2(k)$, ..., $y_n(k)$ are linearly dependent.

11 COROLLARY

Let $y_1(k)$, $y_2(k)$, ..., $y_n(k)$ be n solutions of (2). Then

(i) these solutions are linearly dependent if and only if $\det[C(k)] = \emptyset$ on N;

(ii) these solutions are linearly independent if and only if $\det[C(k)] \neq 0$ for all $k \in N$.

Proof: Exercise.

These results provide the means to characterize the set of all solutions of (2). As remarked previously, it is clear that H is a subspace of V_∞. The next step is to determine the dimension of H.

12 THEOREM

Any solution of the nth-order equation (2) can be represented as a linear combination of n linearly independent solutions of (2).

Proof: Let $y_i(k)$, $i = 1, 2, \ldots, n$, be the unique solutions of (2) satisfying the initial conditions

$$y_1(0) = 1, y_1(1) = y_1(2) = \cdots = y_1(n - 1) = 0$$
$$y_2(0) = 0, y_2(1) = 1, y_2(2) = y_2(3) = \cdots = y_2(n - 1) = 0$$
$$\vdots$$
$$y_n(0) = y_n(1) = \cdots = y_n(n - 2) = 0, y_n(n - 1) = 1$$

or more compactly,

$$E^{i-1}y_i(0) = 1$$

for $i = 1, 2, \ldots, n$, with all other initial values equal to zero.

Evidently, the Casorati matrix for these n solutions evaluated at $k = 0$, that is, $C(0)$, is the identity matrix. Thus $\det[C(0)] = 1$, and this implies that the y_i are independent. Therefore, equation (2) has n linearly independent solutions.

Let $y(k)$ be a solution of (2) and consider

$$z(k) = y(0)y_1(k) + y(1)y_2(k) + \cdots + y(n - 1)y_n(k)$$

Thus $z(k)$ is merely a linear combination of the n independent solutions considered above. Since

$$E^m z(0) = z(m) = y(m)$$

for $m = 0, 1, 2, \ldots, n - 1$, it follows from the uniqueness theorem that $y(k) = z(k)$ on N. Hence any solution of (2) can be expressed as a linear combination of these n independent solutions.

Any set of n linearly independent solutions of (2) is called a *solution basis* of (2). This is the terminology which was introduced in Chapter 2. The proof of Theorem 12 entailed the construction of a solution basis for equation (2). In studying the proof of the theorem, it should be apparent that the solution basis considered is by no means unique. For example, a slight modification of an initial condition, say $y_i(0) = 2$, with all other values unchanged, would have served just as well. In fact, the concept of a solution basis of a linear homogeneous difference equation or differential equation is merely a particular case of the general concept of a basis of a vector space. Thus a change of basis is equally appropriate in this application of linear algebra.

13 Example

It is easy to verify that

$$y_1(k) = 7$$

and

$$y_2(k) = 2^k$$

are solutions of

(A) $(E^2 - 3E + 2I)y(k) = \varnothing$

Since $y_1(k)$ and $y_2(k)$ are linearly independent, these two functions are a solution basis for equation (A), and

$$y(k) = c_1 7 + c_2 2^k$$

is the two-parameter family of solutions.

14 EXERCISES

1. Show that the following sets of functions are, or are not, linearly independent.
   a. $y_1(k) = (-1)^k$, $y_2(k) = \cos(k\pi)$, $y_3(k) = k$
   b. $y_1(k) = 2^k$, $y_2(k) = (-2)^k$, $y_3(k) = (-1)^k$
   c. $y_1(k) = 2^k$, $y_2(k) = k(2)^k$, $y_3(k) = k^{(2)}(2)^k$
   d. $y_1(k) = k^{(2)}$, $y_2(k) = k^{(3)}$, $y_3(k) = k^{(4)}$
   e. $y_1(k) = k^{(-1)}$, $y_2(k) = k^{(-2)}$, $y_3(k) = 1$
   f. $y_1(k) = k^2$, $y_2(k) = k$, $y_3(k) = k^{(2)}$
2. Form the Casorati matrix for each set of functions in Problem 1 and calculate $\det[C(k)]$.
3. Let $C(k)$ be the Casorati matrix for $y_1(k) = \sin(k)$, $y_2(k) = \cos(k)$. Compute $\Delta\{\det[C(k)]\}$.
4. Prove Corollary 11.

5. Show that $y_1(k) = -1$, $y_2(k) = 2^{k+7}$ is a solution basis for equation (A) in Example 13.

6. Show that if $y_1(k)$, $y_2(k)$ is a solution basis for a second-order linear homogeneous difference equation and c_1, c_2, c_3, and c_4 are constants, then $u_1(k) = c_1 y_1(k) + c_2 y_2(k)$, $u_2(k) = c_3 y_1(k) + c_4 y_2(k)$ is a solution basis if and only if

$$\det \begin{bmatrix} c_1 & c_2 \\ c_3 & c_4 \end{bmatrix} \neq 0$$

7. Consider the second-order equation

$$E^2 y(k) + p(k)Ey(k) + q(k)y(k) = \emptyset$$

and let $D(k) = \det[C(k)]$, where $C(k)$ is the Casorati matrix for a solution basis of this equation. Show that $D(k)$ is a solution of the first-order equation

$$Ey(k) - q(k)y(k) = \emptyset$$

and thus

$$\det[C(k)] = \det[C(0)] \prod_{i=0}^{k-1} q(i)$$

8. Use Problem 7 to show that if $y_1(k)$, $y_2(k)$ is solution basis for

$$E^2 y(k) + p(k)Ey(k) + q(k)y(k) = \emptyset$$

$y_1(k) \neq 0$ for all $k \in N$, and $C(k)$ is the Casorati matrix of y_1, y_2, then

$$\Delta \left[\frac{y_2(k)}{y_1(k)} \right] = \frac{\det[C(0)] \prod_{i=0}^{k-1} q(i)}{y_1(k)Ey_1(k)}$$

9. Obtain a theorem that is analogous to Theorem 5, Section 12.

We conclude this development of the general theory of linear difference equations by considering the nonhomogeneous equation. The main results here are identical in form with those of Section 11, and the proofs are left as exercises.

Associated with the linear operator

(15) $\Lambda = E^n + p_1(k)E^{n-1} + \cdots + p_n(k)I$

is the homogeneous equation

(16) $\Lambda[y(k)] = \emptyset$

as well as nonhomogeneous equations of the form

(17) $\Lambda[y(k)] = f(k)$

where $f(k) \neq \emptyset$. Equation (16) is called the *reduced equation* of equation (17).

18 THEOREM

If $u(k)$ and $v(k)$ are solutions of the nonhomogeneous equation (17), then $y(k) = u(k) - v(k)$ is a solution of the homogeneous equation (16).

19 THEOREM

If $z(k)$ is a solution of the nonhomogeneous equation (17) and $y_1(k), y_2(k), \ldots,$ $y_n(k)$ is a solution basis for the homogeneous equation (16), then every solution of (17) can be represented as

$$y(k) = c_1 y_1(k) + c_2 y_2(k) + \cdots + c_n y_n(k) + z(k)$$

where $c_1, c_2, \ldots, c_n$ are constants.

20 Example

Consider the nonhomogeneous equation

(A) $(E^2 - 3E + 2I)y(k) = 3^k$

It is easy to verify that $z(k) = (\frac{1}{2})3^k$ is a solution of this equation. Furthermore, since $y_1(k) = 7$, $y_2(k) = 2^k$ is a solution basis for the reduced equation, it follows that any solution of (A) is of the form

$$y(k) = c_1 \cdot 7 + c_2 \cdot 2^k + (\tfrac{1}{2})3^k$$

Consider, for example, the initial-value problem consisting of (A) and initial values $y(0) = 3$ and $y(1) = 0$. The initial conditions imply that

$$3 = y(0) = 7c_1 + c_2 + \tfrac{1}{2}$$
$$0 = y(1) = 7c_1 + 2c_2 + \tfrac{3}{2}$$

Thus $c_1 = \frac{13}{14}$ and $c_2 = -4$. The unique solution of this initial-value problem is:

$$y(k) = (\tfrac{13}{14})7 - 4(2^k) + (\tfrac{1}{2})3^k$$
$$= \tfrac{1}{2}(13 - 2^{k+3} + 3^k)$$

21 EXERCISES

1. Prove Theorem 18.
2. Prove Theorem 19.
3. Let $y_1(k), y_2(k)$ be a solution basis for the homogeneous equation

$$E^2 y(k) + p_1(k)Ey(k) + p_2(k)y(k) = \varnothing$$

and let $C(k)$ be the associated Casorati matrix. Show that a solution of the nonhomogeneous equation

$$E^2 y(k) + p_1(k)Ey(k) + p_2(k)y(k) = f(k)$$

is $y(k) = u(k)y_1(k) + v(k)y_2(k)$, where

$$\Delta u(k) = -E\left[\frac{y_2(k)}{\det[C(k)]}\right]f(k)$$

and

$$\Delta v(k) = E\left[\frac{y_1(k)}{\det[C(k)]}\right] f(k)$$

This is a form of the method of variation of parameters.

4. Use the method of variation of parameters to determine a solution of $E^2 y(k) - 3Ey(k) + 2y(k) = k$. (Hint: See Example 13.)

5. a. Show that a solution basis for

$$E^2 y(k) - (k + 2)^{(2)} y(k) = \varnothing$$

   is $y_1(k) = k!$, $y_2(k) = (-1)^{k+1} k!$.

   b. Calculate $\det[C(k)]$, where $C(k)$ is the Casorati matrix for $y_1(k)$, $y_2(k)$.

   c. Find a solution of

$$E^2 y(k) - (k + 2)^{(2)} y(k) = 2(k + 2)!$$

6. Does the problem $\Delta^2 y(k) = E^2 y(k)$; $y(0) = 1$, have a solution?

Section 25 Linear Difference Equations with Constant Coefficients

In Section 20, Equidimensional Differential Equations, many of the details were omitted since Sections 14, 15, and 16 could be used as a model for the study of these equations. The topic of this section, that is, linear difference equations with constant coefficients, can be viewed as another application of the ideas developed in Sections 14, 15, and 16, and, again, most of the details will be omitted.

1 DEFINITION

Let $P(t) = a_0 t^n + a_1 t^{n-1} + \cdots + a_{n-1} t + a_n$ be a real polynomial of degree n such that $a_n \neq 0$. The nth-order operator polynomial, $P(E)$, associated with $P(t)$ is defined by

$$P(E) = a_0 E^n + a_1 E^{n-1} + \cdots + a_{n-1} E + a_n I$$

This is the same notational convention as used in the discussion of differential equations with constant coefficients. Thus

$$P(E)y = \varnothing$$

is the abbreviated description of the homogeneous difference equation

$$a_0 E^n y + a_1 E^{n-1} y + \cdots + a_{n-1} Ey + a_n y = \varnothing$$

Notice that in Definition 1 it is specified that $a_n \neq 0$, which is equivalent to $P(0) \neq 0$. This is the same restriction that was made in Definition 6, Section

22. Some additional comments about this restriction are made at the end of this section.

As noted in equation (24), Section 19,

$$Eb^k = b(b^k)$$

which is similar to

$$De^{bx} = be^{bx}$$

Thus the following result compares with Theorem 10, Section 13.

2 THEOREM

If $P(E)$ is an nth-order operator polynomial, then

$$P(E)[b^k] = P(b)[b^k]$$

Proof: Exercise.

Theorem 2 indicates that there is the same sort of connection between roots of the operator polynomial and solutions of the homogeneous difference equation as was the case with differential equations. The obvious consequence of Theorem 2 that is useful in characterizing the solutions of the homogeneous difference equation is given in the following corollary.

3 COROLLARY

If $P(E)$ is an operator polynomial and $P(b) = 0$, then $y(k) = b^k$ is a solution of $P(E)y = \varnothing$.

Of course, if the restriction $P(0) \neq 0$ is dropped, then Theorem 2 and Corollary 3 are still correct. However, if $P(0) = 0$, then it follows from Corollary 3 that $y(k) = 0^k$ is a solution of $P(E)y = \varnothing$. But the function defined by $y(k) = 0^k$ is merely the zero function, and $y = \varnothing$ is always a solution of a homogeneous linear equation. A final word of explanation of this topic is the last item in this section.

4 Example

If $P_1(t) = a_0 t + a_1$, where $a_0 a_1 \neq 0$, then $P_1(-a_1/a_0) = 0$, so that $y(k) = (-a_1/a_0)^k$ is a solution of the first-order equation

(A) $a_0 E y + a_1 y = \varnothing$

Thus

$$y(k) = \left(\frac{-a_1}{a_0}\right)^k y(0)$$

is the form of all solutions of (A). Also, by writing (A) in the equivalent form

(B) $$Ey + \left(\frac{a_1}{a_0}\right) y = \emptyset$$

the above solution can be obtained by using (5), Section 23.

5 EXERCISES

1. Prove Theorem 2.
2. Solve the following first-order equations.
   a. $Ey + 2y = \emptyset$ b. $Ey - ey = \emptyset$ c. $3Ey + y = \emptyset$
3. Let b and c be two nonzero constants. Show that:
   a. no constant function is a solution of $Ey - y = c$, and
   b. the only constant function that is a solution of $Ey - by = c$ is:

   $$y(k) = \frac{c}{1 - b}$$

   provided $b \neq 1$.
4. Solve the initial-value problem

   $$Ey - 4y = 3; \qquad y(0) = 0$$

5. Suppose that

   $$P(t) = a_0 t^n + a_1 t^{n-1} + \cdots + a_n$$

   and

   $$Q(t) = b_0 t^n + b_1 t^{n-1} + \cdots + b_n$$

   are two polynomials such that

   $$P(E) \equiv Q(\Delta)$$

   is an operator identity. What restriction on $Q(t)$ is equivalent to $P(0) \neq 0$?

We now consider second-order homogeneous equations by using the ideas in Section 14. These equations can be represented in terms of the polynomial

$$P_2(t) = a_0 t^2 + a_1 t + a_2$$

where $a_0 a_2 \neq 0$. By applying Corollary 3, we have that

$$y(k) = b^k$$

is a solution of

(6) $$P_2(E)y = \emptyset$$

if and only if

$$P_2(b) = 0$$

that is, if and only if b is a root of P_2.

Rather than expressing the roots of P_2 in terms of the coefficients of P_2, as was done in (2), Section 14, we consider the three possible forms of P_2 directly. Since the equation $P_2(E)y = \varnothing$ is homogeneous, it is convenient to consider P_2 with leading coefficient one, that is,

$$P_2(t) = t^2 + \left(\frac{a_1}{a_0}\right)t + \left(\frac{a_2}{a_0}\right)$$

CASE I. Assume that $P_2(t) = (t - r_1)(t - r_2) = t^2 - (r_1 + r_2)t + r_1 r_2$, where r_1 and r_2 are distinct real numbers and are both nonzero. In this case, $y_1(k) = r_1{}^k$ and $y_2(k) = r_2{}^k$ are solutions of $P_2(E)y = \varnothing$, and the Casorati matrix of this pair of functions is:

$$C(K) = \begin{bmatrix} r_1{}^k & r_2{}^k \\ r_1 r_1{}^k & r_2 r_2{}^k \end{bmatrix}$$

Thus $\det[C(k)] = (r_2 - r_1)(r_1 r_2)^k \neq 0$ for all k, and the two solutions are linearly independent. All solutions of $P_2(E)y = \varnothing$ are of the form

$$y(k) = c_1(r_1)^k + c_2(r_2)^k$$

CASE II. Assume that $P_2(t) = (t - r)^2 = t^2 - 2rt + r^2$, where $r \neq 0$ is a real number. In this case, by using Section 14 as a model, we might speculate that $y_1(k) = r^k$, $y_2(k) = k(r)^k$ is a solution basis for $P_2(E)y = \varnothing$. This conjecture is, in fact, correct and its verification is a straightforward exercise. Thus all solutions of $P_2(E)y = \varnothing$ are of the form

$$y(k) = (c_1 + c_2 k)r^k$$

CASE III. Assume that

$$P_2(t) = (t - r_1 - ir_2)(t - r_1 + ir_2) = t^2 - 2r_1 t + r_1{}^2 + r_2{}^2$$

where r_1 and r_2 are real numbers and $r_2 \neq 0$. This is the case where P_2 has a pair of complex roots, that is, $P_2(r_1 + ir_2) = 0$ and $P_2(r_1 - ir_2) = 0$. In order to obtain solutions that are real-valued functions, it is helpful to represent the complex number $r_1 + ir_2$ in polar form (see Section 3). Thus

$$r_1 + ir_2 = r (\cos s + i \sin s)$$

where

$$r = (r_1{}^2 + r_2{}^2)^{1/2} \quad \text{and} \quad \tan s = \left(\frac{r_2}{r_1}\right)$$

With this form, it can be shown that

$$y_1(k) = r^k \cos(ks)$$

and

$$y_2(k) = r^k \sin(ks)$$

are linearly independent solutions of $P_2(E)y = \varnothing$. Verification of this requires the use of trigonometric identities and is left as an exercise. Thus all real solutions of $P_2(E)y = \emptyset$ are of the form

$$y(k) = r^k[c_1 \cos(ks) + c_2 \sin(ks)]$$

7 Example

If $P_2(t) = 3t^2 - 3t - 6$, then $P(-1) = P(2) = 0$, so that $y_1(k) = (-1)^k$, $y_2(k) = 2^k$ is a solution basis for

(A) $3E^2y - 3Ey - 6y = \emptyset$

All solutions of (A) are of the form

$$y(k) = c_1(-1)^k + c_2 2^k$$

The solution of (A) satisfying the initial conditions

$$y(0) = 4; \qquad y(1) = -1$$

is found by solving for c_1 and c_2 as follows:

$$4 = y(0) = c_1 + c_2$$
$$-1 = y(1) = -c_1 + 2c_2$$

or $c_1 = 3$ and $c_2 = 1$. Thus the solution of the initial-value problem is:

$$y(k) = 3(-1)^k + 2^k$$

8 Example

If $P_2(t) = t^2 + 14t + 49$, then $P(-7) = 0$ and -7 is the only root of this polynomial. Thus all solutions of

$$P_2(E)y = \emptyset$$

are of the form $y(k) = (c_1 + c_2 k)(-7)^k$

9 Example

If $P_2(t) = t^2 - 6t + 12$, then $P_2(3 + i\sqrt{3}) = P_2(3 - i\sqrt{3}) = 0$. The polar form of the complex number $3 + i\sqrt{3}$ is:

$$3 + i\sqrt{3} = (12)^{1/2}\left[\cos\left(\frac{\pi}{6}\right) + i\sin\left(\frac{\pi}{6}\right)\right]$$

Therefore, all solutions of $P_2(E)y = \emptyset$ are of the form

$$y(k) = (12)^{k/2}\left[c_1 \cos\left(\frac{k\pi}{6}\right) + c_2 \sin\left(\frac{k\pi}{6}\right)\right]$$

10 EXERCISES

1. Show that if $P(t) = (t - r)^2$, where $r \neq 0$, then $y_1(k) = r^k$, $y_2(k) = kr^k$ is a solution basis for $P(E)y = \varnothing$.

2. Verify that $y_1(k)$ and $y_2(k)$ in Case III are solutions of the second-order equation and show that this pair of solutions is linearly independent.

3. Solve the following initial-value problems.
   a. $-E^2y - 6Ey + 7y = \varnothing$; $y(0) = 0$, $y(1) = 24$
   b. $4E^2y - 4Ey + y = \varnothing$; $y(0) = 1$, $y(1) = 2$
   c. $E^2y = Ey + y$; $y(0) = 0$, $y(1) = 1$
   d. $E^2y - 6Ey + 18y = \varnothing$; $y(0) = 2$, $y(1) = 9$

4. For each of the following homogeneous equations, determine whether some, all, or none of the nontrivial solutions are bounded functions on N.
   a. $(4E^2 - 2E + I)y = \varnothing$ b. $(E^2 + E + (\frac{1}{4})I)y = \varnothing$
   c. $(E^2 - 2E + I)y = \varnothing$ d. $(E^2 + 3E + 2I)y = \varnothing$
   e. $(E^2 + 4E + 4I)y = \varnothing$

5. Consider the equation $P(E)y = \varnothing$ for the following choices of $P(t)$. Determine what initial conditions imply that a solution has the property

$$\lim_{k \to \infty} y(k) = 0$$

   a. $P(t) = t^2 + 4t + 4$ b. $P(t) = t^2 - 8t + 17$
   c. $P(t) = 6t^2 + 5t + 1$.   d. $P(t) = 2t^2 + 3t + 1$
   e. $P(t) = 9t^2 + 4t + 1$

The study of higher-order homogeneous equations amounts to showing the relationship between the roots of the operator polynomial, together with their multiplicities, and a solution basis of the difference equation. Since Section 15 contained an exhaustive treatment of the analogous situation for differential equations, the proofs of the basic results here are listed as exercises. Some of these are very straightforward and some are a real test of ingenuity.

The easy case is that involving a simple polynomial, that is, a polynomial whose roots are all of multiplicity one.

11 THEOREM

If $P(t)$ is a simple polynomial of degree n, $P(0) \neq 0$, and $b_1, b_2, \ldots, b_n$ are the distinct roots (real or complex) of $P(t)$, then a solution basis for $P(E)y = \varnothing$ is:

$$\{y_i(k) = (b_i)^k\}$$

where $i = 1, 2, \ldots, n$. Thus the n-parameter family of solutions of $P(E)y = \varnothing$ has the form

$$y(k) = \sum_{i=1}^{n} c_i(b_i)^k$$

The effect of a multiple zero of P on the form of the solutions of $P(E)y = \varnothing$ involves factorial powers. Before stating the general result, we consider a third-order illustration.

If $P_3(t) = (t - b)^3 = t^3 - 3bt^2 + 3b^2t - b^3$, where $b \neq 0$, then $P_3(E)$ is an operator polynomial and, by Corollary 3, $y_1(k) = b^k$ is a solution of $P_3(E)y = \varnothing$. From our study of second-order equations, in particular, Case II, we might suspect that $y_2(k) = kb^k$ is another solution. This is easy to verify. A convenient choice for a third solution is $y_3(k) = k^{(2)}b^k$, and these three solutions form a solution basis for $P_3(E)y = \varnothing$.

In order to verify these claims, notice that

$$(E - bI)y_3(k) = (E - bI)[k^{(2)}b^k] = (k + 1)^{(2)}b^{k+1} - k^{(2)}bb^k$$
$$= bb^k[(k + 1)(k) - k(k - 1)] = 2b[kb^k]$$

Thus

$$(E - bI)y_3(k) = 2by_2(k)$$

Next,

$$(E - bI)^2y_3(k) = (E - bI)[(E - bI)y_3(k)] = 2b(E - bI)y_2(k)$$
$$= 2b(E - bI)(kb^k) = 2b[(k + 1)b^{k+1} - kbb^k]$$
$$= 2b^2(b^k) = 2b^2y_1(k)$$

Finally,

$$P(E)y_3(k) = (E - bI)^3y_3(k) = (E - bI)2b^2y_1(k)$$
$$= 2b^2(E - bI)y_1(k) = \varnothing$$

By a careful examination of these calculations, a pattern should become apparent that leads to the next result.

12 THEOREM

If $P(t) = (t - b)^n$, where $b \neq 0$, then a solution basis for $P(E)y = \varnothing$ is:

$$y_1(k) = b^k; \, y_2(k) = kb^k; \, \ldots \, ; \, y_i(k) = k^{(i-1)}b^k; \, \ldots \, ; \, y_n(k) = k^{(n-1)}b^k$$

and all solutions of $P(E)y = \varnothing$ are of the form

$$y(k) = [c_1 + c_2k + c_3k^{(2)} + \cdots + c_nk^{(n-1)}]b^k$$

The final result on homogeneous equations can be stated in a form almost identical to Theorem 14, Section 15.

13 THEOREM

If $P(t) = a_0(t - b_1)^{r_1}(t - b_2)^{r_2} \cdots (t - b_m)^{r_m}$, where $b_1, b_2, \ldots, b_m$ are distinct (real or complex) and nonzero numbers, then a solution basis for

$P(E)y = \emptyset$ is the collection of functions of the form

$$S_j = \{b_j{}^{r_j}, kb_j{}^{r_j}, \ldots, k^{(r_j - 1)}b_j{}^{r_j}\}$$

$j = 1, 2, \ldots, m.$

In the hypothesis of the theorem, the statement that the numbers $b_1, b_2, \ldots, b_m$ are all nonzero is, of course, equivalent to the restriction $P(0) \neq 0$.

14 Example

If $P(t) = 3(t^2 - 6t + 12)^3(t + 7)^4$, then a solution basis for $P(E)y = \emptyset$ is:

$$y_1(k) = 12^{k/2} \cos\left(\frac{k\pi}{6}\right) \qquad y_2(k) = 12^{k/2} \sin\left(\frac{k\pi}{6}\right)$$

$$y_3(k) = ky_1(k) \qquad\qquad y_4(k) = ky_2(k)$$

$$y_5(k) = k^{(2)}y_1(k) \qquad\qquad y_6(k) = k^{(2)}y_2(k)$$

$$y_7(k) = (-7)^k \qquad\qquad y_8(k) = k(-7)^k$$

$$y_9(k) = k^{(2)}(-7)^k \qquad\qquad y_{10}(k) = k^{(3)}(-7)^k$$

This example is a composite of Examples 8 and 9.

15 EXERCISES

1. Let $P(t) = (t - 1)^n$, where n is a positive integer. Show that all nontrivial solutions of $P(E)y = \emptyset$ are polynomials and that every polynomial of degree $(n - 1)$ or less is a solution.

2. Let $P(t) = (t - 3)^3$. Show that $y_1(k) = 3^k$, $y_2(k) = k(3)^k$, $y_3(k) = k^2(3)^k$ is a solution basis for $P(E)y = \emptyset$.

3. Form the Casorati matrix for $y_1(k) = 3^k$, $y_2(k) = (-4)^k$, $y_3(k) = 1$, and show that these three functions are linearly independent.

4. Find an equation of the form $P(E)y = \emptyset$ for which the following functions form a solution basis.
   a. $y_1(k) = 1$, $y_2(k) = k$, $y_3(k) = (-1)^k$
   b. $y_1(k) = 2^k$, $y_2(k) = (-2)^k$, $y_3(k) = k(2)^k$, $y_4(k) = k^2(2)^k$
   c. $y_1(k) = \cos(k)$, $y_2(k) = \sin(k)$, $y_3(k) = 1$

5. Let $P(t)$ be a polynomial of degree $n > 0$. Show that all solutions of $P(E)y = \emptyset$ are bounded on N if and only if
   (i) $P(r) = 0$ implies that $|r| \leq 1$, and
   (ii) $P(r) = 0$ and $|r| = 1$ implies that $DP(r) \neq 0$.

6. Determine the bounded solutions of $P(E)y = \emptyset$ in the following cases.
   a. $P(t) = (t - 1)^3(t + 7)^2$ b. $P(t) = (2t - 1)^3(t + 6)$
   c. $P(t) = (t^2 - 2t + 5)(3t + 1)(t - 1)$ d. $P(t) = (t^2 + 4)(t + 4)^2$
   e. $P(t) = (t^2 + 1)(t + 1)^2$

A method of undetermined coefficients for certain nonhomogeneous difference equations can be modeled after the results in Section 16. The equivalent of Table 3, Section 16, is all that is required.

16 TABLE   If $P(t)$ is a polynomial, j is a nonnegative integer and $P(0) \neq 0$, then

A solution of $P(E)y = \varnothing$ is:	If and only if a factor of $P(t)$ is:
(a) $k^{(j)}r^k$	$(t - r)^{j+1}$
(b) $k^{(j)}a^k \cos(kb); \; a > 0$	$(t^2 - 2a\cos(b)t + a^2)^{j+1}$
(c) $k^{(j)}a^k \sin(kb); \; a > 0$	$(t^2 - 2a\cos(b)t + a^2)^{j+1}$

The form of the factors in entries (b) and (c) of Table 16 are different from the corresponding entries in Table 3, Section 16. This is due to the fact that in Table 16 complex numbers are taken to be in polar form. Thus a polynomial whose roots include the complex numbers

$$a(\cos b + i \sin b)$$

and

$$a(\cos b - i \sin b)$$

where $a > 0$, has the factor

$$(t - a \cos b - ia \sin b)(t - a \cos b + ia \sin b)$$
$$= t^2 - 2a \cos(b)t + a^2$$

Rather than repeat the details in Section 16, with only minor modification, we consider a few examples to indicate how the method of undetermined coefficients works in the case of difference equations.

17 Example

In order to solve the initial-value problem

(A) $(E - 2I)y = 3^k; \qquad y(0) = 0$

we first consider the homogeneous equation

(B) $(E - 2I)y = \varnothing$

Clearly all solutions of (B) are of the form

$$y(k) = c_1 2^k$$

Next, $Q(t) = t - 3$ is a nullifying polynomial for the function 3^k so that all solutions of (A) are solutions of

(C) $(E - 3I)(E - 2I)y = \varnothing$

Equation (C) has solutions of the form

$$z(k) = c_1 3^k + c_2 2^k$$

so that some member of this family of solutions is a solution of (A), that is,

(D) $\quad (E - 2I)(c_1 3^k + c_2 2^k) = 3^k$

for some choice of c_1 and c_2. Since

$$
\begin{aligned}
(E - 2I)(c_1 3^k + c_2 2^k) &= c_1(E - 2I)3^k + c_2(E - 2I)2^k \\
&= c_1 3(3^k) - 2c_1(3^k) + c_2(\varnothing) \\
&= c_1 3^k
\end{aligned}
$$

it follows that

$$c_1 = 1 \quad \text{and} \quad c_2 = 0$$

is a solution of (D), or that $z(k) = 3^k$ is a solution of (A). Thus all solutions of (A) are of the form

$$y(k) = c_1 2^k + 3^k$$

By applying the initial conditions,

$$0 = y(0) = c_1 + 1$$

we have that

$$c_1 = -1$$

Therefore, the unique solution of the initial-value problem is:

$$y(k) = -(2)^k + 3^k$$

18 Example

Consider the second-order equation

(A) $\quad E^2 y + Ey - 2y = 9k^{(2)}$

A nullifying polynomial for $k^{(2)}$ can be found in Table 16 since this function is of the form of entry (a) with $j = 2$ and $r = 1$. Also, since $\Delta^3 k^{(2)} = \varnothing$ and $\Delta^3 \equiv (E - I)^3$, the nullifying polynomial $Q(t) = (t - 1)^3$ can be determined in this manner. Equation (A), in operator form, is:

$$P(E)y = 9k^{(2)}$$

where $P(t) = (t - 1)(t + 2)$. Thus $R(t) = Q(t)P(t) = (t - 1)^4(t + 2)$ is a fifth-degree polynomial, and each solution of (A) is a solution of

(B) $\quad R(E)y = \varnothing$

The solutions of (B) are of the form

$$z(k) = c_1 + c_2 k + c_3 k^{(2)} + c_4 k^{(3)} + c_5(-2)^k$$

Because $R(t)$ and $P(t)$ have $(t - 1)(t + 2)$ as a common factor,

$$P(E)z(k) = P(E)[c_2 k + c_3 k^{(2)} + c_4 k^{(3)}]$$

Thus there is a solution of (A) of the form $c_2 k + c_3 k^{(2)} + c_4 k^{(3)}$, and the specific choices of constants are found by solving

(C) $c_2 P(E)k + c_3 P(E)k^{(2)} + c_4 P(E)k^{(3)} = 9k^{(2)}$

The actual calculations can be simplified by use of the operator identity $E \equiv \Delta + I$. Thus

$$P(E) = E^2 + E - 2I = (\Delta + I)^2 + (\Delta + I) - 2I = \Delta^2 + 3\Delta$$

and equation (C) becomes

$$c_2[\Delta^2 + 3\Delta]k + c_3[\Delta^2 + 3\Delta]k^{(2)} + c_4[\Delta^2 + 3\Delta]k^{(3)} = 9k^{(2)}$$

or

$$3c_2 + c_3[2 + 6k] + c_4[6k + 9k^{(2)}] = 9k^{(2)}$$

or

$$(3c_2 + 2c_3) + (6c_3 + 6c_4)k + 9c_4 k^{(2)} = 9k^{(2)}$$

Therefore, $c_4 = 1$, $c_3 = -1$, and $c_2 = \frac{2}{3}$.
    Finally, all solutions of (A) are of the form

$$y(k) = c_1 + c_2(-2)^k + (\tfrac{2}{3})k - k^{(2)} + k^{(3)}$$

19 EXERCISES

1. Solve the following initial-value problems.
    a. $(E^2 - 2\cos(1)E + I)y = -2\sin(1) + \sin(z)$; $y(0) = y(1) = 0$
    b. $(4E^2 - 4E + I)y = k + 8$; $y(0) = 9$, $y(1) = 11$
    c. $E^2 y - 6Ey + 18y = -13k + 17$; $y(0) = 3$, $y(1) = 9$
    d. $(2E^2 - 3E + 3I)y = 2k^2$; $y(0) = -2$, $y(1) = -2$
    e. $\Delta^3 y = 2^k$; $y(0) = 0$, $y(1) = 1$, $y(2) = 3$
    f. $\Delta^2 y + E^2 y = 13(3)^k + 1$; $y(0) = 2$, $y(1) = 4$
2. Determine which of the following equations have some bounded solutions.
    a. $(E^2 + E - 2I)y = 1$ b. $(E^2 + I)y = 1$
    c. $(9E^2 - 9E + I)y = (\tfrac{3}{4})^k$ d. $(E^2 - I)y = (-1)^k$
    e. $\Delta^2 y = 2^{-k}$ f. $(2E^3 - E^2 - 2E + I)y = (\tfrac{1}{2})^k$
    g. $(2E^3 - E^2 - 2E + I)y = (\tfrac{1}{2})$ h. $(2E^3 - E^2 - 2E + I)y = 10^{-50}$
3. Suppose $P(E)y = f(k)$ is a nonhomogeneous equation, and a solution of this equation is:

$$y(k) = 7 - 3k^2 + 2^k$$

What are the possibilities for the order of this equation?

4. Show that if $P(t)$ is a polynomial of degree $n > 0$ and $P(r) \neq 0$, then

$$y(k) = \frac{r^k}{P(r)}$$

is a solution of $P(E)y = r^k$.

5. Let $P(t) = (t - r)^n$, where $n > 0$. Show that

$$y(k) = r^{k-n} \binom{k}{n}$$

is a solution of $P(E)y = r^k$.

6. Suppose that $P(t)$ is a second-degree polynomial and that

$$P(-1) = 2 + 2 \cos r$$

$$P(0) = 1$$

$$P(1) = 2 - 2 \cos r$$

where r is some constant. Find all solutions of

$$P(E)y = 2 - 2 \cos r$$

As a final item, we consider again the restriction on the form of a difference equation that was first mentioned in Definition 6, Section 22. The simplest equation which does not satisfy the definition of a difference equation is:

(20) $Ey = \emptyset$

Evidently, if $y(k)$ is a solution of (20), then

$$0 = y(1) = y(2) = \cdots = y(k)$$

for all $k \geq 1$, and $y(0)$ can be any real number. In fact, it is easy to show that all solutions of (20) are of the form

$$y(k) = c_1 e_1(k)$$

where

$$e_1(0) = 1 \quad \text{and} \quad e_1(k) = 0$$

for all $k \geq 1$. Thus the solutions of equation (20) are in conflict with Theorem 9, Section 24. However, to really appreciate this situation, it is necessary to consider a second-order example.

21 Example

Consider the equation

(A) $(E^2 + a_1 E)y = \emptyset$

where $a_1 \neq 0$. This equation does not satisfy the definition of a difference equation. However, $y_1(k) = e_1(k)$ is clearly a solution of this equation. In order to find another solution, consider the equation in the form

(B) $E(Ey) + a_1(Ey) = \emptyset$

This is a first-order linear difference equation in (Ey). A solution of this equation is:

$Ey(k) = (-a_1)^k$

and, since a constant multiple of a solution is also a solution of this first-order equation, it follows that

$Ey(k) = (-a_1)(-a_1)^k = (-a_1)^{k+1}$

is also a solution of (B). Thus

$y_2(k) = (-a_1)^k$

is a solution of equation (A).

The Casorati matrix for this pair of solutions is:

$$C(k) = \begin{bmatrix} e_1(k) & (-a_1)^k \\ \emptyset & (-a_1)(-a_1)^k \end{bmatrix}$$

and $\det[C(k)] = e_1(k)(-a_1)(-a_1)^k$. Notice that, from the definition of $e_1(k)$,

$\det[C(k)] = 0$

for $k \geq 1$. This is in apparent conflict with Corollary 11 even though y_1, y_2 is a linearly independent pair of functions.

Finally, while

$y(k) = c_1 e_1(k) + c_2(-a_1)^k$

is the form of all solutions of equation (A), notice that the value of c_1 has no effect on the function y for $k \geq 1$, that is,

$y(k) = c_2(-a_1)^k$

for $k \geq 1$.

The point of Example 21 is to add insight into the nature of the restriction in Definition 6, Section 22. The restriction is essential in order to preserve the analogies with differential equations. However, as indicated in the next theorem, the results of this chapter can be applied to a larger class of equations.

22 THEOREM

If $P(t) = a_0 t^n + a_1 t^{n-1} + \cdots + a_n$, where $P(0) \neq 0$, $Q(t) = t^m P(t)$, and $e_i(k)$, where $i = 1, 2, \ldots, m$, are functions defined on N by

$e_i(k) = 1$

if $k = i - 1$,

$$e_i(k) = 0$$

if $k \neq i - 1$, then all solutions of $Q(E)y = \varnothing$ are of the form

$$y(k) = u(k) + c_1 e_1(k) + c_2 e_2(k) + \cdots + c_m e_m(k)$$

where

$$P(E)u(k) = \varnothing$$

_____

ANALYSIS OF NUMERICAL METHODS*

Section 26 Introduction

The purpose of the chapter is to consider the problems associated with producing approximate solutions of differential equations. As indicated in the discussion in Chapter 2, there is no universal method for producing solutions of differential equations. Thus effective methods to approximate the solutions of a differential equation are both necessary and of practical interest. The approximation methods are usually called *numerical methods*, and the analysis of these methods is, in general, quite difficult.

In order to keep the treatment of numerical methods fairly elementary, only first-order linear differential equations are considered in this chapter. However, some additional results that are applicable to a larger class of differential equations are presented in Sections 42 and 47. Also, only a very few of the possible numerical methods are discussed. While hundreds of numerical methods have appeared in print, the number of possibilities is limitless.

It was shown in Chapter 2 that the initial-value problem

(1) $Dy = q(x)y + f(x); \qquad y(x_0) = a_0$

has the unique solution

(2) $y(x) = \exp\left[\int_{x_0}^{x} q(t)\, dt\right]\left\{a_0 + \int_{x_0}^{x} f(s) \exp\left[-\int_{x_0}^{s} q(t)\, dt\right] ds\right\}$

* This chapter depends upon some of the material from each of the preceding chapters.

In this chapter, it will be convenient to consider first-order linear equations in the form (1) rather than the form

$$Dy + p(x)y = f(x)$$

that was used in Chapter 2.

While it is easy to verify that (2) is the solution of (1), the representation of $y(x)$ as given in (2) has serious limitations in practical work. To illustrate this situation, we consider three particular initial-value problems, each with the same initial condition. In each example, we want to find the value of $y(2)$, where $y(x)$ is the solution of the initial-value problem.

3 Example

The initial-value problem

$$Dy = (x + 1)^{-1}y; \qquad y(0) = 1$$

has the solution $y(x) = x + 1$. This solution is a particularly simple function and clearly,

$$y(2) = 3$$

4 Example

The initial-value problem

$$Dy = y; \qquad y(0) = 1$$

has the solution $y(x) = e^x$. Therefore, $y(2) = e^2$. However, in many applications, a decimal approximation of $y(2)$ would not only be adequate, it would be preferable to e^2. A decimal approximation for e^2 can be determined from a table since $y(x) = e^x$ is a known function. While $y(2) \neq 7.389$, this number is a four-digit decimal approximation of e^2.

5 Example

The initial-value problem

$$Dy = [\sin x^2]y; \qquad y(0) = 1$$

has the solution $y(x) = \exp[\int_0^x \sin(t^2)\,dt]$. Determining a decimal approximation in this case is radically different from the problems presented in Examples 3 and 4. One reason for this is that a more convenient representation of

$$\int_0^x \sin(t^2)\,dt$$

is not readily available. The fundamental theorem of calculus is not applicable here since there is no elementary, or known, function whose derivative is the integrand, that is, $\sin(t^2)$. While it is possible that decimal approximations of $y(x)$ have been tabulated, it is clear that tables of approximations of every function of the form of (2) cannot exist.

These three examples illustrate distinct kinds of computational problems, as well as indicate a need for numerical methods to investigate the quantitative nature of solutions of initial-value problems. One approach to these numerical methods is to consider ways of producing tables of approximate values of the function under consideration, namely, the solution of an initial-value problem. The tables would be of essentially the same form as those with which we are familiar, that is, trigonometric, exponential, square-root, and so forth.

In any table of approximate values of a function (look at a sine table, for example), several features determine its utility. There is the portion of the domain of the function for which approximations are tabulated (usually $[0, \pi/4]$ in trigonometric tables); there are the increments in the domain (usually degrees or minutes, or hundredths of a radian, for example); and there are the number of decimal digits in the approximate function values (that is, a five-place table, or an eight-place table, etc.). In general, the nature of the computational problem influences the choice of the table to be used. Considerations of this sort are important in this chapter.

The idea behind the numerical methods in this chapter is, for a given, or selected, positive number h, to determine approximate values of $y(x_0)$, $y(x_0 + h), \ldots, y(x_0 + nh)$, where $x_0 + kh$ is in the domain of y and y is the solution of the initial-value problem (1). In short, the idea is to produce a table of approximations of y, beginning with the initial condition, in increments of h. In general, the choice of h is decided by practical considerations along with certain limitations imposed by the numerical methods.

In considering approximations of this type, the matters of symbolism, notation, and terminology can easily get out of hand. However, it is important that a clear distinction be made between the value of $y(x_0 + kh)$ and an approximate value of $y(x_0 + kh)$. To do this, an approximate value of $y(x_0 + kh)$ will be denoted by $v(k)$.

If $v(k)$ is an approximate value of $y(x_0 + kh)$, then the next question concerns the quality of $v(k)$. This can best be answered in terms of the error function defined by

$$e(k) = y(x_0 + kh) - v(k)$$

A reasonable goal of any numerical method is to minimize the maximum value of $|e(k)|$, and ideally we want to have the error function be the zero function. The figure shown here illustrates a graphical interpretation of these concepts.

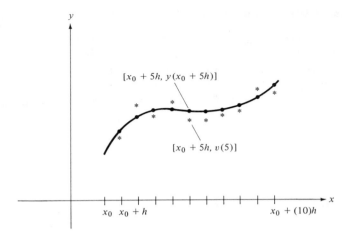

In this figure, the smooth curve represents the graph of $y(x)$, the solution of initial-value problem (1), on the interval $[x_0, x_0 + 10h]$. The dots on the curve represent points for which approximate values are desired, and the points marked * are the approximations. Notice, for example, that $e(0) = 0$, $e(1) > 0$, and $e(2) < 0$.

To be of practical interest, a numerical method must provide an explicit computational procedure for determining the approximate values $v(k)$. Since $y(x_0)$ is the given initial condition, the best choice for $v(0)$ is $v(0) = y(x_0)$. The methods considered in this chapter are called *step-by-step methods* in that $v(1)$ is a computable function of $v(0)$, and $v(2)$ is a computable function of $v(0)$ and $v(1)$, and so forth. The numerical value of h is called the *stepsize*. While step-by-step methods have been used for centuries, the advent of high-speed computers has intensified the interest in methods of this type.

Interest in numerical methods has resulted in the development of an extensive and growing field of new knowledge. This area of study has become quite specialized so that the treatment in this book is necessarily very brief. However, the approach to numerical methods taken here is in the spirit of recent research in this area, and, therefore, this chapter can provide a smooth transition from the classical theory of differential equations to recent results in numerical methods. Also, the material in this chapter is a first instance in which a knowledge of both differential equations and difference equations can be applied to explore a new area of mathematics.

Modern computing systems have the ability of doing millions of arithmetic operations per second. A discriminating computer user can use these machines to great advantage in scientific work. However, the examples in this chapter are clear indications that computational mathematics is not a routine activity.

Section 27 Euler's Method

The simplest numerical method is known as Euler's method. Various ways of generalizing Euler's method are considered in Sections 28 and 29. The first consideration is a derivation of Euler's method, and this is followed by an analysis of the error function associated with this numerical method.

Suppose $y(x)$ is the solution of the first-order linear differential equation

(1) $Dy = q(x)y + f(x)$

that satisfies the initial condition

(2) $y(x_0) = a_0$

If h is a positive number and $x_0 + h$ is in the domain of y, then the simplest form of the mean value theorem (Theorem 2, Section 2) is applicable, and

(3) $y(x_0 + h) = y(x_0) + h \, Dy(\alpha_0)$

for some number $\alpha_0 \in (x_0, x_0 + h)$.

Only the existence of α_0 is known, and, evidently, the value of α_0, along with $Dy(\alpha_0)$, depends upon h. Thus equation (3) cannot be used to determine $y(x_0 + h)$. In fact, the only value of Dy that can be determined is $Dy(x_0)$, and this can be obtained from the differential equation as

(4) $Dy(x_0) = q(x_0)y(x_0) + f(x_0)$

While, in general, $Dy(x_0) \neq Dy(\alpha_0)$, it is true that Dy is a continuous function. Therefore, corresponding to each $\varepsilon > 0$, there is a $\delta > 0$ such that $0 < h < \delta$ implies that $|Dy(x_0) - Dy(\alpha_0)| \leq \varepsilon$. Thus by suitably restricting the value h, the difference between the unknown number $Dy(\alpha_0)$ and the known number $Dy(x_0)$ can be made small.

For these reasons, we consider the effect of replacing $Dy(\alpha_0)$ in equation (3) with $Dy(x_0)$. Of course, if this is done, then equation (3) is no longer correct, so that a new symbol is needed to represent an approximate value of $y(x_0 + h)$. If $v(1)$ is used to represent an approximate value of $y(x_0 + h)$, then (3) becomes

(5) $v(1) = y(x_0) + h \, Dy(x_0)$

Notice that

$$e(1) = y(x_0 + h) - v(1) = y(x_0) + h \, Dy(\alpha_0) - y(x_0) - h \, Dy(x_0)$$

or

$$e(1) = h[Dy(\alpha_0) - Dy(x_0)]$$

Now, the above sequence of reasoning can be applied to the interval $[x_0 + h, x_0 + 2h]$ to obtain an approximate value of $y(x_0 + 2h)$. However, the calculation to determine $Dy(x_0 + h)$ will necessarily be an approximation. From the differential equation (1), we have that

(6) $Dy(x_0 + h) = q(x_0 + h)y(x_0 + h) + f(x_0 + h)$

The correct value of $y(x_0 + h)$ is not available, but the approximate value $v(1)$ could be used. Thus, if $Dv(1)$ is used to denote an approximate value of $Dy(x_0 + h)$, (6) becomes

(7) $Dv(1) = q(x_0 + h)v(1) + f(x_0 + h)$

The extension of (5) to the interval $[x_0 + h, x_0 + 2h]$ becomes

(8) $v(2) = v(1) + h\,Dv(1)$

where $Dv(1)$ is computed according to (7).

Finally, a continuation of this scheme yields the difference equation

(9) $v(k + 1) = v(k) + h\,Dv(k), \qquad k \in N$

where

(10) $Dv(k) = q(x_0 + kh)v(k) + f(x_0 + kh)$

and this is Euler's method for differential equation (1). In order that (9) agree with (5) when $k = 0$, it is necessary to define $v(0) = y(x_0)$ so that $e(0) = 0$.

Before beginning an analysis of the approximations produced by Euler's method, we consider a graphical interpretation of these equations, as shown in the figure here.

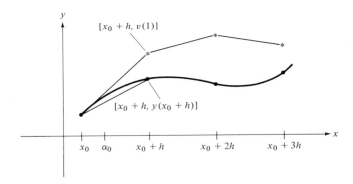

In this figure, the smooth curve represents the graph of $y(x)$, the solution of equation (1) with initial condition (2). The points on the graph of $y(x)$ that are marked • are the values for which approximations are sought. The computation to determine $v(1)$ in equation (5) amounts to finding the point with first coordinate $x_0 + h$ that is on the line with slope $Dy(x_0)$ and through the point $[x_0, y(x_0)]$. In this case, $v(1) \neq y(x_0 + h)$ since the correct slope is $Dy(\alpha_0)$. Notice that in calculating $v(2)$, the same interpretation of Euler's method can be made except that it is with respect to a new initial-value problem, namely, equation (1) together with $y(x_0 + h) = v(1)$. Thus in this picture, there is a solution of equation (1) that goes through the point $[x_0 + h, v(1)]$ but it is not the solution of (1) together with initial condition (2).

Before considering some examples, we formalize the definition of Euler's method.

11 DEFINITION

Euler's method as applied to differential equation

$$Dy = q(x)y + f(x)$$

together with initial condition

$$y(x_0) = a_0$$

is:

$$v(k + 1) = v(k) + h \, Dv(k)$$

where $v(0) = y(x_0) = a_0$, $h > 0$, and

$$Dv(k) = q(x_0 + kh)v(k) + f(x_0 + kh)$$

The error function is:

$$e(k) = y(x_0 + kh) - v(k)$$

---

12 Example

Consider the initial-value problem in Example 4, Section 26, that is,

$$Dy = y; \qquad y(0) = 1$$

Since $Dy(0) = y(0) = 1$, the first step in Euler's method is:

$$v(1) = 1 + h$$

whereas

$$y(0 + h) = y(h) = e^h$$

Thus the error in the first step is:

$$e(1) = y(h) - v(1) = e^h - 1 - h$$

It is easy to show that $e(1) > 0$ for all $h > 0$, but that $\lim_{h \to 0} e(1) = 0$.

13 EXERCISES

1. Verify the two claims made at the end of Example 12.
2. Continue the calculations in Example 12 and determine $v(2)$, $v(3)$, and $v(4)$.
3. Let $Dy = (\cos x)y$; $y(0) = 1$. Apply Euler's method to this initial-value problem. Find $v(1)$, $y(h)$, $e(1)$, and $\lim_{h \to 0} e(1)$.
4. Continue Problem 3 by finding $v(2)$, $v(3)$, and $v(4)$.
5. Consider the initial-value problem $Dy = -y$; $y(0) = 1$. Determine $v(1)$ if $h = 4$, and compare $v(1)$ with $y(4)$.
6. Let c_1 and c_2 be constants and apply Euler's method to $Dy = c_1$; $y(0) = c_2$. Show that for each $k \in N$,

$$v(k) = y(kh)$$

and thus

$$e(k) = \varnothing$$

Since an analysis of any numerical method depends upon the differential equations under investigation, we begin a study of Euler's method with a particular family of differential equations of the form of (1), namely,

(14) $Dy = ay + b$

on $[0, +\infty)$, where $a \neq 0$ and b are constants, together with

(15) $y(0) = 1$

The solution of this initial-value problem is:

(16) $y(x) = \left(\dfrac{a + b}{a}\right) e^{ax} - \dfrac{b}{a}$

Euler's method applied to (14) yields

(17) $v(k + 1) = v(k) + h[av(k) + b]$

and writing this first-order difference equation in standard form, we get

(18) $[E - (1 + ah)I]v = hb$

Since $v(0) = y(0) = 1$, the initial condition associated with (18) is:

$$v(0) = 1$$

Equation (18) together with this initial condition can be solved by using Theorem 9, Section 23, or the method of undetermined coefficients, as illustrated in Section 25. The solution of this initial-value problem is:

(19) $v(k) = \left(\dfrac{a + b}{a}\right)(1 + ah)^k - \dfrac{b}{a}$

where for each k, $v(k)$ is an approximate value of

(20) $y(kh) = \left(\dfrac{a + b}{a}\right) e^{ahk} - \dfrac{b}{a}$

Thus for differential equations of the form (14), the error function is:

(21) $e(k) = y(kh) - v(k) = \left(\dfrac{a + b}{a}\right)\left[e^{ahk} - (1 + ah)^k\right]$

This illustrates what is true in general, namely, that Euler's method transforms a first-order linear differential equation into a first-order linear difference equation. Although, in practice, the values of $v(k)$ would be computed one at a time, or step-by-step, from (17), this example is considered because of the explicit representations of the values of $v(k)$ and $y(kh)$ as given in (19) and (20). Naturally, no one would seriously propose the application of a numerical method to (14) and (15) when the solution (16) is so easy to obtain. This example is merely intended to illustrate some shortcomings and some advantages of Euler's method.

To continue with differential equations of the form (14), one obvious result is that when $a = -b$, $e(k) = 0$ for all k, and this is true independent of h. This is, of course, a trivial case since if $a = -b$, clearly $y(x) = 1$ and $v(k) = 1$ are the solutions of the respective differential and difference equations. Therefore, in what follows, we assume that $a + b \neq 0$.

Another case that is easy to resolve can be based upon the observation that all solutions of differential equation (14) are bounded on $[0, +\infty)$ if and only if $a \leq 0$. When it is possible to characterize the bounded and unbounded solutions of a differential equation, this information can be used in analyzing a numerical method. A typical example of this situation is our first result in analyzing Euler's method as applied to equation (14).

22 THEOREM

If $a < 0$, then for any $h > 0$, $y(kh)$ is bounded for all $k > 0$; whereas if $h > 2/|a|$, then $\lim\limits_{k \to \infty} |e(k)| = +\infty$.

Proof: The claim concerning $y(kh)$ is easy to verify and is left as an exercise. If $a < 0$, then

$$h > \frac{2}{|a|} = \frac{-2}{a}$$

implies that $ah < -2$ or $1 + ah < -1$. Thus in this case, $|1 + ah|^k$ is an unbounded function of k, and clearly,

$$\lim\limits_{k \to \infty} |e(k)| = +\infty$$

Theorem 22 amounts to giving conditions under which a differential equation with bounded solutions can have an approximate solution that is unbounded. This situation is to be avoided in practical work.

23 Example

Applying Euler's method to

$$Dy = -1000y; \qquad y(0) = 1$$

with $h = 0.01$, means that

$$ah = -1000(0.01) = -10$$

and $1 + ah = -9$. Thus in this case,

(A) $e(k) = [e^{-10k} - (-9)^k] = y(kh) - v(k)$

Notice that $e(1) = e^{-10} - (-9)$, or $e(1) > 9$.

With $h = 0.0001$, $ah = -1000(0.0001) = -0.1$, and

$1 + ah = 1 + (-0.1) = 0.9$,

so that

(B) $e(k) = e^{-0.1k} - (0.9)^k$

Notice that $e(100) = e^{-10} - (0.9)^{100}$ and, with the help of a table, we find that $e(100) < 0.00002$.

The two error functions in (A) and (B) correspond to different values of h. However, $y(0.01) = y[(0.01)1] = y[(0.0001)100]$. Thus applying Euler's method with the larger value of h gives a ridiculous approximation for $y(0.01)$, namely, $v(1) = -9$, while using $h = 0.0001$ and $k = 100$ results in a much closer approximation of $y(0.01)$.

24 COROLLARY

If $a < 0$ and $h < 2/|a|$, then both $y(kh)$ and $v(k)$ are bounded functions.

This corollary, which is easy to prove, gives an upper bound for h for the case of $a < 0$. Establishing an upper bound for h is a useful first step in an analysis of a numerical method for a given family of differential equations. The results in Example 23 suggest that the quality of the approximations improves as h goes to zero. In order to investigate this situation, it is convenient to adopt a slightly more cumbersome notation for the error function, namely,

(25) $e(k, h) = y(kh) - v(k)$

The reason for considering the error as a function of both k and h is to emphasize the fact that in order to approximate a specified value of y, say $y(m)$, there is an infinite choice of k and h. For example, using the notation in (25), the error functions (A) and (B) in Example 23 become

$e(1, 0.01) = y(0.01) - v(1)$

and

$e(100, 0.0001) = y(0.01) - v(100)$

Both of these represent the error made in approximating $y(0.01)$; however, on the one hand, only one step, or application, of Euler's method was used because of the large value of h, while for the smaller value of h, it was necessary to use 100 steps to approximate $y(0.01)$, but with much superior results.

In general, if an approximate value of $y(m)$ is needed, then, since $y(m) = y(kh)$, any choice of k and h such that

$kh = m$

can be considered. Since k is an integer, and

$h = \dfrac{m}{k}$

the available choices are $h = m, m/2, m/3, \ldots$, where the smaller the value of h, the larger the value of k. Of course, as k increases, the number of steps in Euler's method, and hence the number of calculations, increases. This should seem reasonable since it implies that to obtain greater accuracy, more calculations are required. The next consideration is to formalize these observations.

We now consider the general problem of approximating $y(m)$, where $y(x)$ is the solution of equation (14) that satisfies (15), and where $m \in (0, \infty)$. We do this by studying the error function

(26) $\qquad e(k, h) = \dfrac{a + b}{a} \left[e^{ahk} - (1 + ah)^k \right]$

Our first result establishes a useful fact about this error function.

27 LEMMA

If $0 < h < 1/|a|$ and $k > 0$, then

$$e^{ahk} - (1 + ah)^k > 0$$

Proof: It follows from Taylor's theorem (Theorem 2, Section 2) that

$$e^{ah} = 1 + ah + \frac{(ah)^2}{2} e^{\alpha}$$

for some number α. Thus

$$e^{ah} > (1 + ah)$$

Since $0 < h < 1/|a|$ implies that $1 + ah > 0$, it follows that

$$e^{ahk} > (1 + ah)^k$$

or

$$e^{ahk} - (1 + ah)^k > 0$$

Another way of stating the above lemma is as follows: If $0 < h < 1/|a|$ and $k > 0$, then

$$|e(k, h)| = \left| \frac{a + b}{a} \right| \{ e^{ahk} - (1 + ah)^k \}$$

In what follows, we assume that $0 < h < 1/|a|$ and we consider the special sequence of error functions specified by

$$|e(k, h)|, \qquad |e(2k, 2^{-1}h)|, \qquad |e(2^2 k, 2^{-2}h)|, \qquad |e(2^3 k, 2^{-3}h)|, \ldots$$

Notice that each term in this sequence is the error in approximating $y(m)$, where

$$m = kh = (2k)(2^{-1}h) = (2^2 k)(2^{-2}h) = (2^3 k)(2^{-3}h) = \cdots$$

Consideration of this sequence is motivated in part by what is commonly done in practice, namely, after calculating one table of approximate values of y using a stepsize h, to compute a second table with stepsize $(h/2)$. The next theorem shows the mathematical advantages of this process.

28 THEOREM

If $0 < h < 1/|a|$ and $k > 0$, then

$$|e(k, h)| > \left| e\left(2k, \frac{h}{2} \right) \right|$$

Proof: Let $T = |e(k, h)| - |e(2k, h/2)|$. We show that $T > 0$. From Lemma 27,

$$T = \left| \frac{a + b}{a} \right| \{e^{ahk} - (1 + ah)^k\} - \left| \frac{a + b}{a} \right| \left\{ e^{ahk} - \left(1 + \frac{ah}{2} \right)^{2k} \right\}$$

This simplifies to

$$T = \left| \frac{a + b}{a} \right| \left\{ \left(1 + \frac{ah}{2} \right)^{2k} - (1 + ah)^k \right\}$$

$$= \left| \frac{a + b}{a} \right| \left(\frac{1}{4^k} \right) \{(2 + ah)^{2k} - (4 + 4ah)^k\}$$

or

$$T = \left| \frac{a + b}{a} \right| \left(\frac{1}{4^k} \right) \{[4 + 4ah + (ah)^2]^k - (4 + 4ah)^k\}$$

and clearly T is positive.

An immediate consequence of Theorem 28 is the following corollary.

29 COROLLARY

If $0 < h < 1/|a|$ and $k > 0$, then the sequence

$$t(n) = |e(2^n k, 2^{-n}h)|$$

where $n = 1, 2, \ldots$, is monotone decreasing.

An inductive proof of this corollary is an easy exercise. Notice that Theorem 28 supplies the first step in the induction. The corollary implies that each successive approximation of $y(m)$ is superior to its predecessor. Of course, in order to reduce the error, it is necessary to decrease the step-size and increase the number of steps.

Finally, there is the question of how small the error in approximating $y(m)$ can be. This can be formulated as follows: what is the value of the following expression?

$$\lim_{n \to \infty} |e(2^n k, 2^{-n}h)|$$

Since $t(n) = |e(2^n k, 2^{-n}h)|$ is a decreasing sequence and $t(n) > 0$ for $n \geq 1$, it follows that

$$\lim_{n \to \infty} t(n)$$

exists.

30 THEOREM

If $m = kh$, then

$$\lim_{n \to \infty} |e(2^n k, 2^{-n}h)| = 0$$

Proof: Since

$$|e(2^n k, 2^{-n}h)| = \left|\frac{a+b}{a}\right| [\exp(a2^n k 2^{-n}h) - (1 + a2^{-n}h)^{2^n k}]$$

$$= \left|\frac{a+b}{a}\right|\left[e^{am} - \left(1 + \frac{ahk}{2^n k}\right)^{2^n k}\right]$$

$$= \left|\frac{a+b}{a}\right|\left[e^{am} - \left(1 + \frac{am}{2^n k}\right)^{2^n k}\right]$$

we need to show that

$$\lim_{n \to \infty} \left(1 + \frac{am}{2^n k}\right)^{2^n k} = e^{am}$$

This is a standard limit problem of a type to be found in most calculus books.

In summary, our study of Euler's method as applied to differential equation (14) resulted in the following information:

(a) an upper bound for the stepsize was determined, namely, $h < 1/|a|$
(b) it was shown that better approximate values of $y(m)$ are obtained by reducing the stepsize
(c) for each $\varepsilon > 0$, it is possible to choose h so that $|y(m) - v(k)| < \varepsilon$.

31 EXERCISES

1. Verify that the function in (16) is the solution of equation (14) that satisfies the initial condition (15).
2. Verify that the function in (19) is the solution of equation (18) that satisfies the initial condition $v(0) = 1$.
3. Complete the proof of Theorem 22 by finding a bound for $|y(kh)|$, where $k \in N$.
4. Prove Corollary 24.

5. Prove Corollary 29.

6. Show that

$$\lim_{n \to \infty} \left(1 + \frac{am}{2^n k}\right)^{2^n k} = e^{am}$$

7. Show that

$$e^{ahk} = \sum_{i=0}^{\infty} \frac{(ah)^i k^i}{i!}$$

and that

$$(1 + ah)^k = \sum_{i=0}^{k} \frac{(ah)^i k^{(i)}}{i!}$$

The remaining problems refer to the initial-value problem (14), (15).

8. Show that if $a < 0$ and $h < 2/|a|$, then

$$\lim_{k \to \infty} e(k, h) = 0$$

9. Show that if $a > 0$, then for each $h > 0$,

$$\lim_{k \to \infty} y(kh) = \lim_{k \to \infty} e(k, h) = \infty$$

10. If $a > 0$, investigate the limit

$$\lim_{k \to \infty} \frac{e(k, h)}{y(kh)}$$

The function $e(k, h)/y(kh)$ is called the *relative error*.

As a second example, we consider

(32) $Dy = ay + \cos x$

with the initial condition

$$y(0) = 1$$

Most of the details in this illustration will be omitted, but they are mentioned in the exercises that follow.

Application of Euler's method to equation (32) results in the difference equation

(33) $v(k + 1) - (1 + ah)v(k) = h \cos(kh)$

The solutions of this first-order nonhomogeneous difference equation can be found by the method of undetermined coefficients in Section 25. The solution of (33) that satisfies the initial condition

$$v(0) = 1$$

is:

(34) $v(k) = (1 - A)(1 + ah)^k + A \cos(kh) + B \sin(kh)$

where

(35)

$$A = \frac{-h(1 + ah - \cos h)}{1 - 2(1 + ah) \cos h + (1 + ah)^2}$$

$$B = \frac{h \sin h}{1 - 2(1 + ah) \cos h + (1 + ah)^2}$$

Since

$$\lim_{h \to 0} A = \frac{-a}{1 + a^2}$$

and

$$\lim_{h \to 0} B = \frac{1}{1 + a^2}$$

it follows that

(36)

$$A = \frac{-a}{1 + a^2} + g_1(h)$$

$$B = \frac{1}{1 + a^2} + g_2(h)$$

where $\lim_{h \to 0} g_1(h) = \lim_{h \to 0} g_2(h) = 0$. The solution of (32) satisfying $y(0) = 1$ is:

(37) $y(kh) = (1 + a^2)^{-1}[(a^2 + a + 1)e^{ahk} - a \cos(kh) + \sin(kh)]$

Therefore, the error function is expressible as

(38)
$$e(k, h) = \left(\frac{a^2 + a + 1}{1 + a^2}\right)[e^{ahk} - (1 + ah)^k]$$
$$+ g_1(h)[(1 + ah)^k - \cos(kh)] - g_2(h) \sin(kh)$$

39 THEOREM

If $m = kh$, and $e(k, h)$ is as defined in (38), then

$$\lim_{n \to \infty} e(2^n k, 2^{-n} h) = 0$$

Proof: We consider the three terms in $e(2^n k, 2^{-n} h)$ independently. First,

$$\lim_{n \to \infty} \left[e^{ahk} - \left(1 + \frac{am}{2^n k}\right)^{2^n k} \right] = 0$$

from Theorem 30. Next,

$$\lim_{n \to \infty} g_1(2^{-n} h) \left[\left(1 + \frac{am}{2^n k}\right)^{2^n k} - \cos(m) \right] = 0$$

since $\lim_{h \to 0} g_1(h) = 0$ and

$$\lim_{n \to \infty} \left(1 + \frac{am}{2^n k}\right)^{2^n k} = e^{am}$$

Finally,

$$\lim_{n \to \infty} g_2(2^{-n}h) \sin(m) = 0$$

since $\lim_{h \to 0} g_2(h) = 0$.

___

Thus it is possible to determine an approximation of $y(m)$, where $y(x)$ is the solution of equation (32), by Euler's method with a nonzero error that is as small as required. Notice that this and all of the results in this section are of a point-wise nature. That is, if m is a number in the domain of the solution $y(x)$ of an initial-value problem and $\varepsilon > 0$, then it is possible to find h and k such that $m = hk$ and $|e(k, h)| < \varepsilon$. However, if $m_1 > m$, then, in general, the same stepsize h will not produce a comparable error for $y(m_1)$. It is for this reason that Theorems 30 and 39 are called point-wise convergence theorems for Euler's method.

40 EXERCISES

1. Verify that (34) is a solution of equation (33).
2. Show that (36) is a correct representation of the coefficients A and B by finding $\lim_{h \to 0} A$ and $\lim_{h \to 0} B$.
3. Verify that (37) is a solution of equation (32).
4. Consider the approximations obtained from Euler's method for the initial-value problem

   $$Dy = ay + x; \qquad y(0) = 1$$

   Analyze the error function for this problem.

A point-wise convergence theorem for the general first-order linear equation

(41) $\qquad Dy = q(x)y + f(x)$

on $[a, b]$ is possible. Since $q(x)$ is assumed to be continuous on $[a, b]$, there is a number M such that

$$|q(x)| \leq M$$

for all $x \in [a, b]$, and it is this fact that leads to the point-wise convergence theorem. However, additional details on this topic are deferred to Chapter 10.

A final point that is treated in numerical analysis texts concerns the effect of arithmetic errors in Euler's method. Naturally, if a calculation is being done, whether with a pocket calculator or a large computer system, only a finite number of digits are used to represent each number. This results in so-called rounding errors. The effects of rounding errors are important and practical considerations, but their analysis is beyond the introductory level.

42 Example

Consider the initial-value problem

$$Dy = \pi y; \qquad y(0) = 1$$

Suppose we want to know $y(1)$. Since $y(x) = e^{\pi x}$ is the solution of the initial-value problem, $y(1) = e^{\pi}$ is approximately 23.14069. Euler's method involves the step-by-step calculation of

$$v(k + 1) = v(k) + h\pi v(k); \qquad v(0) = 1$$

If this calculation is done in two-digit arithmetic with $h = 0.1$, then $h\pi = 0.31$ (to two-digit accuracy) and

$$v(1) = 1 + 0.31 = 1.3$$

$$v(2) = 1.3 + (0.31)(1.3) = 1.7$$

and so on. Here, all arithmetic operations are rounded to two digits, and errors are made at each step. Continuing in this way, we find that

$$v(10) = 14$$

(to two digits). Notice that

$$|e(10, 0.1)| = |y(1) - v(10)| > 9$$

If the same calculations are done in five-digit arithmetic, then

$$v(10) = 15.363$$

and here

$$|e(10, 0.1)| = |y(1) - v(10)| < 8$$

The following exercises illustrate one elementary approach to the analysis of arithmetic errors within a numerical method.

43 EXERCISES

1. Suppose that an incorrect initial condition, namely, $v(0) = 1 + \delta$, where $\delta \neq 0$, is used with the first-order difference equation (18).
   a. Find the solution of (18) that satisfies $v(0) = 1 + \delta$.
   b. If Euler's method together with $v(0) = 1 + \delta$ is used to approximate $y(kh)$, where $y(x)$ is the solution of $Dy = ay + b; y(0) = 1$, what is the error function?
   c. Let $e(k, h)$ be the error function determined above. Show that if $m = kh$, then
   $$\lim_{n \to \infty} |e(2^n k, 2^{-n}h)| = |\delta| e^{am} \neq 0$$

2. Repeat the above analysis with the assumption that an arithmetic error is made at step j in Euler's method, that is,

$$v(j) = \left(\frac{a + b}{a}\right)(1 + ah)^j - \frac{b}{a} + \delta$$

Show that the effects of this error are essentially the same as in Problem 1.

Section 28 Multistep Methods

The remainder of this chapter is devoted to two types of generalizations of Euler's method. The emphasis in this section is on the relationships between a solution of a first-order linear differential equation and the solutions of an approximating difference equation. Also, only first-order linear differential equations are considered here. However, Chapter 9 contains a section devoted to applications of the methods of this section to other types of differential equations.

Suppose $y(x)$ is the solution of the first-order linear differential equation

(1) $Dy = q(x)y + f(x)$

that satisfies the initial condition

(2) $y(x_0) = b$

3 DEFINITION

A *linear two-step* method as applied to differential equation (1) together with initial condition (2) is:

$$v(k + 2) = a_1 v(k + 1) + a_2 v(k) + h[b_1 \, Dv(k + 1) + b_2 \, Dv(k)]$$

where $v(0) = y(x_0) = b$; $h > 0$; a_1, a_2, b_1, and b_2 are constants; and

$$Dv(j) = q(x_0 + jh)v(j) + f(x_0 + jh)$$

The error function is:

$$e(k, h) = y(x_0 + kh) - v(k)$$

In the case of $a_2 = b_2 = 0$, the two-step method is essentially a one-step method, that is, of the form of Euler's method. In this sense, a two-step method is a generalization of Euler's method.

Whereas only a single one-step method was discussed in the previous section, a number of useful two-step methods are possible. It is for this reason that a derivation of a specific two-step method is somewhat more involved than is the case for Euler's method. We first consider specific two-step methods.

The choice of coefficients $a_1 = 1$, $a_2 = 0$, $b_1 = \frac{3}{2}$, and $b_2 = -\frac{1}{2}$ in (3) defines the two-step method

(4) $v(k + 2) = v(k + 1) + h[\frac{3}{2} Dv(k + 1) - \frac{1}{2} Dv(k)]$

As a first illustration, consider the same differential equation (14), Section 27, studied previously, that is,

(5) $Dy = ay + b$

on $[0, \infty)$, where $a \neq 0$ and b are constants, together with

(6) $y(0) = 1$

From (5), we have that

$Dv(k) = av(k) + b$

so that the application of the two-step method (4) to the differential equation (5) gives

(7) $v(k + 2) = v(k + 1) + \left(\dfrac{h}{2}\right)[3av(k + 1) + 3b - av(k) - b]$

along with the initial condition

(8) $v(0) = 1$

Notice that (7) is a second-order difference equation, so that the single initial condition in (8) does not define a unique solution of (7). This is the situation with all multistep methods, that is, in order to use or analyze them, it is necessary to provide one or more additional initial conditions. In the case of a two-step method, it is required that a value of $v(1)$ be specified. Naturally, the value of $v(1)$ will completely determine the solution of (7), so that the choice of $v(1)$ is an important consideration.

In order to keep the analysis of this multistep method specific, Euler's method is used to assign a value to $v(1)$. As was seen in (19), Section 27, if Euler's method is applied to equation (5), then

(9) $v(1) = \left(\dfrac{a + b}{a}\right)(1 + ah) - \dfrac{b}{a} = 1 + (a + b)h$

By using (8) and (9) as initial conditions, it is now possible to compute $v(2)$, $v(3)$, $v(4)$, $v(5)$, ... , in a step-by-step manner from (7).

In order to analyze this step-by-step approximate solution of (5) using the two-step method (4), consider the resultant difference equation (7) in standard form, namely,

(10) $v(k + 2) - \left(1 + \dfrac{3ha}{2}\right)v(k + 1) + \dfrac{ha}{2}v(k) = hb$

This equation, in operator form, is:

$P(E)v(k) = hb$

where

$$P(t) = t^2 - \left(1 + \frac{3ha}{2}\right)t + \frac{ha}{2}$$

Following the methods of Section 25, it is easy to show that the solutions of (10) are of the form

(11) $v(k) = c_1(t_1)^k + c_2(t_2)^k - \dfrac{b}{a}$

where t_1 and t_2 are the two roots of $P(t)$. In fact,

$$t_1 = \frac{2 + 3ha + \sqrt{(2 + 3ha)^2 - 8ha}}{4}$$

(12)

$$t_2 = \frac{2 + 3ha - \sqrt{(2 + 3ha)^2 - 8ha}}{4}$$

and since $(2 + 3ha)^2 - 8ha > 0$ for all h and a, t_1 and t_2 are distinct real numbers.

The coefficients c_1 and c_2 in (11) can be expressed in terms of $v(0)$ and $v(1)$ as

$$c_1 = \frac{t_2[av(0) + b] - [av(1) + b]}{a(t_2 - t_1)}$$

$$= \frac{-t_2[av(0) + b] + [av(1) + b]}{a(t_1 - t_2)}$$

(13)

$$c_2 = \frac{-t_1[av(0) + b] + [av(1) + b]}{a(t_2 - t_1)}$$

$$= \frac{t_1[av(0) + b] - [av(1) + b]}{a(t_1 - t_2)}$$

These expressions for c_1 and c_2 are useful if various options are considered for selecting a value for $v(1)$. However, by using $v(0) = 1$ and $v(1) = 1 + (a + b)h$, as noted in (9), the coefficients can be written as

$$c_1 = \left(\frac{a + b}{a}\right)\left(\frac{-t_2 + (1 + ah)}{t_1 - t_2}\right)$$

(14)

$$c_2 = \left(\frac{a + b}{a}\right)\left(\frac{t_1 - (1 + ah)}{t_1 - t_2}\right)$$

15 EXERCISES

1. Show that $t_1 - t_2 > 0$ for all choices of a and h.
2. Show that $\lim\limits_{ah \to -\infty} t_1 = \tfrac{1}{3}$ and $\lim\limits_{ah \to +\infty} t_2 = \tfrac{1}{3}$.
3. Verify (13) and (14).

Since the solution (11) of the approximating difference equation (10), together with the coefficients (14), all depend on the roots t_1 and t_2 of

$$P(t) = t^2 - \left(1 + \frac{3ah}{2}\right)t + \frac{ah}{2}$$

the next item is to investigate this polynomial. The polynomial equation

$$t^2 - \left(1 + \frac{3ah}{2}\right)t + \frac{ah}{2} = 0$$

can be solved for ah, which gives the equivalent equation

(16) $$ah = \frac{2t(1 - t)}{1 - 3t} = g(t)$$

The important properties of the solution (11) can be determined by examining a graph of $g(t)$, shown in the figure here.

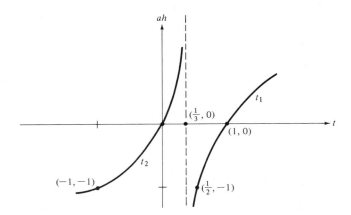

The two branches in the graph of $ah = g(t)$ are labeled t_1 and t_2, where these two functions of ah are defined explicitly in (12). Notice that both t_1 and t_2 are increasing functions of ah. The information contained in this graph can be used to determine the nature of the solution of the approximating difference equation (10).

Since the solution of the differential equation (5) that satisfies the initial condition (6) is:

(17) $$y(x) = \left(\frac{a + b}{a}\right)e^{ax} - \frac{b}{a}$$

it follows that all solutions of (5) are bounded on $[0, \infty)$ if and only if $a \leq 0$. However, if $a \leq 0$, it is possible to have unbounded solutions of the approximating difference equation (10).

18 THEOREM

All solutions of difference equation (10) are bounded if and only if $-1 \leq ah \leq 0$.

Proof: The form of the solutions is given in (11). From the graph, we see that $t_1{}^k$ is a bounded function of k if and only if $ah \leq 0$, and that $t_2{}^k$ is a bounded function of k if and only if $-1 \leq ah$. Thus $-1 \leq ah \leq 0$.

The coefficients c_1 and c_2 as defined in (14) are also functions of ah. The following facts about these coefficients are easy to verify.

19 THEOREM

If c_1 and c_2 are as defined in (14), then

$$\lim_{ah \to 0} c_1 = \left(\frac{a + b}{a} \right) \quad \text{and} \quad \lim_{ah \to 0} c_2 = 0$$

This property of c_1 and c_2 implies that the effect of the term $c_2(t_2)^k$ in (11) is negligible if $|ah|$ is small.

There are a number of important considerations associated with two-step methods, but the preservation of boundedness is our primary interest. The differential equation (5) has bounded solutions if and only if $a \leq 0$, and, as indicated in Theorem 18, it is possible to choose h so that the approximations produced by the two-step method are bounded functions. This property of the two-step method (4) is the motivation for the next definition.

20 DEFINITION

A two-step method is called *stable* if there is a number $d > 0$ such that $-d < ah < 0$ implies that all approximate solutions of $Dy = ay + b$, $a < 0$, are bounded.

Theorem 18 could be reworded to state that the two-step method (4) is stable. Not all linear two-step methods are stable, however. The next topic is an example of an unstable two-step method.

Consider the linear two-step method

(21) $v(k + 2) = -2v(k + 1) + 3v(k) + h[3\,Dv(k + 1) + Dv(k)]$

Application of (21) to the differential equation (5) yields the difference equation

$$v(k + 2) = -2v(k + 1) + 3v(k) + 3ahv(k + 1) + 3hb + ahv(k) + hb$$

In standard form, this difference equation is:

(22) $P(E)v(k) = 4hb$

where

(23) $P(t) = t^2 + (2 - 3ah)t - (3 + ah)$

It is easy to verify that the solutions of (22) are of the form

(24) $v(k) = c_1(t_1)^k + c_2(t_2)^k - \dfrac{b}{a}$

where $P(t_1) = P(t_2) = 0$ and $t_1 \neq t_2$. Rather than solve for t_1 and t_2 explicitly, we consider the graph of the roots of $P(t)$ as functions of ah. Thus $P(t) = 0$ if and only if

(25) $ah = \dfrac{t^2 + 2t - 3}{3t + 1} = \dfrac{(t + 3)(t - 1)}{3t + 1} = g(t)$

and the graph of this function is shown in the figure here.

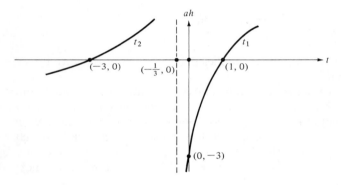

This graph reveals an undesirable feature of the two-step method (21). Notice that if $ah \leq 0$, then $t_2 \leq -3$. Thus while the differential equation has a bounded solution if $a \leq 0$, the solution of the approximating difference equation (22), with suitable initial conditions, is unbounded. This is an example of an *unstable numerical method*.

26 EXERCISES

1. Show that $g(t)$, as defined in (16), is an increasing function on $(-\infty, \tfrac{1}{3})$ and $(\tfrac{1}{3}, +\infty)$.
2. Determine the Maclaurin series expansion of t_1 as defined in (12) and compare it with

    $$e^{ah} = 1 + ah + \frac{(ah)^2}{2!} + \frac{(ah)^3}{3!} + \cdots$$

3. Prove Theorem 19.
4. Show that if $v(1) = 1 + (ah)^2$, then the coefficients c_1 and c_2 as defined in (13) have the limit properties of Theorem 19.
5. Show that if $v(1)$ is defined so that $\lim_{h \to 0} v(1) = v(0)$, then the general coefficients c_1 and c_2 as defined in (13) will have the limit properties of Theorem 19.

6. Show that if $a \neq 0$, then for all $h > 0$, c_2, as defined in (14), is nonzero.
7. Find explicit representations for the roots of the polynomial (23).
8. If $v(0) = 1$ and $v(1) = 1 + ah$ are the initial conditions for equation (22), determine c_1 and c_2 in (24) as functions of ah. Do these coefficients have the limit properties of Theorem 19? For what values of ah is $c_2 = 0$?
9. Determine if the following sets of coefficients define a stable or unstable linear two-step method.
   a. $a_1 = -2, a_2 = 3, b_1 = 0, b_2 = 1$ b. $a_1 = 1, a_2 = 0, b_1 = 1, b_2 = 1$
   c. $a_1 = \frac{1}{2}, a_2 = \frac{1}{2}, b_1 = 1, b_2 = 0$

These two examples illustrate that the choice of coefficients in a linear two-step method is very important. While a complete analysis of linear multistep methods is well beyond the scope of an introductory text, we can give some indication of the principles used to derive particular two-step methods.

The form used to define the family of linear two-step methods in Definition 3 is the most convenient one for the purposes of step-by-step computation. An equivalent form is:

(27) $v(k + 2) - a_1 v(k + 1) - a_2 v(k) - h[b_1 \, Dv(k + 1) + b_2 \, Dv(k)] = \varnothing$

Since $v(k)$ and $Dv(k)$ are approximate values of $y(x_0 + kh)$ and $Dy(x_0 + kh)$, respectively, we consider the operator T defined on $\mathscr{C}^1(I)$ by

(28) $$T(y) = y[(k + 2)h] - a_1 y[(k + 1)h] - a_2 y[kh]$$
$$- h\{b_1 \, Dy[(k + 1)h] + b_2 \, Dy[kh]\}$$

For simplicity, we let $x_0 = 0$, but this only amounts to a change in variable.

Evidently, T can be interpreted as a transformation from $\mathscr{C}^1(I)$ into V_∞, where $I = [0, \infty)$. More properly, T represents a family of such transformations, since to specify T, one must use a specific a_1, a_2, b_1, b_2, and h. It is easy to verify that T is a linear transformation.

29 THEOREM

Let T be a member of the family of linear operators (28). If y is a solution of a first-order initial-value problem and $T(y) = \varnothing$ for all $h > 0$, then the linear two-step method (27) associated with T, with appropriate initial conditions, gives exact values of $y(kh)$ for all $k \in N$.

Theorem 29 is easy to prove, and it is the reason for considering $T(y)$. The method used to determine specific values for a_1, a_2, b_1, and b_2 in a linear two-step method is based upon the functional equation

(30) $T(y) = \varnothing$

In particular, we consider $T(x^m) = \varnothing$, where m is a nonnegative integer. A reason for this choice can be explained by considering $T(x^m)$, which is, in simplified form,

(31)
$$T(x^m) = h^m\{(k + 2)^m - a_1(k + 1)^m - a_2k^m$$
$$- b_1m(k + 1)^{m-1} - b_2mk^{m-1}\}$$

where $m \geq 0$.

Since T is a linear operator, identity (31) implies that T transforms a polynomial of degree m in $\mathscr{C}^1[0, \infty)$ into a polynomial of the same degree in V_∞. In particular,

$$T(x^0) = (1 - a_1 - a_2)$$

$$T(x^1) = h[k + 2 - a_1k - a_1 - a_2k - b_1 - b_2]$$
$$= h[k(1 - a_1 - a_2) + 2 - a_1 - b_1 - b_2]$$

(32)
$$T(x^2) = h^2[k^2 + 4k + 4 - a_1k^2 - 2a_1k$$
$$- a_1 - a_2k^2 - 2b_1k - 2b_1 - 2b_2k]$$
$$= h^2[k^2(1 - a_1 - a_2) + 2k(2 - a_1 - b_1 - b_2)$$
$$+ 4 - a_1 - 2b_1]$$

$$T(x^3) = h^3[k^3(1 - a_1 - a_2) + 3k^2(2 - a_1 - b_1 - b_2)$$
$$+ 3k(4 - a_1 - 2b_1) + 8 - a_1 - 3b_1]$$

33 DEFINITION

A linear two-step method, as defined in (3), has *order r* if

$$T(x^0) = T(x^1) = \cdots = T(x^r) = \varnothing \qquad \text{and} \qquad T(x^{r+1}) \neq \varnothing$$

By using equations (32), the concept of order can be translated directly into equations involving the coefficients of the linear two-step method.

34 THEOREM

A linear two-step method has order r, where $r = 0, 1, 2, 3$, if and only if the coefficients satisfy the first $r + 1$ of the following equations:

(35)
$$1 - a_1 - a_2 = 0$$
$$2 - a_1 - b_1 - b_2 = 0$$
$$4 - a_1 - 2b_1 = 0$$
$$8 - a_1 - 3b_1 = 0$$
$$16 - a_1 - 4b_1 = 0$$

and fail to satisfy the $(r + 2)$nd equation.

The steps required to prove Theorem 34 are indicated in the exercises. Both of the linear two-step methods considered in this section happen to have order two although, as has been shown, they provide radically different approximations. The only linear two-step method of order three is obtained by solving the system consisting of the first four equations in (35). This order-three method is:

(36) $v(k + 2) = -4v(k + 1) + 5v(k) + h[4\, Dv(k + 1) + 2\, Dv(k)]$

Although the order of this method is maximal, there are some serious limitations in its application.

Consider applying (36) to the initial-value problem

(37) $Dy = -y; \quad y(0) = 1$

The resultant difference equation is:

(38) $P(E) = \varnothing$

where $P(t) = t^2 + 4(1 + h)t - (t - 2h)$.

Following the same procedure as before, we note that $P(t) = 0$ if and only if

(39) $h = -\frac{1}{2}\left[\dfrac{(t + 5)(t + 1)}{2t + 1}\right] = g(t)$

An analysis of this function will show that (38) has unbounded solutions for all $h > 0$, whereas the solution of (37) is bounded.

40 EXERCISES

1. Prove Theorem 29. [Hint: The "appropriate initial conditions" are $v(0) = y(0)$ and $v(1) = y(h)$.]
2. Verify that equation (31) is correct.
3. Enlarge the system of equations (32) by calculating $T(x^4)$ and $T(x^5)$.
4. Use equations (32) together with the fact that $\{1, k, k^2, \ldots\}$ is linearly independent to prove Theorem 34.
5. Verify that (4) and (21) are both linear two-step methods of order two.
6. Show that (36) is the only linear two-step method of order three.
7. Find all stable linear two-step methods of order three.
8. Show that $v(k + 2) = v(k) + 2h\, Dv(k + 1)$ is a two-step method of order two.

The examples in this section illustrate how to begin an analysis of numerical methods for differential equations. However, there is little more that can be done using elementary methods, since the deeper problems in this area involve questions in advanced function theory.

Much of what has been treated for linear two-step methods can be extended to include the general n-step methods of the form

$$v(k + n) = \sum_{i=1}^{n} a_i v(k + n - i) + h \sum_{i=1}^{n} b_i\, Dv(k + n - i)$$

An understanding of difference equations is an important first step in approaching these numerical methods.

One final remark on two-step methods might help to unify the observations made in three specific illustrations in this section. By applying the general linear two-step method to

(41) $Dy = ay$

the resultant difference equation is:

(42) $P(E)v(k) = \varnothing$

where $P(t) = t^2 - (a_1 + ahb_1)t - (a_2 + ahb_2)$. Of course, the solutions of (42) are linear combinations of $t_1{}^k$ and $t_2{}^k$ provided $P(t_1) = P(t_2) = 0$ and $t_1 \neq t_2$. The case of multiple roots of P is not really important since they would occur only for a very special choice of h. As before, we note that $P(t) = 0$ if and only if

(43) $ah = \dfrac{t^2 - a_1 t - a_2}{b_1 t + b_2} = g(t)$

Particular cases of $g(t)$ occurred in (16), (25), and (39). Just to give it a name, we will call $g(t)$ the *characterizing function* of a linear two-step method.

If the coefficients satisfy the first of the order equations (35), that is, if $1 - a_1 - a_2 = 0$, then $a_1 = 1 - a_2$. In this case,

$$g(t) = \frac{t^2 - t + a_2 t - a_2}{b_1 t + b_2} = \frac{t(t - 1) + a_2(t - 1)}{b_1 t + b_2}$$

or

(44) $g(t) = \dfrac{(t - 1)(t + a_2)}{b_1 t + b_2}$

Notice that $g(1) = 0$ is the case for each of the characterizing functions (16), (25), (39), and (44).

The general linear two-step method of order two can be determined by solving the first three of the order equations (35). It is easy to verify that the solutions of

$$1 - a_1 - a_2 = 0$$
$$2 - a_1 - b_1 - b_2 = 0$$
$$4 - a_1 - 2b_1 = 0$$

are:

$$a_1 = 1 - a_2$$

(45) $b_1 = \tfrac{1}{2}(3 + a_2)$

$$b_2 = \tfrac{1}{2}(a_2 - 1)$$

where a_2 is any real number.

By using the solution (45) of the order equations in the general characterizing function (43) we get, after simplification,

(46) $\qquad g(t) = \dfrac{2(t - 1)(t + a_2)}{(3 + a_2)t + a_2 - 1}$

Thus (46) is the characterizing function of all linear two-step methods of order two. There is one exception to this last statement and that is that $a_2 \neq 5$. If $a_2 = 5$, then the characterizing function (46) becomes that of the order-three method, as given in (39). Notice also that $a_2 = 0$ in (46) is the characterizing function defined in (16) and that $a_2 = 3$ yields (25).

An appropriate choice of a_2 in the linear two-step methods of order two,

$$v(k + 2) = (1 - a_2)v(k + 1) + a_2 v(k)$$

(47)

$$+ \left(\frac{h}{2}\right)[(3 + a_2)\, Dv(k + 1) + (a_2 - 1)\, Dv(k)]$$

will yield a stable method.

It is a consequence of Theorem 18 that the linear two-step method (4) is stable. In fact, the linear two-step method (4) is both stable and of order two. Since the third-order method (36) is not stable, it follows that the highest-order stable linear two-step methods are those of order two. However, not all methods of order two are stable. In particular, the linear two-step method (21) is not stable. The idea of a stable method along with the consequences of this property is motivated by Theorem 18. By analyzing the characterizing function (46), it is easy to prove a sufficient condition for a two-step method to be stable.

48 THEOREM

If $|a_2| < 1$, then the linear two-step method (47) is stable and has order two.

49 EXERCISES

1. Verify that (46) represents the characterizing function of all linear two-step methods of order two, provided $a_2 \neq 5$.
2. Prove Theorem 48.
3. Consider the linear three-step methods

$$v(k + 3) = a_1 v(k + 2) + a_2 v(k + 1) + a_3 v(k)$$
$$+ h[b_1\, Dv(k + 2) + b_2\, Dv(k + 1) + b_3\, Dv(k)]$$

and determine a set of "order equations" analogous to (35) for these three-step methods.
4. Determine the following.
   a. the unique three-step method of order five

b. the one-parameter family of three-step methods of order four

c. the two-parameter family of three-step methods of order three

5. By considering the characterizing function of a three-step method, show that the highest-order stable three-step method is order three.

6. Show that if $b_2 = b_3 = 0$ in a stable three-step method, then the method cannot have order three.

Section 29 Other Numerical Methods

In this section, we illustrate how some other families of numerical methods are derived. These methods can be viewed as other generalizations of Euler's method.

The first of this new class of numerical methods involves the *trapezoidal rule* for approximating an integral. A derivation of the trapezoidal rule can be found in most calculus texts. This integration formula is:

$$\textbf{(1)} \qquad \int_b^{b+h} f(x)\, dx = (\tfrac{1}{2})h[f(b) + f(b+h)] - \frac{h^3}{12} D^2 f(c)$$

where $b < c < b + h$.

In fact, the trapezoidal rule has a simple geometric interpretation which is illustrated in the graph shown here.

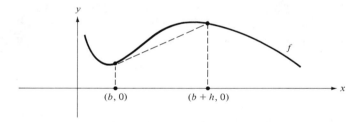

The first term on the right side of (1) is the area of the trapezoid bounded by the line segment from b to $b + h$ and the dotted lines.

Of course, the error term in (1), that is, $(-h^3/12)\, D^2 f(c)$, assumes that $f \in \mathscr{C}^2(I)$, where $[b, b + h] \subset I$. In general, we can consider the formula

$$\textbf{(2)} \qquad \int_b^{b+h} f(x)\, dx = (\tfrac{1}{2})h[f(b) + f(b+h)] + E(h)$$

where f is continuous on $[b, b + h]$. In this case, there is no convenient representation for the error term $E(h)$, but nonetheless, $\lim_{h \to 0} E(h) = 0$.

Now if y is a solution of a first-order differential equation, then an application of the trapezoidal rule (2) to Dy gives

$$\textbf{(3)} \qquad \int_b^{b+h} Dy(x)\, dx = \left(\frac{h}{2}\right)[Dy(b) + Dy(b+h)] + E(h)$$

However,

(4) $\quad \int_b^{b+h} Dy(x)\,dx = y(b + h) - y(b)$

so that the combination of (3) and (4) gives

(5) $\quad y(b + h) = y(b) + (\frac{1}{2})h[Dy(b) + Dy(b + h)] + E(h)$

Finally, by taking $b = x_0 + kh$, we get a form of the trapezoidal rule that is useful in approximating solutions of differential equations, namely,

(6)
$$y[x_0 + (k + 1)h]$$
$$= y(x_0 + kh) + (\tfrac{1}{2})h\{Dy(x_0 + kh) + Dy[x_0 + (k + 1)h]\} + E(h)$$

In order to simplify the symbolism, we will consider first-order differential equations of the form

(7) $\quad Dy = g(x, y)$

with the initial condition

(8) $\quad y(x_0) = b$

Of course, the first-order linear equations are of this form, where $g(x, y) = q(x)y + f(x)$, and only linear differential equations will be considered in this section.

As before, an approximate value of $y(x_0 + kh)$ is denoted by $v(k)$, and an approximate value of $Dy(x_0 + kh)$ is denoted by $Dv(k)$. Also, $Dv(k)$ is computed from the differential equation by

(9) $\quad Dv(k) = g[x_0 + kh, v(k)]$

If the error term in the trapezoidal rule (6) is deleted, then this formula is no longer correct and so v must be used in place of y. This replacement gives the approximation

(10) $\quad v(k + 1) = v(k) + (\tfrac{1}{2})h[Dv(k) + Dv(k + 1)]$

and this is the final form of the trapezoidal rule that we consider.

The first thing to notice about (10) is that in order to use this formula to compute $v(k + 1)$, the value of $v(k + 1)$ must already be known. That is, since $Dv(k + 1)$ is on the right side of (10) and

$$Dv(k + 1) = g[x_0 + (k + 1)h, v(k + 1)]$$

the value of $v(k + 1)$ must be available in order to compute $Dv(k + 1)$. Thus, of itself, the trapezoidal rule would seem to have limited interest in approximating solutions of a differential equation.

However, one approach to the utilization of the trapezoidal rule is to use it in conjunction with Euler's method, that is,

(11) $\quad v(k + 1) = v(k) + h\,Dv(k)$

This arrangement amounts to first using Euler's method (11) to obtain a preliminary value of $v(k + 1)$. This value of $v(k + 1)$ is then used to determine $Dv(k + 1)$, which, in turn, is used in the trapezoidal rule (10) to obtain a final value of $v(k + 1)$. This composition of the trapezoidal rule and Euler's method is called a *Runge-Kutta method*.

The common way to describe this particular Runge-Kutta method is in terms of the function g as used in (9). Thus using the definition of Dv in (9), Euler's method becomes

(12) $v(k + 1) = v(k) + hg[x_0 + kh, v(k)]$

Using this value of $v(k + 1)$ in the trapezoidal rule (10) gives

(13)
$$v(k + 1) = v(k) + \tfrac{1}{2}h(g[x_0 + kh, v(k)]$$
$$+ g\{x_0 + (k + 1)h, v(k) + hg[x_0 + kh, v(k)]\})$$

Equation (13) is a first-order difference equation and is a Runge-Kutta method for the first-order differential equation (7).

It is immaterial whether this Runge-Kutta method is considered in the compact form (13), or as the composition of Euler's method (11) and the trapezoidal rule (10). In either case, our derivation shows that the method amounts to assuming that two error or remainder terms, one associated with Euler's method and the other with the trapezoidal rule, are negligible. Thus this method is of a different type than that of either Section 27 or Section 28.

Consider once again the first-order linear equation

(14) $Dy = ay + b$

on $[0, \infty)$, where $a \neq 0$ and b are constants, together with

(15) $y(0) = 1$

In applying method (13) to this differential equation, we have that $g(x, y) = ay + b$ so that

$$g[kh, v(k)] = av(k) + b$$

Making the appropriate substitutions gives

$$v(k + 1) = v(k) + \tfrac{1}{2}h\{av(k) + b + a[v(k) + hav(k) + hb] + b\}$$

or

$$v(k + 1) = v(k) + \tfrac{1}{2}ahv(k) + \tfrac{1}{2}bh + \tfrac{1}{2}ahv(k)$$
$$+ \tfrac{1}{2}(ah)^2 v(k) + \tfrac{1}{2}abh^2 + \tfrac{1}{2}bh$$

This, in turn, can be written as

(16) $v(k + 1) = [1 + ah + \tfrac{1}{2}(ah)^2]v(k) + (bh + \tfrac{1}{2}abh^2)$

so that the effect of applying the Runge-Kutta method (13) to the linear equation (14) is to transform it into a first-order linear difference equation. Also, the appropriate initial condition associated with (16) is:

(17) $v(0) = 1$

In actual practice, the difference equation (13) would be solved (usually on a computer) in a step-by-step manner starting with the initial value. Notice that we are dealing with a one-step method in the sense that the numerical value of $v(0)$ completely determines all subsequent values of v. However, the particular difference equation (16) is simple enough to solve explicitly by the methods of Chapter 5.

Notice first that $z(k) = -b/a$ is a solution of (16). Since

$$v(k) = [1 + ah + \tfrac{1}{2}(ah)^2]^k$$

is a solution of the homogeneous equation

$$v(k + 1) = [1 + ah + \tfrac{1}{2}(ah)^2]v(k)$$

it follows that all solutions of (16) are of the form

$$v(k) = c[1 + ah + \tfrac{1}{2}(ah)^2]^k - \frac{b}{a}$$

Using the initial value (17), we have that

$$v(0) = 1 = c - \frac{b}{a}$$

or

$$c = \left(\frac{a + b}{a}\right)$$

Thus the solution of (16) and (17) is:

(18) $$v(k) = \left(\frac{a + b}{a}\right)[1 + ah + \tfrac{1}{2}(ah)^2]^k - \frac{b}{a}$$

It is an easy matter to determine which solutions (18) are bounded. We can first show that

(19) $1 + ah + \tfrac{1}{2}(ah)^2 \geq \tfrac{1}{2}$

with equality if and only if $ah = -1$. Since the solutions of

$$1 + ah + \tfrac{1}{2}(ah)^2 = 1$$

are $ah = 0$ and $ah = -2$, it should be clear that the graph of $z = 1 + ah + \tfrac{1}{2}(ah)^2$ is as shown in the figure here.

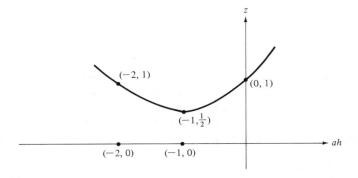

20 THEOREM

The solution $v(k)$ in (18) is a bounded function if and only if

$$-2 \le ah \le 0$$

Since the differential equation is the same as the one in Section 27, the value of $y(kh)$ can be found in (20) of Section 27 and is:

$$y(kh) = \left(\frac{a+b}{a}\right) e^{ahk} - \frac{b}{a}$$

Thus the error function is:

(21) $e(k, h) = \left(\frac{a+b}{a}\right) \{e^{ahk} - [1 + ah + \tfrac{1}{2}(ah)^2]^k\}$

It might be well to compare the above error function with the one that is associated with Euler's method, namely,

(22) $e(k, h) = \left(\frac{a+b}{a}\right) \{e^{ahk} - (1 + ah)^k\}$

Notice that in (22) the expression $(1 + ah)$ is the first two terms of the power series representation of e^{ah}, whereas in (21) the expression $[1 + ah + \tfrac{1}{2}(ah)^2]$ is the first three terms of e^{ah}. It is for this reason that the method defined in (13) is called a *second-order* Runge-Kutta method.

Additional details in an analysis of the error function (21) can be developed in just about the same manner as in Section 27. Some suggestions are given in the exercises.

23 EXERCISES

1. Show that the Runge-Kutta method transforms the general first-order equation $Dy = q(x)y + f(x)$ into a first-order linear difference equation.
2. Verify that the inequality (19) is correct.
3. Let $e_2(k, h)$ be the error function (21) for the Runge-Kutta method and $e_1(k, h)$ the error function (22) for Euler's method. Show that if $|ah| < 1$, then

    $$|e_1(k, h)| \ge |e_2(k, h)| \ge 0$$

4. Apply the Runge-Kutta method to $Dy = ay + x$; $y(0) = 1$, and investigate the error function.

A second-order Runge-Kutta method can be used to obtain another solution to a problem in Section 28. In applying a linear two-step method to the differential equation (14), we observed that the resultant second-order difference equation had but one initial condition associated with it. In particular, if Euler's method is used, then

(24) $v(1) = 1 + (a + b)h$

as was noted in (9), Section 28. However, if the Runge-Kutta method is used to supply a value for $v(1)$, then from (18) we have

(25)
$$v(1) = \left(\frac{a+b}{a}\right)[1 + ah + \tfrac{1}{2}(ah)^2] - \frac{b}{a}$$
$$= 1 + (a+b)h(1 + \tfrac{1}{2}ah)$$

Thus for suitable choices of h, the Runge-Kutta method provides an improved value of $v(1)$, which in turn can be used in conjunction with the two-step method (4) of Section 28.

A reasonable question at this point is this: Why bother with a linear two-step method, especially if a second-order Runge-Kutta method is to be used to determine $v(1)$? Although it is difficult to make comparisons between two radically different approximation methods, it should be clear that the Runge-Kutta methods require two evaluations of Dv per step, whereas only one evaluation of Dv per step is needed with a linear two-step method. This potential for reducing the number of calculations, and hence computer time, is the main attraction of linear multistep methods.

26 EXERCISES

1. Use the initial conditions

$$v(0) = 1$$
$$v(1) = 1 + (a+b)h(1 + \tfrac{1}{2}ah)$$

   to compute the coefficients c_1 and c_2 in (13) of Section 28.
2. Compare the coefficients computed above with (14) of Section 28.
3. Show that the coefficients determined in Problem 1 have the limit properties of Theorem 19, Section 28.

It is possible to employ the trapezoidal rule to obtain approximate solutions of some types of first-order differential equations, in particular, linear equations. For the equation

(27) $\quad Dy = q(x)y + f(x)$

we have that

(28) $\quad Dv(k) = q(kh)v(k) + f(kh)$

By substituting the approximation (28) in the trapezoidal rule (10), we obtain

$$v(k+1) = v(k)$$
$$+ \tfrac{1}{2}h\{q(kh)v(k) + f(kh) + q[(k+1)h]v(k+1) + f[(k+1)h]\}$$

This difference equation can be put into the form

(29)
$$\{1 - \tfrac{1}{2}hq[(k+1)h]\}v(k+1)$$
$$= \{1 + \tfrac{1}{2}hq(kh)\}v(k) + \tfrac{1}{2}h\{f(kh) + f[(k+1)h]\}$$

In this form, (29) is suitable for step-by-step solution.

Notice that (29) is a linear difference equation. We consider a very simple example to illustrate an analysis of this form of application of the trapezoidal rule.

If $Dy = ay$, $y(0) = 1$, then (29) becomes

$$(1 - \tfrac{1}{2}ah)v(k + 1) = (1 + \tfrac{1}{2}ah)v(k); \qquad v(0) = 1$$

or

(30) $(2 - ah)v(k + 1) = (2 + ah)v(k); \qquad v(0) = 1$

Clearly,

(31) $v(k) = \left(\dfrac{2 + ah}{2 - ah}\right)^k, \qquad ah \neq 2$

is the solution of (30). An interesting fact about this solution is that it is a bounded function for all $ah < 0$. In fact, both $v(k)$ and

$$y(kh) = e^{ahk}$$

are bounded for $ah \leq 0$ and both have the property that

$$\lim_{k \to \infty} y(kh) = \lim_{k \to \infty} v(k) = 0$$

for all $ah < 0$.

One simple comparison of Euler's method, the trapezoidal rule, and the Runge-Kutta method is to examine the approximation of e^{ah} that each of these provides. If $Dy = ay$, $y(0) = 1$, then

(32) $y(h) = e^{ah}$

whereas the three approximations are:

(33.a)   *Euler's method:* $v(1) = 1 + ah$

(33.b)   *trapezoidal rule:* $v(1) = \dfrac{2 + ah}{2 - ah}$

(33.c)   *Runge-Kutta method:* $v(1) = 1 + ah + \tfrac{1}{2}(ah)^2$

34 EXERCISES

1. Show that (31) is a bounded function on N for all $ah \leq 0$.

2. Show that if $ah < 0$, then $\lim_{k \to \infty} v(k) = 0$, where $v(k)$ is given by (31).

3. For what values of ah is $v(k)$ both bounded and positive-valued?

4. Determine the Maclaurin series expansion of $v(1) = (2 + ah)/(2 - ah)$ and compare it with

$$e^{ah} = 1 + ah + \frac{(ah)^2}{2!} + \cdots$$

5. Apply the trapezoidal rule to $Dy = ay + x$; $y(0) = 1$, and solve the resultant difference equation.

Another numerical method that employs both Euler's method and the trapezoidal rule is the *iterated Euler-trapezoidal* method. The idea is to find a first value for $v(k + 1)$ by using Euler's method and then compute a sequence of subsequent values of $v(k + 1)$ by using the trapezoidal rule. A proper explanation of this iterated method requires a slightly modified notational system.

The equations that define this iterated method are:

(35) $v(k + 1, 1) = v(k) + h\,Dv(k)$

(36) $v(k + 1, j + 1) = v(k) + \frac{1}{2}h[Dv(k) + Dv(k + 1, j)]$

$j = 1, 2, 3, \ldots$.

In practice, only a finite number of terms in the sequence $\{v[k + 1, j + 1)\}$ can be calculated, so that eventually $v(k + 1, n)$, for some n, is used to define $v(k + 1)$, and the initial choice of $v(k + 2, 1)$ is then determined from (35). In this iterated method, the differential equation, the choice of h, and the number of iterations per step all influence the final result.

As an illustration, consider once again

$$Dy = ay; \qquad y(0) = 1$$

From (35), we have that

$$v(1, 1) = v(0) + h\,Dv(0) = 1 + ah$$

Thus

$$Dv(1, 1) = av(1, 1) = a + a^2h$$

Now, using (36),

$$v(1, 2) = v(0) + \tfrac{1}{2}h[Dv(0) + Dv(1, 1)]$$
$$= 1 + \tfrac{1}{2}h[a + (a + a^2h)]$$

or

$$v(1, 2) = 1 + ah + \tfrac{1}{2}(ah)^2$$

Next,

$$Dv(1, 2) = av(1, 2) = a + a^2h + \tfrac{1}{2}a(ah)^2$$

so that

$$v(1, 3) = 1 + \tfrac{1}{2}h[a + a + a^2h + \tfrac{1}{2}a(ah)^2]$$

or

$$v(1, 3) = 1 + ah + \tfrac{1}{2}(ah)^2 + \tfrac{1}{4}(ah)^3$$

It is an easy inductive proof to show that

$$v(1, n) = 1 + ah + \tfrac{1}{2}(ah)^2 + \tfrac{1}{4}(ah)^3 + \cdots 2^{-n+1}(ah)^n$$

or

$$v(1, n + 1) = 1 + ah\left[1 + \left(\frac{ah}{2}\right) + \left(\frac{ah}{2}\right)^2 + \cdots + \left(\frac{ah}{2}\right)^n\right]$$

Also, since $v(1, n + 1)$ involves a geometric series, it can be represented as

$$v(1, n + 1) = 1 + ah \left[\frac{1 - \left(\dfrac{ah}{2}\right)^{n+1}}{1 - \left(\dfrac{ah}{2}\right)} \right]$$

By using the properties of a geometric series, it is easy to prove the following theorem.

37 THEOREM

If $|ah| < 2$, then

$$\lim_{n \to \infty} v(1, n + 1) = \frac{2 + ah}{2 - ah}$$

If $v(1)$ is defined to be the limiting value, then $|ah| < 2$ implies that $v(1) = (2 + ah)/(2 - ah)$, which is the same value as in (33.b).

38 EXERCISES

1. Supply all of the details to prove Theorem 37.
2. Suppose that in the illustration, $v(1, 3)$ is used to define $v(1)$ and, in general, $v(k, 3) = v(k)$. Continue the approximate solution of $Dy = ay$ by computing $v(2)$, $v(3)$, and $v(4)$.
3. Simpson's rule, as applied to differential equations, is:

$$v(k + 2) = v(k) + \left(\frac{h}{3}\right) [Dv(k + 2) + 4\, Dv(k + 1) + Dv(k)]$$

Apply Simpson's rule to $Dy = ay$, and by considering the characterizing function, show that this method is unstable.

7

SERIES SOLUTIONS OF DIFFERENTIAL EQUATIONS*

Section 30 Introduction

Throughout our treatment of linear differential equations, we have tried to emphasize that only in very special cases is it possible to obtain explicit solutions in terms of elementary functions. Since the solutions of most linear differential equations do not have any simple representation in terms of elementary functions, alternative methods for representing, or approximating, solutions must be sought. Numerical methods for approximating solutions were discussed in the preceding chapter. The purpose of this chapter is to develop techniques for obtaining power series representations of solutions and for approximating solutions using power series.

For a brief account of the appropriate definitions and basic properties of power series, the reader is referred to the review of sequences, series, and power series in Section 6.

We begin our discussion of the power series method by considering some simple first-order linear equations. These examples were chosen because they illustrate the basic power series technique without involving unnecessarily complicated calculations, and because the power series representations of the solutions can be compared with the explicit solutions obtained by using the methods of Section 8.

---

* This chapter depends upon the material in Chapters 2 and 3. Also some of the material in this chapter is presented as an application of results in Chapters 4 and 5.

1 Example

Consider the first-order linear homogeneous differential equation

$$Dy - y = \varnothing$$

on $J = (-\infty, \infty)$. Using the methods of Section 8, the one-parameter family of solutions is:

(A) $y(x) = ce^x$

Assume that the equation has a power series solution of the form

(B) $y(x) = \sum_{n=0}^{\infty} a_n x^n$

which converges on some interval $J = (-R, R), 0 < R \leq \infty$; that is, assume that the equation has a solution which is analytic at $x = 0$. Then, since a power series can be differentiated term-wise, and the differentiated series has the same radius of convergence, we have

(C) $D[y(x)] = \sum_{n=1}^{\infty} n a_n x^{n-1}$

Substituting (B) and (C) into the differential equation yields

$$\sum_{n=1}^{\infty} n a_n x^{n-1} - \sum_{n=0}^{\infty} a_n x^n = \varnothing$$

or

$$(a_1 + 2a_2 x + 3a_3 x^2 + \cdots) - (a_0 + a_1 x + a_2 x^2 + \cdots) = \varnothing$$

which can be written

$$a_1 - a_0 + (2a_2 - a_1)x + (3a_3 - a_2)x^2 + \cdots = \varnothing$$

From the uniqueness theorem (Theorem 23, Section 6), it follows that $a_1 - a_0 = 0, 2a_2 - a_1 = 0, 3a_3 - a_2 = 0$, and, in general, $n a_n - a_{n-1} = 0, n = 1, 2, \ldots$. We can use these equations to express each coefficient $a_n, n \geq 1$, in terms of a_0:

$$a_1 = a_0, \qquad a_2 = \frac{a_1}{4} = \frac{a_0}{2}, \qquad a_3 = \frac{a_2}{3} = \frac{a_0}{2 \cdot 3}, \ldots, a_n = \frac{a_0}{n!}, \ldots$$

By regarding the undetermined coefficient a_0 as a parameter and substituting these equations into (B), we have

(D) $y(x) = \sum_{n=0}^{\infty} \frac{a_0}{n!} x^n = a_0 \sum_{n=0}^{\infty} \frac{x^n}{n!}$

From the ratio test, the series (D) has radius of convergence $R = \infty$. Of course, in this case, we recognize the series (D) as being the

Maclaurin series expansion of e^x, and so (D) coincides with (A) as was to be expected.

The form of the power series (B), that is, powers of x, or about $a = 0$, was selected for convenience. The form of the assumed power series expansion is often dictated by the particular equation, or initial-value problem, whose solution is sought. Also, the calculations involved in substituting the power series into the differential equation can be organized by using summation notation. In Example 1, we have

(2) $$\sum_{n=1}^{\infty} n a_n x^{n-1} - \sum_{n=0}^{\infty} a_n x^n = \varnothing$$

The index of summation, n, is a dummy variable, comparable to the variable of integration in a definite integral, and so changes of indices are possible. In particular, defining m by means of the equation $m = n - 1$ (or $n = m + 1$), the first summation in (2) can be written

$$\sum_{n=1}^{\infty} n a_n x^{n-1} = \sum_{m=0}^{\infty} (m + 1) a_{m+1} x^m$$

Of course, the second summation can also be written

$$\sum_{n=0}^{\infty} a_n x^n = \sum_{m=0}^{\infty} a_m x^m$$

Thus (2) becomes

$$\sum_{m=0}^{\infty} (m + 1) a_{m+1} x^m - \sum_{m=0}^{\infty} a_m x^m = \varnothing$$

or

$$\sum_{m=0}^{\infty} [(m + 1) a_{m+1} - a_m] x^m = \varnothing$$

from which it follows that

(3) $(m + 1) a_{m+1} - a_m = 0, \qquad m = 0, 1, \ldots$

Note that the coefficients in the power series expansion of the solution $y(x)$ are obtained by solving the equation (3). This equation is called a *recursion formula*. Of course, we also recognize (3) as a first-order linear difference equation, and so we can expect the methods and results of Chapter 5 to be applicable in this chapter.

4 Example

Consider the initial-value problem

$$Dy - y = 2e^{-x}; \qquad y(1) = 0$$

From Section 8, the solution of the initial-value problem is $y(x) = e^{x-2} - e^{-x}$.

Assume that the solution has a power series expansion of the form

(A) $$y(x) = \sum_{n=0}^{\infty} a_n(x - 1)^n$$

Note that the expansion is about the point $x = 1$, since the initial value is specified at $x = 1$. From the power series (A), we have

(B) $$Dy(x) = \sum_{n=1}^{\infty} na_n(x - 1)^{n-1}$$

The Taylor series expansion of e^{-x} in powers of $(x - 1)$ is:

(C) $$e^{-x} = \sum_{n=0}^{\infty} \frac{(-1)^n e^{-1}}{n!}(x - 1)^n$$

Substituting (A), (B), and (C) into the differential equation yields

$$\sum_{n=1}^{\infty} na_n(x - 1)^{n-1} - \sum_{n=0}^{\infty} a_n(x - 1)^n = 2\sum_{n=0}^{\infty} \frac{(-1)^n e^{-1}}{n!}(x - 1)^n$$

The first summation on the left-hand side can also be written

$$\sum_{n=0}^{\infty} (n + 1)a_{n+1}(x - 1)^n$$

Thus we obtain

$$\sum_{n=0}^{\infty} [(n + 1)a_{n+1} - a_n](x - 1)^n = 2\sum_{n=0}^{\infty} \frac{(-1)^n e^{-1}}{n!}(x - 1)^n$$

from which we get the recursion formula

(D) $$(n + 1)a_{n+1} - a_n = 2e^{-1}\frac{(-1)^n}{n!}$$

$n = 0, 1, \ldots$, a first-order linear nonhomogeneous difference equation. There are two ways to proceed from this point. First, the recursion formula (D) will allow us to calculate each of the coefficients a_n, $n = 1, 2, \ldots$, in terms of a_0. In particular, successively setting $n = 0, 1, \ldots$, in (D) yields the equations

$$a_1 - a_0 = 2e^{-1}$$

$$2a_2 - a_1 = \frac{-2e^{-1}}{1!}$$

$$3a_3 - a_2 = \frac{2e^{-1}}{2!}$$

$$4a_4 - a_3 = \frac{-2e^{-1}}{3!}$$

and so on, the solutions of which are:

$$a_1 = a_0 + 2e^{-1}$$

$$a_2 = \frac{a_0}{2!}$$

$$a_3 = \frac{a_0}{3!} + \frac{2e^{-1}}{3!}$$

$$a_4 = \frac{a_0}{4!}$$

In general, we will have

$$a_n = \begin{cases} \dfrac{a_0}{n!} & n \text{ even} \\[2ex] \dfrac{a_0}{n!} + \dfrac{2e^{-1}}{n!} & n \text{ odd} \end{cases}$$

We use the initial condition $y(1) = 0$ in (A) to obtain

$$0 = y(1) = \sum_{n=0}^{\infty} a_n(1 - 1)^n = a_0$$

or $a_0 = 0$. Thus

$$a_n = \begin{cases} 0 & n \text{ even} \\[2ex] \dfrac{2e^{-1}}{n!} & n \text{ odd} \end{cases}$$

and

$$y(x) = \sum_{n=0}^{\infty} \frac{2e^{-1}}{(2n + 1)!} (x - 1)^{2n+1}$$

is a series solution. The reader can verify that the Taylor series expansion in powers of $(x - 1)$ of

$$y(x) = e^{x-2} - e^{-x}$$

coincides with the result above. It is easy to show that this series converges for all x.

As an alternative to calculating successive coefficients a_n in terms of a_0, we could have used the general theory of first-order linear difference equations, Section 23. In particular, from Theorem 9 in Section 23, the one-parameter family of solutions is:

$$a_n = t(n) \sum \frac{2e^{-1}(-1)^n}{(n + 1)! \, t(n + 1)} + ct(n)$$

where

$$t(n) = (-1)^n \prod_{i=0}^{n-1} \left(\frac{-1}{i+1} \right) = \frac{1}{n!}$$

Thus

$$a_n = \frac{2e^{-1}}{n!} \sum (-1)^n + \frac{c}{n!}$$

$$= \frac{2e^{-1}}{n!} f(n) + \frac{c}{n!}$$

where $\Delta f(n) = (-1)^n$ and

$$f(n) = \begin{cases} 0 & \text{if } n \text{ is even} \\ 1 & \text{if } n \text{ is odd} \end{cases}$$

Since

$$0 = a_0 = \frac{2e^{-1}}{0!} f(0) + \frac{c}{0!} = c$$

we get that

$$a_n = \frac{2e^{-1}}{n!} f(n)$$

which agrees with the previous calculation.

5 EXERCISES

In each of the following, find the one-parameter family of solutions of the given equation by assuming a power series solution in powers of x. In each case, compare your result with the one-parameter family obtained by using the methods of Section 8.

1. $Dy + 2y = \emptyset$
2. $Dy - 3y = 3$
3. $Dy + y = 2x - 1$
4. $Dy + 2y = e^x$
5. $Dy - y = e^x$
6. $Dy - y = \sin x$

Determine the power series solution of each of the following initial-value problems.

7. $Dy - y = e^x + 1$; $y(0) = 1$
8. $Dy + y = x + 1$; $y(1) = 0$ [Hint: Expand $f(x) = x + 1$ in powers of $(x - 1)$.]
9. Consider the differential equation $Dy + 2xy = 2x$.
   a. Assume a power series solution in powers of x and show that the recursion formula is a second-order linear nonhomogeneous difference equation with nonconstant coefficients.
   b. By using the power series method, determine the one-parameter family of solutions. Compare with the solutions using Corollary 15, Section 8.

c. Determine the solution of the equation satisfying the initial condition $y(0) = 1$.

10. Consider the differential equation $Dy + 3x^2y = \varnothing$.
   a. Assume a power series solution in powers of x and show that the recursion formula is a third-order linear homogeneous difference equation with nonconstant coefficients.
   b. Use the power series method to determine the one-parameter family of solutions of the equation. Compare your result with the solutions obtained by using Corollary 8, Section 8.

11. Consider the differential equation $Dy + p(x)y = f(x)$, where p is a polynomial of degree k and f has a power series expansion in powers of x. Assume a power series solution in powers of x.
   a. Show that the recursion formula is a $(k + 1)$st-order linear difference equation with nonconstant coefficients.
   b. Show that the recursion formula is homogeneous if and only if the differential equation is homogeneous.

12. Find the first five terms of the power series solution of the initial-value problem

$$Dy + xy = \ln x; \qquad y(1) = 1$$

[Hint: Expand $p(x) = x$ and $f(x) = \ln x$ in powers of $(x - 1)$.]

We now illustrate a second power series technique which is sometimes more convenient than the method using the summation notation illustrated in the examples above. The technique shown here is often called the *Taylor series method*. It is particularly useful in solving initial-value problems by power series, and it has the advantage of being adaptable to nonlinear differential equations (see Section 46). Recall that the Taylor series expansion of a function h in powers of $x - a$ is:

$$h(x) = h(a) + \frac{Dh(a)}{1!}(x - a) + \frac{D^2h(a)}{2!}(x - a)^2 + \cdots$$

$$= \sum_{n=0}^{\infty} \frac{D^nh(a)}{n!}(x - a)^n$$

Thus, for example, if we are given the initial-value problem

$$Dy + p(x)y = f(x); \qquad y(a) = b, \quad a, x \in J$$

then we assume that the solution y has the Taylor series expansion

$$y(x) = \sum_{n=0}^{\infty} \frac{D^ny(a)}{n!}(x - a)^n = y(a) + \frac{Dy(a)}{1!}(x - a) + \cdots$$

The initial condition, $y(a) = b$, determines the first coefficient in the series. From the differential equation,

$$Dy(a) = f(a) - p(a)b$$

and this determines the second coefficient. Differentiating the differential equation (assuming that $p(x)$ and $f(x)$ are differentiable functions on the interval J), we get

$$D^2y(x) = Df(x) - y(x)\,Dp(x) + p(x)\,Dy(x)$$

Therefore,

$$D^2y(a) = Df(a) - b\,Dp(a) + p(a)\,Dy(a)$$

We continue by calculating successive derivatives and evaluating at $x = a$. Note that to carry this method out, the functions $p(x)$ and $f(x)$ will have to be infinitely differentiable on J. We illustrate with the initial-value problem of Example 4.

6 Example

Given the initial-value problem

$$Dy - y = 2e^{-x}; \qquad y(1) = 0$$

Here $p(x) = -1$ and $f(x) = 2e^{-x}$ are infinitely differentiable on $(-\infty, \infty)$. Assuming that the solution has the Taylor series expansion

$$y(x) = y(1) + \frac{Dy(1)}{1!}(x - 1) + \frac{D^2y(1)}{2!}(x - 1)^2 + \cdots$$

we have $y(1) = 0$, $Dy(1) = 2e^{-1} + y(1) = 2e^{-1}$.
From the differential equation,

$$D^2y(x) = -2e^{-x} + Dy(x)$$

so that

$$D^2y(1) = -2e^{-1} + 2e^{-1} = 0$$

Calculating successive derivatives and evaluating at $x = 1$ yields

$$D^3y(x) = 2e^{-x} + D^2y(x)$$

and

$$D^3y(1) = 2e^{-1} + 0 = 2e^{-1}$$
$$D^4y(x) = -2e^{-x} + D^3y(x)$$

and

$$D^4y(1) = -2e^{-1} + 2e^{-1} = 0$$

and, in general,

$$D^ny(x) = (-1)^{n+1}2e^{-x} + D^{n-1}y(x)$$

so that

$$D^ny(1) = \begin{cases} 0 & n \text{ even} \\ 2e^{-1} & n \text{ odd} \end{cases}$$

Thus

$$y(x) = \sum_{n=0}^{\infty} \frac{2e^{-1}}{(2n+1)!} (x-1)^{2n+1}$$

as determined in Example 4.

Although our examples illustrating the power series method have been restricted to first-order equations, it should be clear that the techniques are applicable to higher-order equations as well. From the examples, we might be led to believe that the power series approach provides a solution method for all linear equations. This is definitely not true; but the method is applicable in enough important cases to justify its study and use. We conclude this section with a few remarks which indicate when the power series method is applicable as well as some of the shortcomings of the method.

First, as suggested above, not every differential equation has a power series solution in powers of $(x-a)$ for every real number a. For example, if we attempt to find a power series solution of

(7) $x\, Dy + y = \varnothing$

in powers of x, we are led to consider

$$\sum_{n=1}^{\infty} n a_n x^n + \sum_{n=0}^{\infty} a_n x^n = \varnothing$$

or

$$a_0 + \sum_{n=1}^{\infty} (n+1) a_n x^n = \varnothing$$

from which it follows that $a_n = 0$ for all n. Thus the equation does not have a power series solution in powers of x. The equation does have, however, a power series solution in powers of $(x-a)$, $a \neq 0$. This raises the question as to when a given linear differential equation will have a power series expansion in powers of $(x-a)$ for some real number a.

Recall that a necessary condition for a function to have a power series expansion in powers of $(x-a)$ is that the function be infinitely differentiable on some interval containing $x = a$. It is shown in Theorem 5, Section 13, that any solution of a linear homogeneous equation with constant coefficients is infinitely differentiable on $\mathscr{R}$, the real line. From the results in Chapter 2, it follows that if $y = y(x)$ is a solution of

(8) $D^n y + p_1(x)\, D^{n-1} y + \cdots + p_{n-1}(x)\, Dy + p_n(x) y = f(x)$

$x \in J$, and if each of the coefficients $p_1, p_2, \ldots, p_n$, and f are in $\mathscr{C}^{\infty}(J)$, then $y = y(x) \in \mathscr{C}^{\infty}(J)$. In general, this does not imply that y has a power series expansion for the reasons stated in Theorem 18, Section 6. Motivated by these observations, we state the following general theorem specifying when a linear differential equation has a power series solution.

9 THEOREM

Given the nth-order linear differential equation (8). If each of the coefficients $p_1, p_2, \ldots, p_n$, and f are analytic at $x = a$, with $r_1, r_2, \ldots, r_n$, and r the radii of convergence of their respective power series expansions in powers of $(x - a)$, and if $R = \min(r_1, r_2, \ldots, r_n, r)$, then each solution $y = y(x)$ of (8) has a power series expansion

$$y(x) = \sum_{n=0}^{\infty} a_n(x - a)^n$$

which is valid on the interval $(a - R, a + R)$.

A second question which is raised by the power series method is concerned with the convergence of the power series expansion of a solution. In Example 4, we found that the power series expansion of the solution of the initial value problem is:

$$y(x) = \sum_{n=0}^{\infty} \frac{2e^{-1}}{(2n + 1)!} (x - 1)^{2n+1}$$

and that this series converges for all real numbers x, which agrees with Theorem 9. Suppose we wanted to calculate the value of this solution at some point $x = x_0$, $x_0 \neq 1$. If x_0 is sufficiently close to 1, then the resulting series

(10) $$y(x_0) = \sum_{n=0}^{\infty} \frac{2e^{-1}}{(2n + 1)!} (x_0 - 1)^{2n+1}$$

converges rapidly enough that relatively few terms of the series are required to obtain a "good" approximation of $y(x_0)$. For example, if $|x_0 - 1| < 1$, then eight terms will produce an approximation for $y(x_0)$ accurate to four decimal places. On the other hand, if we want to approximate the value of the solution at $x_0 = 4$ by using the first eight terms of the series, then the error would be greater than 0.408. Thus from a computational point of view, the series expansion of the solution in powers of $(x - 1)$ is not particularly useful for calculating values of the solution at points $x = x_0$ which are not "close" to $x = 1$.

As a second example along these lines, consider the initial-value problem

$$Dy + \frac{1}{x} y = \varnothing; \qquad y(1) = 1$$

and suppose we want to find the value of the solution at $x = 10$. The solution is $y(x) = 1/x$, and so $y(10) = \frac{1}{10}$. As indicated above (and seen by examining the solution), we cannot expect to have a power series solution in powers of x. We can, however, obtain a power series in powers of $(x - 1)$. In particular, it can be verified that the power series

$$y(x) = \sum_{n=0}^{\infty} (-1)^n (x - 1)^n$$

is the solution of the initial-value problem. The radius of convergence of this series is $R = 1$, so the series converges only on $J = (0, 2)$, and we cannot use this series to approximate $y(10)$.

11 EXERCISES

Determine the first five terms of the power series solution of each of the following initial-value problems by using the Taylor series method. In each case, use Theorem 9 to specify an interval on which the power series solution will converge.

1. $Dy + xy = \ln x$; $y(1) = 1$
2. $(1 - x^2) Dy - xy = 1$; $y(0) = 0$
3. $Dy + (\cos x)y = \sin x$; $y(\pi/4) = 1$
4. $Dy - (2/x)y = (1 - x/x)$; $y(1) = \frac{1}{2}$
5. Consider the initial-value problem $Dy + (1/x)y = \varnothing$; $y(1) = 1$. Assume a power series solution in powers of $(x - 1)$ and determine the solution. (Hint: Write the equation in the equivalent form $x\,Dy + y = \varnothing$.) Use Theorem 9 to specify an interval on which the power series solution will converge.
6. Assume a power series solution in powers of x and determine the solution of the initial-value problem $(1 - x^2) Dy - xy = 1$; $y(0) = 0$. Compare your result with Problem 2.
7. Assume a power series solution in powers of $(x - 1)$ and determine the solution of the initial-value problem $Dy - (2/x)y = (1 - x)/x$; $y(1) = \frac{1}{2}$.
8. Assume a power series solution in powers of x and determine the one-parameter family of solutions of $Dy - [1/(x - 1)]y = x^2/(x - 1)$.
9. Obtain the one-parameter family of solutions of $Dy + (1/x)y = \varnothing$ in powers of $x - 10$. Apply the initial condition $y(1) = 1$, and then determine $y(10)$.

Section 31 Power Series Solutions at Ordinary Points

In the next two sections, the power series methods are used to obtain solutions of higher-order linear equations. Although the methods are applicable in general, we shall concentrate on second-order linear equations because they occur most frequently in physical applications, and because they exhibit the important phenomena of the method without involving unnecessarily long and complicated calculations.

The first step in characterizing the set of solutions is to determine a solution basis for the solutions of the associated homogeneous equation. If a solution basis is known, then it can be used to construct a solution for the nonhomogeneous equation by the method of variation of parameters. For this reason, homogeneous equations will receive most of our attention.

Consider the second-order linear homogeneous differential equation

(1) $q_0(x) D^2 y + q_1(x) Dy + q_2(x)y = \varnothing$

where q_0, q_1, and q_2 are continuous functions on some interval J. Recall from Section 9 the difficulties that can arise at the points at which $q_0(x) = 0$.

As in Chapter 2, we divide the equation by q_0 in order to put (1) in the form

(2) $$D^2y + p_1(x)\,Dy + p_2(x)y = \varnothing$$

on J. In this form, the zeros of q_0 (if any) become points of discontinuity of $p_1 = q_1/q_0$ and $p_2 = q_2/q_0$. Now according to Theorem 9, Section 30, (2) will have a power series solution in powers of $(x - a)$, $a \in J$, if p_1 and p_2 are analytic at $x = a$, that is, if each of p_1 and p_2 has a power series expansion in powers of $(x - a)$ which is valid on some interval $(a - R, a + R)$, $R > 0$, contained in J.

3 DEFINITION

Given the differential equation (2). The point $x = a$ is an *ordinary point* of (2) if each of p_1 and p_2 is analytic at $x = a$. If one (or both) of the functions fails to be analytic at this point, then $x = a$ is called a *singular point* of (2).

We emphasize that in order to determine whether the point $x = a$ is an ordinary point or a singular point, the differential equation must be written in the form (2), that is, with leading coefficient 1.

4 Examples

a. Consider the differential equation

$$D^2y + [\cos(x)]\,Dy + e^x y = \varnothing$$

on $J = (-\infty, \infty)$. Here, $p_1(x) = \cos(x)$ and $p_2(x) = e^x$. Each of these functions is analytic at every point of J, and so all the points of J are ordinary points of the differential equation.

b. Consider the differential equation

$$(x - 2)\,D^2y + [\sin(2x)]\,Dy + (x^2 + 1)y = \varnothing$$

on $J = (-\infty, \infty)$. Writing this equation in the form (2), we get

$$D^2y + \frac{\sin(2x)}{(x - 2)}\,Dy + \frac{x^2 + 1}{x - 2}\,y = \varnothing$$

Now $p_1(x) = \sin(2x)/(x - 2)$ and $p_2(x) = (x^2 + 1)/(x - 2)$. Clearly, these functions are not analytic at $x = 2$, but they are analytic at every other point of J. Thus $x = 2$ is a singular point of the differential equation, while all other points are ordinary points.

While the zeros of $q_0(x)$ in equation (1) are singular points, Definition 3 actually includes other types of singular points. For example, the points $x = 0$ and $x = -1$ are singular points of the equation

$$D^2y + |x|\,Dy + (x + 1)^{1/3}y = \varnothing$$

on $J = (-\infty, \infty)$, since the coefficient functions fail to be analytic at these points.

In this section, we shall be concerned with power series solutions about ordinary points. Series solutions about singular points are treated in the next section.

The following theorem provides the theoretical basis for the method used in this section. The proof of this result involves complex variable theory and is beyond the scope of the text.

5 THEOREM

If $x = a$ is an ordinary point of (2), then there exist two linearly independent power series solutions

$$y_1(x) = \sum_{n=0}^{\infty} a_n(x - a)^n$$

$$y_2(x) = \sum_{n=0}^{\infty} b_n(x - a)^n$$

and each of these series converges on the interval $(a - R, a + R)$ contained in J, where R is the minimum of the radii of convergence of the power series expansion in powers of $(x - a)$ of p_1 and p_2.

Theorem 5 implies that all solutions of equation (2) on the interval $J = (a - R, a + R)$ can be represented as a linear combination of two power series.

6 Example

Consider the differential equation

$$D^2y - Dy - 6y = \varnothing$$

on $J = (-\infty, \infty)$. The functions $y_1(x) = e^{-2x}$ and $y_2(x) = e^{3x}$ form a solution basis for the equation. As a result of Theorem 5, however, we can obtain a pair of linearly independent power series solutions in powers of $(x - a)$ for any real number a, and each of the power series will converge for all x. In particular, $x = 0$ is an ordinary point, and so we shall assume a power series solution of the form

(A) $\qquad y(x) = \sum_{n=0}^{\infty} a_n x^n$

Differentiating (A) term by term, we obtain

$$Dy = \sum_{n=1}^{\infty} na_n x^{n-1}$$

and

$$D^2 y = \sum_{n=2}^{\infty} n(n-1)a_n x^{n-2}$$

Substituting these series into the differential equation yields

(B) $$\sum_{n=2}^{\infty} n(n-1)a_n x^{n-2} - \sum_{n=1}^{\infty} na_n x^{n-1} - 6\sum_{n=0}^{\infty} a_n x^n = \emptyset$$

The change of index $k = n - 2$ transforms the first series in (B) into

$$\sum_{k=0}^{\infty} (k+2)(k+1)a_{k+2}x^k$$

The change of index $k = n - 1$ transforms the second series into

$$\sum_{k=0}^{\infty} (k+1)a_{k+1}x^k$$

Thus (B) can be written

(C) $$\sum_{k=0}^{\infty} [(k+2)(k+1)a_{k+2} - (k+1)a_{k+1} - 6a_k]x^k = \emptyset$$

From this equation, we conclude that the coefficients a_n in the power series (A) must satisfy the *recursion formula*

(D) $$(k+2)(k+1)a_{k+2} - (k+1)a_{k+1} - 6a_k = 0$$

$k = 0, 1, 2, \ldots$, which we recognize as a second-order linear homogeneous difference equation.

There are several ways to proceed from this point. First, by using the ideas in Section 24, we could attempt to determine two linearly independent solutions of (D). The reader can verify that the functions

$$z_1(n) = \frac{(-2)^n}{n!} \quad \text{and} \quad z_2(n) = \frac{3^n}{n!}$$

are linearly independent solutions of (D). Note that z_1 and z_2 are the coefficients in the Maclaurin series expansions of e^{-2x} and e^{3x}. The two-parameter family of solutions of (D) is:

$$a_k = c_1 \frac{(-2)^k}{k!} + c_2 \frac{3^k}{k!}$$

By substituting this into (A), we have

$$y(x) = \sum_{n=0}^{\infty} \left[c_1 \frac{(-2)^n}{n!} + c_2 \frac{3^n}{n!} \right] x^n$$

or

(E) $$y(x) = c_1 \sum_{n=0}^{\infty} \frac{(-2)^n}{n!} x^n + c_2 \sum_{n=0}^{\infty} \frac{3^n}{n!} x^n$$

It is easy to verify that the power series

$$y_1(x) = \sum_{n=0}^{\infty} \frac{(-2)^n}{n!} x^n$$

and

$$y_2(x) = \sum_{n=0}^{\infty} \frac{3^n}{n!} x^n$$

are linearly independent solutions of the differential equation, and that each series converges for all real numbers x. Finally, we recognize $y_1(x)$ and $y_2(x)$ as being the Maclaurin series expansions of e^{-2x} and e^{3x}.

In general, it is not possible to obtain explicit representations of two linearly independent solutions of the difference equation. A second procedure is to use the difference equation (D) to successively calculate the coefficients a_k, $k \geq 2$, in terms of a_0 and a_1. In particular, for $k = 0$ in (D), we get

$$2a_2 - a_1 - 6a_0 = 0$$

or

$$a_2 = \frac{a_1}{2} + 3a_0$$

For $k = 1$, we have

$$6a_3 - 2a_2 - 6a_1 = 0$$

or

$$a_3 = \frac{a_2}{3} + a_1 = \tfrac{1}{3}\left(\frac{a_1}{2} + 3a_0\right) + a_1 = \frac{7a_1}{6} + a_0$$

For $k = 2$, we have

$$12a_4 - 3a_3 - 6a_2 = 0$$

or

$$a_4 = \frac{a_3}{4} + \frac{a_2}{2} = \tfrac{1}{4}\left[\frac{7a_1}{6} + a_0\right] + \tfrac{1}{2}\left[\frac{a_1}{2} + 3a_0\right]$$

$$= \frac{13a_1}{24} + \frac{7a_0}{4}$$

Clearly, this procedure can be continued, and as many coefficients as desired can be calculated in terms of a_0 and a_1.

The calculations above yield

$$y(x) = a_0 + a_1 x + (\tfrac{1}{2}a_1 + 3a_0)x^2 + (\tfrac{7}{6}a_1 + a_0)x^3$$
$$+ (\tfrac{13}{24}a_1 + \tfrac{7}{4}a_0)x^4 + \cdots$$

or

(F) $$y(x) = a_0(1 + 3x^2 + x^3 + \tfrac{7}{4}x^4 + \cdots)$$
$$+ a_1(x + \tfrac{1}{2}x^2 + \tfrac{7}{6}x^3 + \tfrac{13}{24}x^4 + \cdots)$$

It can be verified that the power series

$$u_1(x) = 1 + 3x^2 + x^3 + \tfrac{7}{4}x^4 + \cdots$$

and

$$u_2(x) = x + \tfrac{1}{2}x^2 + \tfrac{7}{6}x^3 + \tfrac{13}{24}x^4 + \cdots$$

are linearly independent solutions of the differential equation, so that (F) represents the two-parameter family of solutions. According to Theorem 5, each of these series will converge for all real numbers x. Verification that the two families (E) and (F) are the same is left to the reader.

An examination of this step-by-step procedure for calculating the coefficients $a_2, a_3, \ldots$, in terms of a_0 and a_1 shows that we are solving a system of linear algebraic equations. This suggests that vector-matrix notation will provide a means of organizing the calculations. In order to calculate a_2, a_3, and a_4 in terms of a_0 and a_1, as above, the three equations involved are:

$$2a_2 - a_1 - 6a_0 = 0$$

$$6a_3 - 2a_2 - 6a_1 = 0$$

$$12a_4 - 3a_3 - 6a_2 = 0$$

These equations can be represented in the vector-matrix form

$$\begin{bmatrix} -6 & -1 & 2 & 0 & 0 \\ 0 & -6 & -2 & 6 & 0 \\ 0 & 0 & -6 & -3 & 12 \end{bmatrix} \begin{bmatrix} a_0 \\ a_1 \\ a_2 \\ a_3 \\ a_4 \end{bmatrix} = \begin{bmatrix} 0 \\ 0 \\ 0 \end{bmatrix}$$

By partitioning the matrix and the vector on the left as follows

$$\left[\begin{array}{cc|ccc} -6 & -1 & 2 & 0 & 0 \\ 0 & -6 & -2 & 6 & 0 \\ 0 & 0 & -6 & -3 & 12 \end{array}\right] \begin{bmatrix} a_0 \\ a_1 \\ a_2 \\ a_3 \\ a_4 \end{bmatrix} = \begin{bmatrix} 0 \\ 0 \\ 0 \end{bmatrix}$$

and multiplying in this partitioned form, we have

$$
\begin{bmatrix} -6 & -1 \\ 0 & -6 \\ 0 & 0 \end{bmatrix} \begin{bmatrix} a_0 \\ a_1 \end{bmatrix} + \begin{bmatrix} 2 & 0 & 0 \\ -2 & 6 & 0 \\ -6 & -3 & 12 \end{bmatrix} \begin{bmatrix} a_2 \\ a_3 \\ a_4 \end{bmatrix} = \begin{bmatrix} 0 \\ 0 \\ 0 \end{bmatrix}
$$

or

$$
\begin{bmatrix} 2 & 0 & 0 \\ -2 & 6 & 0 \\ -6 & -3 & 12 \end{bmatrix} \begin{bmatrix} a_2 \\ a_3 \\ a_4 \end{bmatrix} = \begin{bmatrix} 6 & 1 \\ 0 & 6 \\ 0 & 0 \end{bmatrix} \begin{bmatrix} a_0 \\ a_1 \end{bmatrix} = \begin{bmatrix} 6a_0 + a_1 \\ 6a_1 \\ 0 \end{bmatrix}
$$

The values of a_2, a_3, and a_4 are now obtained from this vector-matrix equation.

Notice that the power series method transforms the problem of solving a linear differential equation into one of solving a linear difference equation. Thus the theory and results of Chapter 5 are applicable. In many cases, however, the problem of finding explicit solutions of the difference equation will be of the same order of difficulty as solving the differential equation. The following examples illustrate additional aspects of the power series method. Since Example 6 has been treated in great detail, the presentation in the remaining examples will be somewhat brief, with routine manipulations and calculations left to the reader.

7 Example

Consider the differential equation

$$D^2 y - xy = \varnothing$$

on $J = (-\infty, \infty)$. All points of J are ordinary points of the equation. In particular, $x = 1$ is an ordinary point, and we shall determine two linearly independent solutions of the form

(A) $\qquad y(x) = \sum_{n=0}^{\infty} a_n(x - 1)^n$

Calculating $D^2 y(x)$ from (A) and substituting into the differential equation yields

(B) $\qquad \sum_{n=2}^{\infty} n(n - 1)a_n(x - 1)^{n-2} - x \sum_{n=0}^{\infty} a_n(x - 1)^n = \varnothing$

Before we can manipulate the series and combine the coefficients of the corresponding powers of $(x - 1)$, the coefficient $p_2(x) = x$ must be written as $p_2(x) = 1 + (x - 1)$. With this form of p_2, equation (B) becomes

(C)
$$\sum_{n=2}^{\infty} n(n-1)a_n(x-1)^{n-2} - \sum_{n=0}^{\infty} a_n(x-1)^n$$
$$- \sum_{n=0}^{\infty} a_n(x-1)^{n+1} = \varnothing$$

By making the appropriate changes of index, (C) can be written as

(D)
$$(2a_2 - a_0) + \sum_{k=1}^{\infty} [(k+2)(k+1)a_{k+2} - a_k - a_{k-1}](x-1)^k = \varnothing$$

Therefore, we get $2a_2 - a_0 = 0$ and the recursion formula

$$(k+2)(k+1)a_{k+2} - a_k - a_{k-1} = 0$$

$k = 1, 2, \ldots$. This is a third-order linear homogeneous difference equation which, in standard form, is:

(E)
$$(k+3)(k+2)a_{k+3} - a_{k+1} - a_k = \varnothing$$

Although (E) has three linearly independent solutions (for example, the three initial values a_0, a_1, and a_2 can be assigned arbitrarily), the solutions which we seek must also satisfy $2a_2 - a_0 = 0$, and there are only two linearly independent solutions of (E) with this property. That is, a_0 and a_1 can be assigned values arbitrarily, then $a_2 = \frac{1}{2}a_0$, and all the remaining coefficients a_n can be calculated in terms of a_0 and a_1 using (E).

By setting $k = 0, 1, 2$ in (E), together with $2a_2 - a_0 = 0$, the system of equations in vector-matrix form is:

$$\begin{bmatrix} -1 & 0 & 2 & 0 & 0 & 0 \\ -1 & -1 & 0 & 6 & 0 & 0 \\ 0 & -1 & -1 & 0 & 12 & 0 \\ 0 & 0 & -1 & -1 & 0 & 20 \end{bmatrix} \begin{bmatrix} a_0 \\ a_1 \\ a_2 \\ a_3 \\ a_4 \\ a_5 \end{bmatrix} = \begin{bmatrix} 0 \\ 0 \\ 0 \\ 0 \end{bmatrix}$$

which is the same as

(F)
$$\begin{bmatrix} 2 & 0 & 0 & 0 \\ 0 & 6 & 0 & 0 \\ -1 & 0 & 12 & 0 \\ -1 & -1 & 0 & 20 \end{bmatrix} \begin{bmatrix} a_2 \\ a_3 \\ a_4 \\ a_5 \end{bmatrix} = \begin{bmatrix} 1 & 0 \\ 1 & 1 \\ 0 & 1 \\ 0 & 0 \end{bmatrix} \begin{bmatrix} a_0 \\ a_1 \end{bmatrix} = \begin{bmatrix} a_0 \\ a_0 + a_1 \\ a_1 \\ 0 \end{bmatrix}$$

The solution of this system is:

$$a_2 = \frac{a_0}{2}, \qquad a_3 = \frac{a_1}{6} + \frac{a_0}{6}$$

$$a_4 = \frac{a_0}{24} + \frac{a_1}{12}, \qquad a_5 = \frac{a_0}{30} + \frac{a_1}{120}$$

Substituting these coefficients into (A), we get

(G)
$$y(x) = a_0[1 + \tfrac{1}{2}(x - 1)^2 + \tfrac{1}{6}(x - 1)^3$$
$$+ \tfrac{1}{24}(x - 1)^4 + \tfrac{1}{30}(x - 1)^5 + \cdots]$$
$$+ a_1[(x - 1) + \tfrac{1}{6}(x - 1)^3 + \tfrac{1}{12}(x - 1)^4$$
$$+ \tfrac{1}{120}(x - 1)^5 + \cdots]$$

It can be verified that the two series

(H)
$$y_1(x) = 1 + \tfrac{1}{2}(x - 1)^2 + \tfrac{1}{6}(x - 1)^3$$
$$+ \tfrac{1}{24}(x - 1)^4 + \tfrac{1}{30}(x - 1)^5 + \cdots$$
$$y_2(x) = (x - 1) + \tfrac{1}{6}(x - 1)^3 + \tfrac{1}{12}(x - 1)^4 + \tfrac{1}{120}(x - 1)^5 + \cdots$$

are linearly independent. Thus (G) represents the two-parameter family of solutions of the differential equation. Finally, according to Theorem 5, each of the series y_1 and y_2 will converge for all real numbers x.

At the end of Example 6 we noted that the power series method transforms a differential equation into a difference equation. From this example we see that the respective orders of these equations may be different. It is easy to verify that the order of the difference equation which the power series method produces will always be at least as great as the order of the differential equation. The order of the difference equation, however, depends upon both the differential equation and the coefficients $p_i(x)$ in the equation. A second observation concerns the vector-matrix representation of the system of equations which occur in the step-by-step calculation of the coefficients a_n. Note that the coefficient matrix of the "unknown" vector $(a_2, a_3, a_4, a_5)^T$ in (F) is lower triangular with nonzero entries on the main diagonal. Such a matrix is always invertible, and in some cases the inverse is easy to determine.

8 Example

Consider the differential equation

$$(1 - x^2)\, D^2 y - x\, Dy + y = \emptyset$$

on $J = (-\infty, \infty)$. Here,

$$p_1(x) = \frac{-x}{1 - x^2}$$

$$p_2(x) = \frac{1}{1 - x^2}$$

and these functions are not analytic at $x = 1$ and $x = -1$. They are analytic at all other points. In particular, $x = 0$ is an ordinary point of the differential equation; and so we seek a pair of linearly independent solutions of the form

(A) $$y(x) = \sum_{n=0}^{\infty} a_n x^n$$

According to Theorem 5, the series solutions will converge on the interval $(-1, 1)$ at least. They may be convergent on a larger interval. By differentiating (A) twice and substituting into the differential equation, we obtain

$$(1 - x^2) \sum_{n=2}^{\infty} n(n - 1)a_n x^{n-2} - x \sum_{n=1}^{\infty} na_n x^{n-1} + \sum_{n=0}^{\infty} a_n x^n = \emptyset$$

or

(B) $$\sum_{n=2}^{\infty} n(n - 1)a_n x^{n-2} - \sum_{n=2}^{\infty} n(n - 1)a_n x^n$$

$$- \sum_{n=1}^{\infty} na_n x^n + \sum_{n=0}^{\infty} a_n x^n = \emptyset$$

It is important to note that we used the given form of the equation rather than

$$D^2 y - \frac{x}{1 - x^2} Dy + \frac{x}{1 - x^2} y = \emptyset$$

To use this form would require the expansion of $x/(1 - x^2)$ and $1/(1 - x^2)$ into power series and would entail two multiplications of power series. Clearly these calculations would be difficult.

Now (B) can be simplified to

(C) $$a_0 + \sum_{n=2}^{\infty} n(n - 1)a_n x^{n-2} - \sum_{n=2}^{\infty} [n(n - 1) + n - 1]a_n x^n = \emptyset$$

With appropriate changes of index, (C) can be written

(D) $$(2a_2 + a_0) + 6a_3 x + \sum_{k=2}^{\infty} [(k + 2)(k + 1)a_{k+2}$$

$$- (k^2 - 1)a_k]x^k = \emptyset$$

It now follows that $2a_2 + a_0 = 0$, $a_3 = 0$, and

$$(k + 2)(k + 1)a_{k+2} - (k^2 - 1)a_k = 0$$

$k = 2, 3, \ldots$. These equations are represented by the recursion formula

(E) $$(k + 2)a_{k+2} - (k - 1)a_k = 0$$

$k = 0, 1, 2, \ldots$, a second-order linear homogeneous difference equation. Since $a_3 = 0$, it follows from (E) that $a_5 = a_7 = \cdots = a_{2k+1} = \cdots = 0$. Thus a step-by-step calculation of the coefficients will be convenient in this case. We have

$$a_2 = -\frac{a_0}{2}$$

$$a_4 = \frac{a_2}{4} = -\frac{a_0}{2 \cdot 4}$$

$$a_6 = \frac{3a_4}{6} = -\frac{3a_0}{2 \cdot 4 \cdot 6}$$

$$a_8 = \frac{5a_0}{8} = -\frac{3 \cdot 5 \cdot a_0}{2 \cdot 4 \cdot 6 \cdot 8}$$

In general, we conclude that

$$a_{2k} = -\frac{1 \cdot 3 \cdot 5 \cdots \cdots (2n - 3)}{2 \cdot 4 \cdot 6 \cdots \cdots (2n)} a_0$$

$k = 1, 2, \ldots$. We leave as an exercise the verification that a_{2k} can also be written

(F) $\qquad a_{2k} = -\frac{4^{-k}}{(2k - 1)}\binom{2k}{k}$

Thus we obtain

(G) $\qquad y(x) = a_0 \sum_{n=0}^{\infty} \frac{4^{-n}}{(2n - 1)}\binom{2n}{n} x^{2n} + a_1 x$

Let $y_1(x)$ be the power series

$$y_1(x) = \sum_{n=0}^{\infty} \frac{4^{-n}}{(2n - 1)}\binom{2n}{n} x^{2n}$$

and let $y_2(x)$ be the (finite) series $y_2(x) = x$. It is clear that y_1 and y_2 are linearly independent solutions of the differential equation. Thus (G) represents the two-parameter family of solutions. Note that the "series" $y_2(x) = x$ converges for all real numbers x, while, using the ratio test, the series $y_1(x)$ has radius of convergence equal to 1.

9 EXERCISES

1. Determine the ordinary points and the singular points of each of the following differential equations.

   a. $D^2 y - 2 Dy + y = \varnothing$
   b. $(x + 1) D^2 y + (x - 2) Dy + y = \varnothing$
   c. $(1 + x^2) D^2 y + 2x Dy = \varnothing$
   d. $(1 - x^2) D^2 y - 6x Dy - 4y = \varnothing$
   e. $x(x + 1) D^2 y + (x + 2) Dy + 4y = \varnothing$
   f. $xe^x D^2 y - (x^2 - 2) Dy + x^3 y = \varnothing$

g. $(e^x - 1) D^2 y + y = \varnothing$

h. $(x^2 - 5x + 6) D^2 y + e^x Dy - 7y = \varnothing$

In each of the following, verify that the indicated point $x = a$ is an ordinary point; determine two linearly independent power series solutions in powers of $(x - a)$; and indicate the interval of convergence. Express your power series solutions in terms of elementary functions whenever possible.

2. $D^2 y - y = \varnothing, x = 0$

3. $D^2 y + 4y = \varnothing, x = 0$

4. $D^2 y - 2 Dy = \varnothing, x = 0$

5. $D^2 y + xy = \varnothing, x = 0$

6. $D^2 y - x^2 Dy - y = \varnothing, x = 0$

7. $D^2 y - (x + 1) Dy - y = \varnothing, x = -1$

8. $D^2 y + (x - 1) Dy + 3y = \varnothing, x = 1$

9. $(x + 1) D^2 y + (x - 2) Dy + y = \varnothing, x = 0$

10. $(2 + x) D^2 y - y = \varnothing, x = -1$

11. $D^2 y + (x + 2) Dy + (6x + 8)y = \varnothing, x = -2$

12. $(1 - x^2) D^2 y - 6x Dy - 4y = \varnothing, x = 0$

13. $(1 + x^2) D^2 y - 4x Dy + 6y = \varnothing, x = 0$

14. $(1 - x^2) D^2 y - 4x Dy - 2y = \varnothing, x = 0$

15. $(1 + x^2) D^2 y + 2x Dy = \varnothing, x = 0$

16. $(2 + x) D^2 y = y, x = 0$

Use the power series method to find three linearly independent solutions in powers of x of each of the following third-order equations.

17. $D^3 y + x^2 D^2 y + 6x Dy + 6y = \varnothing$

18. $D^3 y - x^2 D^2 y + 5x Dy - 3y = \varnothing$

19. $D^3 y - x Dy - 3y = \varnothing$

20. Consider the second-order homogeneous equation

(A) $(ax^2 + 1) D^2 y + bx Dy + cy = \varnothing$

where a, b, and c are constants.

a. Give a necessary and sufficient condition for (A) to have a polynomial solution of degree k.

b. Give a necessary and sufficient condition for (A) to have a solution basis consisting of exactly one polynomial solution, say of degree k, and one power series (infinite) solution.

c. Give a necessary and sufficient condition for (A) to have only polynomial solutions.

21. Verify that the two series solutions y_1 and y_2 as given in (H) in Example 7 are linearly independent.

22. Verify in Example 8 that (F) is a solution of (E).

The power series method can also be used to obtain series solutions in powers of $(x - a)$ of the nonhomogeneous equation

(10) $D^2 y + p_1(x) Dy + p_2(x)y = f(x)$

when each of the functions $p_1(x)$, $p_2(x)$, and $f(x)$ is analytic at $x = a$, that is, when $x = a$ is an ordinary point of the reduced equation and $f(x)$ has a power series expansion in powers of $(x - a)$. As with homogeneous equations, a power series solution of (10) will have a radius of convergence which is at least as large as the minimum of the radii of convergence of the expansions in powers of $(x - a)$ of p_1, p_2, and f. The following example illustrates the power series method applied to nonhomogeneous equations.

11 Example

Consider the differential equation

$$D^2 y + y = \ln(x)$$

on $J = (0, \infty)$. The functions $p_1 = \varnothing$, $p_2 = 1$, and $f(x) = \ln(x)$ are analytic on J and thus we can determine a power series solution in powers of $(x - a)$ for any $a \in J$. In particular, letting $a = 1$, we seek a power series solution of the form

(A) $\qquad y(x) = \sum_{n=0}^{\infty} a_n(x - 1)^n$

By differentiating (A) twice and substituting into the differential equation, we get

$$\sum_{n=2}^{\infty} n(n - 1)a_n(x - 1)^{n-2} + \sum_{n=0}^{\infty} a_n(x - 1)^n = \ln(x)$$

Of course, before we can calculate the coefficients a_n we must expand the function $\ln(x)$ in powers of $(x - 1)$. The Taylor series expansion of $\ln(x)$ in powers of $(x - 1)$ is:

$$\ln(x) = \sum_{n=1}^{\infty} \frac{(-1)^{n+1}}{n}(x - 1)^n$$

The radius of convergence of this series is 1. Replacing $\ln(x)$ by its Taylor series at $x = 1$, we obtain

(B) $\qquad \sum_{n=2}^{\infty} n(n - 1)a_n(x - 1)^{n-2} + \sum_{n=0}^{\infty} a_n(x - 1)^n = \sum_{n=1}^{\infty} \frac{(-1)^{n+1}}{n}(x - 1)^n$

Now, by making the appropriate changes of index, (B) can be written

(C) $\qquad \sum_{k=0}^{\infty} [(k + 2)(k + 1)a_{k+2} + a_k](x - 1)^k = \sum_{k=1}^{\infty} \frac{(-1)^{k+1}}{k}(x - 1)^k$

In order to equate the coefficients of the corresponding powers of $(x - 1)$, we write (C) as

$$(2a_2 + a_0) + \sum_{k=1}^{\infty} [(k + 2)(k + 1)a_{k+2} + a_k](x - 1)^k$$

$$= \sum_{k=1}^{\infty} \frac{(-1)^{k+1}}{k}(x - 1)^k$$

Thus

$$2a_2 + a_0 = 0$$

(D)

$$(k + 2)(k + 1)a_{k+2} + a_k = \frac{(-1)^{k+1}}{k}$$

$k = 1, 2, \ldots$. The latter equation is the recursion formula; it is a second-order linear nonhomogeneous difference equation.

By doing a step-by-step calculation of $a_2, a_3, \ldots$, we conjecture that

$$a_{2k} = \frac{(-1)^k a_0}{(2k)!} + \frac{-(2k-3)! + (2k-5)! - \cdots + (-1)^{k+1}}{(2k)!}$$

$$k = 2, 3, \ldots$$

$$a_{2k+1} = \frac{(-1)^k a_1}{(2k+1)!} + \frac{(2k-2)! - (2k-4)! + \cdots + (-1)^{k+1}}{(2k+1)!}$$

$$k = 1, 2, \ldots$$

The validity of these solutions can be established by induction. We substitute these values for a_k into (A) to obtain the series solution

(E)

$$y(x) = a_0 \left[1 - \frac{(x-1)^2}{2!} + \frac{(x-1)^4}{4!} - \cdots \right]$$

$$+ a_1 \left[(x-1) - \frac{(x-1)^3}{3!} + \frac{(x-1)^5}{5!} - \cdots \right]$$

$$+ \frac{(x-1)^3}{3!} - \frac{(x-1)^4}{4!} + \frac{(2! - 1)}{5!}(x-1)^5$$

$$+ \frac{(1 - 3!)}{6!}(x-1)^6 + \cdots$$

We recognize the coefficients of a_0 and a_1 in (E) as being the Taylor series expansions of $\cos(x - 1)$ and $\sin(x - 1)$, respectively, in powers of $(x - 1)$. Thus

(F)

$$y(x) = a_0 \cos(x - 1) + a_1 \sin(x - 1) + z(x)$$

where z is the power series

$$z(x) = \frac{(x-1)^3}{3!} - \frac{(x-1)^4}{4!} + \frac{(2! - 1)}{5!}(x-1)^5$$

$$+ \frac{(1 - 3!)}{6!}(x-1)^6 + \cdots$$

Direct substitution of z into the given nonhomogeneous equation shows that z is a solution of the equation. It can be verified that $R = 1$ is the radius of convergence of $z(x)$.

We conclude this section by considering an example in which it is required to solve an initial-value problem. We shall illustrate both the power series method and the Taylor series method.

12 Example

Consider the initial-value problem

$$x\, D^2y + Dy + 2y = \varnothing; \qquad y(-1) = 2, \quad Dy(-1) = 4$$

on $J = (-\infty, \infty)$. The initial point $x = -1$ is an ordinary point, and so we assume that the solution has a power series expansion of the form

(A) $$y(x) = \sum_{n=0}^{\infty} a_n(x + 1)^n$$

From the initial conditions, it follows that $a_0 = 2$ and $a_1 = 4$. A straightforward application of the power series method yields

$$\sum_{k=0}^{\infty} [(k + 2)(k + 1)a_{k+2} - (k + 1)^2 a_{k+1} - 2a_k](x + 1)^k = \varnothing$$

Since $a_0 = 2$ and $a_1 = 4$, we must solve the initial-value problem

(B) $$(k + 2)(k + 1)a_{k+2} - (k + 1)^2 a_{k+1} - 2a_k = \varnothing; \qquad a_0 = 2, \quad a_1 = 4$$

From a step-by-step calculation of the solution of (B), we get

$$a_2 = 4, \qquad a_3 = 4, \qquad a_4 = \tfrac{11}{3}, \qquad a_5 = \tfrac{10}{3}$$

Thus the first six terms of the power series solution of the given initial-value problem are:

$$y(x) = 2 + 4(x + 1) + 4(x + 1)^2 + 4(x + 1)^3$$
$$+ \tfrac{11}{3}(x + 1)^4 + \tfrac{10}{3}(x + 1)^5 + \cdots$$

We now solve the initial-value problem by using the Taylor series method. The Taylor series expansion of the solution is:

(C) $$y(x) = \sum_{n=0}^{\infty} \frac{D^n y(-1)}{n!}(x + 1)^n$$

From the initial conditions, $y(-1) = 2$, $Dy(-1) = 4$, and from the differential equation

(D) $$x\, D^2y(x) = -Dy(x) - 2y(x)$$

so that

$$D^2y(-1) = 8$$

Next, we differentiate (D) to get

(E) $$x\, D^3y(x) + D^2y(x) = -D^2y(x) - 2\, Dy(x)$$

and this is used to obtain $D^3y(-1) = 24$. By continuing in this manner, we can calculate as many coefficients as desired. We leave it as an exercise to show that

$$y(x) = 2 + 4(x + 1) + \frac{8}{2!}(x + 1)^2 + \frac{24}{3!}(x + 1)^3$$

(F)
$$+ \frac{88}{4!}(x + 1)^4 + \frac{400}{5!}(x + 1)^5 + \cdots$$

$$= 2 + 4(x + 1) + 4(x + 1)^2 + 4(x + 1)^3$$

$$+ \tfrac{11}{3}(x + 1)^4 + \tfrac{10}{3}(x + 1)^5 + \cdots$$

A concluding remark motivated by Example 12 is appropriate. Note that we determined a power series solution of the initial-value problem in powers of $(x + 1)$. This was an obvious choice since the initial conditions were specified at $x = -1$. However, because of notational considerations, it is sometimes more convenient to work with power series expansions about $x = 0$. By means of the simple change in independent variable $t = x - a$, it is always possible to transform a problem involving power series in powers of $(x - a)$ into a problem involving powers series in powers of t. In particular, if we let $t = x + 1$, then the initial-value problem of Example 12 is transformed into the initial-value problem

$$(t - 1)\frac{d^2y}{dt^2} + \frac{dy}{dt} + 2y = \varnothing; \qquad y(0) = 2, \quad \frac{dy}{dt}(0) = 4$$

Of course, this same observation also applies to Examples 7 and 11.

13 EXERCISES

In each of the following, verify that the indicated point $x = a$ is an ordinary point of the differential equation; determine the two-parameter family of solutions as a power series in powers of $(x - a)$; and give the interval of convergence of the power series.

1. $D^2y - y = e^x$; $x = 0$
2. $D^2y - 2Dy + 2y = x + 1$; $x = 0$
3. $x^2 D^2y - 2x Dy + 2y = x + 1$; $x = -1$
4. $D^2y - x Dy + 2y = 1/(1 - x)$; $x = 0$
5. $D^2y - (x - 1) Dy = 1n\ x$; $x = 1$
6. $(1 - x^2) D^2y - 4x Dy - 2y = -3(x + 1)$; $x = 0$

Find a power series solution of each of the following initial-value problems.

7. $D^2y + y = \cos x$; $y(0) = 0, Dy(0) = 1$
8. $D^2y - (x - 1) Dy = 2x^2 - 4x$; $y(1) = 0, Dy(1) = 2$
9. $D^2y + 2x Dy - 8y = \varnothing$; $y(0) = 1, Dy(0) = 0$
10. $(1 - x^2) D^2y - 3x Dy - y = 4(x + 1)$; $y(0) = 1, Dy(0) = 1$

Use the Taylor series method described in Section 30 to find the first five terms in the power series solution of each of the following initial-value problems.

11. $D^2 y + 2x\, Dy + y = e^{-x}$; $y(0) = 0$, $Dy(0) = 1$

12. $D^2 y - [x - (\pi/2)]^2 y = \cos x$; $y(\pi/2) = 1$, $Dy(\pi/2) = 1$

13. $(1 - x^2)\, D^2 y - 4x\, Dy - 2y = \varnothing$; $y(0) = 0$, $Dy(0) = 0$

In the following two problems verify that the indicated point $x = a$ is an ordinary point of the differential equation. Determine the two-parameter family of solutions as a power series in powers of $(x - a)$ by using the indicated change of variable $t = x - a$ and finding the two-parameter family of solutions of the transformed equation.

14. $D^2 y - (x^2 + 2x + 1)\, Dy - (2x + 2)y = \varnothing$; $a = -1$, $t = x + 1$

15. $(2x - x^2)\, D^2 y - 5(x - 1)\, Dy - 3y = \varnothing$; $a = 1$, $t = x - 1$

16. Solve the initial-value problem using the method specified in Problems 14 and 15:

$$(x^2 - 2x + 2)\, D^2 y + (2x - 2)\, Dy = x^2 - 2x + 1; \qquad y(1) = Dy(1) = 0$$

where $a = 1$ and $t = x - 1$.

17. Supply the missing details in each of the examples of this section.

Section 32 Power Series Solutions at Singular Points

In this section, we develop techniques for obtaining power series solutions of

$$(1) \qquad D^2 y + p_1(x)\, Dy + p_2(x)y = \varnothing$$

in powers of $(x - a)$, where $x = a$ is a singular point of the differential equation. Recall that $x = a$ is a singular point of (1) if either (or both) of the coefficient functions p_1 and p_2 fails to be analytic at $x = a$. In this case, Theorem 5, Section 31, and the power series methods of the previous section cannot be applied, and, consequently, additional techniques must be developed.

As a means of motivating the classification of singular points and the power series methods to be developed in this section, consider the second-order Cauchy equation

$$(2) \qquad x^2\, D^2 y - 2x\, Dy + \tfrac{9}{4} y = \varnothing$$

A solution technique for Cauchy equations was introduced in Chapter 3, and Cauchy equations were studied in detail in Section 20. It is easy to see that $x = 0$ is a singular point of the equation. If we assume a series solution of the form

$$(3) \qquad y(x) = \sum_{n=0}^{\infty} a_n x^n$$

and substitute into (2) using the techniques of the previous section, we obtain

$$x^2 \sum_{n=2}^{\infty} n(n - 1)a_n x^{n-2} - 2x \sum_{n=1}^{\infty} na_n x^{n-1} + \tfrac{9}{4} \sum_{n=0}^{\infty} a_n x^n = \varnothing$$

This equation can be simplified to

$$\tfrac{9}{4}a_0 + \tfrac{1}{4}a_1 x + \sum_{n=2}^{\infty} [n^2 - 3n + \tfrac{9}{4}]a_n x^n = \varnothing$$

Therefore, $a_0 = 0$, $a_1 = 0$, and, since $n^2 - 3n + \tfrac{9}{4} \neq 0$ for $n = 2, 3, \ldots$, it follows that $a_n = 0$ for all n. Thus (2) does not have a power series solution of the form (3) (except for the zero function). It can be shown, however, that

$$y_1(x) = x^{3/2} \qquad \text{and} \qquad y_2(x) = x^{3/2} \ln(x)$$

are linearly independent solutions of (2). These functions are not analytic at 0, and they cannot be expressed as a power series in powers of x.

In general, the Cauchy equation

(4) $b_0 x^2\, D^2 y + b_1 x\, Dy + b_2 y = \varnothing, \qquad b_0 \neq 0$

and the form of its solutions (see Section 20) suggests a means of classifying the singular points of a differential equation, and this leads to a solution technique which is a straightforward modification of the power series method developed in the previous section.

5 DEFINITION

Consider the differential equation (1) and assume that at least one of the functions p_1 and p_2 is not analytic at $x = a$. If each of the functions

$$(x - a)p_1(x) \qquad \text{and} \qquad (x - a)^2 p_2(x)$$

is analytic at $x = a$, then a is called a *regular singular point* of (1). If either (or both) of these products fails to be analytic at $x = a$, then a is called an *irregular singular* point of (1).

---

6 Examples

   a. To see that the Cauchy equation has a regular singular point at $x = 0$, consider equation (4) in the form

$$D^2 y + \frac{b}{x} Dy + \frac{c}{x^2} y = \varnothing$$

   where b and c are the numbers b_1/b_0 and b_2/b_0, respectively. Since

$$xp_1(x) = \frac{xb}{x} = b$$

   and

$$x^2 p_2(x) = \frac{x^2 c}{x^2} = c$$

are constants, and, consequently, are analytic for all x, $x = 0$ is a regular singular point.

b. Consider the differential equation

$$x^2(x - 1)^2 D^2 y + 3(x - 1) Dy + (x^2 + x)y = \emptyset$$

By writing this equation in the form (1), we obtain

$$D^2 y + \frac{3}{x^2(x - 1)} Dy + \frac{x^2 + x}{x^2(x - 1)^2} y = \emptyset$$

Here,

$$p_1(x) = \frac{3}{x^2(x - 1)}$$

and

$$p_2(x) = \frac{x^2 + x}{x^2(x - 1)^2} = \frac{x + 1}{x(x - 1)^2}$$

It is clear that $x = 0$ and $x = 1$ are singular points of the equation. We consider these points separately. First, at $x = 0$, we form the products

$$x p_1(x) = \frac{3}{x(x - 1)}$$

and

$$x^2 p_2(x) = \frac{x(x + 1)}{(x - 1)^2}$$

Although the second product $x^2 p_2(x)$ is analytic at $x = 0$, the first product $x p_1(x)$ is *not* analytic at $x = 0$. Thus $x = 0$ is an *irregular* singular point of the equation.

Now consider the point $x = 1$. Forming the products

$$(x - 1) p_1(x) = \frac{3}{x^2}$$

and

$$(x - 1)^2 p_2(x) = \frac{x + 1}{x}$$

we find that each is analytic at $x = 1$. Thus $x = 1$ is a regular singular point of the equation.

A general treatment of irregular singular points requires mathematical techniques which are beyond the scope of this text. Consequently, our discussion in this section will be confined to a treatment of differential equations with regular singular points.

A Cauchy equation has a regular singular point at $x = 0$, and the form of the solutions suggests the possibility of obtaining a series solution of the form

$$x^r \sum_{n=0}^{\infty} a_n x^n$$

where r is a constant which may be either a real number or a complex number. The following theorem provides the basic information concerning series solutions at a regular singular point.

7 THEOREM

If $x = a$ is a regular singular point of the differential equation

$$D^2 y + p_1(x)\, Dy + p_2(x)y = \varnothing$$

then the equation has at least one nontrivial solution of the form

(8) $$(x - a)^r \sum_{n=0}^{\infty} a_n(x - a)^n$$

where r is a constant. This solution is valid on the interval $(a, a + R)$ for some $R > 0$. The solution can be extended to the interval $(a - R, a)$ by replacing $(x - a)^r$ by $|x - a|^r$.

If $x = a$ is a regular singular point of the differential equation (1), then the theorem assures us of at least one solution of the form (8). Note that we may assume $a_0 \neq 0$ in (8), for if $a_0 = 0$ and $a_1 \neq 0$, then

$$(x - a)^r \sum_{n=0}^{\infty} a_n(x - a)^n = (x - a)^r[a_1(x - a) + a_2(x - a)^2 + \cdots]$$

$$= (x - a)^{r+1}[a_1 + a_2(x - a) + \cdots]$$

$$= (x - a)^{r+1} \sum_{n=0}^{\infty} a_{n+1}(x - a)^n$$

which is an expression of the form (8) with the constant term in the power series portion of the expression being nonzero.

Suppose $x = a$ is a regular singular point of (1). The immediate question is this: How do we determine the number r and coefficients a_n in the series solution (8)? The procedure for doing this is an extension of the power series method of the previous section, and it is usually called the *method of Frobenius*. We shall illustrate the technique by considering some examples in detail.

9 Example

Consider the differential equation

$$2(x - 1)^2\, D^2 y - (x - 1)\, Dy + (2 - x)y = \varnothing$$

It is easy to verify that $x = 1$ is a regular singular point of the equation.

According to Theorem 7, there exists a solution of the form

(A) $$y(x) = (x - 1)^r \sum_{n=0}^{\infty} a_n(x - 1)^n, \qquad a_0 \neq 0$$

By writing (A) as

$$y(x) = \sum_{n=0}^{\infty} a_n(x - 1)^{n+r}$$

and differentiating twice, we obtain

$$Dy(x) = \sum_{n=0}^{\infty} (n + r)a_n(x - 1)^{n+r-1}$$

$$D^2 y(x) = \sum_{n=0}^{\infty} (n + r)(n + r - 1)a_n(x - 1)^{n+r-2}$$

Substituting these into the differential equation, we obtain

$$2(x - 1)^2 \sum_{n=0}^{\infty} (n + r)(n + r - 1)a_n(x - 1)^{n+r-2}$$

$$- (x - 1) \sum_{n=0}^{\infty} (n + r)a_n(x - 1)^{n+r-1}$$

$$+ (2 - x) \sum_{n=0}^{\infty} a_n(x - 1)^{n+r} = \emptyset$$

Since $2 - x = 1 - (x - 1)$, this equation can be written

(B) $$\sum_{n=0}^{\infty} [2(n + r)(n + r - 1) - (n + r) + 1]a_n(x - 1)^{n+r}$$

$$- \sum_{n=0}^{\infty} a_n(x - 1)^{n+r+1} = \emptyset$$

With the change of index $k = n + 1$ in the second series of (B), and by letting $k = n$ in the first series, we get

$$\sum_{k=0}^{\infty} [2(k + r)(k + r - 1) - (k + r) + 1]a_k(x - 1)^{k+r}$$

$$- \sum_{k=1}^{\infty} a_{k-1}(x - 1)^{k+r} = \emptyset$$

or

(C) $$[2r(r - 1) - r + 1]a_0(x - 1)^r + \sum_{k=1}^{\infty} \{[2(k + r)(k + r - 1)$$

$$- (k + r) + 1]a_k - a_{k-1}\}(x - 1)^{k+r} = \emptyset$$

Since the left side of (C) can be written as a product of $(x - 1)^r$ and a power series, and since $a_0 \neq 0$, we must have

(D) $2r(r - 1) - r + 1 = 0$

This equation determines the values of r for which the differential equation has a solution of the form (A). Equation (D), a quadratic in r, is called the *indicial equation* of the given differential equation, and its roots are called the *indicial roots*, or *exponents*, of the differential equation. From (D) we obtain the indicial roots $r_1 = 1$ and $r_2 = \frac{1}{2}$. From (C) we must also have

(E) $[2(k + r)(k + r - 1) - (k + r) + 1]a_k - a_{k-1} = 0$

$k = 1, 2, \ldots$, which can be used to calculate the coefficients a_k in terms of a_0. In particular, for $r = 1$, we obtain the recursion formula

$[2k(k + 1) - (k + 1) + 1]a_k - a_{k-1} = 0$

$k = 1, 2, \ldots$, or

$k(2k + 1)a_k - a_{k-1} = \emptyset$

This is a first-order linear homogeneous difference equation which, in the standard form (3) of Section 23, is:

(F) $a_{k+1} - \dfrac{1}{(k + 1)(2k + 3)} a_k = \emptyset$

We can use (F) for a step-by-step calculation of the coefficients $a_1, a_2, \ldots$, in terms of a_0. We get

$$a_1 = \frac{a_0}{3}, \qquad a_2 = \frac{a_0}{30}, \qquad a_3 = \frac{a_0}{630}, \ldots$$

However, in Section 23 we derived the explicit representation (5) of the one-parameter family of solutions of a first-order linear homogeneous difference equation, and this result can be applied here. The one-parameter family of solutions of (F) is:

$$a_k = (-1)^k \prod_{i=0}^{k-1} \frac{(-1)a_0}{(i + 1)(2i + 3)} = \frac{2^k}{(2k + 1)!} a_0$$

Substituting $r = 1$ and these coefficients into (A), we obtain the series solution

$$y(x) = a_0(x - 1)\left[1 + \frac{(x - 1)}{3} + \frac{(x - 1)^2}{30} + \frac{(x - 1)^3}{630} + \cdots\right]$$

$$= a_0(x - 1) \sum_{n=0}^{\infty} \frac{2^n}{(2n + 1)!}(x - 1)^n$$

By letting $a_0 = 1$, we get the particular solution

$$y_1(x) = (x - 1) \sum_{n=0}^{\infty} \frac{2^n}{(2n + 1)!} (x - 1)^n$$

Theorem 7 implies that this series will converge on an interval of the form $(1, 1 + R)$ for some $R > 0$. From the form of the solution and the ratio test, however, we find that this series converges for all real numbers.

We now go through the same procedure with $r = \frac{1}{2}$. From (E), with $r = \frac{1}{2}$, we have

$$[2(k + \tfrac{1}{2})(k - \tfrac{1}{2}) - (k + \tfrac{1}{2}) + 1]a_k - a_{k-1} = 0$$

$k = 1, 2, \ldots$, which in standard form is:

$$a_{k+1} - \frac{1}{(k + 1)(2k + 1)} a_k = \varnothing$$

From (5) of Section 23, the one-parameter family of solutions is:

$$a_k = \frac{2^k}{(2k)!} a_0$$

Substituting $r = \frac{1}{2}$ and these coefficients into (A), we obtain the series solutions

$$y(x) = a_0(x - 1)^{1/2} \sum_{n=0}^{\infty} \frac{2^n}{(2n)!} (x - 1)^n$$

If we let $a_0 = 1$, we get the particular solution

$$y_2(x) = (x - 1)^{1/2} \sum_{n=0}^{\infty} \frac{2^n}{(2n)!} (x - 1)^n$$

Note that y_2 is not real-valued when $x < 1$, and that y_2 is not differentiable at $x = 1$. It is easy to verify using the ratio test that the power series

$$\sum_{n=0}^{\infty} \frac{2^n}{(2n)!} (x - 1)^n$$

converges for all real numbers. Thus y_2 is analytic on $(1, \infty)$. To obtain a solution which is valid for values of $x < 1$, we replace $(x - 1)^{1/2}$ by $|x - 1|^{1/2}$. Finally, it is easy to show that y_1 and y_2 are linearly independent on $(1, \infty)$. Thus

$$y(x) = c_1 y_1(x) + c_2 y_2(x)$$

represents the two-parameter family of solutions.

Example 9 illustrates a straightforward application of the method of Frobenius to obtain a series solution of a differential equation at a regular singular point. We now give a more formal discussion of the method and apply the results in an example. Suppose $x = a$ is a regular singular point of the differential equation

$$D^2y + p_1(x)\, Dy + p_2(x)y = \varnothing$$

If we multiply the equation by $(x - a)^2$, we get

$$(x - a)^2\, D^2y + (x - a)^2 p_1(x)\, Dy + (x - a)^2 p_2(x)y = \varnothing$$

which can be written

(10) $(x - a)^2\, D^2y + (x - a)q_1(x)\, Dy + q_2(x)y = \varnothing$

where $q_1(x) = (x - a)p_1(x)$ and $q_2(x) = (x - a)^2 p_2(x)$ are analytic on some interval J containing $x = a$. According to Theorem 7, the equation has a series solution of the form

(11) $y(x) = (x - a)^r \displaystyle\sum_{n=0}^{\infty} a_n(x - a)^n, \qquad a_0 \neq 0$

Let $z(x - a)$ be the power series part of $y(x)$; that is,

(12) $z(x - a) = \displaystyle\sum_{n=0}^{\infty} a_n(x - a)^n$

Then

$$y(x) = (x - a)^r z(x - a)$$

$$Dy(x) = r(x - a)^{r-1}z(x - a) + (x - a)^r\, Dz(x - a)$$

$$D^2y(x) = r(r - 1)(x - a)^{r-2}z(x - a) + 2r(x - a)^{r-1}\, Dz(x - a)$$
$$+ (x - a)^r\, D^2z(x - a)$$

Substituting these equations into (10) and combining common terms, we get

$$(x - a)^{r+2}\, D^2z + (x - a)^{r+1}[2r + q_1(x)]\, Dz$$
$$+ (x - a)^r[r(r - 1) + rq_1(x) + q_2(x)]z = \varnothing$$

or

(13) $(x - a)^2\, D^2z + (x - a)[2r + q_1(x)]\, Dz$
$$+ [r(r - 1) + rq_1(x) + q_2(x)]z = \varnothing$$

Notice that if we let $x = a$ in (13), we get

$$[r(r - 1) + rq_1(a) + q_2(a)]a_0 = 0$$

Since $a_0 \neq 0$, it follows that

(14) $r(r - 1) + rq_1(a) + q_2(a) = 0$

and this is the *indicial equation*. The roots, r_1 and r_2, of (14) are the *indicial roots*.

Equation (14) is quadratic in r. According to Theorem 7, at least one of the two indicial roots can be used in equation (13) to determine the coefficients of the power series (12). This in turn provides a solution of the form (11). In the next examples, we use this formal approach to obtain a power series solution at a regular singular point. In contrast to Example 9, these examples also illustrate that the method of Frobenius does not always produce two linearly independent solutions.

15 Examples

a. Consider the differential equation

$$x^2 D^2 y + x(3 - x) Dy + (1 - 2x)y = \varnothing$$

The point $x = 0$ is a regular singular point. Thus the equation has a series solution of the form

(A) $\qquad y(x) = x^r \sum_{n=0}^{\infty} a_n x^n = x^r z(x), \qquad a_0 \neq 0$

Differentiating (A) twice and substituting into the differential equation, we obtain

(B) $\qquad x^2 D^2 z + x[2r + 3 - x] Dz$
$$+ [r(r - 1) + r(3 - x) + (1 - 2x)]z = \varnothing$$

as the analogue of (13). Since $q_1(x) = 3 - x$ and $q_2(x) = 1 - 2x$, it follows from (14) that the indicial equation is:

$$r(r - 1) + 3r + 1 = (r + 1)^2 = 0$$

Thus $r_1 = r_2 = -1$ are the indicial roots. Set $r_1 = -1$ in (B) to obtain the equation

$$x^2 \sum_{n=2}^{\infty} n(n - 1)a_n x^{n-2} + x(1 - x) \sum_{n=1}^{\infty} na_n x^{n-1} - x \sum_{n=0}^{\infty} a_n x^n = \varnothing$$

By making the appropriate changes of index, this equation is transformed into

$$\sum_{k=1}^{\infty} [k^2 a_k - ka_{k-1}]x^k = \varnothing$$

which yields the recursion formula

$$ka_k - a_{k-1} = 0$$

$k = 1, 2, \ldots$. This is the same as the first-order difference equation

$$a_{k+1} - \frac{1}{k + 1} a_k = \varnothing$$

whose solution is $a_k = (1/k!)$. Thus for $r_1 = -1$, we get the series solution

$$y(x) = a_0 x^{-1} \sum_{n=0}^{\infty} \frac{1}{n!} x^n$$

and for $a_0 = 1$, we have the particular solution

$$y_1(x) = x^{-1} \sum_{n=0}^{\infty} \frac{1}{n!} x^n = x^{-1} e^x$$

Finally, since the roots of the indicial equation are equal, it is clear that the method of Frobenius produces only one solution of the equation. Compare this with Example 9 where the two distinct roots yielded two linearly independent solutions.

b. Consider the differential equation

$$x^2 D^2 y - x \, Dy + 8(x^2 - 1)y = \varnothing$$

The point $x = 0$ is a regular singular point, and so the equation has a series solution of the form

(C) $$y(x) = x^r \sum_{n=0}^{\infty} a_n x^n, \qquad a_0 \neq 0$$

Equations (13) and (14) in this case are:

(D) $$x^2 \sum_{n=2}^{\infty} n(n-1)a_n x^{n-2} + x(2r-1) \sum_{n=1}^{\infty} na_n x^{n-1}$$

$$+ [r(r-1) - r + 8(x^2 - 1)] \sum_{n=0}^{\infty} a_n x^n = \varnothing$$

and

$$r(r-1) - r - 8 = r^2 - 2r - 8 = 0$$

respectively. Thus $r_1 = 4$ and $r_2 = -2$ are the indicial roots. By setting $r = 4$ in (D) and simplifying, we obtain

$$\sum_{n=2}^{\infty} [(n^2 + 6n)a_n + 8a_{n-2}]x^n + 7a_1 x = \varnothing$$

Therefore, we have $a_1 = 0$ and the recursion formula

$$(n^2 + 6n)a_n + 8a_{n-2} = 0$$

$n = 2, 3, \ldots$. A step-by-step calculation of these coefficients yields

$$a_1 = a_3 = a_5 = \cdots = 0$$

and

$$a_2 = (-\tfrac{1}{2})a_0, \qquad a_4 = (\tfrac{1}{10})a_0, \qquad a_6 = (-\tfrac{1}{90})a_0, \ldots$$

Thus with $r = 4$ and $a_0 = 1$ in (C), we get the series solution

(E) $$y_1(x) = x^4 \left[1 - \frac{x^2}{2} + \frac{x^4}{10} - \frac{x^6}{90} + \cdots \right]$$

corresponding to the larger indicial root.

Now consider the smaller indicial root, $r_2 = -2$. If we set $r = -2$ in (D) and simplify, we obtain

$$\sum_{n=2}^{\infty} [(n^2 - 6n)a_n + 8a_{n-2}]x^n - 5a_1 x = \varnothing$$

Therefore, we have $a_1 = 0$ and the recursion formula

$$(n^2 - 6n)a_n + 8a_{n-2} = 0$$

$n = 2, 3, \ldots$, from which it follows that $a_1 = a_3 = a_5 = \cdots = 0$. If we begin a step-by-step calculation of the even-ordered coefficients, we find that

$$-8a_2 + 8a_0 = 0 \qquad \text{or} \qquad a_2 = a_0$$

$$-8a_4 + 8a_2 = 0 \qquad \text{or} \qquad a_4 = a_2 = a_0$$

$$0a_6 + 8a_4 = 0$$

From the last equation, $a_4 = 0$, which implies that $a_0 = 0$ and contradicts the fact that $a_0 \neq 0$. Thus there does not exist a series solution of the form (C) corresponding to the smaller indicial root.

16 EXERCISES

1. Determine the ordinary points and the singular points of each of the following differential equations. In the case of a singular point, indicate whether it is regular or irregular.

   a. $(x - 1)^2 D^2 y + (x - 1) Dy + x^2 y = \varnothing$
   b. $D^2 y - 2 Dy + y = \varnothing$
   c. $(x - 1)^2 D^2 y + Dy + x^2 y = \varnothing$
   d. $(x - 1)^2 D^2 y + (x - 1) Dy + [x^2/(x - 1)]y = \varnothing$
   e. $x^2(x - 1) D^2 y + (x - 1) Dy + 2xy = \varnothing$
   f. $x^2(x^2 + 1) D^2 y + (x + 1)x Dy + (x - 1)y = \varnothing$
   g. $x^3(x^2 + 1) D^2 y + x(x + 1) Dy + (x - 1)y = \varnothing$
   h. $(1 - x^2) D^2 y + 4x Dy - 2y = \varnothing$
   i. $x^4(x + 1)^3 D^2 y + 4x^2 y = \varnothing$
   j. $(x^4 - 1) D^2 y + (x - 1) Dy + xy = \varnothing$
   k. $e^x D^2 y + x \sin(x) Dy + (x - 1)y = \varnothing$
   l. $D^2 y + [1/(x - 1)] Dy + (x - 1)^2 y = \varnothing$
   m. $x^2(2x + 1)^3 D^2 y + x(2x + 1) Dy + (2x + 1)y = \varnothing$

2. Consider the differential equation

(A) $\qquad D^2y + p(x)\, Dy + q(x)y = \varnothing$

on $[a, \infty)$.

a. Show that the change in independent variable defined by $t = 1/x$ transforms (A) into

(B) $\qquad t^4 \dfrac{d^2y}{dt^2} + \left[2t^3 - t^2 p\left(\dfrac{1}{t}\right) \right] \dfrac{dy}{dt} + q\left(\dfrac{1}{t}\right) y = \varnothing$

b. In applications, it is often necessary to investigate $\lim\limits_{x\to\infty} y(x)$, where y is a solution of (A). The *point at infinity* is an ordinary point if $t = 0$ is an ordinary point of (B). Otherwise, the point at infinity is a singular point of (A). If the point at infinity is a singular point of (A), then it is called regular if $t = 0$ is a regular singular point of (B), otherwise it is called an irregular singular point. Show that the point at infinity is a regular singular point if at least one of the functions

$$2t^{-1} - t^{-2} p\left(\frac{1}{t}\right) \qquad \text{and} \qquad t^{-4} q\left(\frac{1}{t}\right)$$

fails to be analytic at $t = 0$, and each of the functions

$$2 - t^{-1} p\left(\frac{1}{t}\right) \qquad \text{and} \qquad t^{-2} q\left(\frac{1}{t}\right)$$

is analytic at $t = 0$.

3. Determine whether the point at infinity is an ordinary point, a regular singular point, or an irregular singular point for each of the following differential equations.
   a. $(1 - x^2)\, D^2y - 2x\, Dy + by = \varnothing$ (Legendre's equation), b constant
   b. $x^2\, D^2y + x\, Dy + (x^2 - b^2)y = \varnothing$ (Bessel's equation), b constant
   c. $x^4\, D^2y + x(x^2 + 1)\, Dy + (x + 1)y = \varnothing$

In each of the following, verify that the given point $x = a$ is a regular singular point; determine the indicial equation and the two indicial roots; determine a series solution of the form $y(x) = x^r \sum_{n=0}^{\infty} a^n(x - a)^n$; indicate whether there is a second (independent) solution of this form, and if there is, find it; specify the interval of convergence in each case.

4. $2x^2\, D^2y - x\, Dy + (x^2 + 1)y = \varnothing; \ x = 0$
5. $3(x + 1)\, D^2y + 2\, Dy + (x + 1)y = \varnothing; \ x = -1$
6. $x\, D^2y + (x - 1)\, Dy - y = \varnothing; \ x = 0$
7. $x^2\, D^2y - x\, Dy + (1 - x)y = \varnothing; \ x = 0$
8. $3x^2\, D^2y + 8x\, Dy + (x^2 - 2)y = \varnothing; \ x = 0$
9. $4(x - 2)^2\, D^2y - (x - 2)\, Dy + (x - 1)y = \varnothing; \ x = 2$
10. $(x + 1)^2\, D^2y - (x + 1)\, Dy + 8(x^2 + 2x)y = \varnothing; \ x = -1$
11. $x^2\, D^2y - x\, Dy + (x^2 - \frac{1}{4})y = \varnothing; \ x = 0$
12. $(x - 1)^2\, D^2y + 5(x - 1)\, Dy + (x + 3)y = \varnothing; \ x = 1$
13. $4x^2\, D^2y + (2x^2 - 5x)\, Dy + (x^2 + 2)y = \varnothing; \ x = 0$
14. $x^2\, D^2y + x\, Dy + (x - 1)y = \varnothing; \ x = 0$
15. $2x^2\, D^2y + x(x + 1)\, Dy + (x - 3)y = \varnothing; \ x = 0$

16. Show that the point at infinity is a regular singular point of the equation

$$x^3 D^2 y + 2x^2 Dy + y = \varnothing$$

Determine a series solution of this equation which is valid in a neighborhood of ∞, that is, valid for large x, by transforming the equation using the method suggested in Problem 2 and finding a solution of the transformed equation in a neighborhood of 0.

Examples 9 and 15 illustrate that the method of Frobenius may or may not produce two linearly independent solutions of the form

(17) $$y(x) = (x - a)^r \sum_{n=0}^{\infty} a_n(x - a)^n, \qquad a_0 \neq 0$$

The next theorem gives a complete characterization of a solution basis for a second-order differential equation at a regular singular point.

18 THEOREM

Assume that $x = a$ is a regular singular point of the differential equation (1). Let r_1 and r_2 be the two roots of the corresponding quadratic indicial equation. Then (1) has two linearly independent solutions y_1 and y_2, valid on $(a, a + R)$, for some $R > 0$, whose form is determined by r_1 and r_2 as follows.

(i)   If $r_1 - r_2$ is not an integer, then

$$y_1(x) = (x - a)^{r_1} \sum_{n=0}^{\infty} a_n(x - a)^n, \qquad a_0 = 1$$

$$y_2(x) = (x - a)^{r_2} \sum_{n=0}^{\infty} b_n(x - a)^n, \qquad b_0 = 1$$

(ii) If $r_1 = r_2 = r$, then

$$y_1(x) = (x - a)^r \sum_{n=0}^{\infty} a_n(x - a)^n, \qquad a_0 = 1$$

and

$$y_2(x) = (x - a)^{r+1} \sum_{n=0}^{\infty} b_n(x - a)^n + y_1(x) \ln(x - a), \qquad b_0 \neq 0$$

(iii) If $r_1 - r_2$ is an integer and $r_1 > r_2$, then

$$y_1(x) = (x - a)^{r_1} \sum_{n=0}^{\infty} a_n(x - a)^n, \qquad a_0 = 1$$

and

$$y_2(x) = (x - a)^{r_2} \sum_{n=0}^{\infty} b_n(x - a)^n + Cy_1(x) \ln(x - a)$$

where C is a constant.

Some remarks concerning the theorem are appropriate. First, if the indicial roots r_1 and r_2 are complex conjugates, then case (i) of the theorem applies. Although the corresponding series solutions will not be real-valued functions in this situation, real-valued solutions can be obtained in the manner indicated in Section 20. Next, the constant C in case (iii) may or may not be zero. This will be illustrated in Examples 20 and 21. Finally, the solutions can be extended to the interval $(a - R, a)$ by replacing $(x - a)^{r_i}$ by $|x - a|^{r_i}$ in each case.

Example 9 illustrates the first case of Theorem 18. In Example 15a, the roots of the indicial equation were equal, and so it can be used to illustrate the second case of the theorem.

19 Example

(Example 15a continued.) The point $x = 0$ is a regular singular point of the differential equation

$$x^2 D^2 y + (3x - x^2) Dy + (1 - 2x)y = \emptyset$$

The method of Frobenius leads to the indicial equation

$$(r + 1)^2 = 0$$

and to the series solution

$$y_1(x) = x^{-1} \sum_{n=0}^{\infty} \frac{1}{n!} x^n = x^{-1} e^x$$

To find a second solution y_2 of the equation which is independent of y_1, we can use the deflation rule for second-order linear homogeneous equations discussed in Section 12. In particular, according to Theorem 5, Section 12,

(A) $$y_2(x) = y_1(x) \int_1^x \frac{\exp\left\{ -\int_1^t \frac{3s - s^2}{s^2}\, ds \right\}}{y_1{}^2(t)}\, dt$$

where $x = 1$ is chosen as a convenient lower limit of integration. The right side of (A) simplifies to

$$y_2(x) = y_1(x) \int_1^x \frac{e^{t-1} t^{-3}\, dt}{y_1{}^2(t)}$$

At this point in a typical problem, we would have to take the series expansion y_1, square it, and divide the result into the series expansion of the numerator. The resulting series would then be integrated term by term and multiplied by the series y_1. In this particular example, however, we can take advantage of the closed-form representation $y_1(x) = x^{-1}e^x$ to evaluate the integral. We have

$$y_2(x) = x^{-1}e^x \int_1^x \frac{e^{t-1}}{t^3 t^{-2} e^{2t}} \, dt = e^{-1} x^{-1} e^x \int_1^x \frac{e^{-t}}{t} \, dt$$

$$= e^{-1} x^{-1} e^x \int_1^x \frac{1}{t} \left[1 - t + \frac{t^2}{2!} - \frac{t^3}{3!} + \cdots \right] dt$$

$$= e^{-1} x^{-1} e^x \left[\ln(t) - t + \frac{t^2}{2 \cdot 2!} - \frac{t^3}{3 \cdot 3!} + \cdots \right] \Big|_1^x$$

Since evaluation at the lower limit will merely produce a constant multiple of $y_1(x)$, we can ignore it and set

$$y_2(x) = x^{-1} e^x \ln(x) + x^{-1} e^x \sum_{n=0}^{\infty} \frac{(-1)^{n+1}}{(n+1)(n+1)!} x^{n+1}$$

$$= x^{-1} e^x \ln(x) + x^{-1} \left(\sum_{n=0}^{\infty} \frac{1}{n!} x^n \right) \left(\sum_{n=0}^{\infty} \frac{(-1)^{n+1}}{(n+1)(n+1)!} x^{n+1} \right)$$

$$= x^{-1} e^x \ln(x) + x^{-1} \left[-x - \tfrac{3}{4} x^2 - \tfrac{11}{36} x^3 - \tfrac{25}{288} x^4 - \cdots \right]$$

The solution y_2 has the form specified in case (ii) of Theorem 18.

The next examples illustrate the third case of Theorem 18.

20 Examples

a. Consider the differential equation

$$x^2 D^2 y + x(1 - x) Dy - y = \varnothing$$

The point $x = 0$ is a regular singular point. From the method of Frobenius, we assume a series solutions of the form

(A) $$y(x) = x^r \sum_{n=0}^{\infty} a_n x^n = x^r z(x), \qquad a_0 \neq 0$$

Differentiating (A) twice, substituting into the differential equation, and using the manipulations illustrated in either Example 9 or Example 15, we obtain the equation

(B) $$\sum_{n=1}^{\infty} \{[(n+r)^2 - 1]a_n - (n+r-1)a_{n-1}\} x^n + (r^2 - 1)a_0 = \varnothing$$

Thus the indicial equation is $r^2 - 1 = 0$, and the indicial roots are $r_1 = 1$ and $r_2 = -1$. We let $r = 1$ in (B) to get the recursion formula

$$(n^2 + 2n)a_n - na_{n-1} = 0$$

$n = 1, 2, \ldots$, a first-order difference equation which, in standard form, is:

$$a_{k+1} + \frac{-1}{k+3} a_k = \varnothing$$

From Section 23, equation (5), the one-parameter family of solutions of this difference equation is:

$$a_k = \frac{1}{(k+2)!} a_0$$

We let $r = 1$, $a_0 = 1$, and $a_k = 1/(k+2)!$ in (A) to obtain the series solution

$$y_1(x) = x \sum_{n=0}^{\infty} \frac{1}{(n+2)!} x^n$$

Now, since the difference of the two indicial roots $r_1 - r_2 = 2$, we are in case (iii) of Theorem 18. As illustrated by Example 19, the deflation rule can be used to determine a second solution y_2 which is independent of y_1. However, as indicated in the remarks following the statement of Theorem 18, the constant C, the coefficient of $y_1(x) \ln(x - a)$, may or may not be zero. If it should turn out that $C = 0$, then y_2 also has the form (A), and this suggests the possibility that y_2 can be determined directly from (B) using the procedure which yielded y_1. Since the deflation rule is difficult to apply when series are involved, this possibility is worth investigating.

We put $r = r_2 = -1$ in (B) and obtain the recursion formula

$$(n^2 - 2n)a_n - (n - 2)a_{n-1} = 0$$

$n = 1, 2, \ldots$, a first-order difference equation which, in standard form, is:

$$a_{k+1} - \frac{1}{k+1} a_k = \varnothing$$

Again, from Section 23, the one-parameter family of solutions is:

$$a_k = \frac{1}{k!} a_0$$

We set $r = -1$, $a_0 = 1$, and $a_k = 1/k!$ in (A) to obtain the series solution

$$y_2(x) = x^{-1} \sum_{n=0}^{\infty} \frac{1}{n!} x^n = x^{-1} e^x$$

It is easy to verify that each of the solutions y_1 and y_2 converge for all x on $(0, \infty)$. In fact, $y_1(x)$ converges for all x on

$(-\infty, \infty)$. Finally, by calculating the Wronskian of y_1 and y_2, it can be shown that the functions are linearly independent. Thus

$$y(x) = c_1 x \sum_{n=0}^{\infty} \frac{1}{(n+2)!} x^n + c_2 x^{-1} e^x$$

is the two-parameter family of solutions on $(0, \infty)$.

b. The differential equation

$$(x+1)^2 D^2 y + (3x^2 + 8x + 5) Dy - 2y = \varnothing$$

can be written equivalently as

(C) $(x+1)^2 D^2 y + (x+1)[2 + 3(x+1)] Dy - 2y = \varnothing$

and it is easy to see that $x = -1$ is a regular singular point. Thus the equation has a series solution of the form

(D) $$y(x) = (x+1)^r \sum_{n=0}^{\infty} a_n (x+1)^n, \qquad a_0 \neq 0$$

Substitution of this series into the differential equation (C), followed by simplification, yields

(E) $$\sum_{n=1}^{\infty} \{[(n+r)^2 + n + r - 2]a_n + [3(n+r) - 3]a_{n-1}\}(x+1)^n$$
$$+ (r^2 + r - 2)a_0 = \varnothing$$

Thus the indicial equation is $r^2 + r - 2 = 0$, and the indicial roots are $r_1 = 1$ and $r_2 = -2$. By letting $r = 1$ in (E), we obtain the recursion formula

$$n(n+3)a_n + 3na_n = 0$$

$n = 1, 2, \ldots$, which, in the form of a first-order difference equation, is:

(F) $$a_{k+1} + \frac{3}{k+4} a_k = \varnothing$$

The one-parameter family of solutions of this difference equation is:

$$a_k = \frac{(-1)^k 2 \cdot 3^{k+1}}{(k+3)!} a_0$$

Thus by using these coefficients in (D), together with $r = 1$ and $a_0 = 1$, we get the series solution

(G) $$y_1(x) = (x+1) \sum_{n=0}^{\infty} \frac{(-1)^n 2 \cdot 3^{n+1}}{(n+3)!} (x+1)^n$$

Since the indicial roots differ by an integer, we are again in case (iii) of Theorem 18. As suggested by part (a) of this example, it is worthwhile to determine whether the smaller indicial root also corresponds to a series solution of the form (D). If we let $r = -2$ in (E), we obtain the recursion formula

(H) $n(n - 3)a_n + 3(n - 3)a_{n-1} = 0$

$n = 1, 2, \ldots$. We cannot divide (H) by $(n - 3)$ since such a division would amount to dividing by 0 when $n = 3$. From (H), it follows that

$$-2a_1 - 6a_0 = 0$$

or

$$a_1 = -3a_0$$

and

$$-2a_2 - 3a_1 = 0$$

or

$$a_2 = -\tfrac{3}{2}a_1 = \tfrac{9}{2}a_0$$

When $n = 3$ in (H), we have $0a_3 + 0a_2 = 0$. Since this equation holds regardless of the value assigned to a_3, we conclude that a_3 is independent of a_0. We shall treat a_3 as an arbitrary constant (parameter) and continue the calculation of the coefficients. For $n > 3$, (H) can be written

$$na_n + 3a_{n-1} = 0$$

$n = 4, 5, \ldots$. This is a first-order difference equation which can be written

$$(k + 4)b_{k+1} + 3b_k = 0$$

$k = 0, 1, 2, \ldots$, by letting $k = n - 4$ and $b_k = a_{k+3}$. Since this equation is the same as (F), it follows that

$$b_k = \frac{(-1)^k 2 \cdot 3^{k+1}}{(k + 3)!} b_0$$

or

$$a_{k+3} = \frac{(-1)^k 2 \cdot 3^{k+1}}{(k + 3)!} a_3$$

$k = 0, 1, 2, \ldots$.
We now have all of the coefficients a_n in the series (D) in terms of a_0 and a_3. These coefficients, together with $r = -2$ in (D), yield the series solutions

$$y(x) = (x + 1)^{-2} \left(a_0[1 - 3(x + 1) + \tfrac{9}{2}(x + 1)^2] \right.$$
$$\left. + a_3[(x + 1)^3 - \tfrac{3}{4}(x + 1)^4 + \tfrac{9}{20}(x + 1)^5 - \cdots] \right)$$

which can be written as

(I) $\qquad y(x) = a_0[(x + 1)^{-2} - 3(x + 1)^{-1} + \frac{9}{2}]$

$$+ a_3(x + 1) \sum_{n=0}^{\infty} \frac{(-1)^n 2 \cdot 3^{n+1}}{(n + 3)!} (x + 1)^n$$

If we let $a_0 = 1$ and $a_3 = 0$ in (I), we get the solution

$$y_2(x) = (x + 1)^{-2} - 3(x + 1)^{-1} + \frac{9}{2}$$

The solution y_1 given in (G) results from (I) by letting $a_0 = 0$ and $a_3 = 1$. It is easy to show that y_1 and y_2 are linearly independent. Thus (I) represents the two-parameter family of solutions of the differential equation.

The practical conclusion which can be drawn from the two examples in Examples 20 is that when the indicial roots differ by an integer, it is worthwhile working with the smaller indicial root first.

The final example of this section uses the equation in Example 15b to illustrate the case where the constant C in part (iii) of Theorem 18 is nonzero.

21 Example

(Example 15b continued.) As determined in Example 15b, the differential equation

$$x^2 D^2 y - x Dy + 8(x^2 - 1)y = \varnothing$$

does not have a series solution of the form

$$y(x) = x^{-2} \sum_{n=0}^{\infty} b_n x^n, \qquad a_0 \neq 0$$

corresponding to the smaller indicial root. Thus from part (iii) of Theorem 18, the second solution will have the form

(A) $\qquad y_2(x) = x^{-2} \sum_{n=0}^{\infty} b_n x^n + C y_1(x) \ln x$

where $C \neq 0$ and $y_1(x)$ is the series solution determined in Example 15b.

The use of the deflation rule for obtaining the second solution y_2 has been illustrated in Example 19, and so we will use an alternate approach here. Let $z(x) = \sum_{n=0}^{\infty} b_n x^n$. Then from (A),

$$y_2 = x^{-2}z + C y_1 \ln x$$

$$Dy_2 = -2x^{-3}z + x^{-2} Dz + C \ln(x) Dy_1 + Cx^{-1}y_1$$

$$D^2 y_2 = 6x^{-4}z - 4x^{-3} Dz + x^{-2} D^2 z + C \ln(x) D^2 y_1$$
$$+ 2Cx^{-1} Dy_1 - Cx^{-2} y_1$$

By substituting y_2 and its derivatives into the differential equation, we get

$$D^2 z - 5x^{-1} Dz - 8z + 2Cx Dy_1 - 2Cy_1 = \emptyset$$

or

(B) $\quad \sum_{n=2}^{\infty} n(n-1)b_n x^{n-2} - 5x^{-1} \sum_{n=1}^{\infty} nb_n x^{n-1} + 8 \sum_{n=0}^{\infty} b_n x^n$

$$= 2C(y_1 - x\, Dy_1)$$

Now, by simplifying the left side of (B) in the usual manner, and by using the series expansion of y_1 given in Example 15b to evaluate the right side of (B), we obtain

(C) $\quad 5b_1 x^{-1} + \sum_{n=0}^{\infty} [(n+2)(n-4)b_{n+2} + 8b_n]x^n$

$$= 2C(-3x^4 + \tfrac{5}{2}x^6 - \tfrac{7}{10}x^8 + \cdots)$$

It follows from (C) that $b_1 = b_3 = b_5 = \cdots = 0$;

$$-8b_2 + 8b_0 = 0 \quad \text{or} \quad b_2 = b_0$$

$$-8b_4 + 8b_2 = 0 \quad \text{or} \quad b_4 = b_2 = b_0$$

and

$$0b_6 + 8b_4 \quad = -6C$$

From this last equation we see that $C = -\tfrac{4}{3}b_0$ and that b_6 can be assigned arbitrarily. If we set $b_6 = 0$, then, since we have C in terms of b_0, the remaining coefficients can be calculated in terms of b_0. The solution which will result from using these coefficients in (A) with $b_0 = 1$ is:

$$y_2(x) = x^{-2}(1 + x^2 + x^4 + \tfrac{5}{12}x^8 + \cdots) - \tfrac{4}{3}y_1(x)\ln(x)$$

In general, if b_0 and b_6 are left as arbitrary constants in the calculations above, then it can be verified that the resulting solutions are $y(x) = b_0 y_2(x) + b_6 y_1(x)$, which we recognize as the two-parameter family of solutions of the given equation. This is similar to Example 20b.

22 EXERCISES

In each of the following, verify that the given point $x = a$ is a regular singular point; determine the indicial equation and the two indicial roots r_1 and r_2 $[Re(r_1) \geq Re(r_2)]$; determine a series solution of the form $y(x) = x^{r_1} \sum_{n=0}^{\infty} a_n(x - a)^n$ corresponding to the larger root; indicate whether there is a second independent solution of this form corresponding to the second root, and if there is, find it; if not, indicate the form of a second solution corresponding to the smaller root; specify the interval of convergence in each case.

1. $2x\,D^2y + 5\,Dy + 3y = \varnothing; x = 0$
2. $x^2\,D^2y + x(3 - x)\,Dy - xy = \varnothing; x = 0$
3. $(x + 2)^2\,D^2y - 3(x + 2)\,Dy + (x + 6)y = \varnothing; x = -2$
4. $2x^2\,D^2y - x\,Dy + (1 - x^2)y = \varnothing; x = 0$
5. $x(1 - x)\,D^2y - 3\,Dy + 2y = \varnothing; x = 0$
6. $(x - 2)^2\,D^2y - (x - 2)\,Dy + 8(x^2 - 4x + 3)y = \varnothing; x = 2$
7. $3x^2\,D^2y + 7x\,Dy + (1 + x^2)y = \varnothing; x = 0$
8. $(x - 1)^2\,D^2y + 3(x - 1)\,Dy + xy = \varnothing; x = 1$
9. $4x\,D^2y + 2\,Dy + y = \varnothing; x = 0$
10. $(x - 1)\,D^2y - 3\,Dy + y = \varnothing; x = 1$
11. $x^2\,D^2y - x(2 + x^2)\,Dy + x^2y = \varnothing; x = 0$
12. Show that $x = -1$ is a regular singular point of the differential equation

$$(x + 1)\,D^2y + Dy + 2y = \varnothing$$

Determine two linearly independent solutions of the equation.

13. Show that $x = 0$ is a regular singular point of the differential equation

$$x^2\,D^2y + x(x - 2)\,Dy + 2y = \varnothing$$

Determine two linearly independent solutions of the equation.

14. Show that the differential equation

$$\ln(x + 1)\,D^2y + 3\,Dy + y = \varnothing$$

has a regular singular point at $x = 0$. Determine the indicial equation and the two indicial roots. Determine the first three terms of the series solution corresponding to the larger root. Specify an interval of convergence.

15. Consider the differential equation

$$x^3\,D^2y + ax\,Dy + by = \varnothing$$

a. Show that $x = 0$ is an irregular singular point.
b. Show that if we assume a solution of the form

$$y(x) = x^r \sum_{n=0}^{\infty} a_nx^n, \qquad a_0 \neq 0$$

then r satisfies a linear equation. Thus there can be at most one solution of this form.

16. Consider the differential equation

$$x^3\,D^2y + x\,Dy - y = \varnothing$$

a. Assume a series solution of the form

(A) $$y(x) = x^r \sum_{n=0}^{\infty} a_nx^n, \qquad a_0 \neq 0$$

and determine the indicial equation and the indicial root.
b. Obtain a series solution of the form (A).
c. Determine a solution basis for the equation.

17. Consider the differential equation

$$x^2 D^2 y + (ax + b) Dy + cy = \emptyset$$

a. Show that $x = 0$ is an irregular singular point.

b. Show that if we assume a series solution of the form

$$y(x) = x^r \sum_{n=0}^{\infty} a_n x^n, \qquad a_0 \neq 0$$

then $r = 0$. Thus there can be at most one solution of this form.

18. Consider the differential equation

$$x^2 D^2 y + (3x - 1) Dy + y = \emptyset$$

a. Show that a series solution is

$$y(x) = \sum_{n=0}^{\infty} n! \, x^n$$

b. Show that the radius of convergence of this series is $R = 0$. Thus the equation does not have a power series solution in powers of x.

Section 33 Classical Second-Order Equations

We conclude this chapter by considering certain special classes of differential equations which are of primary importance in applications and whose solutions have been studied in detail by using power series methods.

1 DEFINITION

A differential equation of the form

$$(1 - x^2) D^2 y - 2x \, Dy + by = \emptyset$$

where b is a constant, is called *Legendre's equation*.

Legendre's equation occurs in a variety of problems in science and engineering. It is easy to see that $x = 1$ and $x = -1$ are singular points of Legendre's equation, and that all other points are ordinary points. In particular, $x = 0$ is an ordinary point, and so, from Theorem 5, Section 31, we can obtain two linearly independent power series solutions of the form

(2) $$y(x) = \sum_{n=0}^{\infty} a_n x^n$$

By differentiating (2) twice and substituting into (1), we obtain

$$(1 - x^2) \sum_{n=2}^{\infty} n(n - 1)a_n x^{n-2} - 2x \sum_{n=1}^{\infty} na_n x^{n-1} + b \sum_{n=0}^{\infty} a_n x^n = \emptyset$$

With the appropriate changes of index, this equation simplifies to

$$\sum_{k=0}^{\infty} \{(k + 2)(k + 1)a_{k+2} - [k(k + 1) - b]a_k\}x^k = \emptyset$$

Thus the recursion formula is:

(3) $(k + 2)(k + 1)a_{k+2} - [k(k + 1) - b]a_k = 0$

$k = 0, 1, 2, \ldots,$ a second-order linear homogeneous difference equation. Calculating the coefficients a_k in terms of a_0 and a_1 using (3), we get

$$a_2 = \frac{(0 \cdot 1 - b)a_0}{2 \cdot 1} \qquad a_3 = \frac{(1 \cdot 2 - b)a_1}{3 \cdot 2}$$

$$a_4 = \frac{(2 \cdot 3 - b)a_2}{4 \cdot 3} = \frac{(0 \cdot 1 - b)(2 \cdot 3 - b)a_0}{4!}$$

and, in general,

$$a_{2k} = \frac{(0 \cdot 1 - b)(2 \cdot 3 - b) \cdots [(2k - 2)(2k - 1) - b]a_0}{(2k)!}$$

$$a_{2k+1} = \frac{(1 \cdot 2 - b)(3 \cdot 4 - b) \cdots [(2k - 1)(2k) - b]a_1}{(2k + 1)!}$$

Therefore, the two-parameter family of solutions of Legendre's equation is:

$$y(x) = a_0 z_0(x) + a_1 z_1(x)$$

where z_0 and z_1 are the linearly independent power series solutions:

(4) $z_0(x) = 1 + \displaystyle\sum_{k=1}^{\infty} \frac{(0 \cdot 1 - b)(2 \cdot 3 - b) \cdots [(2k - 2) \cdot (2k - 1) - b]}{(2k)!} x^{2k}$

(5) $z_1(x) = x + \displaystyle\sum_{k=1}^{\infty} \frac{(1 \cdot 2 - b)(3 \cdot 4 - b) \cdots [(2k - 1) \cdot (2k) - b]}{(2k + 1)!} x^{2k+1}$

By the ratio test, each of the series (4) and (5) has radius of convergence $R = 1$. It is important to notice that if the constant b in Legendre's equation has the form $b = m(m + 1)$, where m is a nonnegative integer, then one (and only one) of the two series z_0 and z_1 terminates, that is, is a polynomial. This is the case which we want to consider in more detail.

6 DEFINITION

Let m be a nonnegative integer. The differential equation

$$(1 - x^2) D^2 y - 2x \, Dy + m(m + 1)y = \emptyset$$

is called *Legendre's equation of order m.*

With $b = m(m + 1)$, m a nonnegative integer, the series specified by (4) and (5) take the form

(7) $\quad y_0(x) = 1 + \displaystyle\sum_{k=1}^{\infty} \frac{[0 \cdot 1 - m(m + 1)][2 \cdot 3 - m(m + 1)] \cdots [(2k - 2) \cdot (2k - 1) - m(m + 1)]}{(2k)!} x^{2k}$

(8) $\quad y_1(x) = x + \displaystyle\sum_{k=1}^{\infty} \frac{[1 \cdot 2 - m(m + 1)][3 \cdot 4 - m(m + 1)] \cdots [(2k - 1)(2k) - m(m + 1)]}{(2k + 1)!} x^{2k+1}$

Note that since y_0 involves only even powers of x, y_0 is an even function on $(-1, 1)$, and since y_1 involves only odd powers of x, y_1 is an odd function on $(-1, 1)$.

The proof of the following theorem is left as an exercise.

9 THEOREM

Let m be a nonnegative integer and consider Legendre's equation of order m. If m is even, then the power series y_0 given by (7) terminates and y_0 has degree m. If m is odd, then the power series y_1 given by (8) terminates and y_1 has degree m. Finally, if p is any polynomial solution of Legendre's equation of order m, then either p is a multiple of y_0 or a multiple of y_1, depending upon whether m is even or odd.

According to this theorem, Legendre's equation of order m has a one-parameter family of solutions, each nontrivial member of which is a polynomial of degree m. We leave it as an exercise to show that if $q(x)$ is a polynomial solution of Legendre's equation of order m, then $q(1) \neq 0$. This fact is used in the next definition.

10 DEFINITION

Let m be a nonnegative integer. *Legendre's polynomial of degree m*, denoted P_m, is the polynomial solution of Legendre's equation of order m which satisfies $P_m(1) = 1$.

We can use (7) and (8) to calculate the Legendre polynomials of degrees $0, 1, 2, \ldots$. In particular, the first four Legendre polynomials are:

$m = 0$: $P_0(x) = 1$

$m = 1$: $P_1(x) = x$

$m = 2$: $P_2(x) = \frac{3}{2}x^2 - \frac{1}{2}$

$m = 3$: $P_3(x) = \frac{5}{2}x^3 - \frac{3}{2}x$

The graphs of these polynomials on $[-1, 1]$ are shown in the figure here.

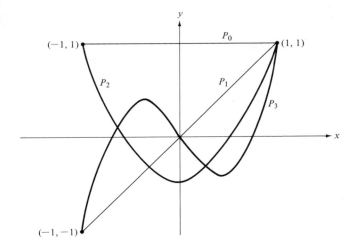

It can be shown that the Legendre polynomials satisfy the recursion relation

(11) $mP_m(x) - (2m - 1)xP_{m-1}(x) + (m - 1)P_{m-2}(x) = \varnothing$

Thus, having $P_0(x) = 1$ and $P_1(x) = x$, this recursion relation provides an alternative method for calculating Legendre's polynomial P_m for any $m \geq 2$.

Another important property of the Legendre polynomials is specified by the following theorem.

12 THEOREM

If m and n are nonnegative integers, $m \neq n$, then

$$\int_{-1}^{1} P_m(x)P_n(x) \, dx = 0$$

Proof: The polynomial P_m satisfies the equation

$$(1 - x^2) \, D^2 P_m(x) - 2x \, DP_m(x) + m(m + 1)P_m(x) = \varnothing$$

which can be written

(A) $D[(1 - x^2) \, DP_m(x)] + m(m + 1)P_m(x) = \varnothing$

Similarly, $P_n(x)$ is a solution of

(B) $D[(1 - x^2) \, DP_n(x)] + n(n + 1)P_n(x) = \varnothing$

Multiplying (A) by $P_n(x)$, (B) by $P_m(x)$, and subtracting the resulting equations, we get

$$P_n \cdot D[(1 - x^2) \, DP_m] - P_m \cdot D[(1 - x^2) \, DP_n]$$
$$+ [m(m + 1) - n(n + 1)]P_m P_n = \varnothing$$

This equation is the same as

$$D\{(1 - x^2)[P_n \cdot DP_m - P_m \cdot DP_n]\}$$
$$+ [m(m + 1) - n(n + 1)]P_m P_n = \emptyset$$

Now we integrate from -1 to 1 to obtain

$$[m(m + 1) - n(n + 1)] \int_{-1}^{1} P_m(x) P_n(x) \, dx = 0$$

and since $m \neq n$, the theorem follows.

Two functions f and g defined on an interval (a, b) are said to be *orthogonal* on (a, b) if $\int_a^b f(x)g(x) \, dx = 0$. Thus Theorem 12 states that the Legendre polynomials P_m and P_n are orthogonal on $(-1, 1)$ whenever $m \neq n$.

Two other important properties of the Legendre polynomials are:

(13) $P_m(x) = \dfrac{1}{2^m m!} D^m[(x^2 - 1)^m]$

$m = 0, 1, 2, \ldots$, and

(14) $\displaystyle\int_{-1}^{1} P_m^2(x) \, dx = \dfrac{2}{2m + 1}$

The representation (13) of the Legendre polynomial P_m is known as *Rodrigues' formula*. Hints for establishing (13) and (14) are contained in the exercises.

15 EXERCISES

1. Prove Theorem 9.
2. Use the recursion relation (11) to determine P_2, P_3, and P_4.
3. Show that the set of Legendre polynomials is linearly independent.
4. Given the polynomial $q(x) = 2x^3 - 3x^2 + 5x - 1$. Express q as a linear combination of Legendre polynomials.
5. Let q be a polynomial of degree n. Prove that

    $$\int_{-1}^{1} P_m(x)q(x) \, dx = 0$$

    whenever $m > n$.
6. Establish Rodrigues' formula (13). [Hint: Let $z = (x^2 - 1)^m$ and differentiate $m + 2$ times to show that

    $$(x^2 - 1) D^{m+2}z + 2x D^{m+1} z - m(m + 1) D^m z = \emptyset$$

7. Show that the Legendre polynomials satisfy the recursion relation

    $$D[P_{m+1}(x)] = x D[P_m(x)] + (m + 1)P_m(x)$$

    $m = 0, 1, \ldots$. [Hint: Use Rodrigues' formula for $P_{m+1}(x)$ to show that

    $$P_{m+1}(x) = \dfrac{1}{m! \, 2^m} D^m[x(x^2 - 1)^m]$$

so that

(A) $$D[P_{m+1}(x)] = \frac{1}{m! \, 2^m} D^{m+1}[x(x^2 - 1)^m]$$

Calculate this derivative using Leibnitz's rule, Theorem 8, Section 2.]

8. Show that the Legendre polynomials satisfy the recursion relation

$$D[P_{m+1}(x)] = (2m + 1)P_m(x) + D[P_{m-1}(x)]$$

(Hint: Use (A) in Problem 7 and compute D^{m+1} as $D^m D$.)

9. Derive the recursion relation (11) using the relations in Problems 7 and 8.
10. Establish (14). (Hint: Use (13) and integrate by parts n times.)
11. Prove that

$$(2m + 1) \int_x^1 P_m(x) \, dx = P_{m-1}(x) - P_{m+1}(x)$$

12. For $m > 0$, show that
   a. $\int_0^1 P_m(x) \, dx = [1/(m + 1)]P_{m-1}(0)$
   b. $\int_0^1 P_{2m}(x) \, dx = 0$
   c. $\int_0^1 P_{2m+1}(x) \, dx = (-1)^m [\prod_{i=1}^m (2i - 1)/\prod_{i=1}^m (2i - 2)]$
13. Verify the statement that precedes Definition 10.
14. Consider the Legendre equation

$$(1 - x^2) D^2 y - 2x \, Dy + \alpha(\alpha + 1)y = \emptyset$$

where α is any real number.
   a. Show that $x = 1$ and $x = -1$ are regular singular points.
   b. Determine a series solution of the form

$$y(x) = (x - 1)^r \sum_{n=0}^{\infty} a_n(x - 1)^n$$

   c. Show that if α is a nonnegative integer, then the equation has a polynomial solution.

The next equation which we shall consider is one of the most important differential equations from the standpoint of applications.

16 DEFINITION

A differential equation of the form

$$x^2 D^2 y + x \, Dy + (x^2 - b^2)y = \emptyset$$

where b is a nonnegative constant, is called *Bessel's equation of order b*.

The solutions of this equation are known as *Bessel functions*, and they are useful in a wide variety of problems in mechanics, thermodynamics, elasticity, and acoustics. The point $x = 0$ is a regular singular point of

Bessel's equation. Thus we shall determine a series solution of the form

(17) $y(x) = x^r \sum_{n=0}^{\infty} a_n x^n, \qquad a_0 \neq 0$

using the methods of Section 32. In particular, differentiating (17) twice and substituting into Bessel's equation, we get the equation

$$\sum_{n=0}^{\infty} (n + r)(n + r - 1)a_n x^{n+r} + \sum_{n=0}^{\infty} (n + r)a_n x^{n+r} + \sum_{n=0}^{\infty} a_n x^{n+r+2}$$

$$- b^2 \sum_{n=0}^{\infty} a_n x^{n+r} = \varnothing$$

which simplifies to

(18) $(r^2 - b^2)a_0 + [(r + 1)^2 - b^2]a_1 x + \sum_{k=2}^{\infty} \{[(k + r)^2 - b^2]a_k$

$$+ a_{k-2}\}x^k = \varnothing$$

Thus the indicial equation is $r^2 - b^2 = 0$, and the indicial roots are $r_1 = b$ and $r_2 = -b$. By substituting $r_1 = b$ into (18), we see that $a_1 = 0$, and we obtain the recursion formula

(19) $(k^2 + 2kb)a_k + a_{k-2} = 0$

$k = 2, 3, \ldots$. Since $a_1 = 0$, we conclude from (19) that $a_{2k+1} = 0$, $k = 1, 2, \ldots$. Calculating the coefficients a_{2k}, $k = 1, 2, \ldots$, in terms of a_0 using (19), we get that

$$a_2 = \frac{-a_0}{4(1 + b)}, \qquad a_4 = \frac{-a_2}{2 \cdot 4(2 + b)} = \frac{a_0}{2! \, 4^2(2 + b)(1 + b)}$$

and, in general,

$$a_{2k} = \frac{(-1)^k a_0}{k! \, 4^k(k + b)(k - 1 + b) \cdots (2 + b)(1 + b)}$$

$k = 1, 2, \ldots$. We now substitute $r_1 = b$ and these coefficients into (17) to obtain the one-parameter family of series solutions

(20) $y(x) = a_0 x^b \sum_{k=0}^{\infty} \frac{(-1)^k x^{2k}}{k! \, 4^k(k + b)(k - 1 + b) \cdots (2 + b)(1 + b)}$

It is easy to verify that the power series portion of $y(x)$ in (20) converges for all real numbers x. Thus, in general, the series (20) is valid on the interval $(0, \infty)$. The Bessel function is obtained from (20) by assigning a particular value to the parameter a_0. This assignment, however, involves a generalization of the factorial function known as the *gamma function*, and consequently a digression from our treatment of Bessel's equation is necessary here.

21 DEFINITION

The *gamma function*, denoted by Γ, is given by

$$\Gamma(s) = \int_0^\infty x^{s-1} e^{-x} \, dx$$

Since the gamma function is defined by an improper integral, the reader is referred to Section 2 for a brief review of improper integrals and methods for evaluating them. It can be shown that $\Gamma(s)$, when s is a positive integer, can be calculated directly from Definition 21 using integration by parts. In general, however, the following recursion relation provides a more efficient means of determining values of Γ.

22 THEOREM

Let $s > 0$. Then

$$\Gamma(s + 1) = s\Gamma(s)$$

Proof: Fix any $s > 0$. From the definition of Γ,

$$\Gamma(s + 1) = \int_0^\infty x^s e^{-x} \, dx = \lim_{c \to \infty} \int_0^c x^s e^{-x} \, dx$$

Now, using integration by parts,

$$\int_0^c x^s e^{-x} \, dx = -x^s e^{-x} \Big|_0^c + s \int_0^c x^{s-1} e^{-x} \, dx$$

$$= -c^s e^{-c} + s \int_0^c x^{s-1} e^{-x} \, dx$$

Therefore,

$$\Gamma(s + 1) = \lim_{c \to \infty} (-c^s e^{-c}) + s \lim_{c \to \infty} \int_0^c x^{s-1} e^{-x} \, dx = s\Gamma(s)$$

It is easy to verify that $\Gamma(1) = \int_0^\infty e^{-x} \, dx = 1$. Thus by repeated use of the recursion relation, it follows that $\Gamma(n + 1) = n!$ for each nonnegative integer n. This observation supports the remark that Γ is a generalized factorial.

The recursion relation given by Theorem 22 can also be used to define Γ for certain negative values of s. From the recursion relation,

(23) $\qquad \Gamma(s) = \dfrac{\Gamma(s + 1)}{s}$

and it is clear that $\Gamma(0)$ is undefined. However, we can use (23) to determine values of $\Gamma(s)$ for $-1 < s < 0$, since for $s \in (-1, 0)$, we have $s + 1 \in (0, 1)$. Now, having defined Γ on $(-1, 0)$ using (23), we can repeat the procedure to obtain values of $\Gamma(s)$ for $s \in (-2, -1)$. Continuing in this manner, the domain of Γ is extended to all negative values of s except for $s = -1, -2, \ldots$. Thus we may assume that the domain of Γ is all real numbers except for $s = 0, -1, -2, \ldots$. The graph of Γ is shown in the figure here.

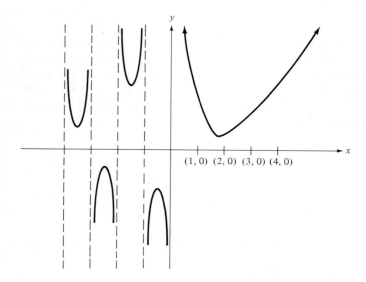

Returning to the discussion of Bessel's equation, consider the one-parameter family of series solutions (20) and let

$$a_0 = \frac{1}{2^b \Gamma(b + 1)}$$

We leave it as an exercise to show that the resulting series solution can be written as

(24) $$J_b(x) = \left(\frac{x}{2}\right)^b \sum_{k=0}^{\infty} \frac{(-1)^k}{k!\,\Gamma(k + 1 + b)} \left(\frac{x}{2}\right)^{2k}$$

25 DEFINITION

The solution J_b, given by (24), of Bessel's equation of order b is called *Bessel's function of the first kind of order b*.

We now investigate the possibility of obtaining a second series solution of the form (17) of Bessel's equation. Recall from Theorem 18, Section 32, that if the difference of the indicial roots, $r_1 - r_2 = 2b$, is not a nonnegative integer, then the differential equation has a second series solution of the form (17) which is independent of J_b. We assume, therefore, that $2b$ is not a nonnegative integer and substitute $r_2 = -b$ into (18). This yields $a_1 = 0$ and the recursion formula

(26) $(k^2 - 2kb)a_k + a_{k-2} = 0$

$k = 2, 3, \ldots$. Since $a_1 = 0$, it follows from (26) that $a_{2k+1} = 0, k = 1, 2, \ldots$. The even-numbered coefficients a_{2k}, $k = 1, 2, \ldots$, are given in terms of a_0 by

$$(27) \qquad a_{2k} = \frac{(-1)^k a_0}{k! \, 4^k (k - b)(k - 1 - b) \cdots (2 - b)(1 - b)}$$

We now substitute $r_2 = -b$ and these coefficients into (17) and obtain the one-parameter family of series solutions

$$(28) \qquad y(x) = a_0 x^{-b} \sum_{k=0}^{\infty} \frac{(-1)^k x^{2k}}{k! \, 4^k (k - b)(k - 1 - b) \cdots (2 - b)(1 - b)}$$

Notice that the series (28) is merely the series (20) with b replaced by $-b$. If we set

$$a_0 = \frac{1}{2^{-b} \Gamma(1 - b)}$$

in (28), then the resulting solution

$$(29) \qquad J_{-b}(x) = \left(\frac{x}{2}\right)^{-b} \sum_{k=0}^{\infty} \frac{(-1)^k}{k! \, \Gamma(k + 1 - b)} \left(\frac{x}{2}\right)^{2k}$$

is a second solution of Bessel's equation and is independent of $J_b(x)$. Again, it is easy to verify that the power series portion of J_{-b} converges for all real numbers x, so that J_{-b} is valid on the interval $(0, \infty)$. The solution (29) is also called a Bessel function of the first kind.

If $2b = 0$, that is, if $b = 0$, then we are in case (ii) of Theorem 18, Section 32, and some other technique must be used to find a second solution which is independent of J_0. If $2b$ is a positive integer, then case (iii) of Theorem 18, Section 32, applies. Note that in case (iii) it is possible for the equation to have a second series solution of the form (17). An examination of the recursion formula (26) shows that if $2b = 2m$ is an even positive integer, that is, if $b = m$ is a positive integer, then a_{2m} is undefined and there does not exist a second solution of the form (17). If, on the other hand, $2b = 2m - 1$ is an odd positive integer, that is, $b = (2m - 1)/2$ for some positive integer m, then the coefficients a_{2k} are defined for all k and are given by (27). As a result, the solution J_{-b} given by (29) is valid when $2b$ is an odd positive integer. These remarks are summarized by the following theorem, with the proof being left as an exercise.

30 THEOREM

If b is not a nonnegative integer, then the functions J_b and J_{-b}, given by (24) and (29), respectively, are linearly independent solutions of Bessel's equation on $(0, \infty)$. These solutions can be extended to the interval $(-\infty, 0)$ by replacing x by $|x|$.

31 EXERCISES

1. Calculate $\Gamma(2)$ using integration by parts.
2. Show that $\Gamma(\frac{1}{2}) = \sqrt{\pi}$. (Hint: With the change of variable $x = u^2$, $\Gamma(\frac{1}{2}) = 2 \int_0^\infty e^{-u^2} du$.)
3. Let n be a positive integer. Determine $\Gamma(n + \frac{1}{2})$.
4. Verify that if $a_0 = 1/[2^b \Gamma(b + 1)]$ in (20), then the resulting solution is given by (24). Show also that if $a_0 = 1/[2^{-b} \Gamma(1 - b)]$ in (28), then the resulting solution is given by (29).
5. Show that $\Gamma(k + b + 1) = (k + b)^{(k)} \Gamma(b + 1)$, where the superscript (k) denotes factorial power. Use this result to express the coefficients in (20) in terms of factorial powers.
6. Prove Theorem 30.
7. Let z be a solution of Bessel's equation. Show that the function v defined by $v(x) = x^{1/2}z(x)$ on $(0, \infty)$ is a solution of

$$D^2 y + \left[1 + \frac{\frac{1}{4} - b^2}{x^2}\right] y = \varnothing$$

8. Use the results in Problems 2, 3, and 7 to show that

$$J_{1/2}(x) = \left(\frac{2}{\pi x}\right)^{1/2} \sin x$$

$$J_{-1/2}(x) = \left(\frac{2}{\pi x}\right)^{1/2} \cos x$$

9. Show that

$$J_{3/2}(x) = \left(\frac{2}{\pi x}\right)^{1/2} \left(\frac{\sin x}{x} - \cos x\right)$$

$$J_{-3/2}(x) = \left(\frac{2}{\pi x}\right)^{1/2} \left(\frac{\cos x}{x} - \sin x\right)$$

The Bessel functions of the first kind satisfy a number of useful recursion relations. One of the most important of these relations is given by the following theorem.

32 THEOREM

For every positive number b,

$$J_{b+1}(x) = \frac{2b}{x} J_b(x) - J_{b-1}(x)$$

Proof: Since J_b is an analytic function on $(0, \infty)$, the series

$$x^b J_b(x) = \sum_{k=0}^\infty \frac{(-1)^k x^{2k+2b}}{2^{2k+b} k! \, \Gamma(k + 1 + b)}$$

can be differentiated term-wise. This yields

$$D[x^b J_b(x)] = \sum_{k=0}^{\infty} \frac{(-1)^k 2(k+b)x^{2k+2b-1}}{2^{2k+b}k!\,\Gamma(k+1+b)}$$

$$= x^b \left(\frac{x}{2}\right)^{b-1} \sum_{k=0}^{\infty} \frac{(-1)^k}{k!\,\Gamma(k+b)} \left(\frac{x}{2}\right)^{2k}$$

Thus

(A) $\qquad D[x^b J_b(x)] = x^b J_{b-1}(x)$

In the same manner, it can be shown that

(B) $\qquad D[x^{-b} J_b(x)] = -x^{-b} J_{b+1}(x)$

Of course, using the product rule for differentiation,

$$D[x^b J_b(x)] = bx^{b-1} J_b(x) + x^b D[J_b(x)]$$

and

$$D[x^{-b} J_b(x)] = -bx^{-b-1} J_b(x) + x^{-b} D[J_b(x)]$$

Combining these expressions with (A) and (B), respectively, we get

(C) $\qquad D[J_b(x)] = J_{b-1}(x) - \frac{b}{x} J_b(x)$

and

(D) $\qquad D[J_b(x)] = -J_{b+1}(x) + \frac{b}{x} J_b(x)$

The recursion relation is now an immediate consequence.

As a result of the recursion relation specified in Theorem 32, it is only necessary to calculate values of the Bessel functions J_b for $0 \le b < 2$. In particular, the Bessel functions J_m, $m > 1$ a positive integer, can be expressed in terms of J_0 and J_1. Extensive tables of these two functions have been compiled.

The conclusions of Theorem 30 can be used to characterize the set of solutions of Bessel's equation of order b when b is not a nonnegative integer. Thus the case where $b = m$ is a nonnegative integer remains to be treated, and so we will make this restriction on b for the remainder of this discussion of Bessel's equation. Since $\Gamma(k+1+m) = (k+m)!$, Bessel's function of the first kind of integral order m has the simpler form

(33) $\qquad J_m(x) = \left(\frac{x}{2}\right)^m \sum_{k=0}^{\infty} \frac{(-1)^k}{k!\,(k+m)!} \left(\frac{x}{2}\right)^{2k}$

In particular, J_0 and J_1 are given by

(34) $\qquad J_0(x) = \sum_{k=0}^{\infty} \frac{(-1)^k}{(k!)^2} \left(\frac{x}{2}\right)^{2k}$

and

(35) $$J_1(x) = \sum_{k=0}^{\infty} \frac{(-1)^k}{k! \, (k+1)!} \left(\frac{x}{2}\right)^{2k+1}$$

From these series representations, note the similarity of J_0 and J_1 to the trigonometric functions $\cos x$ and $\sin x$. This analogy is further reinforced by the fact that $D[J_0(x)] = -J_1(x)$, which follows from equation (D) in the proof of Theorem 32. In more advanced treatments of Bessel's equation, it is shown that each of J_0 and J_1 has infinitely many zeros on $(0, \infty)$, and that the functions are almost periodic in the sense that as x approaches infinity, the distance between consecutive zeros approaches π as a limit. The graphs of J_0 and J_1 are shown in the figure here.

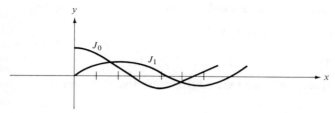

To obtain a second solution of Bessel's equation of order m, we can use the deflation rule, Theorem 5, Section 12. Use of the deflation rule yields the solution

(36) $$z(x) = J_m(x) \int_{x_0}^{x} \frac{dt}{t J_m{}^2(t)}$$

where x_0 is chosen as a convenient lower limit of integration.

As an alternative to the deflation rule, we can appeal directly to Theorem 18, Section 32, case (ii) or case (iii). Case (ii) corresponds to $m = 0$ and the second solution has the form

(37) $$y(x) = x \sum_{n=0}^{\infty} c_n x^n + J_0(x) \ln(x)$$

Case (iii) corresponds to m a positive integer, and the form of the second solution can be assumed to be

(38) $$y(x) = x^{-m} \sum_{n=0}^{\infty} c_n x^n + J_m(x) \ln(x)$$

Consider the case $m = 0$. Differentiating (37) twice and substituting into Bessel's equation of order 0, we get

$$\sum_{n=1}^{\infty} (n+1)n c_n x^{n+1} + \sum_{n=0}^{\infty} (n+1)c_n x^{n+1} + \sum_{n=0}^{\infty} c_n x^{n+3}$$
$$+ 2x \, D[J_0(x)] + \{x^2 \, D^2[J_0(x)]$$
$$+ x \, D[J_0(x)] + x^2 J_0(x)\} \ln(x) = \varnothing$$

Now, using the fact that $J_0(x)$ is a solution of Bessel's equation of order 0, and by making the appropriate changes in index, this equation simplifies to

$$(39) \qquad c_0 x + 4c_1 x^2 + \sum_{k=2}^{\infty} [(k+1)^2 c_k + c_{k-2}] x^{k+1} = -2 \sum_{k-1}^{\infty} \frac{(-1)^k 2k x^{2k}}{2^{2k}(k!)^2}$$

where the expression on the right side of (39) is the series expansion of $-2x\, D[\, J_0(x)]$. By equating coefficients and observing that the right side contains only even powers of x, we have

$$0 = c_0 = c_2 = c_4 = \cdots$$

$$c_1 = \tfrac{1}{4}$$

$$c_3 = \frac{-1}{2^4 (2!)^2} (1 + \tfrac{1}{2})$$

and, in general,

$$c_{2k+1} = \frac{(-1)^{k-1}}{2^{2k}(k!)^2} \left(1 + \frac{1}{2} + \cdots + \frac{1}{k} \right)$$

$k = 1, 2, \ldots$. Thus a second solution of Bessel's equation of order 0 which is independent of J_0 is:

$$(40) \qquad K_0(x) = \sum_{k=1}^{\infty} \frac{(-1)^{k+1}}{(k!)^2} \left(1 + \frac{1}{2} + \cdots + \frac{1}{k} \right) \left(\frac{x}{2}\right)^{2k} + J_0(x) \ln(x)$$

It can be verified that K_0 is defined on the interval $(0, \infty)$.

It will be left as an exercise to show that when m is a positive integer, a second solution of Bessel's equation of order m is given by

$$(41) \qquad K_m(x) = -\frac{1}{2} \left(\frac{x}{2}\right)^{-m} \left[\sum_{k=0}^{m-1} \frac{(m-k-1)!}{k!} \left(\frac{x}{2}\right)^{2k} \right.$$

$$\left. + \frac{1}{m!} \left(1 + \frac{1}{2} + \cdots + \frac{1}{m} \right) \left(\frac{x}{2}\right)^{m} \right]$$

$$- \frac{1}{2} \left(\frac{x}{2}\right)^{m} \sum_{k=0}^{\infty} \frac{(-1)^k}{k!\,(k+m)!} \left\{ \left(1 + \frac{1}{2} + \cdots + \frac{1}{k} \right) \right.$$

$$\left. + \left(1 + \frac{1}{2} + \cdots + \frac{1}{k+m} \right) \right\} \left(\frac{x}{2}\right)^{2k} + J_m(x) \ln(x)$$

42 DEFINITION

The solutions K_m, $m = 0, 1, 2, \ldots$, given by (40) and (41), of Bessel's equation of integral order m are called *Bessel functions of the second kind of order m*.

The discussion of Bessel's equation of integral order m is summarized by the following theorem.

43 THEOREM

If $m = 0$, then the functions J_0 and K_0, given by (34) and (40), respectively, are linearly independent solutions of Bessel's equation of order 0, on the interval $(0, \infty)$. If m is a positive integer, then the functions J_m and K_m, given by (33) and (41), respectively, are linearly independent solutions of Bessel's equation of order m on $(0, \infty)$. These solutions can be extended to the interval $(-\infty, 0)$ by replacing x by $|x|$.

44 EXERCISES

1. a. Express J_3 and J_4 in terms of J_0 and J_1.

    b. Express J_4 in terms of J_0 and DJ_0.

2. Let m and n be nonnegative integers.

    a. Show that if $m > n$, then

    $$\int x^m J_n(x)\, dx = x^m J_{n+1}(x) - (m - n - 1) \int x^{m-1} J_{n+1}(x)\, dx$$

    b. Show that if $m \le n$, then

    $$\int x^m J_n(x)\, dx = - x^m J_{n-1}(x) + (m + n - 1) \int x^{m-1} J_{n-1}(x)\, dx$$

3. Show that

    $$DJ_b(x) = \tfrac{1}{2}[J_{b-1}(x) - J_{b+1}(x)]$$

    for all $b > 0$.

4. Show that

    $$D[x^{-b} J_b(x)] = -x^{-b} J_{b+1}(x)$$

    for all $b > 0$.

5. a. Let n be a positive integer. Show that

    $$\lim_{r \to -n} |\Gamma(r)| = \infty$$

    b. Define $1/\Gamma(-n) = 0$ when n is a positive integer and show that $J_{-n}(x) = (-1)^n J_n(x)$.

6. Show that:

    a. $\int x^3 J_0(x)\, dx = x^3 J_1(x) - 2x^2 J_2(x) + c$

    b. $\int x^4 J_1(x)\, dx = x^4 J_2(x) - 2x^3 J_3(x) + c$

    c. $\int x J_1(x)\, dx = -x J_0 + \int J_0(x)\, dx$

    d. $\int J_2(x)\, dx = \int J_0(x)\, dx - 2J_1(x)$.

7. Establish the recursion relation

    $$J_{b-1}(x) + J_{b+1}(x) = 2D[J_b(x)]$$

    for all $b > 0$.

8. a. Prove that

    $$\int_0^\infty J_{n+1}(x)\, dx = \int_0^\infty J_{n-1}(x)\, dx$$

b. Show that

$$\int_0^\infty J_n(x)\,dx = 1$$

for all positive integers n.

9. Verify that the functions defined in (40) and (41) are solutions of Bessel's equation.

10. Show that the Wronskian of J_b and J_{-b} is $-2(\pi x)^{-1} \sin bx$.

11. Show that

$$\int_0^\infty \frac{J_n(x)}{x}\,dx = \frac{1}{n}$$

for all positive integers n. (Hint: Use Theorem 32 and Problem 8.)

12. a. Show that if $0 < x_1 < x_2$ are zeros of $J_b(x)$, then $J_{b+1}(x)$ has a zero on (x_1, x_2) for any $b \geq 0$.

    b. Show that if $0 < y_1 < y_2$ are zeros of $J_{b+1}(x)$, then $J_b(x)$ has a zero on (y_1, y_2) for any $b > 0$.

13. Show that $J_b(sx)$ is a solution of

$$x^2 D^2 y + x\,Dy + (s^2 x^2 - b^2)y = \varnothing, \qquad b \geq 0$$

14. a. Show that $\int J_0(x) \sin(x)\,dx = x J_0(x) \sin(x) - x J_1(x) \cos(x) + c$.

    b. Show that $\int x J_0(x) \cos(x)\,dx = x J_0(x) \cos(x) + x J_1(x) \sin(x) + c$.

We conclude this section with a brief description of some additional second-order differential equations which occur frequently in applications, and which have been studied in detail using series methods. Since the techniques required to verify the results which follow have been illustrated by our treatment of the Legendre and Bessel equations, the proofs will be left as exercises.

Each of the equations to be considered here has polynomial solutions, and these polynomial solutions satisfy an orthogonality relation which is a generalization of the orthogonality property of the Legendre polynomials, Theorem 12.

45 DEFINITION

Let w be a continuous function on the interval (a, b) such that $w(x) \geq 0$ for all x on (a, b) and $w \neq \varnothing$. Let $\{f_n\}$ be a sequence of functions defined on (a, b). If

$$\int_a^b w(x) f_m(x) f_n(x)\,dx = 0$$

whenever $m \neq n$, then the sequence is said to be *orthogonal on* (a, b) *with respect to the weight function* w.

___

In the terminology of Definition 45, the sequence $\{P_m\}$ of Legendre polynomials is orthogonal on $(-1, 1)$ with respect to the weight function $w(x) = 1$.

46 DEFINITION

A differential equation of the form

$$(1 - x^2)\, D^2 y - x\, Dy + m^2 y = \varnothing$$

where m is a nonnegative integer, is called *Chebyshev's equation of order m.*

The points $x = 1$ and $x = -1$ are regular singular points of Chebyshev's equation, and all other points are ordinary points. In particular, $x = 0$ is an ordinary point, and the equation has two linearly independent power series solutions of the form

(47) $$y(x) = \sum_{n=0}^{\infty} a_n x^n$$

48 THEOREM

Exactly one of the two power series solutions of Chebyshev's equation of order m terminates. The resulting polynomial solution p has the following properties:

(i) p has degree m;

(ii) p is an even function if m is even, p is an odd function if m is odd;

(iii) if q is any polynomial solution of the equation, then $q(x) = c \cdot p(x)$ for some constant c.

---

49 DEFINITION

The *Chebyshev polynomial of degree m*, denoted T_m, is the polynomial solution of Chebyshev's equation of order m which satisfies $T_m(1) = 1$.

The first three Chebyshev polynomials are:

$$T_0(x) = 1, \qquad T_1(x) = x, \qquad T_2(x) = 2x^2 - 1$$

The Chebyshev polynomials $T_m, m = 0, 1, 2, \ldots$, are useful in approximating continuous functions on $[-1, 1]$. These polynomials satisfy the following orthogonality relation.

50 THEOREM

The sequence $\{T_m\}$ of Chebyshev polynomials is orthogonal on $(-1, 1)$ with respect to the weight function $w(x) = (1 - x^2)^{-1/2}$, that is,

$$\int_{-1}^{1} (1 - x^2)^{-1/2} T_m(x) T_n(x)\, dx = 0$$

whenever $m \neq n$.

The change in independent variable determined by $x = \cos \theta$, $-1 \leq x \leq 1$, transforms Chebyshev's equation of order m into the equation

(51) $\qquad \dfrac{d^2 y}{d\theta^2} + m^2 y = \varnothing$

which has $z_1(\theta) = \cos(m\theta)$ and $z_2(\theta) = \sin(m\theta)$ as a pair of linearly independent solutions. This relationship between Chebyshev's equation and equation (51) can be used to establish a recursion relation.

52 THEOREM

For every positive integer m,

$$T_{m+1}(x) = 2x T_m(x) - T_{m-1}(x)$$

on $[-1, 1]$.

53 EXERCISES

1. Prove Theorem 48.
2. Prove Theorem 50.
3. By using the change of variable $x = \cos \theta$, show that Chebyshev's equation is transformed into equation (51).
4. Prove Theorem 52.
5. Compute the Chebyshev polynomials T_3, T_4, and T_5.
6. Show that $T_{m+n}(x) + T_{m-n}(x) = 2T_m(x)T_n(x)$. (Hint: Consider a corresponding identity for cosines.)
7. Show that $T_m(x) = \cos m[\arccos x]$ for each nonnegative integer m.
8. Show that $T_m(T_n(x)) = T_n(T_m(x)) = T_{mn}(x)$ for all nonnegative integers m and n.
9. Show that $|T_m(x)| \leq 1$ for $-1 \leq x \leq 1$ and for all nonnegative integers m.
10. Show that

$$\int_{-1}^{1} (1 - x^2)^{-1/2} T_m{}^2(x) \, dx = \begin{cases} \dfrac{\pi}{2} & m > 0 \\[2mm] \pi & m = 0 \end{cases}$$

11. Verify the recursion relation

$$DT_{m+1}(x) = 2(m+1)T_m(x) + \frac{m+1}{m-1} DT_{m-1}(x), \qquad m \geq 2$$

53 DEFINITION

A differential equation of the form

$$D^2 y - 2x \, Dy + 2my = \varnothing$$

where m is a nonnegative integer, is called *Hermite's equation of order* m.

It is clear that all points on $(-\infty, \infty)$ are ordinary points of the equation. Hence $x = 0$ is an ordinary point, and there exist two linearly independent power series solutions of the form

(55) $y(x) = \displaystyle\sum_{n=0}^{\infty} a_n x^n$

each of which will converge for all x on $(-\infty, \infty)$.

56 THEOREM

Exactly one of the two power series solutions of Hermite's equation of order m terminates. The resulting polynomial solution p has the following properties:

(i) p has degree m;

(ii) p is an even function if m is even, p is an odd function if m is odd;

(iii) if q is any polynomial solution of the equation, then $q(x) = c \cdot p(x)$ for some constant c.

57 DEFINITION

The Hermite polynomial of degree m, denoted by H_m, is the polynomial solution of Hermite's equation of order m whose leading coefficient is 2^m, that is,

$H_m(x) = 2^m x^m + \cdots$

The first three Hermite polynomials are:

$H_0(x) = 1, \qquad H_1(x) = 2x, \qquad H_2(x) = 4x^2 - 2$

The Hermite polynomials are used in the theory of statistics for the representation of frequency functions over the interval $(-\infty, \infty)$. The orthogonality relation satisfied by the Hermite polynomials is given by the following theorem.

58 THEOREM

The sequence $\{H_m\}$ of Hermite polynomials is orthogonal on $(-\infty, \infty)$ with respect to the weight function $w(x) = e^{-x^2}$, that is,

$$\int_{-\infty}^{\infty} e^{-x^2} H_m(x) H_n(x)\, dx = 0$$

whenever $m \neq n$.

Let $w(x) = e^{-x^2}$ on $(-\infty, \infty)$. Since $Dw(x) = -2xw(x)$, it follows that w satisfies the first-order differential equation

$Dw + 2xw = \varnothing$

on $(-\infty, \infty)$. By repeated differentiation of this equation, we find

(59) $D^{m+2}w + 2x\, D^{m+1}w + (2m + 2)\, D^m w = \varnothing$

$m = 0, 1, 2, \ldots$. Equation (59) can be used to establish the following Rodrigues' formula for Hermite polynomials.

60 THEOREM

For each nonnegative integer m,

$$H_m(x) = (-1)^m e^{x^2} D^m[e^{-x^2}]$$

Equation (59) can also be written as

(61) $\quad D^{m+1}w + 2x D^m w + 2m D^{m-1}w = \emptyset$

$m = 1, 2, \ldots$. Multiplying (61) by $(-1)^{m+1}w^{-1}$ and using Theorem 60, we obtain the recursion relation

(62) $\quad H_{m+1}(x) = 2xH_m(x) - 2mH_{m-1}(x)$

Another useful recursion relation which can be obtained from Rodrigues' formula for Hermite polynomials is:

(63) $\quad H_{m+1}(x) = 2xH_m(x) - D[H_m(x)]$

The advantage of the recursion relation (63) over (62) is that it allows the computation of $H_{m+1}(x)$ from $H_m(x)$ alone.

64 EXERCISES

1. Prove Theorem 56.
2. Prove Theorem 58.
3. Prove Theorem 60.
4. Verify the recursion relations (62) and (63).
5. Compute the Hermite polynomials H_3, H_4, and H_5.
6. Show that $\int_{-\infty}^{\infty} e^{-x^2} H_m{}^2(x)\,dx = 2^m m! \sqrt{\pi}, \quad m = 0, 1, 2, \ldots$.
7. Verify the recursion relation

$$DH_{m+1}(x) = 2(m+1)H_m(x)$$

$m = 0, 1, 2, \ldots$.

8. Show that

$$\int_0^x e^{-t^2} H_m(t)\,dt = H_{m-1}(0) - e^{-x^2} H_{m-1}(x)$$

$m = 1, 2, \ldots$.

65 DEFINITION

A differential equation of the form

$$x D^2 y + (1 - x) Dy + my = \emptyset$$

where m is a nonnegative integer, is called *Laguerre's equation of order m.*

The point $x = 0$ is a regular singular point of Laguerre's equation, and all other points are ordinary points. By using the theory and methods of Section 32, we can obtain a series solution of the form

(66) $y(x) = x^r \sum\limits_{n=0}^{\infty} a_n x^n, \qquad a_0 \neq 0$

Substitution of the series (66) into Laguerre's equation yields the power series equation

$$r^2 a_0 + \sum_{k=1}^{\infty} [(k + r)^2 a_k - (k - 1 + r - m)a_{k-1}]x^k = \emptyset$$

Thus the indicial equation is $r^2 = 0$ and the indicial roots are $r_1 = r_2 = 0$. These remarks provide a basis for a proof of the following theorem.

67 THEOREM

A solution of the form (66) of Laguerre's equation of order m is a poly-nomial p of degree m. Any solution of the equation which is independent of p involves $\ln(x)$.

68 DEFINITION

The *Laguerre polynomial of degree m*, denoted by L_m, is the polynomial solution of Laguerre's equation of order m whose leading coefficient is $(-1)^m/m$, that is,

$$L_m(x) = \frac{(-1)^m}{m} x^m + \cdots$$

The first three Laguerre polynomials are:

$L_0(x) = 1, \qquad L_1(x) = 1 - x, \qquad L_2(x) = 1 - 2x + \tfrac{1}{2}x^2$

The Laguerre polynomials satisfy the following orthogonality relation.

69 THEOREM

The sequence $\{L_m\}$ of Laguerre polynomials is orthogonal on $(0, \infty)$ with respect to the weight function $w(x) = e^{-x}$, that is,

$$\int_0^{\infty} e^{-x} L_m(x) L_n(x) \, dx = 0$$

whenever $m \neq n$.

Let $u(x) = x^m e^{-x}$. By using either Leibnitz's rule (Theorem 8, Section 2) or Lemma 6, Section 15,

$$D^m[u(x)] = e^{-x} \sum_{k=0}^{\infty} (-1)^k \binom{m}{k} \frac{m!}{k!} x^k$$

Thus $q(x) = e^x D^m u(x)$ is a polynomial of degree m with leading coefficient $a_m = (-1)^m$. It can be verified that q is a solution of Laguerre's equation of order m, and this fact yields a Rodrigues' formula for the Laguerre polynomials.

70 THEOREM

For each nonnegative integer m,

$$L_m(x) = \frac{1}{m!} e^x D^m \left[x^m e^{-x} \right]$$

The following recursion relation is an easy consequence of Rodrigues' formula for the Laguerre polynomials:

(71) $L_{m+1}(x) = x D[L_m(x)] + (m + 1)L_m(x)$

It can also be verified that the Laguerre polynomials satisfy the recursion relation

(72) $(m + 1)L_{m+1}(x) = (2m + 1 - x)L_m(x) - mL_{m-1}(x)$

73 EXERCISES

1. Prove Theorem 67.
2. Prove Theorem 69.
3. Prove Theorem 70.
4. Verify the recursion relation 71.
5. Verify the recursion relation 72.
6. Compute the Laguerre polynomials L_3, L_4, and L_5.
7. Show that $\int_0^\infty e^{-x} L_m^2(x)\, dx = 1, m = 0, 1, 2, \ldots$.
8. Show that $\int_0^x L_{m+n}(t)\, dt = L_{m+n}(x) - L_{m+n+1}(x)$, m, n nonnegative integers.
9. Show that $|L_m(x)| \le e^{x/2}, 0 \le x < \infty, m = 0, 1, 2, \ldots$.

THE LAPLACE TRANSFORM*

Section 34 Introduction

The Laplace transform is defined in terms of a linear operator $\mathscr{L}$, and it provides an effective method for solving initial-value problems for linear differential equations with constant coefficients. The usefulness of Laplace transforms, however, is by no means restricted to this class of problems, and some understanding of the basic theory is an essential part of the mathematical background required of engineers, scientists, and mathematicians.

The Laplace transform is defined in terms of an integral over the interval $[0, \infty)$. Integrals over an infinite interval are called *improper integrals*, and a brief review of the basic properties of such integrals is contained in Section 2.

If the function f is defined on the interval $[b, \infty)$, and if the improper integral $\int_b^\infty |f(x)|\,dx$ is convergent, then the improper integral $\int_b^\infty f(x)\,dx$ is said to be *absolutely convergent*. That every absolutely convergent improper integral is convergent is a fact which is analogous to a similar result for infinite series. Another fact which will be useful in the development of the properties of the Laplace transform, and which also has an infinite series analogue, is the following comparison theorem.

1 THEOREM

Let the functions f and g be defined on $[b, \infty)$ and integrable on $[b, c]$ for each $c \geq b$. If $|f(x)| \leq g(x)$ on $[b, \infty)$ and $\int_b^\infty g(x)\,dx$ converges, then $\int_b^\infty f(x)\,dx$ is absolutely convergent, and hence convergent. On the other hand, if $0 \leq g(x) \leq f(x)$ on $[b, \infty)$ and $\int_b^\infty g(x)\,dx$ is divergent, then $\int_b^\infty f(x)\,dx$ is divergent.

---

* This chapter depends upon the material in Chapters 2 and 3.

Although we shall not give a formal proof of this theorem, the result can be made intuitively clear by considering the areas represented by $\int_b^\infty g(x)\,dx$ and $\int_b^\infty |f(x)|\,dx$. The comparison theorem is useful in establishing the convergence or divergence of improper integrals $\int_b^\infty f(x)\,dx$ in which the function f is difficult (or impossible) to integrate explicitly in terms of elementary functions.

We now begin our discussion of the Laplace transform.

2 DEFINITION

Let $f \in \mathscr{C}[0, \infty)$. The Laplace transform of f, denoted by $\mathscr{L}[f(x)]$, or by $F(s)$, is given by

(3) $\mathscr{L}[f(x)] = F(s) = \displaystyle\int_0^\infty e^{-sx} f(x)\,dx$

The domain of F is the set of all real numbers s for which this improper integral converges.

In more advanced treatments of the Laplace transform, the parameter s assumes complex values, but the restriction to real values of s is sufficient for our purposes. Clearly, the operator $\mathscr{L}$ transforms the function f into a function F of the parameter s. In the definition, note that we have assumed the function f is continuous on $[0, \infty)$. This will be the assumption throughout the first three sections of this chapter, and it is made for convenience in presenting the basic properties of $\mathscr{L}$ and applying the Laplace transform method in solving initial-value problems. In the final section of this chapter, we extend the definition of $\mathscr{L}$ to a larger class of functions, the piecewise continuous functions on $[0, \infty)$, and we apply $\mathscr{L}$ to the problem of solving a nonhomogeneous equation in which the nonhomogeneous term is piecewise continuous. This will involve some extension of our concepts of a differential equation, solutions, and existence and uniqueness.

The primary application of Laplace transforms of interest to us is in solving initial-value problems for linear differential equations with constant coefficients. Referring to the work in Chapter 3, the functions which arise naturally in the treatment of equations with constant coefficients are those specified in the Table 3, Section 16. The Laplace transforms of some of these functions are calculated in the following examples.

4 Examples

a. Consider the constant function $f = 1$ on $[0, \infty)$. Using Definition 2,

$$\mathscr{L}[1] = \int_0^\infty e^{-sx} \cdot 1\,dx = \lim_{c \to \infty} \int_0^c e^{-sx}\,dx$$

$$= \lim_{c \to \infty} \left[\frac{e^{-sx}}{-s} \Big|_0^c \right] = \lim_{c \to \infty} \left[\frac{e^{-sc}}{-s} \right] + \frac{1}{s}$$

Now $\lim_{c \to \infty} (e^{-sc}/-s)$ exists if and only if $s > 0$. Thus $\mathscr{L}[1] = 1/s$, $s > 0$.

b. Consider the function $f(x) = e^{rx}, r \neq 0$, on $[0, \infty)$. Then

$$\mathscr{L}[e^{rx}] = \int_0^\infty e^{-sx} e^{rx} \, dx = \lim_{c \to \infty} \int_0^c e^{-(s-r)x} \, dx$$

$$= \lim_{c \to \infty} \left[\frac{e^{-(s-r)x}}{-(s-r)} \Big|_0^c \right]$$

$$= \lim_{c \to \infty} \left[\frac{e^{-(s-r)c}}{-(s-r)} \right] + \frac{1}{s-r}$$

Since the limit exists (and has the value 0) if and only if $s - r > 0$, we have

$$\mathscr{L}[e^{rx}] = \frac{1}{s-r}, \qquad s > r$$

Note that if $r = 0$, then we have the result in (a).

c. Let $f(x) = \sin(ax), a \neq 0$, on $[0, \infty)$. Then

$$\mathscr{L}[\sin(ax)] = \int_0^\infty e^{-sx} \sin(ax) \, dx = \lim_{c \to \infty} \int_0^c e^{-sx} \sin(ax) \, dx$$

$$= \lim_{c \to \infty} \frac{[s \sin(ax) - a \cos(ax)]e^{-sx}}{s^2 + a^2} \Big|_0^c$$

(Note: The integral is calculated using integration by parts.)

$$\mathscr{L}[\sin(ax)] = \lim_{c \to \infty} \frac{[s \sin(ac) - a \cos(ac)]e^{-sc}}{s^2 + a^2} + \frac{a}{s^2 + a^2}$$

Since $[s \sin(ac) - a \cos(ac)]/(s^2 + a^2)$ is bounded, the limit exists (and has the value 0) if and only if $s > 0$. Thus

$$\mathscr{L}[\sin(ax)] = \frac{a}{s^2 + a^2}, \qquad s > 0$$

The following table completes the list of the Laplace transforms of the functions specified by Table 3, Section 16. The entries in the table can be obtained using Definition 3. However, these entries can be verified more efficiently by using the basic properties of the Laplace transform which are to be developed in the next section. For a more extensive table see the *Handbook of Mathematical Functions*, N.B.S. Applied Mathematics Series No. 55.

5 TABLE

$f(x)$	$F(s) = \mathscr{L}[f(x)]$
c, c a constant	$\dfrac{c}{s}$, $s > 0$
e^{rx}, r a real number	$\dfrac{1}{s - r}$, $s > r$
$\sin(ax)$, a a real number	$\dfrac{a}{s^2 + a^2}$, $s > 0$
$\cos(ax)$, a a real number	$\dfrac{s}{s^2 + a^2}$, $s > 0$
x^n, $n = 1, 2, \ldots$	$\dfrac{n!}{s^{n+1}}$, $s > 0$
$x^n e^{rx}$, $n = 1, 2, \ldots$	$\dfrac{n!}{(s - r)^{n+1}}$, $s > r$
$x \sin(ax)$	$\dfrac{2as}{(s^2 + a^2)^2}$, $s > 0$
$x \cos(ax)$	$\dfrac{s^2 - a^2}{(s^2 + a^2)^2}$, $s > 0$
$e^{ax} \sin(bx)$	$\dfrac{b}{(s - a)^2 + b^2}$, $s > a$
$e^{ax} \cos(bx)$	$\dfrac{s - a}{(s - a)^2 + b^2}$, $s > a$

6 EXERCISES

1. The first three entries in Table 5 were verified in Example 3. Verify the entries in Table 5 for $\cos ax$, x^n, and $e^{ax} \sin bx$.

2. Show that
$$\mathscr{L}[\sin x] = \frac{\int_0^{2\pi} e^{-sx} \sin x \, dx}{1 - e^{-2\pi s}}$$

3. Show that
$$\mathscr{L}[\cos 2\pi x] = \frac{\int_0^1 e^{-sx} \cos 2\pi x \, dx}{1 - e^{-s}}$$

4. Let f be a continuous function on $[0, \infty)$ and have the property $f(x + b) = f(x)$, $b > 0$, for all x; that is, f is periodic. Show that
$$\mathscr{L}[f(x)] = \frac{\int_0^b e^{-sx} f(x) \, dx}{1 - e^{-bs}}$$

5. Let f be the function defined by

$$f(x) = \begin{cases} \sin x & 0 \le x \le \pi \\ 0 & \pi < x < 2\pi \end{cases}$$

and extended periodically, that is, $f(x) = \sin x$ on $[2\pi, 3\pi]$, $f(x) = \varnothing$ on $[3\pi, 4\pi]$, and so on.

a. Sketch the graph of f.

b. Show that

$$\mathscr{L}[f(x)] = \frac{1}{(1 - e^{-\pi s})(s^2 + 1)}$$

6. Let f be the function defined by

$$f(x) = \begin{cases} x & 0 \le x \le 1 \\ 2 - x & 1 \le x \le 2 \end{cases}$$

and extended periodically. The graph of f is shown in the figure here.

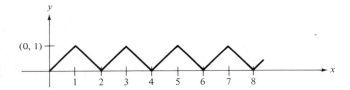

The graph of f is called a *triangular wave*. Show that

$$\mathscr{L}[f(x)] = \frac{1}{s^2} \tanh \left(\frac{s}{2} \right)$$

7. Let f be the function defined by

$$f(x) = \begin{cases} x & 0 \le x \le 3 \\ 3 & 3 < x < \infty \end{cases}$$

Find $\mathscr{L}[f(x)]$.

Section 35 Basic Properties of the Laplace Transform

In the preceding section, we defined the Laplace transform and calculated the Laplace transforms of the functions which arise naturally in solving differential equations with constant coefficients. In this section, we consider the basic question of existence of the Laplace transform of a given function f, and we develop the properties of the Laplace transform operator $\mathscr{L}$ which will be used in solving initial-value problems. All of the functions under consideration in this section will be members of $\mathscr{C}[0, \infty)$.

To motivate the development which follows, consider the second-order equation

(1) $P_2(D)[y] = a_0 D^2 y + a_1 Dy + a_2 y = f(x)$

where a_0, a_1, and a_2 are numbers, and $f \in \mathscr{C}[0, \infty)$. If we assume that $y = y(x)$ is a solution of (1), and if we formally apply the Laplace transform operator, we obtain

(2) $\qquad \mathscr{L}\{a_0 D^2 y(x) + a_1 Dy(x) + a_2 y(x)\} = \mathscr{L}[f(x)]$

The right-hand side of this equation suggests the basic question of the existence of $\mathscr{L}[f(x)]$. That is, for what functions $f \in \mathscr{C}[0, \infty)$ does the function $\mathscr{L}[f]$ exist, and what is its domain?

3 DEFINITION

Let $f \in \mathscr{C}[0, \infty)$. The function f is said to be of *exponential order r*, r a real number, if there exist nonnegative constants M and A such that

$$|f(x)| \le Me^{rx}$$

on $[A, \infty)$.

---

4 Examples

    a. If f is a bounded function [e.g., $f(x) = \sin(ax)$, or $f(x) = \cos(ax)$], then f is of exponential order 0 since f bounded implies

$$|f(x)| \le M = M \cdot e^{0x}$$

on $[0, \infty)$, for some $M \ge 0$. [If $f(x) = \sin(ax)$, or $f(x) = \cos(ax)$, then we could take $M = 1$.]

    b. Consider $f(x) = x$ on $[0, \infty)$. For any positive number r,

$$\lim_{x \to \infty} \frac{x}{e^{rx}} = 0$$

using L'Hôpital's rule. Therefore, there exists a nonnegative number A such that

$$\frac{x}{e^{rx}} \le 1$$

on $[A, \infty)$, or

$$|f(x)| \le 1 \cdot e^{rx}$$

on $[A, \infty)$. Thus $f(x) = x$ is of exponential order r for any positive number r. The same procedure can be used to show that $f(x) = x^p$, $p > 0$, is of exponential order r for any positive number r.

    c. Consider the function $f(x) = e^{x^2}$. If f is of exponential order r for some $r > 0$, then

$$e^{-rx}e^{x^2} \le M$$

on $[A, \infty)$, for some nonnegative constants M and A. But

$$\lim_{x \to \infty} e^{-rx}e^{x^2} = \lim_{x \to \infty} e^{x(x-r)} = \infty$$

for each real number r, and we have a contradiction. Thus $f(x) = e^{x^2}$ is not of exponential order r for any number r.

We now show that a sufficient condition for the Laplace transform of a function $f \in \mathscr{C}[0, \infty)$ to exist is that f be of exponential order r for some real number r.

5 THEOREM

Let $f \in \mathscr{C}[0, \infty)$. If f is of exponential order r, then the Laplace transform $\mathscr{L}[f(x)] = F(s)$ exists for $s > r$.

Proof: We must show that the improper integral

(B) $\int_0^{\infty} e^{-sx}f(x)\, dx$

exists for all $s > r$. In particular, we shall show that this integral is absolutely convergent for $s > r$.

Since f is of exponential order r, there exist constants M and A such that

$$|f(x)| \le Me^{rx}$$

on $[A, \infty)$. Now, for any $c > A$, we have

$$\int_0^c e^{-sx}|f(x)|\, dx = \int_0^A e^{-sx}|f(x)|\, dx + \int_A^c e^{-sx}|f(x)|\, dx$$
$$\le \int_0^A e^{-sx}|f(x)|\, dx + M \int_A^c e^{-sx} \cdot e^{rx}\, dx$$

Thus

$$\int_0^c e^{-sx}|f(x)|\, dx \le \int_0^A e^{-sx}|f(x)|\, dx + M \left.\frac{e^{-(s-r)x}}{-(s-r)}\right|_A^c$$

and

$$\int_0^c e^{-sx}|f(x)|\, dx \le \int_0^A e^{-sx}|f(x)|\, dx + \frac{Me^{-(s-r)A}}{(s-r)} - \frac{Me^{-(s-r)c}}{(s-r)}$$

Notice that $\lim_{c \to \infty} M(s-r)^{-1}e^{-(s-r)c}$ exists if and only if $s > r$. Thus we conclude that

$$\lim_{c \to \infty} \int_0^c e^{-sx}|f(x)|\, ds$$

exists for all $s > r$. It now follows that the improper integral (B) converges for all $s > r$, and this proves the theorem.

The result above answers, in part, the problem of the existence of the right side of equation (2). Since the left side of (2) involves the application of $\mathscr{L}$ to a linear combination of functions, our next result is concerned with this concept in general.

6 THEOREM

The operator $\mathscr{L}$ is linear. That is, if g and h are in $\mathscr{C}[0, \infty)$, and each of $\mathscr{L}[g(x)]$ and $\mathscr{L}[h(x)]$ exists for $s > r$, and if c_1 and c_2 are real numbers, then $\mathscr{L}[c_1 g(x) + c_2 h(x)]$ exists for $s > r$ and

$$\mathscr{L}[c_1 g(x) + c_2 h(x)] = c_1 \mathscr{L}[g(x)] + c_2 \mathscr{L}[h(x)]$$

Proof: Exercise.

7 Example

Consider the function

$$f(x) = 2e^{-3x} + 3 \sin(2x)$$

From Table 5, Section 34, $\mathscr{L}[e^{-3x}] = 1/(s + 3)$, $s > -3$, and $\mathscr{L}[\sin(2x)] = 2/(s^2 + 4)$, $s > 0$. Thus, since $\mathscr{L}$ is a linear operator, we have, for $s > 0$,

$$\mathscr{L}[f(x)] = \mathscr{L}[2e^{-3x} + 3 \sin(2x)] = 2 \left(\frac{1}{s + 3} \right) + 3 \left(\frac{2}{s^2 + 4} \right)$$

$$= \frac{2}{s + 3} + \frac{6}{s^2 + 4} = \frac{2s^2 + 6s + 26}{(s + 3)(s^2 + 4)}$$

Using the linearity of $\mathscr{L}$ in (2), we obtain

(8) $\qquad a_0 \mathscr{L}[D^2 y(x)] + a_1 \mathscr{L}[Dy(x)] + a_2 \mathscr{L}[y(x)] = \mathscr{L}[f(x)]$

This equation suggests the next property of $\mathscr{L}$, namely, a relationship between the Laplace transform of the derivative of a function and the Laplace transform of the function itself.

9 THEOREM

Let $g \in \mathscr{C}^1[0, \infty)$. If g is of exponential order r, then $\mathscr{L}[Dg(x)]$ exists for $s > r$ and

$$\mathscr{L}[Dg(x)] = s\mathscr{L}[g(x)] - g(0)$$

Proof: By Theorem 5, $\mathscr{L}[g(x)]$ exists for $s > r$. From the definition of $\mathscr{L}$, we have

$$\mathscr{L}[Dg(x)] = \int_0^\infty e^{-sx} Dg(x)\, dx = \lim_{c \to \infty} \int_0^c e^{-sx} Dg(x)\, dx$$

Now, for each positive number c,

$$\int_0^c e^{-sx} Dg(x)\, dx = e^{-sx}g(x)\Big|_0^c + s\int_0^c e^{-sx}g(x)\, dx$$
$$= e^{-sc}g(c) - g(0) + s\int_0^c e^{-sx}g(x)\, dx$$

using integration by parts. Therefore,

$$\mathscr{L}[Dg(x)] = s\mathscr{L}[g(x)] - g(0) + \lim_{c\to\infty} e^{-sc}g(c)$$

provided that the last limit exists. Since g is of exponential order r, there exist constants M and A such that

$$|g(x)| \le Me^{rx}$$

on $[A, \infty]$. Thus

$$|e^{-sc}g(c)| \le |e^{-sc}Me^{rc}| = Me^{-(s-r)c}$$

for all $c > A$, and so $\lim_{c\to\infty} e^{-sc}g(c) = 0$ for $s > r$. The theorem now follows.

By repeated application of this rule, that is, by induction, we have the general result given in the following corollary.

10 COROLLARY

Let $g \in \mathscr{C}^n(0, \infty)$. If each of the functions $g, Dg, \ldots, D^{n-1}g$ is of exponential order r, then $\mathscr{L}[D^n g(x)]$ exists for $s > r$ and

$$\mathscr{L}[D^n g(x)] = s^n \mathscr{L}[g(x)] - s^{n-1}g(0) - \cdots - s D^{n-2}g(0) - D^{n-1}g(0)$$

Although the main use of Corollary 10 is in solving linear differential equations, the result can also be used as a means of determining entries in a table of Laplace transforms. For example, if f is a differentiable function and $\mathscr{L}[f(x)]$ is known, then $\mathscr{L}[Df(x)]$ can be determined.

11 Example

From Example 4, Section 34, $\mathscr{L}[\sin(ax)] = a/(s^2 + a^2)$, $s > 0$. Since $D[\sin(ax)] = a\cos(ax)$,

$$\mathscr{L}[a\cos(ax)] = s\mathscr{L}[\sin(ax)] - \sin(0)$$
$$= \frac{sa}{s^2 + a^2}$$

Using the fact that $\mathscr{L}$ is a linear operator, we can solve for $\mathscr{L}[\cos(ax)]$ to obtain

$$\mathscr{L}[\cos(ax)] = \frac{s}{s^2 + a^2}, \qquad s > 0$$

as indicated in Table 5, Section 34.

We now use these properties of $\mathscr{L}$ to determine the Laplace transform of a solution of (1). While this has no significance in itself, it will be shown that this procedure leads to a solution method for certain linear differential equations.

Let $y = y(x)$ be a solution of equation (1). Applying $\mathscr{L}$, we obtain

$$\mathscr{L}[a_0 D^2 y(x) + a_1 Dy(x) + a_2 y(x)] = \mathscr{L}[f(x)]$$

Using the linearity of $\mathscr{L}$, this becomes

$$a_0 \mathscr{L}[D^2 y(x)] + a_1 \mathscr{L}[Dy(x)] + a_2 \mathscr{L}[y(x)] = \mathscr{L}[f(x)]$$

and from Corollary 10, we have

$$a_0 s^2 \mathscr{L}[y(x)] - a_0 sy(0) - a_0 Dy(0) + a_1 s\mathscr{L}[y(x)] - a_1 y(0)$$
$$+ a_2 \mathscr{L}[y(x)] = \mathscr{L}[f(x)]$$

Solving for $\mathscr{L}[y(x)]$ gives

(12) $$\mathscr{L}[y(x)] = \frac{\mathscr{L}[f(x)] + a_0 y(0)s + a_0 Dy(0) + a_1 y(0)}{a_0 s^2 + a_1 s + a_2}$$

Implicit in the derivation of (12) is the assumption that $\mathscr{L}[y(x)]$ exists. Assuming that this is the case, notice that (12) would be especially useful for getting the Laplace transform of the solution of a second-order initial-value problem with initial conditions specified at $x = 0$. Notice also that the denominator in (12) is $P_2(s)$, where $P_2(D)$ is the operator polynomial which defines the equation (1). It should be clear that the same procedure can be applied in the case of an nth-order equation with constant coefficients.

13 Examples

a. Given the initial-value problem

$$D^2 y + 4y = e^{2x} - x; \qquad y(0) = 1, \quad Dy(0) = -2$$

If $y = y(x)$ is the solution, then

$$\mathscr{L}[D^2 y(x) + 4y(x)] = \mathscr{L}[e^{2x} - x] = \frac{1}{s - 2} - \frac{1}{s^2}$$

and

$$s^2 \mathscr{L}[y(x)] - sy(0) - Dy(0) + 4\mathscr{L}[y(x)] = \frac{s^2 - s + 2}{s^2(s - 2)}$$

Using the initial conditions, and solving for $Y(s) = \mathscr{L}[y(x)]$, we obtain

$$Y(s) = \frac{1}{s^2 + 4}\left[\frac{s^2 - s + 2}{s^2(s - 2)} + s - 2\right] = \frac{s^4 - 4s^3 + 5s^2 - s + 2}{(s^2 + 4)(s^2)(s - 2)}$$

b. Given the initial-value problem

$$D^3 y - 5 D^2 y + 6 Dy = \varnothing; \qquad y(0) = 2, \quad Dy(0) = 0, \quad D^2 y(0) = -1$$

If $y = y(x)$ is the solution, then

$$\mathcal{L}[D^3 y(x) - 5 D^2 y(x) + 6 Dy(x)] = \mathcal{L}[\varnothing] = \varnothing$$

and

$$s^3 \mathcal{L}[y(x)] - s^2 y(0) - s Dy(0) - D^2 y(0)$$
$$- 5\{s^2 \mathcal{L}[y(x)] - sy(0) - Dy(0)\} + 6\{s\mathcal{L}[y(x)] - y(0)\} = \varnothing$$

Using the initial conditions and solving for $Y(s) = \mathcal{L}[y(x)]$, we have

$$Y(s) = \frac{2s^2 - 10s + 11}{s^3 - 5s^2 + 6s}$$

There is still the general question of the existence of $\mathcal{L}[y(x)]$, where $y(x)$ is a solution of equation (1). It can be shown that if $f(x)$ is continuous and of exponential order r, then $\mathcal{L}[y(x)]$ does exist. The domain of $\mathcal{L}[y(x)]$ depends upon f and on the operator polynomial $P_2(D)$. This also holds for higher-order linear equations with constant coefficients. Verification of these facts is contained in the exercises.

14 EXERCISES

1. Determine which of the following functions is of exponential order.
   a. $f(x) = x^2 e^x$
   b. $f(x) = x^{3/2}$
   c. $f(x) = x^3 \sin x$
   d. $f(x) = \ln(1 + x)$
   e. $f(x) = x^2 e^{4x} \cos x$
   f. $f(x) = x^2 \ln(1 + x)$

2. Determine the Laplace transform of the solution of each of the following initial-value problems.
   a. $Dy - 2y = \varnothing; y(0) = 1$
   b. $4 D^2 y - 4 Dy + y = \varnothing; y(0) = 0, Dy(0) = 2$
   c. $D^3 y - D^2 y + Dy - y = \varnothing; y(0) = -1, Dy(0) = 1, D^2 y(0) = -2$

3. Determine the Laplace transform of the solution of each of the following initial-value problems.
   a. $Dy - 2y = 1; y(0) = 1$
   b. $4 D^2 y - 4 Dy + y = e^{x/2} + 2 \sin x; y(0) = 0, Dy(0) = 2$
   c. $D^3 y - D^2 y + Dy - y = xe^x; y(0) = -1, Dy(0) = 1, D^2 y(0) = -2$

4. Let P be a polynomial of degree $n \geq 1$, and consider the homogeneous equation $P(D)y = \varnothing$. Show that if $y(x)$ is a solution of this equation, then

$$\mathcal{L}[y(x)] = \frac{Q(s)}{P(s)}$$

where Q is a polynomial of degree $< n$.

5. Consider the nonhomogeneous equation $P(D)y = f(x)$. Show that

$$\mathcal{L}[y(x)] = \frac{\mathcal{L}[f(x)]}{P(s)}$$

if and only if $y(0) = Dy(0) = D^2 y(0) = \cdots = D^{n-1} y(0) = 0$.

6. Verify Corollary 10.
7. Let $f(x)$ be continuous on $[0, \infty)$. Prove that f is of exponential order r if and only if

$$f(x) = g(x)e^{rx}$$

where g is bounded and continuous on $[0, \infty)$.
8. Use the relationship between the hyperbolic functions and the exponential function to show that

$$\mathscr{L}[\sinh(ax)] = \frac{a}{s^2 - a^2}$$

$$\mathscr{L}[\cosh(ax)] = \frac{s}{s^2 - a^2}$$

9. Prove that if f_1 and f_2 are each of exponential order, then $f_1 + f_2$ and $f_1 \cdot f_2$ are of exponential order.
10. Prove that if $f(x)$ is of exponential order and $y(x)$ is a solution of

$$Dy + ay = f(x)$$

then $y(x)$ and $Dy(x)$ are of exponential order and hence $y(x)$ and $Dy(x)$ have Laplace transforms. (Hint: See Section 8 for a representation of $y(x)$, and use Problem 7.)
11. Prove that if $f(x)$ is of exponential order and $y(x)$ is a solution of

$$a\,D^2y + b\,Dy + cy = f(x)$$

then $y(x)$, $Dy(x)$, and $D^2y(x)$ are of exponential order. (Hint: See Theorem 8, Section 12, for a representation of $y(x)$.)
12. Prove Theorem 6.
13. Suppose that y is a solution of equation (1) and that $\mathscr{L}[f]$ exists on (r, ∞) for some r. Let c be the maximum of $\{r, Re(r_1), Re(r_2)\}$, where r_1 and r_2 are the roots of P_2. Show that $\mathscr{L}[y]$ exists on (c, ∞).

We conclude this section with a discussion of two additional properties of the Laplace transform. These properties are useful in determining the entries in a table of Laplace transforms, and they increase the flexibility of such tables.

15 THEOREM

Let $f \in \mathscr{C}[0, \infty)$ and be of exponential order r. Then $F(s) = \mathscr{L}[f(x)]$ has derivatives of all orders, and for $s > r$,

$$\frac{d^n F(s)}{ds^n} = \mathscr{L}[(-x)^n f(x)]$$

$n = 1, 2, \ldots$.

Proof: Exercise.

Assuming that $F(s)$ is differentiable, we give an intuitive justification of the formula for the nth derivative of $F(s)$. In particular, from the definition of $\mathcal{L}[f(x)]$,

$$F(s) = \int_0^\infty e^{-sx} f(x)\, dx$$

Differentiation with respect to s under the integral sign can be justified and yields

$$\frac{dF(s)}{ds} = \int_0^\infty -xe^{-sx} f(x)\, dx = \int_0^\infty e^{-sx}[-xf(x)]\, dx$$

$$= \mathcal{L}[-xf(x)] = -\mathcal{L}[xf(x)]$$

Differentiating a second time, we obtain

$$\frac{d^2 F(s)}{ds^2} = \int_0^\infty x^2 e^{-sx} f(x)\, dx = \mathcal{L}[x^2 f(x)]$$

Thus the formula specified will follow using induction.

Theorem 15 can be used to determine a number of entries in Table 5, Section 34.

16 Example

From Table 5, Section 34, $\mathcal{L}[e^{rx}] = 1/(s - r), s > r$. By Theorem 15,

$$\mathcal{L}[xe^{rx}] = -\frac{d\left(\dfrac{1}{s - r}\right)}{ds} = \frac{1}{(s - r)^2}, \qquad s > r$$

$$\mathcal{L}[x^2 e^{rx}] = \frac{d^2\left(\dfrac{1}{s - r}\right)}{ds^2} = \frac{2}{(s - r)^3}, \qquad s > r$$

In general,

$$\mathcal{L}[x^n e^{rx}] = \frac{n!}{(s - r)^{n+1}}, \qquad s > r$$

n a positive integer. Note that if $r = 0$ in this formula, then we have

$$\mathcal{L}[x^n] = \frac{n!}{s^{n+1}}, \qquad s > 0$$

The final property which we present is called the translation property of $\mathcal{L}$.

17 THEOREM

If $f \in \mathcal{C}[0, \infty)$ and $\mathcal{L}[f(x)] = F(s)$ exists for $s > c$, then for any real number r,

$$\mathcal{L}[e^{rx} f(x)] = F(s - r), \qquad s > c + r$$

Proof: From the definition of the Laplace transform,

$$F(s - r) = \int_0^\infty e^{-(s-r)x}f(x)\,dx = \int_0^\infty e^{-sx}e^{rx}f(x)\,dx$$
$$= \mathcal{L}[e^{rx}f(x)]$$

18 Examples

a. Since $\mathcal{L}[x] = F(s) = 1/s^2$, $s > 0$, we have

$$\mathcal{L}[xe^{rx}] = F(s - r) = \frac{1}{(s - r)^2}, \qquad s > r$$

as also seen in Example 16.

b. Since $\mathcal{L}[\sin(bx)] = b/(s^2 + b^2)$, $s > 0$,

$$\mathcal{L}[e^{ax}\sin(bx)] = \frac{b}{(s - a)^2 + b^2}, \qquad s > a$$

as indicated in Table 5, Section 34.

19 EXERCISES

1. Verify the entries x^n, $x^n e^{rx}$, $x \sin(ax)$, $x \cos(ax)$, and $e^{ax} \cos(bx)$ in Table 5, Section 34, using Theorems 15 and 17.
2. Determine $\mathcal{L}[xe^{ax}\sin(bx)]$ by two methods, namely, by using Theorem 15, and by using Theorem 17.
3. Show that $\mathcal{L}[x \cosh(ax)] = (s^2 + a^2)/(s^2 - a^2)^2$.
4. Show that $\mathcal{L}[\sin(ax)\sinh(ax)] = 2a^2 s/(s^4 + a^4)$.
5. Show that $\mathcal{L}[(1 + x^2)\sin(x) - x\cos(x)] = 8s^2/(s^2 + 1)^3$.
6. Suppose that $\mathcal{L}[f(x)] = F(s)$ and $c > 0$. Show that $\mathcal{L}[f(x/c)] = cF(cs)$. This result is useful for scaling.
7. The combination of scaling and translating is given by

$$\mathcal{L}\left[e^{bx/cf}\left(\frac{x}{c}\right)\right] = cF(cs - b)$$

where $\mathcal{L}[f(x)] = F(s)$ and $c > 0$. Verify this formula.
8. The gamma function Γ is defined by

$$\Gamma(x) = \int_0^\infty t^{x-1}e^{-t}\,dt$$

(See Section 33.)

a. Prove that for any number $r > -1$,

$$\mathcal{L}(x^r) = \frac{\Gamma(r + 1)}{s^{r+1}}, \qquad s > 0$$

b. Show that $\mathcal{L}[x^{-1/2}] = \sqrt{\pi/s}$. (Hint: $\Gamma(\tfrac{1}{2}) = \sqrt{\pi}$.)
9. Show that if f is of exponential order r, then $F(s) = \mathcal{L}[f(x)]$ is differentiable. (Hint: Use the definition of the derivative.)
10. Prove Theorem 15. (Hint: See Exercises 14, Problem 7.)

Section 36 Inverse Transforms and Initial Value Problems

In the preceding section, we used the differential equation

(1) $P_2(D)[y] = a_0 D^2 y + a_1 Dy + a_2 y = f(x)$

to motivate the basic properties of the Laplace transform operator $\mathcal{L}$. Assuming that the Laplace transform of f exists, and using the results of the preceding section, we can transform (1) into

(2) $[a_0 s^2 + a_1 s + a_2]\mathcal{L}[y(x)] - a_0 y(0)s - a_0 Dy(0) - a_1 y(0)$
$$= \mathcal{L}[f(x)]$$

which we can solve for $Y(s) = \mathcal{L}[y(x)]$ to obtain

(3) $Y(s) = \dfrac{1}{a_0 s^2 + a_1 s + a_2} \{\mathcal{L}[f(x)] + a_0 y(0)s + a_0 Dy(0) + a_1 y(0)\}$

After calculating the Laplace transform of the solution of an initial-value problem, as in (3), we would next want to determine the solution from its Laplace transform. The general problem of finding a function with a given Laplace transform is called the *inversion problem*. The inversion problem and its application in solving initial-value problems is the topic of this section.

If $f \in \mathscr{C}[0, \infty)$ and the Laplace transform, $\mathcal{L}[f(x)] = F(s)$, of f exists for $s > r$, then the function F is uniquely determined by f; that is, $\mathcal{L}$ is itself a function. Our first result shows that $\mathcal{L}$ is a one-to-one function on $\mathscr{C}[0, \infty)$. We shall omit the proof of this theorem since it involves methods which are beyond the scope of this introductory treatment.

4 THEOREM

If f and g are functions in $\mathscr{C}[0, \infty)$ and $\mathcal{L}[f(x)] = \mathcal{L}[g(x)]$, then $f = g$; that is, $f(x) = g(x)$ for all $x \in [0, \infty)$.

The terminology and notation used in treating the inversion problem are contained in the following definition.

5 DEFINITION

If $F(s)$ is a given transform and $f \in \mathscr{C}[0, \infty)$ has the property $\mathcal{L}[f(x)] = F(s)$, then f is called the *inverse Laplace transform* of $F(s)$, and is denoted by

$$f(x) = \mathcal{L}^{-1}[F(s)]$$

The operator $\mathcal{L}^{-1}$ is called the *inverse operator* of $\mathcal{L}$.

There is a general formula for the inverse operator $\mathscr{L}^{-1}$ corresponding to (3), Section 34, but use of the formula requires a knowledge of functions of a complex variable, and, therefore, we will not include it in our treatment. The relationship between $\mathscr{L}$ and $\mathscr{L}^{-1}$ is specified by the following equations:

(6) $\mathscr{L}^{-1}\{\mathscr{L}[f(x)]\} = f(x)$

(7) $\mathscr{L}\{\mathscr{L}^{-1}[F(s)]\} = F(s)$

for all $f \in \mathscr{C}[0, \infty)$ such that $\mathscr{L}[f(x)] = F(s)$.

By reading from right to left, Table 5, Section 34, can be used as a table of inverse transforms. We consider some specific examples.

8 Examples

a. Given the transform $F(s) = 1/(s + 2)$, $s > 2$. From Table 5, Section 34,

$$\mathscr{L}[e^{rx}] = \frac{1}{s - r}$$

Thus

$$\mathscr{L}^{-1}\left[\frac{1}{s + 2}\right] = e^{-2x}$$

b. Consider the transform

$$F(s) = \frac{2}{s^2 - 2s + 5}, \qquad s > 1$$

While it may seem that this function does not appear in Table 5, Section 34, notice that $s^2 - 2s + 5 = (s - 1)^2 + 4$. Thus

$$F(s) = \frac{2}{(s - 1)^2 + 2^2}, \qquad s > 1$$

and

$$\mathscr{L}^{-1}[F(s)] = e^x \sin 2x$$

The properties of the Laplace transform operator $\mathscr{L}$ can be used to derive corresponding properties for its inverse operator $\mathscr{L}^{-1}$. The most important property is that of linearity, and it is easy to prove.

9 THEOREM

The operator $\mathscr{L}^{-1}$ is linear, that is, if $F_1(s)$ and $F_2(s)$ are Laplace transforms, and c_1 and c_2 are real numbers, then

$$\mathscr{L}^{-1}[c_1 F_1(s) + c_2 F_2(s)] = c_1 \mathscr{L}^{-1}[F(s)] + c_2 \mathscr{L}^{-1}[F_2(s)]$$

The translation property (Theorem 17, Section 35) of the operator $\mathscr{L}$ is also useful in determining inverse transforms. The analogue of Theorem 17, Section 35, for $\mathscr{L}^{-1}$ is given by the following theorem.

10 THEOREM

If $f \in \mathscr{C}[0, \infty)$ and $\mathscr{L}[f(x)] = F(s)$ exists for $s > c$, then for any real number r,

$$\mathscr{L}^{-1}[F(s - r)] = e^{rx}f(x)$$

11 Example

Consider the function

$$F(s) = \frac{2}{s + 2} + \frac{1}{s^2 - 2s + 5}$$

From Theorem 9,

$$\mathscr{L}^{-1}[F(s)] = 2\mathscr{L}^{-1}\left[\frac{1}{s + 2}\right] + \mathscr{L}^{-1}\left[\frac{1}{s^2 - 2s + 5}\right]$$

From Table 5, Section 34, $\mathscr{L}^{-1}[1/(s + 2)] = e^{-2x}$, and

$$\mathscr{L}^{-1}\left[\frac{1}{s^2 - 2s + 5}\right] = \tfrac{1}{2}\mathscr{L}^{-1}\left[\frac{2}{(s - 1)^2 + 2^2}\right] = \tfrac{1}{2}e^x \sin 2x$$

Thus

$$\mathscr{L}^{-1}[F(s)] = 2e^{-2x} + \tfrac{1}{2}e^x \sin 2x$$

12 EXERCISES

1. Determine the inverse Laplace transform of each of the following.
   \a. $F(s) = 6/(s + 7)$
   b. $F(s) = 1/(2s - 1)$
   \c. $F(s) = 1/(s^2 + 25)$
   d. $F(s) = [4/(s - 3)^3] - [3/(s - 3)^2] + [4/(s - 3)]$
   \e. $F(s) = (s + 4)/(s^2 + 8s + 17)$
   f. $F(s) = 4/(s^2 - 6s + 13)$
   g. $F(s) = (s^2 + 4s - 4)/(s^2 + 4)^2$

2. Suppose $f \in \mathscr{C}[0, \infty)$ and $\mathscr{L}[f(x)] = F(s)$. Show that

$$\mathscr{L}^{-1}\left[\frac{F(s)}{s - a}\right] = \int_0^x e^{-at}f(t)\, dt$$

3. Is there a function $f \in \mathscr{C}[0, \infty)$ such that

$$\mathscr{L}[f(x)] = \ln(s + a)$$

(Hint: Apply Theorem 15, Section 35, to show $\mathcal{L}(x \cdot f(x)) = -1/(s + a)$ and consider $\lim\limits_{x \to 0^+} x \cdot f(x)$.)

We now consider the Laplace transform method for solving initial-value problems by treating several examples in detail.

13 Examples

a. Given the initial-value problem

$$D^2 y + 4y = 5e^x; \qquad y(0) = 1, \quad Dy(0) = 4$$

Applying the operator $\mathcal{L}$, and using the properties presented in the previous section, we have

$$s^2 \mathcal{L}[y(x)] - sy(0) - Dy(0) + 4\mathcal{L}[y(x)] = \frac{5}{s - 1}$$

Using the initial conditions and solving for $Y(s) = \mathcal{L}[y(x)]$, we obtain

$$Y(s) = \frac{1}{s^2 + 4}\left[\frac{5}{s - 1} + s + 4 \right]$$

$$= \frac{5}{(s - 1)(s^2 + 4)} + \frac{s}{s^2 + 4} + \frac{4}{s^2 + 4}$$

To find the solution $y(x)$ of the initial-value problem, we apply the inverse operator to the equation defining $Y(s)$. In particular, we have

$$y(x) = \mathcal{L}^{-1}[Y(s)] = \mathcal{L}^{-1}\left[\frac{5}{(s-1)(s^2+4)} + \frac{s}{s^2+4} + \frac{4}{s^2+4} \right]$$

$$= \mathcal{L}^{-1}\left[\frac{5}{(s-1)(s^2+4)} \right] + \mathcal{L}^{-1}\left[\frac{s}{s^2+4} \right] + 2\mathcal{L}^{-1}\left[\frac{2}{s^2+4} \right]$$

The inverse transforms

$$\mathcal{L}^{-1}\left[\frac{s}{s^2 + 4} \right] = \cos(2x)$$

and

$$\mathcal{L}^{-1}\left[\frac{2}{s^2 + 4} \right] = \sin(2x)$$

are found from Table 5, Section 34, but the function

$$F(s) = \frac{5}{(s - 1)(s^2 + 4)}$$

does not appear in the table.

Since the function F is a rational function, the possibility of using the partial fraction decomposition of F is suggested. In particular, the partial fraction decomposition of F will have the form

$$F(s) = \frac{A}{s-1} + \frac{Bs+C}{s^2+4} = \frac{A}{s-1} + \frac{Bs}{s^2+4} + \frac{C}{s^2+4}$$

for some real numbers, A, B, and C, and we can find the inverse transform of each of these functions from Table 5, Section 34. From the equation

$$\frac{5}{(s-1)(s^2+4)} = \frac{A}{s-1} + \frac{Bs+C}{s^2+4}$$

we obtain $A = 1$, $B = -1$, and $C = -1$. Thus

$$F(s) = \frac{1}{s-1} - \frac{s}{s^2+4} - \frac{1}{s^2+4} = \frac{1}{s-1} - \frac{s}{s^2+4} - \left(\frac{1}{2}\right) \cdot \frac{2}{(s^2+4)}$$

and

$$\mathscr{L}^{-1}[F(s)] = e^x - \cos(2x) - \tfrac{1}{2}\sin(2x)$$

Finally,

$$y(x) = \mathscr{L}^{-1}[Y(s)] = e^x - \cos(2x) - \tfrac{1}{2}\sin(2x) + \cos(2x) + 2\sin(2x)$$
$$= e^x + \tfrac{3}{2}\sin(2x)$$

It is easy to verify that $y(x)$ is the solution of the initial-value problem.

b. Consider the initial-value problem

$$D^3y - 3D^2y + 2Dy = 4 + \sin(x); \qquad y(0) = 1, Dy(0) = D^2y(0) = 0$$

By taking the Laplace transform, we obtain

$$s^3\mathscr{L}[y(x)] - s^2 - 3\{s^2\mathscr{L}[y(x)] - s\} + 2\{s\mathscr{L}[y(x)] - 1\}$$
$$= \frac{4}{s} + \frac{1}{s^2+1}$$

Solving for $Y(s) = \mathscr{L}[y(x)]$, we have

$$Y(s) = \frac{4}{s^2(s-1)(s-2)} + \frac{1}{s(s-1)(s-2)(s^2+1)} + \frac{s^2-3s+2}{s(s-1)(s-2)}$$

$$= \frac{4}{s^2(s-1)(s-2)} + \frac{1}{s(s-1)(s-2)(s^2+1)} + \frac{1}{s}$$

The partial fraction decompositions of the first two functions on the right-hand side are

$$\frac{4}{s^2(s-1)(s-2)} = \frac{3}{s} + \frac{2}{s^2} - \frac{4}{s-1} + \frac{1}{s-2}$$

$$\frac{1}{s(s-1)(s-2)(s^2+1)} = \frac{\frac{1}{2}}{s} - \frac{\frac{1}{2}}{s-1} + \frac{\frac{1}{10}}{s-2} - \frac{(\frac{1}{10}s - \frac{3}{10})}{s^2+1}$$

Thus we have

$$Y(s) = \frac{\frac{9}{2}}{s} + \frac{2}{s^2} - \frac{\frac{9}{2}}{s-1} + \frac{\frac{11}{10}}{s-2} - \frac{1}{10}\left(\frac{s}{s^2+1}\right) + \frac{3}{10}\left(\frac{1}{s^2+1}\right)$$

and, from Table 5, Section 34,

$$y(x) = \mathscr{L}^{-1}[Y(s)] = \tfrac{9}{2} + 2x - \tfrac{9}{2}e^x + \tfrac{11}{10}e^{2x} - \tfrac{1}{10}\cos(x) + \tfrac{3}{10}\sin(x)$$

is the solution of the initial-value problem.

In using the Laplace transform method to solve initial-value problems, it is often the case that the transform $Y(s)$ of the solution is a rational function. Thus, as illustrated by the Example 13, a familiarity with the partial fraction decomposition of a rational function is an essential step in obtaining the inverse transform. The partial fraction decomposition of a rational function is normally encountered first in calculus, where it is used as an integration technique for rational functions. For purposes of completeness here, we shall discuss briefly the partial fraction decomposition of a rational function and give a few illustrative examples.

Recall first that a rational function $r(s)$ is a quotient of two polynomials, that is,

$$r(s) = \frac{q(s)}{p(s)}$$

where

$$p(s) = a_0 s^n + a_1 s^{n-1} + \cdots + a_n$$

and

$$q(s) = b_0 s^k + b_1 s^{k-1} + \cdots + b_k$$

are polynomials. If $n = \deg[p(s)] > k = \deg[q(s)]$, then $r(s)$ is said to be a *proper* rational function. If $n \le k$, then $r(s)$ is said to be *improper* (this corresponds to the terminology for rational numbers). Since an improper rational function can be expressed as a polynomial plus a proper rational function by dividing $q(s)$ by $p(s)$, we shall assume in the remainder of this discussion that $r(s)$ is proper. The essential step in the partial fraction decomposition of a (proper) rational function is the factorization of the denominator $p(s)$ into the product of its linear and quadratic factors corresponding to its real and complex roots.

The primary idea behind the technique is the assumption that the rational function

$$r(s) = \frac{q(s)}{p(s)}$$

$$= \frac{q(s)}{(s-r_1)^{k_1}\cdots(s-r_j)^{k_j}(s^2+a_1 s+b_1)^{m_1}\cdots(s^2+a_t s+b_t)^{m_t}}$$

has been obtained by adding "simple" rational functions of the form

$$\frac{A}{(s - r_i)^k} \quad \text{and} \quad \frac{Bs + C}{(s^2 + a_i s + b_i)^k}$$

We illustrate with some examples.

14 Example

Let $r(s) = (s^2 + 2s + 1)/(s - 1)(s - 2)(s^2 + 1)$. We assume that $r(s)$ has been obtained by adding the proper rational functions

$$\frac{A}{s - 1}, \quad \frac{B}{s - 2}, \quad \text{and} \quad \frac{Cs + D}{s^2 + 1}$$

where A, B, C, and D are unknown numbers which are to be determined so that the identity

$$\frac{s^2 + 2s + 1}{(s - 1)(s - 2)(s^2 + 1)} = \frac{A}{s - 1} + \frac{B}{s - 2} + \frac{Cs + D}{s^2 + 1}$$

holds. To calculate the values of A, B, C, and D, we formally add the terms on the right-hand side to obtain

$$\frac{s^2 + 2s + 1}{(s - 1)(s - 2)(s^2 + 1)}$$

$$= \frac{A(s - 2)(s^2 + 1) + B(s - 1)(s^2 + 1) + (Cs + D)(s - 1)(s - 2)}{(s - 1)(s - 2)(s^2 + 1)}$$

In order for this equation to be an identity, we must have

(A)
$$s^2 + 2s + 1 = A(s - 2)(s^2 + 1) + B(s - 1)(s^2 + 1)$$
$$+ (Cs + D)(s - 1)(s - 2)$$

for all s. Expanding the terms on the right-hand side, we get

$$s^2 + 2s + 1 = (A + B + C)s^3 + (-2A - B - 3C + D)s^2$$
$$+ (A + B + 2C - 3D)s + (-2A - B + 2D)$$

from which it follows that

$$A + B + C = 0$$
$$-2A - B - 3C + D = 1$$
$$A + B + 2C - 3D = 2$$
$$-2A - B + 2D = 1$$

Solving this system of equations, we obtain

$$A = -2, \quad B = \tfrac{9}{5}, \quad C = \tfrac{1}{5}, \quad D = -\tfrac{3}{5}$$

A technique for calculating A, B, C, and D, which is often more efficient than solving the system of equations obtained above, is based on the fact that (A) must hold for *all* values of s. Some judicious choices for s, then, lead to a rapid calculation of the unknown constants. In particular, setting $s = 1$ in (A) yields

$$4 = -2A + 0{\cdot}B + 0{\cdot}(Cs + D), \quad \text{or} \quad A = -2$$

Setting $s = 2$ in (A) yields

$$9 = 0{\cdot}A + 5B + 0{\cdot}(Cs + D), \quad \text{or} \quad B = \tfrac{9}{5}$$

Setting $s = 0$ in (A) yields

$$1 = -2A - B + 2D, \quad \text{or} \quad D = -\tfrac{3}{5}$$

using $A = -2$ and $B = \tfrac{9}{5}$. Setting $s = -1$ in (A) yields

$$0 = -6A - 4B - 6C + 6D, \quad \text{or} \quad C = \tfrac{1}{5}$$

using $A = -2$, $B = \tfrac{9}{5}$, and $D = -\tfrac{3}{5}$.

The case where $p(s)$ has linear or quadratic factors, or both, of multiplicity greater than one is illustrated by the following example.

15 Example

Let $r(s) = (2s - 3)/s(s + 1)^3(s^2 - 2s + 5)^2$. The partial fraction decomposition of $r(s)$ is:

$$\frac{2s - 3}{s(s + 1)^3(s^2 - 2s + 5)^2} = \frac{A}{s} + \frac{B}{s + 1} + \frac{C}{(s + 1)^2} + \frac{D}{(s + 1)^3}$$

$$+ \frac{Es + F}{s^2 - 2s + 5} + \frac{Gs + H}{(s^2 - 2s + 5)^2}$$

and the unknowns $A, B, \ldots, H$ can be calculated using either of the methods illustrated in Example 14.

An examination of the Laplace transform method for solving an initial-value problem of the form

$$P(D)[y] = f(x); \quad y(0) = b_0, \, Dy(0) = b_1, \ldots, D^{n-1}y(0) = b_{n-1}$$

where $P(D)$ is the operator polynomial associated with the polynomial $P(t) = a_0t^n + a_1t^{n-1} + \cdots + a_n$, shows that if the Laplace transform $F(s)$ of f is a rational function, then the Laplace transform $Y(s)$ of the solution is also a rational function. Moreover, the denominator of $Y(s)$ always involves the polynomial $P(s)$. Since the partial fraction decomposition technique requires the complete factorization of the denominator of the rational function, we find that it is necessary to factor the polynomial $P(t)$ (or $P(s)$)

corresponding to the operator polynomial. Of course, this was also an essential step in the methods described in Chapter 3, but an advantage of the Laplace transform method over the methods of Chapter 3 is that the solution is obtained directly, without first finding the n-parameter famil of solutions.

It is obvious from the properties of $\mathscr{L}$, and the illustrative examples, that the transform method requires the initial conditions to be specified at $x = 0$. If we are given an initial-value problem in which the initial conditions are specified at some point x_0, $x_0 \neq 0$, then a simple change of independent variable, in fact, a translation, can be used to transform the given initial-value problem into an initial-value problem with the initial conditions at 0.

16 Example

Given the initial-value problem

$$D^2y - 5\,Dy + 6y = x - 2; \qquad y(1) = 0, \quad Dy(1) = 1$$

Introduce a new independent variable t by means of the equation $t = x - 1$. Then $t = 0$ when $x = 1$. To rewrite the differential equation in terms of t, we use $x = t + 1$ and the chain rule to obtain

$$Dy = \frac{dy}{dx} = \frac{dy}{dt} \cdot \frac{dt}{dx} = \frac{dy}{dt} \cdot 1 = \frac{dy}{dt}$$

and

$$D^2y = \frac{d^2y}{dt^2}$$

In terms of t, the initial-value problem is:

$$\frac{d^2y}{dt^2} - 5\frac{dy}{dt} + 6y = t - 1; \qquad y(0) = 0, \quad \frac{dy}{dt}(0) = 1$$

Using the Laplace transform method, we get

$$s^2Y(s) - 1 - 5Y(s) + 6Y(s) = \frac{1}{s^2} - \frac{1}{s} = \frac{1 - s}{s^2}$$

Solving for $Y(s)$, we have

$$Y(s) = \frac{1 - s}{s^2(s^2 - 5s + 6)} + \frac{1}{s^2 - 5s + 6}$$

The partial fraction decompositions of the two functions are:

$$\frac{1 - s}{s^2(s - 2)(s - 3)} = -\frac{1}{36}\left(\frac{1}{s}\right) + \frac{1}{6}\left(\frac{1}{s^2}\right) + \frac{1}{4}\left(\frac{1}{s - 2}\right) - \frac{2}{9}\left(\frac{1}{s - 3}\right)$$

$$\frac{1}{(s - 2)(s - 3)} = \frac{-1}{s - 2} + \frac{1}{s - 3}$$

Therefore,

$$Y(s) = -\frac{1}{36}\left(\frac{1}{s}\right) + \frac{1}{6}\left(\frac{1}{s^2}\right) - \frac{3}{4}\left(\frac{1}{s-2}\right) + \frac{7}{9}\left(\frac{1}{s-3}\right)$$

and

$$y(t) = -\tfrac{1}{36} + \tfrac{1}{6}t - \tfrac{3}{4}e^{2t} + \tfrac{7}{9}e^{3t}$$

so that

$$y(x) = -\tfrac{1}{36} + \frac{x-1}{6} - \tfrac{3}{4}e^{2(x-1)} + \tfrac{7}{9}e^{3(x-1)}$$

17 EXERCISES

Use the Laplace transform method to solve the following initial-value problems.

1. $Dy - 2y = e^x$; $y(0) = 2$
2. $Dy + y = 2(\sin x + \cos x)$; $y(0) = 1$
3. $4 D^2y - 37 Dy + \sqrt{2}y = \varnothing$; $y(0) = Dy(0) = 0$
4. $D^2y - Dy - 2y = \sin(2x)$; $y(0) = 1, Dy(0) = 1$
5. $D^2y + 2 Dy + y = x + e^x$; $y(0) = -\tfrac{7}{4}, Dy(0) = \tfrac{9}{4}$
6. $D^2y - Dy - 2y = 4x^2$; $y(0) = 1, Dy(0) = 4$
7. $D^2y + 2 Dy + 2y = 2e^{-x}\cos x$; $y(0) = 2, Dy(0) = -2$
8. $D^3y - Dy = \varnothing$; $y(0) = 3, Dy(0) = -1, D^2y(0) = 1$
9. $D^2y - 3 Dy + 2y = e^{x-2}$; $y(2) = 1, Dy(2) = -1$
10. $D^2y - 4 Dy + 5y = e^{2(x+1)}\cos(x+1)$; $y(-1) = Dy(-1) = 0$

In many applications of differential equations, it is not required to determine the solutions explicitly. In some of these cases, desired information about the solutions can be obtained by analyzing their Laplace transforms. This use of the Laplace transform is illustrated in the next example.

18 Example

Consider the differential equation

$$D^2y - Dy - 6y = 7\sin x + \cos x$$

together with an incomplete set of initial conditions, namely, $y(0) = -1$. Is it possible to choose a value for $Dy(0)$ so that the resulting initial-value problem has a bounded solution on $[0, \infty)$? From the theory in Section 16, the solutions of the differential equation have the form

$$y(x) = c_1 e^{-2x} + c_2 e^{3x} + c_3 \sin x + c_4 \cos x$$

Thus it is clear that we would want to choose $Dy(0)$ so that $c_2 = 0$. The Laplace transform method gives an easy way to make this calculation.

Using Laplace transforms,

$$(s^2 - s - 6)\mathcal{L}[y(x)] + s - Dy(0) - 1 = \frac{s + 7}{s^2 + 1}$$

or

(A)

$$\mathcal{L}[y(s)] = \frac{s + 7 + [1 - s + Dy(0)](s^2 + 1)}{(s - 3)(s + 2)(s^2 + 1)}$$

$$= \frac{As + B}{s^2 + 1} + \frac{C}{s + 2} + \frac{D}{s - 3}$$

for some choice of A, B, C, and D. This implies that

(B)

$$s + 7 + [1 - s + Dy(0)](s^2 + 1)$$
$$= (As + B)(s + 2)(s - 3) + C(s^2 + 1)(s - 3)$$
$$+ D(s^2 + 1)(s + 2)$$

Inspecting equation (A), we note that the first two terms are Laplace transforms of bounded functions regardless of the values of A, B, and C, while this is true for the third term only in the case $D = 0$. By putting $D = 0$ and $s = 3$ in (B), we get

$$10 + [-2 + Dy(0)]10 = 0$$

or

$$Dy(0) = 1$$

19 EXERCISES

1. Use the equation in Example 18.
   a. If $Dy(0) = -1$, what value should be assigned to $y(0)$ so that the resulting solution is bounded?
   b. If neither $y(0)$ nor $Dy(0)$ are specified, determine a relationship between them so that the resulting solutions are bounded.
2. Consider the equation

   $$D^2y + 2\,Dy + 2y = 10\sin 2x$$

   together with $y(0) = 14$. For what values of $Dy(0)$ will the resulting solutions be bounded?
3. Consider the equation

   $$D^2y - 5\,Dy + 6y = x$$

   together with $Dy(0) = -20$. For what values of $y(0)$ will the resulting solutions be bounded?
4. What initial conditions should be assigned with the differential equation

   $$D^2y + y = e^{-x}$$

   so that $\lim_{x \to \infty} y(x) = 0$, where $y(x)$ is the solution?

The treatment in this section shows that in applying the Laplace transform method to solve differential equations, it is often necessary to find the inverse transform of a product of two functions. This suggests the general question of a relationship between the inverse transform of a product $F \cdot G$ and the inverse transforms of each of the functions F and G. We conclude this section with a brief discussion of this topic.

Let f and g be continuous functions on $[0, \infty)$, and suppose that each of $\mathscr{L}[f(x)] = F(s)$ and $\mathscr{L}[g(x)] = G(s)$ exists for $s > r$. We seek a representation of the function h such that $\mathscr{L}[h(x)] = F(s)G(s)$, assuming that such a function exists.

From the definition of the Laplace transform,

$$F(s) = \int_0^\infty e^{-su}f(u)\,du$$

$$G(s) = \int_0^\infty e^{-st}g(t)\,dt$$

Thus

$$F(s)G(s) = \left[\int_0^\infty e^{-su}f(u)\,du\right]\left[\int_0^\infty e^{-st}g(t)\,dt\right]$$

Since the integrand of the first integral above does not depend on the variable of integration in the second integral, we can write the product $F \cdot G$ as the iterated integral

$$F(s)G(s) = \int_0^\infty \left[\int_0^\infty e^{-s(u+t)}f(u)g(t)\,du\right]dt$$

This iterated integral can be interpreted as the integral of the function $K(u, t) = e^{-s(u+t)}f(u)g(t)$ over the first quadrant in the u-t plane. By making the change of variable $u + t = x$, $t = t$, we obtain

(20) $$F(s)G(s) = \int_0^\infty \left[\int_t^\infty e^{-sx}f(x - t)g(t)\,dx\right]dt$$

which is the integral of $K(x, t) = e^{-sx}f(x - t)g(t)$ over the wedge-shaped region in the x-t plane shown in the figure here.

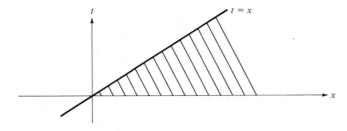

A change in the order of integration in (20) can be justified, and we get

$$\mathscr{L}[h(x)] = F(s)G(s) = \int_0^\infty \left[\int_0^x e^{-sx}f(x - t)g(t)\,dx\right]dt$$

$$= \int_0^\infty e^{-sx}\left[\int_0^x f(x - t)g(t)\,dt\right]dx$$

Therefore, from Theorem 4,

$$h(x) = \int_0^x f(x - t)g(t)\, dt$$

21 DEFINITION

If f and g are functions in $\mathscr{C}[0, \infty)$, then the function $f * g$ defined by

$$(f * g)(x) = \int_0^x f(x - t)g(t)\, dt, \qquad 0 \le x < \infty.$$

is called the *convolution* of f and g.

The convolution of two functions f and g can be regarded as a generalized multiplication of f and g. In fact, it can be verified that convolution has the following properties analogous to ordinary multiplication of functions:

$$f * g = g * f \qquad \text{(commutative)}$$
(22) $$f * (g * h) = (f * g) * h \qquad \text{(associative)}$$
$$f * (g + h) = f * g + f * h \qquad \text{(distributive)}$$

where f, g, and h are continuous functions. The analogy with ordinary multiplication, however, is not complete. For example,

$$(\sin x) * 1 = \int_0^x \sin(x - t) \cdot 1\, dt = \cos(x - t)\Big|_0^x$$

$$= 1 - \cos x \ne \sin x$$

so that $f * 1$ is not, in general, equal to f.

23 Example

In Example 13a, it was necessary to find the inverse transform of

$$F(s) = \frac{5}{(s - 1)(s^2 + 4)} = \left(\frac{5}{s - 1}\right)\left(\frac{1}{s^2 + 4}\right)$$

Since $\mathscr{L}[5e^x] = 5/(s - 1)$ and $\mathscr{L}[\tfrac{1}{2} \sin 2x] = 1/(s^2 + 4)$, we have, using convolutions,

$$\mathscr{L}^{-1}\left[\left(\frac{5}{s - 1}\right)\left(\frac{1}{s^2 + 4}\right)\right] = \int_0^x 5e^{(x - t)}(\tfrac{1}{2}) \sin(2t)\, dt$$

$$= \tfrac{5}{2}e^x \int_0^x e^{-t} \sin(2t)\, dt$$

$$= \tfrac{5}{2}e^x[\tfrac{1}{5}e^{-t}\{-\sin(2t) - 2\cos(2t)\}]\Big|_0^x$$

$$= e^x - \tfrac{1}{2}\sin(2x) - \cos(2x)$$

We leave it as an exercise to show that

$$\tfrac{5}{2} \int_0^x e^t \sin 2(x - t) \, dt = e^x - \tfrac{1}{2} \sin(2x) - \cos(2x)$$

The next example illustrates the use of convolutions in solving initial-value problems.

24 Example

Consider the initial-value problem

$$D^2 y + k^2 y = f(x); \qquad y(0) = Dy(0) = 0$$

k a constant, where f is assumed to have a Laplace transform F. By applying the Laplace transform method, we get

$$(s^2 + k^2)\mathcal{L}[y] = F(s)$$

Therefore,

$$\mathcal{L}[y] = \left(\frac{1}{s^2 + k^2}\right) F(s)$$

and

$$y(x) = \mathcal{L}^{-1}\left[\left(\frac{1}{s^2 + k^2}\right) F(s)\right]$$

or

$$y(x) = \left(\frac{1}{k}\right) \int_0^x f(t) \sin[k(x - t)] \, dt$$

$$= \left(\frac{1}{k}\right) (\sin kx) * f(x)$$

In particular, if $k = 1$ and $f(x) = 1/x$, we get the representation given in Example 16a, Section 16. An equivalent form of the solution given above is:

$$y(x) = \left(\frac{1}{k}\right) \int_0^x \sin(kt) f(x - t) \, dt$$

25 EXERCISES

1. Verify the commutative, associative, and distributive properties of convolution postulated in (22).
2. Let $f(x) = \sin x$ and show that $f * f$ is not, in general, nonnegative.
3. Verify the claim made at the end of Example 23.
4. Use convolutions to find the inverse Laplace transform of each of the following functions.

   a. $1/(s + 2)(s - 3)$ b. $1/(s - 1)^2(s + 1)$ c. $s^2/(s^2 + 1)^2$

5. Solve the initial-value problem

$$D^2y - k^2y = f(x); \qquad y(0) = Dy(0) = 0$$

where k is a constant and f is a function which has the Laplace transform F.

6. A functional equation of the form

$$y(x) = f(x) + \int_0^x h(x - t)y(t) \, dt$$

where f and h are given functions and y is to be determined, is known as an *integral equation*. Assume that f has the Laplace transform F and that for each fixed t, h has the Laplace transform H. Obtain the Laplace transform of y in terms of F and H.

7. Solve each of the following integral equations by using Laplace transforms.
   a. $y(x) = x + \int_0^x y(t) \sin(x - t) \, dt$ b. $y(x) = x^2 + \int_0^x (x - t)y(t) \, dt$
   c. $y(x) = e^{-x}[1 + \int_0^x e^t y(t) \, dt]$

Section 37 Discontinuous Functions

In the preceding sections, we assumed, for simplicity and convenience in presenting the basic properties of the Laplace transform, that the functions under consideration were continuous. In this section, we consider Laplace transforms of certain types of discontinuous functions. Discontinuous functions of the type considered here occur in equations that arise in various applications. This treatment amounts to a slight generalization of the concept of a linear differential equation as in Definition 1, Section 9.

A review of the concept of continuity is in Section 2. A particular example of the type of discontinuous function which we will be considering in this section is the *Heaviside function H* defined on $J = (-\infty, \infty)$ by

(1) $\qquad H(x) = \begin{cases} 0 & x < 0 \\ 1 & x \geq 0 \end{cases}$

The graph of H is shown in the figure here.

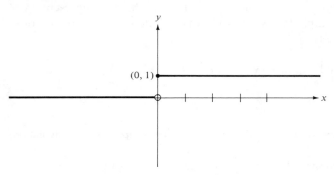

It is clear that H is not continuous at $x = 0$ since $\lim_{x \to 0} H(x)$ does not exist. Notice, however, that if we consider $\lim_{x \to 0} H(x)$ and restrict x to be greater than 0, then the limit does exist, and it has the value 1. Similarly, if x is restricted to negative values only, then the limit exists and has the value 0. These restricted limits are called the *right* and *left limits of H*, respectively, and are denoted by

$$\lim_{x \to 0^+} H(x) \quad \text{and} \quad \lim_{x \to 0^-} H(x)$$

In general, if the function f is defined on an interval J and $x_0 \in J$, x_0 not an endpoint of J, then the right and left limits of f at x_0 are denoted by

$$\lim_{x \to x_0^+} f(x) \quad \text{and} \quad \lim_{x \to x_0^-} f(x)$$

respectively. The concept of continuity of f at x_0 can be defined in terms of the right and left limits of f at x_0 as follows: f is continuous at x_0 if and only if the right and left limits of f at x_0 exist and have the common value $f(x_0)$.

2 DEFINITION

Let f be continuous on an interval J except at a point $x_0 \in J$, x_0 not an endpoint of J. If the right and left limits of f at x_0 exist and are not equal, then f is said to have a *jump* (or *finite*) *discontinuity* at x_0.

3 Examples

    a. The Heaviside function H defined by (1) is continuous on $(-\infty, \infty)$ except at $x = 0$, where it has a jump discontinuity.

    b. Consider the function f defined on $[-5, 5]$ by

$$f(x) = \begin{cases} x + 5 & -5 \leq x < -2 \\ -\dfrac{1}{x} & -2 \leq x < 0 \\ 2x & 0 \leq x < 2 \\ -(x - 5) & 2 \leq x \leq 5 \end{cases}$$

The graph of f is shown in the figure here.

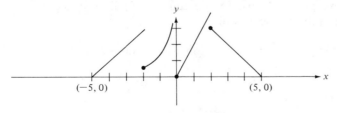

        This function is discontinuous at $x = -2$, $x = 0$, and $x = 2$. Since

$$\lim_{x \to -2^-} f(x) = 3, \quad \lim_{x \to -2^+} f(x) = \tfrac{1}{2}$$

and

$$\lim_{x \to 2^-} f(x) = 4, \qquad \lim_{x \to 2^+} f(x) = 3$$

f has jump discontinuities at these points. The discontinuity at $x = 0$ is not a jump discontinuity since $\lim_{x \to 0^-} f(x)$ does not exist.

c. Consider the function g on $[0, \infty)$ defined by

$$g(x) = \begin{cases} x & 0 < x < 1 \\ 2 & x = 1 \\ \frac{3}{2} & 1 < x < 5 \\ e^{-(x-5)} & x \geq 5 \end{cases}$$

The graph of g is shown in the figure here.

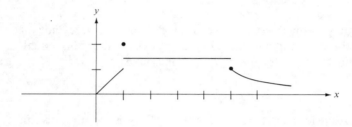

This function has jump discontinuities at $x = 1$ and $x = 5$. Note that the value of g at the jump discontinuity $x = 1$ is independent of the right and left limits of g at $x = 1$.

We now define the class of functions which are considered in this section.

4 DEFINITION

A function f defined on an interval J is *piecewise continuous* provided f is continuous on J, except for at most a finite number of points $x_1, x_2, \ldots, x_n$ of J, at which f has a jump discontinuity.

Note that any continuous function is also piecewise continuous.

5 Examples

a. The Heaviside function H is piecewise continuous on $(-\infty, \infty)$.
b. The function g in Example 3c is piecewise continuous on $[0, \infty)$.
c. The function f in Example 3b is not piecewise continuous on $[-5, 5]$ since the discontinuity at $x = 0$ is not a jump dis-

continuity. Note, however, that f is piecewise continuous on $[-5, 0)$ as well as on $[0, 5]$.

Let c be a real number. The translation of the Heaviside function H by the number c is the function $H(x - c)$ defined on $J = (-\infty, \infty)$ by

(6) $\qquad H(x - c) = \begin{cases} 0 & x < c \\ 1 & x \geq c \end{cases}$

The graph of $H(x - c)$, for $c > 0$, is shown in the figure here.

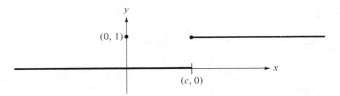

Clearly, $H(x - c)$ is piecewise continuous on $(-\infty, \infty)$, having a jump discontinuity at $x = c$. Since the problem of calculating the Laplace transform of a piecewise continuous function will involve the integration of such a function, we include here a reminder of the necessary concept from integration theory.

Suppose that g is piecewise continuous on $[a, b]$ with jump discontinuities at $c_1, c_2, \ldots, c_n$; $a < c_1 < c_2 < \cdots < c_n < b$. It follows from the definition of the definite integral that

(7) $\qquad \int_a^b g(x)\, dx = \int_a^{c_1} g(x)\, dx + \int_{c_1}^{c_2} g(x)\, dx + \cdots + \int_{c_n}^b g(x)\, dx$

8 Examples

a. Consider the function g defined on $(0, 2)$ by

$$g(x) = \begin{cases} x & 0 < x \leq 1 \\ 4 - x^2 & 1 < x < 2 \end{cases}$$

The graph of g is shown in the figure here.

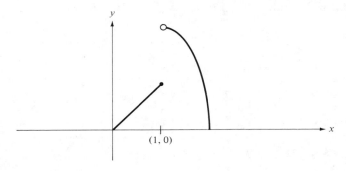

We have that

$$\int_0^2 g(x)\,dx = \int_0^1 x\,dx + \int_1^2 (4 - x^2)\,dx$$

$$= \frac{x^2}{2}\bigg|_0^1 + \left(4x - \frac{x^3}{3}\right)\bigg|_1^2 = \frac{13}{6}$$

Notice that if g_1 is the function defined by

$$g_1(x) = \begin{cases} -1 & x = 0 \\ x & 0 < x < 1 \\ 3 & x = 1 \\ 4 - x^2 & 1 < x < 2 \\ -1 & x = 2 \end{cases}$$

then

$$\int_0^2 g_1(x)\,dx = \int_0^2 g(x)\,dx = \frac{13}{6}$$

b. Equation (7) can be extended to include the (improper) integral of a piecewise continuous function on an infinite interval. Consider the function g of Example 3c. Then

$$\int_0^\infty g(x)\,dx = \int_0^1 x\,dx + \int_1^5 \tfrac{3}{2}\,dx + \int_5^\infty e^{-(x-5)}\,dx$$

$$= \frac{x^2}{2}\bigg|_0^1 + \tfrac{3}{2}x\bigg|_1^5 + \lim_{c\to\infty}\left[-e^{-(x-5)}\bigg|_5^c\right]$$

$$= \tfrac{1}{2} + \left[\tfrac{15}{2} - \tfrac{3}{2}\right] + [0 + 1]$$

$$= \tfrac{15}{2}$$

9 EXERCISES

In each of the following, determine whether the given function is piecewise continuous and sketch the graph.

1. $f_1(x) = \begin{cases} 1 & 0 \le x < 1 \\ -1 & 1 \le x < 2 \end{cases}$

   and f_1 is extended periodically for $x \ge 2$; that is, $f_1(x + 2) = f_1(x)$ for $x \ge 2$.

2. $f_2(x) = \begin{cases} x & 0 \le x < 1 \\ x - 1 & 1 \le x < 2 \\ 0 & x \ge 2 \end{cases}$

3. $f_3(x) = \begin{cases} \dfrac{1}{x - 1} & 0 \le x < 1 \\ 0 & x = 1 \\ \dfrac{1}{x - 1} & 1 < x \le 5 \end{cases}$

4. $f_4(x) = \begin{cases} -1 & 0 \le x \le 1 \\ x - 1 & x > 1 \end{cases}$

5. $f_5(x) = \begin{cases} 0 & 0 \le x \le 2 \\ (x - 2)^2 & 2 \le x \le 3 \\ 2 & 3 < x \le 6 \end{cases}$

6. $f_6(x) = \begin{cases} 1 & 0 \le x \le 2 \\ x - 2 & 2 < x \le 4 \\ e^{-(x-4)} & x > 4 \end{cases}$

7. $f_7(x) = \begin{cases} x^{-1/2} & 0 < x \le 4 \\ \dfrac{8}{x^2} & x \ge 4 \end{cases}$

Using the functions defined above, determine each of the following integrals.

8. a. $\int_0^3 f_1(x)\,dx$ b. $\int_0^4 f_1(x)\,dx$
   c. $\int_0^\infty f_1(x)\,dx$
9. a. $\int_0^2 f_2(x)\,dx$ b. $\int_0^\infty f_2(x)\,dx$
10. a. $\int_0^1 f_3(x)\,dx$ b. $\int_0^5 f_3(x)\,dx$
11. a. $\int_0^{10} f_4(x)\,dx$ b. $\int_0^\infty f_4(x)\,dx$
12. $\int_0^6 f_5(x)\,dx$
13. $\int_0^\infty f_6(x)\,dx$
14. a. $\int_0^4 f_7(x)\,dx$ b. $\int_4^\infty f_7(x)\,dx$
   c. $\int_0^\infty f_7(x)\,dx$

Since a piecewise continuous function defined on a finite interval is integrable on that interval, it now makes sense to examine the existence of the Laplace transform of a piecewise continuous function. The following theorem is an extension of Theorem 5, Section 35. The proof is left as an exercise.

10 THEOREM

If the function f is piecewise continuous on each interval $[0, c]$, $c \ge 0$, and if f is of exponential order r, then the Laplace transform $\mathscr{L}[f(x)]$ exists for $s > r$.

11 Examples

a. Consider the translation of the Heaviside function, $H(x - c)$, $c > 0$:

$$\mathscr{L}[H(x - c)] = \int_0^\infty e^{-sx} H(x - c)\,dx = \int_c^\infty e^{-sx} H(x - c)\,dx$$

$$= \lim_{b \to \infty} \int_c^b e^{-sx}\,dx = \lim_{b \to \infty} \frac{e^{-sx}}{-s}\Big|_c^b = \frac{e^{-sc}}{s}$$

$s > 0$. [Note: If $c = 0$, then $H \equiv 1$ on $[0, \infty)$, and we have the result indicated in Table 5, Section 34.]

b. Let f be defined by

$$f(x) = \begin{cases} 1 & 0 \leq x < 5 \\ 2 & 5 \leq x < \infty \end{cases}$$

Then

$$\mathcal{L}[f(x)] = \int_0^\infty e^{-sx} f(x) \, dx = \int_0^5 e^{-sx} \, dx + \int_5^\infty 2e^{-sx} \, dx$$

$$= \frac{-e^{-5s}}{s} + \frac{1}{s} + \frac{2e^{-5s}}{s} = \frac{1}{s} + \frac{e^{-5s}}{s}$$

$s > 0$.

c. Consider the function g defined in Example 3c.

$$\mathcal{L}[g(x)] = \int_0^\infty e^{-sx} g(x) \, dx$$

$$= \int_0^1 e^{-sx} x \, dx + \int_1^5 e^{-sx} \tfrac{3}{2} \, dx + \int_5^\infty e^{-sx} \cdot e^{-(x-5)} \, dx$$

$$= \frac{-e^{-s}}{s} - \frac{e^{-s}}{s^2} + \frac{1}{s^2} + \frac{-3e^{-5s}}{2s} + \frac{3e^{-s}}{2s} + \frac{e^{-5s}}{s+1}$$

$$= \frac{1}{s^2} - \frac{e^{-s}}{s^2} + \frac{e^{-s}}{2s} - \frac{3e^{-5s}}{2s} + \frac{e^{-5s}}{s+1}$$

The Heaviside function and its translations can be used to obtain translations of arbitrary functions. For example, if f is defined on $[0, \infty]$ and $c > 0$, then the function

$$f_c(x) = f(x - c)H(x - c) = \begin{cases} f(x - c) & \text{for } x \geq c \\ 0 & \text{for } x < c \end{cases}$$

is the function f shifted c units to the right, as shown in the figure.

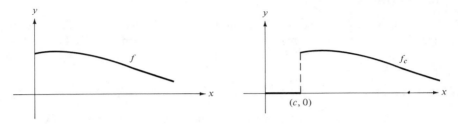

12 THEOREM

Let f be defined on $[0, \infty)$ and suppose $\mathcal{L}[f(x)] = F(s)$ exists for $s > r$. If $c \geq 0$, then $\mathcal{L}[f_c(x)]$ exists for $s > r$ and is given by

$$\mathcal{L}[f_c(x)] = e^{-cs} F(s)$$

Proof: By definition,

$$\mathscr{L}[f_c(x)] = \mathscr{L}[f(x-c)H(x-c)] = \int_0^\infty e^{-sx}f(x-c)H(x-c)\,dx$$

$$= \lim_{b\to\infty} \int_c^b e^{-sx}f(x-c)\,dx$$

Now define a new variable of integration t by $t = x - c$. Then $x = t + c$, $dx = dt$, and $t = 0$ when $x = c$. Thus we have

$$\lim_{b\to\infty} \int_c^b e^{-sx}f(x-c)\,dx = \lim_{b\to\infty} \int_0^{b-c} e^{-s(t+c)}f(t)\,dt$$

$$= e^{-cs}\left(\lim_{b\to\infty} \int_0^{b-c} e^{-st}f(t)\,dt\right)$$

Finally, it is clear that

$$\lim_{b\to\infty} \int_0^{b-c} e^{-st}f(t)\,dt = \lim_{d\to\infty} \int_0^d e^{-sx}f(x)\,dx = \mathscr{L}[f(x)]$$

This proves the theorem.

This theorem can also be expressed equivalently using the inverse Laplace transform.

13 COROLLARY

If $\mathscr{L}^{-1}[F(s)] = f(x)$ and $c \geq 0$, then

$$\mathscr{L}^{-1}[e^{-cs}F(s)] = f(x-c)H(x-c) = f_c(x)$$

We now illustrate an alternate procedure for calculating Laplace transforms of piecewise continuous functions using the Heaviside function and Theorem 12.

14 Examples

a. Consider the function f given in Example 11b,

$$f(x) = \begin{cases} 1 & 0 \leq x < 5 \\ 2 & 5 \leq x < \infty \end{cases}$$

Since f has a jump discontinuity at $x = 5$, we can write f in terms of $H(x - 5)$. Put $f_1(x) = 1 - H(x - 5)$. Then

$$f_1(x) = \begin{cases} 1 & 0 \leq x < 5 \\ 0 & 5 \leq x < \infty \end{cases}$$

Let $f_2(x) = 2H(x - 5)$, or

$$f_2(x) = \begin{cases} 0 & 0 \leq x < 5 \\ 2 & 5 < x < \infty \end{cases}$$

Thus

$$f(x) = f_1(x) + f_2(x) = 1 - H(x - 5) + 2H(x - 5)$$
$$= 1 + H(x - 5)$$

and

$$\mathcal{L}[f(x)] = \mathcal{L}[1 + H(x - 5)] = \frac{1}{s} + \frac{e^{-5s}}{s}$$

$s > 0$.

b. Consider the function h defined by

$$h(x) = \begin{cases} x & 0 \le x < 3 \\ 4 & 3 \le x < \infty \end{cases}$$

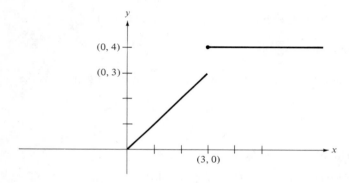

The graph of h is shown in the figure here. Let $h_1(x) = x - xH(x - 3)$ and $h_2(x) = 4H(x - 3)$. Thus

$$h(x) = x - xH(x - 3) + 4H(x - 3) = x - (x - 4)H(x - 3)$$

and

$$\mathcal{L}[h(x)] = \mathcal{L}[x - (x - 4)H(x - 3)]$$

$$= \frac{1}{s^2} - \mathcal{L}[(x - 4)H(x - 3)]$$

To use Theorem 12, however, we must have the coefficient of $H(x - 3)$ expressed as a function of $(x - 3)$. Since $x - 4 = (x - 3) - 1$, we have

$$\mathcal{L}[(x - 4)H(x - 3)] = \mathcal{L}[(x - 3)H(x - 3) - H(x - 3)]$$

$$= \frac{e^{-3s}}{s^2} - \frac{e^{-3s}}{s}$$

It now follows that

$$\mathcal{L}[h(x)] = \frac{1}{s^2} + \frac{e^{-3s}}{s^2} - \frac{e^{-3s}}{s}$$

c. Consider the function g defined by

$$g(x) = \begin{cases} x & 0 \le x < 1 \\ \frac{3}{2} & 1 \le x < 5 \\ e^{-(x-5)} & 5 \le x < \infty \end{cases}$$

See Example 11c. If

$$g_1(x) = x - xH(x - 1)$$
$$g_2(x) = \tfrac{3}{2}H(x - 1) - \tfrac{3}{2}H(x - 5)$$
$$g_3(x) = e^{-(x-5)}H(x - 5)$$

then

$$g(x) = x - xH(x - 1) + \tfrac{3}{2}H(x - 1)$$
$$\quad - \tfrac{3}{2}H(x - 5) + e^{-(x-5)}H(x - 5)$$
$$= x - (x - 1)H(x - 1) + \tfrac{1}{2}H(x - 1)$$
$$\quad - \tfrac{3}{2}H(x - 5) + e^{-(x-5)}H(x - 5)$$

Calculating the Laplace transform of g, we obtain

$$\mathscr{L}[g(x)] = \frac{1}{s^2} - \frac{e^{-s}}{s^2} + \frac{e^{-s}}{2s} - \frac{3e^{-5s}}{2s} + \frac{e^{-5s}}{s + 1}$$

Corollary 13 is used to calculate inverse Laplace transforms.

15 Examples

a. Suppose

$$G(s) = \frac{1}{s^2} + \frac{e^{-3s}}{s - 2}$$

Then

$$\mathscr{L}^{-1}[G(s)] = \mathscr{L}^{-1}\left[\frac{1}{s^2}\right] + \mathscr{L}^{-1}\left[\frac{e^{-3s}}{s - 2}\right]$$

From Table 5, Section 34, $\mathscr{L}^{-1}[1/s^2] = x$. To calculate $\mathscr{L}^{-1}[e^{-3s}/(s - 2)]$, let $F(s) = 1/(s - 2)$. Then $\mathscr{L}^{-1}[F(s)] = e^{2x}$. Thus, by Corollary 13,

$$\mathscr{L}^{-1}\left[\frac{e^{-3s}}{s - 2}\right] = e^{2(x-3)}H(x - 3)$$

Therefore,

$$\mathscr{L}^{-1}[G(s)] = x + e^{2(x-3)}H(x - 3)$$

or

$$\mathscr{L}^{-1}[G(s)] = \begin{cases} x & 0 \le x < 3 \\ x + e^{2(x-3)} & 3 \le x < \infty \end{cases}$$

b. Suppose

$$Y(s) = \frac{1}{s} + \frac{e^{-s}}{(s-1)^2}$$

To calculate $\mathscr{L}^{-1}[e^{-s}/(s-1)^2]$, let $F(s) = 1/(s-1)^2$. Then, from Table 5, Section 34, $\mathscr{L}^{-1}[F(s)] = xe^x$. Thus

$$\mathscr{L}^{-1}\left[\frac{e^{-s}}{(s-1)^2}\right] = (x-1)e^{(x-1)}H(x-1)$$

and

$$\mathscr{L}^{-1}[Y(s)] = 1 + (x-1)e^{(x-1)}H(x-1)$$

16 EXERCISES

1. Prove Theorem 10.

Determine the Laplace transform of each of the following functions.

2. $f(x) = \begin{cases} x & 0 \le x < 1 \\ x-1 & 1 \le x < 2 \\ 0 & x > 2 \end{cases}$

3. $f(x) = \begin{cases} -1 & 0 \le x \le 1 \\ x-1 & x \ge 1 \end{cases}$

4. $f(x) = \begin{cases} 0 & 0 \le x < 5 \\ 2 & x \ge 5 \end{cases}$

5. $f(x) = \begin{cases} 2x & 0 \le x \le 3 \\ 1 & x > 3 \end{cases}$

6. $f(x) = \begin{cases} 1 & 0 \le x \le 2 \\ x-2 & 2 < x \le 4 \\ e^{-(x-4)} & x > 4 \end{cases}$

7. $f(x) = \begin{cases} 1 & 0 \le x < 1 \\ -1 & 1 \le x < 2 \end{cases}$

and extended periodically on $[2, \infty)$. (See Exercises 9, Problem 1.)

8. $f(x) = x - n$ on $n \le x < n + 1$, $n = 0, 1, 2, \ldots$ (Sketch the graph of f.)

9. $f(x) = \begin{cases} 2x & 0 \le x < 2 \\ 2 & 2 \le x < 4 \end{cases}$

and extended periodically on $[4, \infty)$. (Sketch the graph of f.)

Express each of the following functions in terms of the Heaviside function and its translations. Calculate the Laplace transform in each case using Theorem 12.

10. $f(x) = \begin{cases} x^2 & 0 \le x < 3 \\ 3x & x > 3 \end{cases}$

11. $f(x) = \begin{cases} \sin x & 0 \le x < \pi \\ \sin 2x & \pi \le x < 2\pi \\ \sin 3x & x \ge 2\pi \end{cases}$

Sketch the graph of f.

12. $f(x) = \begin{cases} 0 & 0 \le x < \pi \\ 1 + \cos x & \pi \le x < 2\pi \\ 2 \cos x & x \ge 2\pi \end{cases}$

13. $f(x) = \begin{cases} x & 0 \le x < 2 \\ 0 & 2 \le x < 4 \\ (x - 4)^2 & x \ge 4 \end{cases}$

14. $f(x) = \begin{cases} x & 0 \le x \le 1 \\ -x + 2 & 1 \le x \le 3 \\ x - 4 & 3 \le x \le 4 \\ 0 & x > 4 \end{cases}$

Sketch the graph of f.

Calculate the inverse Laplace transform of each of the following functions.

15. $F(s) = (1/s)(e^{-s} + 2e^{-3s} - 6e^{-4s})$
16. $F(s) = (1 + e^{-\pi s})/(s^2 + 1)$
17. $F(s) = (1 - e^{-2(s+1)})/(s + 1)$
18. $F(s) = [s + (s - 1)e^{-\pi s}]/(s^2 + 1)$
19. $F(s) = -e^{\pi s/2}/(s^2 + 1)$
20. $F(s) = (2/s^3) - e^{-2s}[(4/s^2) + (2/s^3)] - \dfrac{4}{s}e^{-4s}$

We now consider equations of the form

$$a_0 D^n y + a_1 D^{n-1}y + \cdots + a_{n-1} Dy + a_n y = f(x)$$

where f is piecewise continuous on some interval J. We will refer to an equation of this type as a differential equation even though it does not meet the requirements of Definition 1, Section 9. This more general type of equation requires a modification of the concept of a solution. We shall use a simple example to illustrate this more general interpretation of the phrase "solution of a differential equation."

17 Example

Given the initial-value problem

$$D^2 y + 4y = f(x); \qquad y(0) = 0, \quad Dy(0) = 1$$

where

$$f(x) = \begin{cases} 1 & 0 \le x < \dfrac{\pi}{2} \\ -1 & \dfrac{\pi}{2} \le x < \infty \end{cases}$$

First consider the differential equation on the interval $[0, \pi/2)$. Since f is continuous on this interval, the initial-value problem

$$D^2 y + 4y = 1; \qquad y(0) = 0, \quad Dy(0) = 1$$

on $[0, \pi/2)$ has a unique solution. In fact, this solution is readily found to be

$$y_1(x) = \tfrac{1}{2} \sin 2x - \tfrac{1}{4} \cos 2x + \tfrac{1}{4}$$

on $[0, \pi/2)$. Evaluating the left limits of $y_1(x)$ and $D[y_1(x)]$ at $x = \pi/2$, we have

$$\lim_{x \to (\pi/2)^-} y_1(x) = y_1 \left(\frac{\pi}{2} \right) = \tfrac{1}{2}$$

$$\lim_{x \to (\pi/2)^-} D[y_1(x)] = D\left[y_1 \left(\frac{\pi}{2} \right) \right] = -1$$

Now, using these values as the intial conditions for the initial-value problem:

$$D^2 y + 4y = -1; \qquad y\left(\frac{\pi}{2} \right) = \tfrac{1}{2}, \quad Dy\left(\frac{\pi}{2} \right) = -1$$

on $[\pi/2, \infty)$, we obtain the unique solution

$$y_2(x) = \tfrac{1}{2} \sin 2x - \tfrac{3}{4} \cos 2x - \tfrac{1}{4}$$

on $[\pi/2, \infty)$. Put

$$y(x) = \begin{cases} y_1(x) & 0 \le x < \dfrac{\pi}{2} \\[2mm] y_2(x) & \dfrac{\pi}{2} \le x < \infty \end{cases}$$

Since

$$\lim_{x \to (\pi/2)^-} y_1(x) = \tfrac{1}{2} = \lim_{x \to (\pi/2)^+} y_2(x)$$

and

$$\lim_{x \to (\pi/2)^-} D[y_1(x)] = -1 = \lim_{x \to (\pi/2)^+} D[y_2(x)]$$

the functions $y(x)$ and $Dy(x)$ are continuous on $[0, \infty)$. Also,

$$\lim_{x \to (\pi/2)^-} D^2[y_1(x)] = -1$$

$$\lim_{x \to (\pi/2)^+} D^2[y_2(x)] = -3$$

so that $D^2[y(x)]$ has a jump discontinuity at $x = \pi/2$.

This example illustrates the procedure for constructing the unique solution of an initial-value problem in which the nonhomogeneous term is piecewise continuous. It also illustrates the generalization of the concept of a solution which we shall use.

18 THEOREM

Consider the initial-value problem

$$P(D)[y] = a_0 D^n y + a_1 D^{n-1} y + \cdots + a_n y = f(x)$$

on J,

$$y(0) = b_0, Dy(0) = b_1, \ldots, D^{n-1} y(0) = b_{n-1}$$

where f is piecewise continuous on J with jump discontinuities at the points $x_1, x_2, \ldots, x_k$, in J. This initial-value problem has a unique solution $y = y(x)$ such that $y(x) \in \mathscr{C}^{n-1}(J)$ and $D^n[y(x)]$ is piecewise continuous on J with jump discontinuities at the points $x_1, x_2, \ldots, x_k$.

We conclude this section with an example illustrating the Laplace transform method for solving initial-value problems with piecewise continuous nonhomogeneous terms.

19 Example

Consider the initial-value problem of Example 17,

$$D^2 y + 4y = f(x); \qquad y(0) = 0, \quad Dy(0) = 1$$

where

$$f(x) = \begin{cases} 1 & 0 \le x < \dfrac{\pi}{2} \\ -1 & \dfrac{\pi}{2} \le x < \infty \end{cases}$$

We first express f in terms of $H(x - \pi/2)$:

$$f(x) = 1 - 2H\left(x - \frac{\pi}{2}\right)$$

By taking Laplace transforms, we have

$$\mathscr{L}[D^2 y + 4y] = s^2 \mathscr{L}[y] - sy(0) - D[y(0)] + 4\mathscr{L}[y]$$

$$= \frac{1}{s} - \frac{2e^{-\pi s/2}}{s}$$

Using the initial conditions and solving for $\mathscr{L}[y] = Y(s)$, we obtain

$$Y(s) = \frac{1}{s^2 + 4}\left[\frac{1}{s} - \frac{2e^{-\pi s/2}}{s} + 1\right]$$

The partial fraction decomposition of $1/s(s^2 + 4)$ is:

$$\frac{1}{s(s^2 + 4)} = \frac{1}{4s} - \frac{s}{4(s^2 + 4)}$$

Therefore,

$$Y(s) = \frac{1}{4s} - \frac{s}{4(s^2 + 4)} - \frac{e^{-\pi s/2}}{2s} + \frac{e^{-\pi s/2} \cdot s}{2(s^2 + 4)} + \frac{1}{s^2 + 4}$$

and

$$y(x) = \mathscr{L}^{-1}[Y(s)]$$

$$= \tfrac{1}{4} - \tfrac{1}{4}\cos(2x) - \tfrac{1}{2}H\left(x - \frac{\pi}{2}\right)$$

$$+ \tfrac{1}{2}\cos\left[2\left(x - \frac{\pi}{2}\right)\right]H\left(x - \frac{\pi}{2}\right) + \tfrac{1}{2}\sin(2x)$$

$$= \tfrac{1}{4} - \tfrac{1}{4}\cos(2x) - \tfrac{1}{2}H\left(x - \frac{\pi}{2}\right)$$

$$- \tfrac{1}{2}\cos(2x)H\left(x - \frac{\pi}{2}\right) + \tfrac{1}{2}\sin(2x)$$

$$= \begin{cases} \tfrac{1}{2}\sin(2x) - \tfrac{1}{4}\cos(2x) + \tfrac{1}{4} & 0 \le x < \dfrac{\pi}{2} \\[2mm] \tfrac{1}{2}\sin(2x) - \tfrac{3}{4}\cos(2x) - \tfrac{1}{4} & \dfrac{\pi}{2} \le x < \infty \end{cases}$$

as determined in Example 17.

20 EXERCISES

Solve each of the following initial-value problems.

1. $D^2y + y = f(x)$; $y(0) = 0$, $Dy(0) = 1$, where

$$f(x) = \begin{cases} 1 & 0 \le x \le 1 \\ 0 & x > 1 \end{cases}$$

2. $D^2y + 4y = f(x)$; $y(0) = 1$, $Dy(0) = 0$, where

$$f(x) = \begin{cases} 0 & 0 \le x \le 2 \\ 1 & x > 2 \end{cases}$$

3. $D^2y + 2\,Dy + y = f(x)$; $y(0) = Dy(0) = 0$, where

$$f(x) = \begin{cases} 1 & 0 \le x \le 2 \\ x - 1 & x > 2 \end{cases}$$

4. $D^2y + 4y = f(x)$; $y(0) = Dy(0) = 0$, where

$$f(x) = \begin{cases} \sin x & 0 \le x \le 2\pi \\ 0 & x > 2\pi \end{cases}$$

5. $D^2y + y = f(x)$, $y(0) = 2$, $Dy(0) = 1$, where

$$f(x) = \begin{cases} 2 & 0 \le x \le 2 \\ x - 2 & x > 2 \end{cases}$$

6. $D^2y - 5\,Dy + 6y = f(x)$; $y(0) = Dy(0) = 0$, where

$$f(x) = \begin{cases} 1 & 0 \le x \le 2 \\ -1 & x > 2 \end{cases}$$

7. $D^2y + y = f(x)$; $y(0) = 0$, $Dy(0) = 0$, where $f(x) = (x - n)$ on $[n, n + 1)$, $n = 0, 1, 2, \dots$. (See Exercises 16, Problem 8.)

8. $D^2y + 4y = f(x)$, $y(1) = Dy(1) = 0$, where

$$f(x) = \begin{cases} 1 & 0 \le x \le 4 \\ 0 & x > 4 \end{cases}$$

(Hint: Use the change of variable $t = x - 1$.)

SYSTEMS OF LINEAR EQUATIONS*

Section 38 Introduction

This chapter is concerned with a study of systems of linear equations. This section and Sections 39 and 40 involve systems of linear differential equations, while the final two sections deal with systems of linear difference equations and numerical methods. The underlying motivation and theory for the material in this chapter are the study of systems of linear algebraic equations, vectors, and matrices, and so the reader is referred to the discussion of this supporting information in Sections 4 and 5.

Systems of linear differential equations, or linear differential systems, arise naturally in a variety of problems involving several dependent variables and one independent variable. For some specific applications of the theory and use of differential systems, the reader is referred to Chapter 11. As a mathematical application, linear differential systems offer an alternative means for representing and studying nth-order linear differential equations. For example, consider the second-order linear differential equation

(1) $\qquad D^2 y + Dy - 6y = \varnothing$

of Example 4, Section 14. If we introduce new (dependent) variables y_1 and y_2 by means of the equations

$$y_1 = y$$

$$y_2 = Dy_1 = Dy$$

---

* Sections 38, 39, and 40 depend upon the material in Chapters 2 and 3. Section 41 is an extension of the results in Chapter 5, and Section 42 is an extension of the results in Chapter 6.

then $Dy_2 = D^2y$, and (1) can be written equivalently as the system of two first-order equations

$$Dy_1 = y_2$$
$$Dy_2 + y_2 - 6y_1 = \varnothing$$

or

(2)
$$Dy_1 = y_2$$
$$Dy_2 = 6y_1 - y_2$$

This same procedure can be used to convert the general second-order linear differential equation

(3) $\quad D^2y + p_1(x)\,Dy + p_2(x)y = f(x)$

where p_1, p_2, and f are continuous functions on an interval J, into an equivalent system of two first-order equations. In particular, we introduce two (dependent) variables, y_1 and y_2, by means of the equations

$$y_1 = y$$
$$y_2 = Dy_1 = Dy$$

In terms of these variables, (3) can be written as the system

(4)
$$Dy_1 = y_2$$
$$Dy_2 = -p_2(x)y_1 - p_1(x)y_2 + f(x)$$

In general, the nth-order linear differential equation

(5) $\quad D^ny + p_1(x)\,D^{n-1}y + \cdots + p_{n-1}(x)\,Dy + p_n(x)y = f(x)$

can be represented by the system of n first-order equations

(6)
$$Dy_1 = y_2$$
$$Dy_2 = y_3$$
$$\vdots$$
$$Dy_{n-1} = y_n$$
$$Dy_n = -p_n(x)y_1 - p_{n-1}(x)y_2 - \cdots - p_1(x)y_n + f(x)$$

by letting

$$y_1 = y;\ y_2 = Dy_1 = Dy;\ \ldots;\ y_n = Dy_{n-1} = D^{n-1}y$$

These examples represent special cases of a system of n first-order equations of the form

(7)
$$Dy_1 = q_{11}(x)y_1 + q_{12}(x)y_2 + \cdots + q_{1n}(x)y_n + f_1(x)$$
$$Dy_2 = q_{21}(x)y_1 + q_{22}(x)y_2 + \cdots + q_{2n}(x)y_n + f_2(x)$$
$$\vdots \qquad \vdots \qquad \vdots \qquad \vdots \qquad \vdots$$
$$Dy_n = q_{n1}(x)y_1 + q_{n2}(x)y_2 + \cdots + q_{nn}(x)y_n + f_n(x)$$

where we assume that the functions $q_{11}, q_{12}, \ldots, q_{nn}, f_1, \ldots, f_n$ are continuous on an interval J.

8 DEFINITION

The system (7) of n first-order linear differential equations is called a *linear differential system*. If $f_1(x) = f_2(x) = \cdots = f_n(x) = \emptyset$, then (7) is called *homogeneous*; otherwise (7) is called *nonhomogeneous*.

9 EXERCISES

Transform the following linear differential equations into equivalent linear differential systems.

1. $D^2 y + 2\,Dy + y = \emptyset$
2. $D^2 y + 5\,Dy + 4y = \ln(x)$
3. $2\,D^2 y + Dy - y = e^x$
4. $D^3 y + 2\,D^2 y - 11\,Dy - 12 = \emptyset$
5. $D^2 y - (\cos x)\,Dy + xy = x^2 + 4$
6. $D^2 y - (\tan x)\,Dy + x^{-2}y = \text{arc}\cos x$

As emphasized in Section 5, vector-matrix notation is an especially convenient means of representing and manipulating a system of linear algebraic equations [see (4), (5), and (6) in Section 5]. Vector-matrix notation is also very useful in studying linear differential systems, but before converting (7) into a vector-matrix form, some preliminary terminology and definitions have to be provided.

10 DEFINITION

Let each of $h_1, h_2, \ldots, h_k$ be a real-valued function defined on an interval J. The vector ψ defined by

$$\psi(x) = \begin{bmatrix} h_1(x) \\ h_2(x) \\ \vdots \\ h_k(x) \end{bmatrix}$$

is called a *vector-valued function* on J. The vector-valued function ψ will be termed *continuous, differentiable*, or *integrable* on J if each of its components $h_1, h_2, \ldots, h_k$ has the indicated property.

11 DEFINITION

Let $q_{11}, q_{12}, \ldots, q_{1n}, q_{21}, \ldots, q_{nn}$ be real-valued functions defined on an interval J. The $n \times n$ matrix Q defined by

$$Q(x) = \begin{bmatrix} q_{11}(x) & q_{12}(x) & \cdots & q_{1n}(x) \\ q_{21}(x) & q_{22}(x) & \cdots & q_{2n}(x) \\ \vdots & \vdots & & \vdots \\ q_{n1}(x) & q_{n2}(x) & \cdots & q_{nn}(x) \end{bmatrix}$$

is called a *matrix-valued function on J*. The matrix-valued function Q will be termed *continuous, differentiable,* or *integrable* if each of its entries q_{ij} has the indicated property.

The calculus of vector-valued functions is defined as follows.

12 DEFINITION

Let

$$\psi = \begin{bmatrix} h_1 \\ h_2 \\ \vdots \\ h_k \end{bmatrix}$$

be a vector-valued function on the interval J. If ψ is differentiable on J, then the derivative of ψ, $D\psi$, is given by

$$D\psi = \begin{bmatrix} Dh_1 \\ Dh_2 \\ \vdots \\ Dh_k \end{bmatrix}$$

Similarly, if ψ is integrable on J and $a \in J$, then the integral of ψ, $\int_a^x \psi(t)\, dt$, is given by

$$\int_a^x \psi(t)\, dt = \begin{bmatrix} \int_a^x h_1(t)\, dt \\ \int_a^x h_2(t)\, dt \\ \vdots \\ \int_a^x h_k(t)\, dt \end{bmatrix}$$

Higher-ordered derivatives of ψ and the calculus of matrix-valued functions are defined in exactly the same manner.

Since the derivative of a vector-valued function is defined component-wise, and in terms of derivatives of real-valued functions, the standard rules of differentiation and integration can be carried over to vector-valued functions. In particular, let σ and ψ be vector-valued functions on the interval J and let c_1 and c_2 be real numbers.

(i) If σ and ψ are differentiable on J, then

$$D(c_1\sigma + c_2\psi) = c_1\, D\sigma + c_2\, D\psi$$

(ii) If σ and ψ are integrable on J, and $a \in J$, then

$$\int_a^x \left[c_1\sigma(t) + c_2\psi(t) \right] dt = c_1 \int_a^x \sigma(t)\, dt + c_2 \int_a^x \psi(t)\, dt$$

The proofs of these rules, as well as additional properties of differentiation and integration of vector-valued functions are contained in the exercises. Note that the differentiation rule (i) implies that the operator D is a linear operator on the set of differentiable vector-valued functions. But we have not explicitly defined the vector spaces which would serve naturally as the domain and range of D. Throughout the text we have been using the notation $\mathscr{C}^n(J)$ to denote the vector space of n-times continuously differentiable functions on the interval J. A natural extension of this notation to vector-valued functions on J is:

$$\mathscr{C}_k{}^0(J) = \mathscr{C}_k(J) = \left\{ \psi = \begin{bmatrix} h_1 \\ h_2 \\ \vdots \\ h_k \end{bmatrix} : \psi \text{ is continuous on } J \right\}$$

$$\mathscr{C}_k{}^1(J) = \left\{ \psi = \begin{bmatrix} h_1 \\ h_2 \\ \vdots \\ h_k \end{bmatrix} : \psi \text{ is continuously differentiable on } J \right\}$$

The reader should note that the subscript k denotes the number of components of the vector-valued function ψ, and the superscript (0 or 1) denotes the assumed order of differentiability.

Now consider the set $\mathscr{C}_k(J)$ and define addition and scalar multiplication of vector-valued functions as follows:

$$(\sigma + \psi)(x) = \sigma(x) + \psi(x) = \begin{bmatrix} h_1(x) \\ h_2(x) \\ \vdots \\ h_k(x) \end{bmatrix} + \begin{bmatrix} g_1(x) \\ g_2(x) \\ \vdots \\ g_k(x) \end{bmatrix}$$

$$= \begin{bmatrix} h_1(x) + g_1(x) \\ h_2(x) + g_2(x) \\ \vdots \\ h_k(x) + g_k(x) \end{bmatrix}$$

and

$$(c\psi)(x) = c\psi(x) = c \begin{bmatrix} h_1(x) \\ h_2(x) \\ \vdots \\ h_k(x) \end{bmatrix} = \begin{bmatrix} ch_1(x) \\ ch_2(x) \\ \vdots \\ ch_k(x) \end{bmatrix}$$

It is easy to verify that $\mathscr{C}_k(J)$ is a vector space and that $\mathscr{C}_k{}^1(J)$ is a subspace of $\mathscr{C}_k(J)$. Note that the zero vector in $\mathscr{C}_k(J)$ is the vector

$$\begin{bmatrix} \varnothing \\ \varnothing \\ \vdots \\ \varnothing \end{bmatrix}$$

It is convenient to denote this vector of zero functions by $\varnothing$, as the proper interpretation of the zero vector will always be clear from the context.

We now return to the linear differential system (7). Let ψ denote the unknown vector-valued function

$$\psi = \begin{bmatrix} y_1 \\ y_2 \\ \vdots \\ y_n \end{bmatrix}$$

let ζ denote the continuous vector-valued function

$$\zeta(x) = \begin{bmatrix} f_1(x) \\ f_2(x) \\ \vdots \\ f_n(x) \end{bmatrix}$$

and let Q be the continuous $n \times n$ matrix-valued function

$$Q(x) = \begin{bmatrix} q_{11}(x) & q_{12}(x) & \cdots & q_{1n}(x) \\ q_{21}(x) & q_{22}(x) & \cdots & q_{2n}(x) \\ \vdots & \vdots & & \vdots \\ q_{n1}(x) & q_{n2}(x) & \cdots & q_{nn}(x) \end{bmatrix}$$

on the interval J. Then (7) can be written in the vector-matrix form

(13) $\qquad D\psi = Q(x)\psi + \zeta(x)$

14 Examples

a. The linear system (6) in vector-matrix form is:

$$D\psi = \begin{bmatrix} \varnothing & 1 & \cdots & \varnothing \\ \varnothing & \varnothing & \cdots & \varnothing \\ \vdots & \vdots & & \vdots \\ \varnothing & \varnothing & \cdots & 1 \\ -p_n(x) & -p_{n-1}(x) & \cdots & -p_1(x) \end{bmatrix} \psi + \begin{bmatrix} \varnothing \\ \varnothing \\ \vdots \\ \varnothing \\ f(x) \end{bmatrix}$$

on J. This system is equivalent to the nth-order linear equation (5).

b. Consider the linear differential system

$Dy_1 = 3y_1 - 2y_2 - y_3$

$Dy_2 = -y_1 + 4y_2 + y_3$

$Dy_3 = y_1 + 2y_2 + 5y_3$

This system can be represented in the vector-matrix form

$$D\psi = \begin{bmatrix} 3 & -2 & -1 \\ -1 & 4 & 1 \\ 1 & 2 & 5 \end{bmatrix} \psi$$

on $(-\infty, \infty)$, where

$$\psi = \begin{bmatrix} y_1 \\ y_2 \\ y_3 \end{bmatrix}$$

15 EXERCISES

1. Transform each of the nonhomogeneous equations in Exercises 9 into a linear system of the form $D\psi = Q(x)\psi + \zeta(x)$. On what interval J are these nonhomogeneous systems defined?
2. Show that $\mathscr{C}_k(J)$ is a vector space with $\mathscr{C}_k{}^1(J)$ as a subspace.
3. Show that D is a linear operator on $\mathscr{C}_k{}^1(J)$.
4. Let

$$Q = \begin{bmatrix} 1 & -3 \\ 2 & 4 \end{bmatrix} \quad \text{and} \quad \psi(x) = \begin{bmatrix} e^x \\ x^2 \end{bmatrix}$$

Calculate $D[Q\psi(x)]$ and compare it with $Q[D\psi(x)]$.
5. Show that if Q is an $n \times n$ matrix of constants and $\psi(x) \in \mathscr{C}_n{}^1(J)$, then $D[Q\psi(x)] = Q[D\psi(x)]$.
6. Let

$$Q(x) = \begin{bmatrix} x & x^2 \\ e^{-x} & 1 \end{bmatrix} \quad \text{and} \quad \psi(x) = \begin{bmatrix} e^x \\ x^2 \end{bmatrix}$$

Calculate $Q(x)[D\psi(x)]$, $[DQ(x)]\psi(x)$, and $D[Q(x)\psi(x)]$, and show that

$$D[Q(x)\psi(x)] = Q(x)[D\psi(x)] + [DQ(x)]\psi(x)$$

7. Generalize the result in Problem 6.
8. Show that the linear differential system

$$\textbf{(A)} \qquad \begin{aligned} Dy &= q_{11}(x)y + q_{12}(x)z + f_1(x) \\ Dz &= q_{21}(x)y + q_{22}(x)z + f_2(x) \end{aligned}$$

can be written in vector-matrix form as

$$D\begin{bmatrix} y \\ z \end{bmatrix} = \begin{bmatrix} q_{11}(x) & q_{12}(x) \\ q_{21}(x) & q_{22}(x) \end{bmatrix}\begin{bmatrix} y \\ z \end{bmatrix} + \begin{bmatrix} f_1(x) \\ f_2(x) \end{bmatrix}$$

or

$$D\begin{bmatrix} z \\ y \end{bmatrix} = \begin{bmatrix} q_{22}(x) & q_{21}(x) \\ q_{12}(x) & q_{11}(x) \end{bmatrix}\begin{bmatrix} z \\ y \end{bmatrix} + \begin{bmatrix} f_2(x) \\ f_1(x) \end{bmatrix}$$

9. In how many ways can the linear differential system (7) be expressed in a vector-matrix form (13)?

In the next section, we shall develop the general theory of linear differential systems. Since (13) is identical in form to the first-order linear differential equation

(16) $Dy = q(x)y + f(x)$

we shall expect some analogies between (13) and (16). The development in the next section closely parallels the general theory of linear differential equations presented in Sections 9, 10, and 11, and so the reader is referred to this material for appropriate background information and facts.

At the beginning of this section, we saw how to transform an nth-order linear differential equation into a linear system. A natural question to ask, then, is this: Can any linear differential system be converted into an equivalent nth-order linear differential equation? We conclude this section with some exercises which will answer this question.

17 EXERCISES

1. Let

$$Q = \begin{bmatrix} 3 & -2 & -1 \\ -1 & 4 & 1 \\ 1 & 2 & 5 \end{bmatrix}$$

and transform the system $D\psi = Q\psi$ into a third-order equation $P(D)y = \emptyset$, where $P(t)$ is a polynomial. Show that $P(2) = P(4) = P(6) = 0$.

2. Let

$$Q(x) = \begin{bmatrix} \sin(x^2) & \ln x \\ 3 & 2x + 1 \end{bmatrix} \quad \text{and} \quad \zeta(x) = \begin{bmatrix} e^{-x} \\ (x-1)^{-1} \end{bmatrix}$$

On what intervals J is $D\psi = Q(x)\psi + \zeta(x)$ defined? Transform this system into a second-order equation.

3. If

$$Q(x) = \begin{bmatrix} q_{11}(x) & q_{12}(x) \\ q_{21}(x) & q_{22}(x) \end{bmatrix}$$

on J, what restrictions can be placed on the coefficient functions $q_{ij}(x)$ so that $D\psi = Q(x)\psi$ can be transformed into a second-order linear differential equation with no singular points?

Section 39 General Theory

In this section, we develop the general theory of the linear differential system

$$Dy_1 = q_{11}(x)y_1 + q_{12}(x)y_2 + \cdots + q_{1n}(x)y_n + f_1(x)$$

$$Dy_2 = q_{21}(x)y_1 + q_{22}(x)y_2 + \cdots + q_{2n}(x)y_n + f_2(x)$$

(1) $\vdots \qquad \vdots \qquad \vdots \qquad \qquad \vdots \qquad \vdots$

$$Dy_n = q_{n1}(x)y_1 + q_{n2}(x)y_2 + \cdots + q_{nn}(x)y_n + f_n(x)$$

where the n^2 functions $q_{11}, q_{12}, \ldots, q_{nn}$, and the functions $f_1, f_2, \ldots, f_n$ are continuous on a given interval J. As we saw at the end of the last section, (1) can be written in the vector-matrix form

(2) $D\psi = Q(x)\psi + \zeta(x)$

on J, where $Q(x) = (q_{ij}(x))$ is a continuous $n \times n$ matrix-valued function, called the *coefficient matrix*, ζ is a continuous vector-valued function with components $f_1, f_2, \ldots, f_n$, and ψ is the (unknown) vector-valued function having components $y_1, y_2, \ldots, y_n$.

Paralleling the presentation in Sections 9, 10, and 11, we begin with definitions of the pertinent terms.

3 DEFINITION

Given the linear differential system (2). A *solution* of (2) is a continuously differentiable vector-valued function ψ on J, that is, $\psi \in \mathscr{C}_n{}^1(J)$, which satisfies (2) on J. If $a \in J$, and β is a constant vector, then the problem of finding a solution ψ of (2) such that $\psi(a) = \beta$ is called an *initial-value problem*. The constant vector β is called the initial condition, or the *initial value* of ψ.

As in the case of initial-value problems associated with linear differential equations, we have a fundamental existence and uniqueness theorem for initial-value problems involving linear differential systems. Although this theorem is essential in the work which follows, we are omitting the proof since it is beyond the scope of this introductory treatment.

4 THEOREM

Given the linear differential system (2). Let $a \in J$ and let β be a constant vector. There is a unique vector-valued function $\psi \in \mathscr{C}_n{}^1(J)$ such that ψ is a solution of (2) and $\psi(a) = \beta$.

5 Examples

a. Consider the homogeneous system $D\psi = Q(x)\psi$, where

$$Q(x) = \begin{bmatrix} 0 & 1 \\ -x^{-2} & x^{-1} \end{bmatrix}$$

The vector-valued function

$$\psi_1(x) = \begin{bmatrix} x \\ 1 \end{bmatrix}$$

is a solution since

$$Q(x)\psi_1(x) = \begin{bmatrix} 1 \\ 0 \end{bmatrix}$$

and

$$D\psi_1(x) = \begin{bmatrix} 1 \\ 0 \end{bmatrix}$$

Also,

$$\psi_1(1) = \begin{bmatrix} 1 \\ 1 \end{bmatrix}$$

so that $\psi_1(x)$ is the solution of the initial-value problem

$$D\psi = Q(x)\psi; \qquad \psi(1) = \begin{bmatrix} 1 \\ 1 \end{bmatrix}$$

b. Consider the nonhomogeneous system $D\psi = Q(x)\psi + \zeta(x)$, where

$$Q(x) = \begin{bmatrix} 0 & 1 \\ 6 & -1 \end{bmatrix} \qquad \text{and} \qquad \zeta(x) = \begin{bmatrix} 0 \\ e^x \end{bmatrix}$$

and the vector-valued function

$$\psi_1(x) = e^{-3x} \begin{bmatrix} 1 \\ -3 \end{bmatrix} - \tfrac{1}{4}e^x \begin{bmatrix} 1 \\ 1 \end{bmatrix}$$

The calculations needed to show that $\psi_1(x)$ is a solution of $D\psi = Q(x)\psi + \zeta(x)$ are:

$$Q(x)\psi_1(x) + \zeta(x) = e^{-3x} \begin{bmatrix} 0 & 1 \\ 6 & -1 \end{bmatrix} \begin{bmatrix} 1 \\ -3 \end{bmatrix}$$

$$- \tfrac{1}{4}e^x \begin{bmatrix} 0 & 1 \\ 6 & -1 \end{bmatrix} \begin{bmatrix} 1 \\ 1 \end{bmatrix} + \begin{bmatrix} 0 \\ e^x \end{bmatrix}$$

$$= e^{-3x} \begin{bmatrix} -3 \\ 9 \end{bmatrix} - \tfrac{1}{4}e^x \begin{bmatrix} 1 \\ 5 \end{bmatrix} + \begin{bmatrix} 0 \\ e^x \end{bmatrix}$$

$$= -3e^{-3x} \begin{bmatrix} 1 \\ -3 \end{bmatrix} - \tfrac{1}{4}e^x \begin{bmatrix} 1 \\ 1 \end{bmatrix}$$

while

$$D\psi_1(x) = -3e^{-3x} \begin{bmatrix} 1 \\ -3 \end{bmatrix} - \tfrac{1}{4}e^x \begin{bmatrix} 1 \\ 1 \end{bmatrix}$$

Thus $\psi_1(x)$ is a solution of $D\psi = Q(x)\psi + \zeta(x)$, and $\psi_1(x)$ is the solution that satisfies the initial condition

$$\psi(0) = \tfrac{1}{4} \begin{bmatrix} 3 \\ -13 \end{bmatrix}$$

6 EXERCISES

1. Verify that

$$\psi_2(x) = \ln(x)\begin{bmatrix} x \\ 1 \end{bmatrix} + \begin{bmatrix} 0 \\ 1 \end{bmatrix}$$

is a solution of the system in Example 5a.

2. Is there a constant c such that

$$\psi(x) = e^x \begin{bmatrix} c \\ c \end{bmatrix}$$

is a solution of the system in Example 5b?

In each of the following, transform the linear differential equation $P(D)y = \emptyset$ into the linear system $D\psi = Q\psi$. Find a solution basis for $P(D)y = \emptyset$ and use these functions to obtain some solutions of $D\psi = Q\psi$.

3. $P(t) = (t + 4)(t + 1)$
4. $P(t) = t^3 + 2t^2 - 11t - 12$
5. $P(t) = t^2 + 2t + 1$
6. Compare each of the vector-valued functions obtained in Problems 3, 4, and 5 with the columns of the Wronski matrix associated with each linear differential equation.

Our objective in presenting the general theory of linear differential systems is to give a characterization of the set of solutions. Following Chapter 2, we shall first characterize the set of solutions of the linear homogeneous differential system

(7) $D\psi = Q(x)\psi$

Let $\mathscr{H}$ denote the set of solutions of (7). Our first result is an exact analogue of Theorem 2, Section 10, and is easy to prove.

8 THEOREM

Let ψ_1 and ψ_2 be members of $\mathscr{H}$ and let c_1 and c_2 be real numbers. Then $\psi = c_1\psi_1 + c_2\psi_2$ is a member of $\mathscr{H}$.

In the terminology of linear algebra, this theorem tells us that $\mathscr{H}$ is a vector space, a subspace of $\mathscr{C}_n^{1}(J)$. If $\psi_1, \psi_2, \ldots, \psi_k$ are members of $\mathscr{H}$ and $c_1, c_2, \ldots, c_k$ are real numbers, then, as an obvious extension of the theorem,

(9) $\psi = c_1\psi_1 + c_2\psi_2 + \cdots + c_k\psi_k = \sum_{i=1}^{k} c_i\psi_i$

is a member of $\mathscr{H}$. The expression (9) is called a *linear combination* of the vector-valued functions $\psi_1, \psi_2, \ldots, \psi_k$. For the special case $c_1 = c_2 = \cdots = c_k = 0$ in (9), the linear combination, called the *trivial linear combination*, is the zero vector. Thus we can conclude that $\emptyset$ is a solution of (7). This can be verified directly, and the zero vector is called the *trivial solution* of $D\psi = Q(x)\psi$.

In many cases, there is an advantage in writing the linear combination (9) in vector-matrix form. Thus if the vector-valued functions ψ_1, $\psi_2, \ldots, \psi_k$ are expressed in terms of their components as

$$\psi_1 = \begin{bmatrix} y_{11} \\ y_{21} \\ \vdots \\ y_{n1} \end{bmatrix}; \psi_2 = \begin{bmatrix} y_{12} \\ y_{22} \\ \vdots \\ y_{n2} \end{bmatrix}; \cdots \psi_k = \begin{bmatrix} y_{1k} \\ y_{2k} \\ \vdots \\ y_{nk} \end{bmatrix}$$

then the linear combination (9) can be written as

$$
(10) \qquad \begin{bmatrix} y_{11} & y_{12} & \cdots & y_{1k} \\ y_{21} & y_{22} & \cdots & y_{2k} \\ \vdots & \vdots & & \vdots \\ y_{n1} & y_{n2} & \cdots & y_{nk} \end{bmatrix} \cdot \begin{bmatrix} c_1 \\ c_2 \\ \vdots \\ c_k \end{bmatrix} = \psi
$$

Continuing with the analogy between differential systems and differential equations, our objective now is to show that the subspace $\mathscr{H}$ of solutions of $D\psi = Q(x)\psi$ has dimension n. For this purpose, we shall need the concepts of linear independence and linear dependence in the vector space $\mathscr{C}_n(J)$.

11 DEFINITION

A set $\psi_1, \psi_2, \ldots, \psi_k$ of vector-valued functions in $\mathscr{C}_n(J)$ is *linearly dependent* if there exists a set of k real numbers $c_1, c_2, \ldots, c_k$, not all of which are zero, such that

$$c_1\psi_1(x) + c_2\psi_2(x) + \cdots + c_k\psi_k(x) = \varnothing$$

on J. The set $\psi_1, \psi_2, \ldots, \psi_k$ is *linearly independent* if it is not linearly dependent.

12 Examples

a. Consider the two vector-valued functions

$$\psi_1(x) = \begin{bmatrix} x \\ 1 \end{bmatrix} \quad \text{and} \quad \psi_2(x) = \begin{bmatrix} x \ln(x) \\ 1 + \ln(x) \end{bmatrix}$$

If $c_1\psi_1(x) + c_2\psi_2(x) = \varnothing$ on $J = [1, \infty)$, then the first components satisfy

$$c_1 x + c_2 x \ln(x) = \varnothing$$

which implies that

$$c_1 + c_2 \ln(x) = \varnothing$$

since $x > 0$. By differentiating this last equation, we have that $c_2/x = \varnothing$ and thus $c_2 = c_1 = 0$. Therefore, $\psi_1(x)$ and $\psi_2(x)$ are linearly independent on $J = [1, \infty)$.

b. If

$$\psi_1(x) = \begin{bmatrix} \cos^2 x \\ \cos x \\ 4 \end{bmatrix}, \qquad \psi_2(x) = \begin{bmatrix} 2\sin^2 x \\ \cos x \\ -1 \end{bmatrix}, \qquad \psi_3(x) = \begin{bmatrix} 2 \\ 3\cos x \\ 7 \end{bmatrix}$$

then, since $2\psi_1(x) + \psi_2(x) - \psi_3(x) = \varnothing$, these vector-valued functions are linearly dependent.

For additional examples of the concepts of linear independence and dependence in a variety of settings, the reader is referred to Sections 4 and 10.

Since our objective is to show that $\mathscr{H}$ has dimension n, we need to show that $D\psi = Q(x)\psi$ has n linearly independent solutions, and that every solution of $D\psi = Q(x)\psi$ can be expressed as a linear combination of these n solutions. This will be accomplished in exactly the same manner as in Section 10. In particular, recall that if $y_1, y_2, \ldots, y_n$ are n solutions of the nth-order homogeneous equation

$$D^n y + p_1(x)\, D^{n-1} y + \cdots + p_{n-1}(x)\, Dy + p_n(x)y = \varnothing$$

on J, then the Wronski matrix, $W(x)$, of these solutions is the $n \times n$ matrix-valued function

$$W(x) = \begin{bmatrix} y_1(x) & y_2(x) & \cdots & y_n(x) \\ Dy_1(x) & Dy_2(x) & \cdots & Dy_n(x) \\ \vdots & \vdots & & \vdots \\ D^{n-1}y_1(x) & D^{n-1}y_2(x) & \cdots & D^{n-1}y_n(x) \end{bmatrix}$$

on J. Consider the columns of this matrix. It is easy to verify that each column of W is a solution of the linear differential system

$$D\psi = \begin{bmatrix} \varnothing & 1 & \varnothing & \cdots & \varnothing \\ \varnothing & \varnothing & 1 & \cdots & \varnothing \\ \vdots & \vdots & \vdots & & \vdots \\ \varnothing & \varnothing & \varnothing & \cdots & 1 \\ -p_n & -p_{n-1} & -p_{n-2} & \cdots & -p_1 \end{bmatrix} \psi$$

This observation provides the motivation for the next definition.

13 DEFINITION

Let $\psi_1, \psi_2, \ldots, \psi_n$ be n solutions of $D\psi = Q(x)\psi$, where

$$\psi_i = \begin{bmatrix} y_{1i} \\ y_{2i} \\ \vdots \\ y_{ni} \end{bmatrix}$$

$i = 1, 2, \ldots, n$. The $n \times n$ matrix-valued function W given by

$$W(x) = \begin{bmatrix} y_{11}(x) & y_{12}(x) & \cdots & y_{1n}(x) \\ y_{21}(x) & y_{22}(x) & \cdots & y_{2n}(x) \\ \vdots & \vdots & & \vdots \\ y_{n1}(x) & y_{n2}(x) & \cdots & y_{nn}(x) \end{bmatrix}$$

is called the *Wronski matrix* of $\psi_1, \psi_2, \ldots, \psi_n$. The determinant, $\det[W(x)]$, of the Wronski matrix is called the *Wronskian* of $\psi_1, \psi_2, \ldots, \psi_n$.

Notice that by using the observation made in (10) together with Theorem 8, it follows that if γ is a vector of constants, then $W(x)\gamma$ is a solution of $D\psi = Q(x)\psi$.

As emphasized by the discussion preceding the definition, the two notions of the Wronski matrix, that is, Definition 8, Section 10, and Definition 13, coincide when (7) is the system representation of an nth-order equation. Note, in addition, that the Wronskian of n solutions of (7) is a continuously differentiable function on J, that is, $\det[W(x)] \in \mathscr{C}^1(J)$.

14 Examples

a. The Wronski matrix for the two vector-valued functions in Example 12a is:

$$W(x) = \begin{bmatrix} x & x \ln(x) \\ 1 & 1 + \ln(x) \end{bmatrix}$$

on $J = [1, \infty)$. Therefore, the Wronskian of this pair of functions is:

$$\det[W(x)] = x + x \ln(x) - x \ln(x) = x$$

and $\det[W(x)] > 0$ on $J = [1, \infty)$.

b. Consider the three vector-valued functions

$$\psi_1(x) = e^{6x} \begin{bmatrix} -1 \\ 1 \\ 1 \end{bmatrix}, \quad \psi_2(x) = e^{4x} \begin{bmatrix} 1 \\ -1 \\ 1 \end{bmatrix}, \quad \psi_3(x) = e^{2x} \begin{bmatrix} -1 \\ -1 \\ 1 \end{bmatrix}$$

The Wronski matrix is:

$$W(x) = \begin{bmatrix} -e^{6x} & e^{4x} & -e^{2x} \\ e^{6x} & -e^{4x} & -e^{2x} \\ e^{6x} & e^{4x} & e^{2x} \end{bmatrix}$$

$$= \begin{bmatrix} -1 & 1 & -1 \\ 1 & -1 & -1 \\ 1 & 1 & 1 \end{bmatrix} \begin{bmatrix} e^{6x} & 0 & 0 \\ 0 & e^{4x} & 0 \\ 0 & 0 & e^{2x} \end{bmatrix}$$

and $\det[W(x)] = -4e^{12x}$.

The following theorem is the analogue of Theorem 10, Section 10, and the proof of that theorem can be used here.

15 THEOREM

Let $\psi_1, \psi_2, \ldots, \psi_n$ be n solutions of $D\psi = Q(x)\psi$, and let W be their Wronski matrix. These functions are linearly dependent on J if and only if their Wronskian is the zero function on J.

The following corollary is the composite of Corollaries 11 and 12 of Section 10. It shows that the Wronskian of n solutions of $D\psi = Q(x)\psi$ determines whether the solutions are linearly dependent or linearly independent. The proof is left as an exercise.

16 COROLLARY

Let $\psi_1, \psi_2, \ldots, \psi_n$ be n solutions of $D\psi = Q(x)\psi$ and let $W(x)$ be their Wronski matrix. Then:

(i) The vector-valued functions are linearly dependent on J if and only if $\det[W(x)] = \varnothing$ on J.

(ii) The vector-valued functions are linearly independent on J if and only if $\det[W(x)] \neq 0$ for all $x \in J$. Moreover, since $\det[W(x)] \in \mathscr{C}^1(J)$, either $\det[W(x)] > 0$ for all $x \in J$, or $\det[W(x)] < 0$ for all $x \in J$.

We are now in a position to establish that the vector space $\mathscr{H}$ of solutions of (7) has dimension n and to provide a characterization of $\mathscr{H}$. The next theorem is the extension of Section 10 to the homogeneous system $D\psi = Q(x)\psi$.

17 THEOREM

The vector space $\mathscr{H}$ of solutions of (7) has dimension n. If $\psi_1, \psi_2, \ldots, \psi_n$ are n linearly independent solutions of (7), then each solution ψ of (7) has a unique representation as a linear combination of $\psi_1, \psi_2, \ldots, \psi_n$; that is, if $\psi \in \mathscr{H}$, then

$$\psi(x) = c_1\psi_1(x) + c_2\psi_2(x) + \cdots + c_n\psi_n(x) = W(x)\gamma$$

where $\gamma = (c_1, c_2, \ldots, c_n)^T$ and $W(x)$ is the Wronski matrix of $\psi_1(x)$, $\psi_2(x), \ldots, \psi_n(x)$.

Proof: Choose any point $b \in J$, and let $\varepsilon_1, \varepsilon_2, \ldots, \varepsilon_n$ be the solutions of (7) determined by the initial conditions

$$\varepsilon_1(b) = \begin{bmatrix} 1 \\ 0 \\ 0 \\ \vdots \\ 0 \end{bmatrix}, \quad \varepsilon_2(b) = \begin{bmatrix} 0 \\ 1 \\ 0 \\ \vdots \\ 0 \end{bmatrix}, \quad \ldots, \quad \varepsilon_n(b) = \begin{bmatrix} 0 \\ 0 \\ 0 \\ \vdots \\ 1 \end{bmatrix}$$

Let W be the Wronski matrix of $\varepsilon_1, \varepsilon_2, \ldots, \varepsilon_n$. Then it is easy to see that $W(b) = I$, the $n \times n$ identity matrix, so that $\det[W(b)] = 1$. Thus the solutions are linearly independent on J.

Now let ψ be any solution of $D\psi = Q(x)\psi$ and evaluate ψ at $x = b$,

$$\psi(b) = \begin{bmatrix} s_1 \\ s_2 \\ \vdots \\ s_n \end{bmatrix}$$

Define the vector-valued function ρ by

$$\rho(x) = s_1\varepsilon_1(x) + s_2\varepsilon_2(x) + \cdots + s_n\varepsilon_n(x)$$

Then ρ is a solution of (7) and

$$\rho(b) = \begin{bmatrix} s_1 \\ s_2 \\ \vdots \\ s_n \end{bmatrix} = \psi(b)$$

Consequently, by the uniqueness theorem, Theorem 4

$$\psi(x) = \rho(x) = \sum_{i=1}^{n} s_i\varepsilon_i(x)$$

We can now conclude that $\mathcal{H}$ has dimension n.

To establish the second part of the theorem, let $\psi_1(x), \psi_2(x), \ldots, \psi_n(x)$ be any set of n linearly independent solutions of $D\psi = Q(x)\psi$, and let $W(x)$ be their Wronski matrix. If $\sigma(x)$ is any solution of $D\psi = Q(x)\psi$ on J and $a \in J$, then there is a unique vector γ such that

$$W(a)\gamma = \sigma(a)$$

The uniqueness of γ is a consequence of $\det[W(a)] \neq 0$. By using this vector γ, we have that

$$\psi(x) = W(x)\gamma$$

is a solution of $D\psi = Q(x)\psi$ since it is a linear combination of $\psi_1(x), \psi_2(x), \ldots, \psi_n(x)$. Finally, $\psi(a) = W(a)\gamma = \sigma(a)$, and thus, by Theorem 4, $\psi(x) = \sigma(x)$.

In the terminology of Section 10, a set of n linearly independent solutions of $D\psi = Q(x)\psi$ is called a *solution basis*. As was the case in Section 10, a solution basis can be characterized as any set of n solutions whose Wronskian at some point of J (and hence throughout J) is nonzero. The particular solution basis used in the proof of Theorem 17 was selected for convenience only.

Other facts about the Wronski matrix of a solution basis of $D\psi = Q(x)\psi$ lead to a number of points of similarity between the theory of linear systems and first-order linear differential equations. The verification of the next result is an easy exercise.

18 THEOREM

If $W(x)$ is the Wronski matrix of a solution basis for $D\psi = Q(x)\psi$, then

$$DW(x) = Q(x)W(x)$$

Theorem 18 could be a starting point in an analysis of yet another kind of differential equation, since it follows from this result that $W(x)$ is a solution of

(19) $DY = Q(x)Y$

where Y is an unknown matrix-valued function.

The solutions of $Dy + p(x)y = \emptyset$ are the one-parameter family $y(x) = c \exp[-\int_a^x p(t)\,dt]$, while the solutions of $D\psi = Q(x)\psi$ are the one-parameter family $\psi(x) = W(x)\gamma$. Also, $u(x) = \exp[-\int_a^x p(t)\,dt]$ is a particular solution of $Dy + p(x)y = \emptyset$ for which $u(x) \neq 0$ for all $x \in J$. This is analogous to Theorem 18 in the case of linear systems where $\det[W(x)] \neq 0$ for $x \in J$. Some related topics are stated in the exercises.

20 Example

If

$$Q = \begin{bmatrix} 3 & -2 & -1 \\ -1 & 4 & 1 \\ 1 & 2 & 5 \end{bmatrix}$$

then it is easy to verify that the three vector-valued functions in Example 14b are solutions of $D\psi = Q(x)\psi$. Again, in Example 14b, the Wronskian was given as $\det[W(x)] = -4e^{12x}$ so that the solutions are a solution basis.

If $\gamma = (c_1, c_2, c_3)^T$ is any constant vector, then all solutions of $D\psi = Q(x)\psi$ are of the form

$$\psi(x) = W(x)\gamma = \begin{bmatrix} -1 & 1 & -1 \\ 1 & -1 & -1 \\ 1 & 1 & 1 \end{bmatrix} \begin{bmatrix} c_1 e^{6x} \\ c_2 e^{4x} \\ c_3 e^{2x} \end{bmatrix}$$

where $W(x)$ was displayed in Example 14b. This is an especially convenient form for solving an initial-value problem at $x = 0$ since

$$\psi(0) = \begin{bmatrix} -1 & 1 & -1 \\ 1 & -1 & -1 \\ 1 & 1 & 1 \end{bmatrix} \begin{bmatrix} c_1 \\ c_2 \\ c_3 \end{bmatrix}.$$

If the initial vector $\psi(0)$ is given, then the above equation is easy to solve for $\gamma = (c_1, c_2, c_3)^T$. For example, if

$$\psi(0) = \begin{bmatrix} -1 \\ 3 \\ 1 \end{bmatrix}$$

is the initial condition, then

$$\gamma = \begin{bmatrix} c_1 \\ c_2 \\ c_3 \end{bmatrix} = \begin{bmatrix} 2 \\ 0 \\ -1 \end{bmatrix}$$

or

$$\psi(x) = 2e^{6x} \begin{bmatrix} -1 \\ 1 \\ 1 \end{bmatrix} - e^{2x} \begin{bmatrix} -1 \\ -1 \\ 1 \end{bmatrix}$$

is the unique solution of $D\psi = Q(x)\psi$ that satisfies this initial condition.

21 EXERCISES

1. Write each of the following linear combinations in the vector-matrix form (10).

a. $7 \begin{bmatrix} 1 \\ x \end{bmatrix} + 4 \begin{bmatrix} e^x \\ x^2 \end{bmatrix}$

b. $-3 \begin{bmatrix} 1 \\ 0 \\ x \end{bmatrix} + 8 \begin{bmatrix} e^x \\ x \\ x \end{bmatrix}$

c. $2 \begin{bmatrix} -1 \\ x \end{bmatrix} + 3 \begin{bmatrix} \sin x \\ x \end{bmatrix} + 8 \begin{bmatrix} e^x \\ 1 \end{bmatrix} - 4 \begin{bmatrix} 2 \\ 6 \end{bmatrix}$

d. $13 \begin{bmatrix} e^{-x} \\ 3e^{-x} \end{bmatrix} - 9 \begin{bmatrix} e^x \\ 7e^x \end{bmatrix}$

2. Determine which of the following sets of vector-valued functions are linearly independent on J.

a. $\left\{ e^x \begin{bmatrix} 1 \\ 1 \end{bmatrix}, e^x \begin{bmatrix} 1 \\ 0 \end{bmatrix} \right\}, J = (-\infty, \infty)$

b. $\left\{ \begin{bmatrix} x \\ 1 \\ -1 \end{bmatrix}, \begin{bmatrix} 1 \\ x \\ -1 \end{bmatrix}, \begin{bmatrix} 1 \\ -1 \\ x \end{bmatrix} \right\}, J = (-\infty, \infty)$

c. $\left\{ e^{-x} \begin{bmatrix} 1 \\ -4 \end{bmatrix}, xe^{-x} \begin{bmatrix} 1 \\ -4 \end{bmatrix} \right\}, J = (-\infty, \infty)$

d. $\left\{ e^{4x} \begin{bmatrix} 1 \\ 1 \\ 7 \end{bmatrix}, e^{4x} \begin{bmatrix} 2 \\ 0 \\ 3 \end{bmatrix}, e^{4x} \begin{bmatrix} -3 \\ 1 \\ 1 \end{bmatrix} \right\}, J = (-\infty, \infty)$

e. $\left\{ \begin{bmatrix} \ln x \\ x \ln x \end{bmatrix}, \begin{bmatrix} 1 \\ x \end{bmatrix} \right\}, J = [1, \infty)$

3. Show that the Wronski matrix in Example 14b is a solution of $D[W(x)] = QW(x)$, where

$$Q = \begin{bmatrix} 3 & -2 & -1 \\ -1 & 4 & 1 \\ 1 & 2 & 5 \end{bmatrix}$$

4. Determine which of the following vector-valued functions can be used to form a solution basis for

$$D\psi = \begin{bmatrix} 3 & 1 \\ 1 & 3 \end{bmatrix} \psi$$

$$\psi_1(x) = e^{2x} \begin{bmatrix} -1 \\ 1 \end{bmatrix}, \qquad \psi_2(x) = e^{3x} \begin{bmatrix} 2 \\ 1 \end{bmatrix}, \qquad \psi_3(x) = \begin{bmatrix} x \\ 1 \end{bmatrix}, \qquad \psi_4(x) = \begin{bmatrix} 0 \\ 0 \end{bmatrix}$$

$$\psi_5(x) = e^{2x} \begin{bmatrix} 1 \\ -1 \end{bmatrix}, \qquad \psi_6(x) = -e^{4x} \begin{bmatrix} 1 \\ 1 \end{bmatrix}, \qquad \psi_7(x) = xe^{2x} \begin{bmatrix} 1 \\ 1 \end{bmatrix}$$

5. Which of the following vector-valued functions form a solution basis for

$$D\psi = \begin{bmatrix} x + \dfrac{1}{2x} & x - \dfrac{1}{2x} \\ x - \dfrac{1}{2x} & x + \dfrac{1}{2x} \end{bmatrix} \psi$$

on $J = [1, \infty)$?

$$\psi_1(x) = \begin{bmatrix} -\frac{1}{2} & \frac{1}{2} \\ -\frac{1}{2} & -\frac{1}{2} \end{bmatrix} \begin{bmatrix} e^{x^2} \\ x \end{bmatrix}, \qquad \psi_2(x) = x \begin{bmatrix} 1 \\ -1 \end{bmatrix}$$

$$\psi_3(x) = \begin{bmatrix} e^{x^2} \\ e^{x^2} \end{bmatrix}, \qquad \psi_4(x) = \begin{bmatrix} e^{x^2} \\ 4x \end{bmatrix}, \qquad \psi_5(x) = \begin{bmatrix} \cos x \\ x^2 \end{bmatrix}$$

$$\psi_6(x) = \begin{bmatrix} x^2 \\ \sin x \end{bmatrix}, \qquad \psi_7(x) = \begin{bmatrix} 2 & -2 \\ 2 & 2 \end{bmatrix} \begin{bmatrix} e^{x^2} \\ x \end{bmatrix}$$

6. Let $W(x)$ be the Wronski matrix of any solution basis for the linear system in Problem 5. Show that

$$\det[W(x)] = \det[W(1)] \exp\left[\int_1^x (2t + t^{-1})\, dt \right]$$

7. Let $g_{11}(x), g_{12}(x), g_{21}(x)$, and $g_{22}(x)$ be members of $\mathscr{C}^1(J)$. Show that

$$D\left[\det \begin{bmatrix} g_{11}(x) & g_{12}(x) \\ g_{21}(x) & g_{22}(x) \end{bmatrix} \right] = \det \begin{bmatrix} Dg_{11}(x) & g_{12}(x) \\ Dg_{21}(x) & g_{22}(x) \end{bmatrix} + \det \begin{bmatrix} g_{11}(x) & Dg_{12}(x) \\ g_{21}(x) & Dg_{22}(x) \end{bmatrix}$$

and

$$D\left[\det\begin{bmatrix} g_{11}(x) & g_{12}(x) \\ g_{21}(x) & g_{22}(x) \end{bmatrix}\right] = \det\begin{bmatrix} Dg_{11}(x) & Dg_{12}(x) \\ g_{21}(x) & g_{22}(x) \end{bmatrix}$$
$$+ \det\begin{bmatrix} g_{11}(x) & g_{12}(x) \\ Dg_{21}(x) & Dg_{22}(x) \end{bmatrix}$$

8. Can you extend the results in Problem 7 to higher-order matrices?
9. Let $W(x)$ be the Wronski matrix of any pair of solutions of

$$D\psi = \begin{bmatrix} q_{11}(x) & q_{12}(x) \\ q_{21}(x) & q_{22}(x) \end{bmatrix}\psi$$

on J. Show that if $a \in J$, then

$$\det[W(x)] = \det[W(a)] \exp\left\{\int_a^x [q_{11}(t) + q_{22}(t)]\, dt\right\}$$

(Hint: Use a formula in Problem 7.)

10. Let

$$D\psi = \begin{bmatrix} 1 - \cos^2 x & \ln(1 + x^2) \\ x^2 + 7 & -\sin^2 x \end{bmatrix}\psi$$

on $J = [0, \infty)$. Show that the Wronskian of any solution basis of this linear system is a constant function.

11. If $Q(x) = (q_{ij}(x))$ is an $n \times n$ matrix-valued function, then

$$\text{tr}[Q(x)] = \sum_{i=1}^{n} q_{ii}(x)$$

is called the *trace of* $Q(x)$. Generalize the result in Problem 9 by showing that if $W(x)$ is the Wronski matrix of n solutions of

$$D\psi = Q(x)\psi$$

on J, then

$$\det[W(x)] = \det[W(a)] \exp\left[\int_a^x \text{tr}[Q(t)]\, dt\right]$$

where $a \in J$.

12. Let $\det[W(x)]$ be the Wronskian of a solution basis of

$$D\psi = Q(x)\psi$$

on J. Show that $\det[W(x)]$ is a constant function on J if and only if $\text{tr}[Q(x)] = \emptyset$ on J.

13. Prove Theorem 15 and Corollary 16.
14. Prove Theorem 18.

There are cases in which the coefficient matrix $Q(x)$ in

$$D\psi = Q(x)\psi$$

on J has a special form which can be used to advantage in finding solutions of the differential system. We introduce two of them here.

First, consider the linear homogeneous system (7) and assume that the coefficient matrix $Q = (q_{ij})$ is a diagonal matrix, that is,

$$Q(x) = \text{diag}[q_1(x), q_2(x), \ldots, q_n(x)]$$

or more explicitly,

$$Q(x) = \begin{bmatrix} q_1(x) & \varnothing & \varnothing & \cdots & \varnothing \\ \varnothing & q_2(x) & \varnothing & \cdots & \varnothing \\ \vdots & \vdots & \vdots & & \vdots \\ \varnothing & \varnothing & \varnothing & \cdots & q_n(x) \end{bmatrix}$$

Here, it is more convenient to write $D\psi = Q(x)\psi$ as the system of n first-order equations

(22)
$$\begin{aligned} Dy_1 &= q_1(x)y_1 \\ Dy_2 &= q_2(x)y_2 \\ &\vdots \\ Dy_n &= q_n(x)y_n \end{aligned}$$

on J. Note that the ith equation depends only on the "unknown" y_i. A linear differential system having this special form is said to be *uncoupled*. Since each equation in (22) is simply a first-order linear homogeneous equation, it is easy to obtain solutions of this system. In particular, for each positive integer i, $1 \leq i \leq n$, $y_i(x) = e^{u_i(x)}$, where $u_i(x) = \int_a^x q_i(t)\, dt$, $a \in J$, is a solution of the ith equation. The vector-valued functions

$$\psi_1(x) = \begin{bmatrix} e^{u_1(x)} \\ \varnothing \\ \varnothing \\ \vdots \\ \varnothing \end{bmatrix}; \quad \psi_2(x) = \begin{bmatrix} \varnothing \\ e^{u_2(x)} \\ \varnothing \\ \vdots \\ \varnothing \end{bmatrix}; \quad \ldots; \quad \psi_n(x) = \begin{bmatrix} \varnothing \\ \varnothing \\ \varnothing \\ \vdots \\ e^{u_n(x)} \end{bmatrix}$$

form a solution basis for $D\psi = Q(x)\psi$ since their Wronski matrix W is the diagonal matrix

$$\begin{bmatrix} e^{u_1(x)} & 0 & 0 & \cdots & 0 \\ 0 & e^{u_2(x)} & 0 & \cdots & 0 \\ \vdots & \vdots & \vdots & & \vdots \\ 0 & 0 & 0 & \cdots & e^{u_n(x)} \end{bmatrix}$$

whose value at $x = a$ is the $n \times n$ identity matrix. Thus $\det[W(x)] \neq 0$ for all $x \in J$ and $\det[W(a)] = 1$. The n-parameter family of solutions is given by

$$\psi(x) = W(x)\gamma = \begin{bmatrix} c_1 e^{u_1(x)} \\ c_2 e^{u_2(x)} \\ \vdots \\ c_n e^{u_n(x)} \end{bmatrix}$$

An $n \times n$ matrix $T = (t_{ij})$ is *upper triangular* if $t_{ij} = \varnothing$ whenever $i > j$. This is equivalent to saying that all the entries of T below the main diagonal are $\varnothing$. If we assume that the coefficient matrix Q in (7) is upper triangular, then, writing (7) as a system of n first-order equations, we have

$$
\textbf{(23)}\qquad
\begin{aligned}
Dy_1 &= q_{11}(x)y_1 + q_{12}(x)y_2 + \cdots + q_{1n}(x)y_n \\
Dy_2 &= \qquad\qquad\quad q_{22}(x)y_2 + \cdots + q_{2n}(x)y_n \\
&\ \ \vdots \\
Dy_n &= \qquad\qquad\qquad\qquad\qquad\qquad q_{nn}(x)y_n
\end{aligned}
$$

on J. Clearly, the last equation of the system (23) is a first-order linear homogeneous equation, and

$$
y_n(x) = \exp\left[\int_a^x q_{nn}(t)\, dt\right] = e^{u_n(x)}, \qquad a \in J
$$

is a solution. Now, the $(n - 1)$st equation in (23) is:

$$
Dy_{n-1} = q_{n-1,\,n-1}(x)y_{n-1} + q_{n-1,\,n}(x)y_n
$$

Since we have found a solution $y_n(x) = e^{u_n(x)}$, this can be substituted into the above equation to produce the first-order linear nonhomogeneous equation

$$
Dy_{n-1} = q_{n-1,\,n-1}(x)y_{n-1} + q_{n-1,\,n}(x)e^{u_n(x)}
$$

A solution of this equation can be determined using the techniques of Section 8. Then the solutions y_{n-1} and y_n are substituted into the $(n - 2)$nd equation, producing a first-order linear nonhomogeneous equation in y_{n-2}. The continuation of this procedure will produce a solution

$$
\psi(x) = \begin{bmatrix} y_1(x) \\ y_2(x) \\ \vdots \\ y_n(x) \end{bmatrix}
$$

of (23). The possibility of obtaining a solution basis and the n-parameter family of solutions in this case is examined in the exercises.

24 EXERCISES

1. Solve the initial-value problem

$$
D\psi = \begin{bmatrix} 1 & -3 & 1 \\ 0 & -2 & -5 \\ 0 & 0 & 3 \end{bmatrix}\psi; \qquad \psi(0) = \begin{bmatrix} 7 \\ 1 \\ 1 \end{bmatrix}
$$

2. Solve the initial-value problem

$$
D\psi = \begin{bmatrix} 3x^2 & x \\ \varnothing & \dfrac{1}{x} \end{bmatrix}\psi; \qquad \psi(1) = \begin{bmatrix} 9e + 1 \\ -3 \end{bmatrix}
$$

3. Let

$$B = \frac{1}{2}\begin{bmatrix} -1 & 1 \\ -1 & -1 \end{bmatrix}, \qquad C = \begin{bmatrix} -1 & -1 \\ 1 & -1 \end{bmatrix}, \qquad R(x) = \begin{bmatrix} 2x & \varnothing \\ \varnothing & x^{-1} \end{bmatrix}$$

a. Show that $C = B^{-1}$.

b. Form the matrix $Q(x) = BR(x)B^{-1}$.

c. Show that the change of variable $\theta(x) = B^{-1}\psi(x)$ transforms

$$D\psi = Q(x)\psi$$

on $J = [1, \infty)$ into

$$D\theta = R(x)\theta$$

on $J = [1, \infty)$.

d. Find all solutions of $D\psi = Q(x)\psi$ by first solving

$$D\theta = R(x)\theta$$

4. Show that if $Q(x) = BT(x)B^{-1}$, then a solution of $D\psi = Q(x)\psi$ is $\psi(x) = B\theta(x)$, where $\theta(x)$ is a solution of $D\theta = T(x)\theta$.

5. Let

$$Q = \begin{bmatrix} -2 & 0 & -5 \\ -3 & 1 & 1 \\ 0 & 0 & 3 \end{bmatrix}$$

Although Q is not upper triangular, show that it is possible to solve $D\psi = Q\psi$ by solving three first-order linear differential equations.

6. Notice that the matrix in Problem 5, namely Q, is merely a rearrangement of the entries of the upper triangular matrix in Problem 1. Find a nonsingular matrix B such that $Q = BTB^{-1}$ and T is upper triangular. Use Problem 4 to find all solutions of $D\psi = Q\psi$.

7. Let

$$Q = \begin{bmatrix} 4 & -1 & 0 & 0 & 0 \\ 0 & 2 & 0 & 0 & 0 \\ 0 & 0 & 1 & -3 & 1 \\ 0 & 0 & 0 & -2 & -5 \\ 0 & 0 & 0 & 0 & 3 \end{bmatrix}$$

$$Q_1 = \begin{bmatrix} 4 & -1 \\ 0 & 2 \end{bmatrix}, \qquad Q_2 = \begin{bmatrix} 1 & -3 & 1 \\ 0 & -2 & -5 \\ 0 & 0 & 3 \end{bmatrix}$$

Find a solution of $D\psi = Q_1\psi$ and a solution of $D\psi = Q_2\psi$, and show how to use them to obtain a solution of $D\psi = Q\psi$.

8. The matrix Q in Problem 7 is said to be in *block diagonal form* since

$$Q = \begin{bmatrix} Q_1 & 0 \\ 0 & Q_2 \end{bmatrix}$$

where 0 denotes zero matrices of appropriate rows and columns. Both Q_1 and Q_2 are square submatrices of Q. Generalize the observation made in Problem 7 for matrices in block diagonal form.

We complete our treatment of the basic theory of linear differential systems by considering the nonhomogeneous system

(25) $D\psi = Q(x)\psi + \zeta(x)$

on J. Associated with (25) is the homogeneous system

(26) $D\psi = Q(x)\psi$

on J, which we shall call the *reduced system* of (25) corresponding to the terminology introduced in Section 11.

Our objective here is to characterize the set of solutions of (25), and our approach parallels that in Section 11. The next theorem is an extension of Theorem 13, Section 9, and Theorem 4, Section 11.

27 THEOREM

If σ_1 and σ_2 are each solutions of (25), then $\psi = \sigma_1 - \sigma_2$ is a solution of the equation (26). If σ is a solution of (25) and $\psi_1, \psi_2, \ldots, \psi_n$ is a solution basis for the reduced equation (26), then each solution ψ of (25) has a unique representation of the form

(28) $\psi(x) = c_1\psi_1(x) + c_2\psi_2(x) + \cdots + c_n\psi_n(x) + \sigma(x)$

on J.

Proof: Exercise.

Theorem 27 provides a characterization of the set of solutions of (25). If we let W be the Wronski matrix of the solution basis $\psi_1, \psi_2, \ldots, \psi_n$, and γ be the constant vector whose components are the parameters c_1, $c_2, \ldots, c_n$, then (28) can be written in the vector-matrix form

(29) $\psi(x) = [W(x)]\gamma + \sigma(x)$

The family of solutions in (29) can be used to provide, in compact form, the solutions of a nonhomogeneous system in terms of an initial condition.

30 COROLLARY

If $W(x)$ is the Wronski matrix of a solution basis for $D\psi = Q(x)\psi$ and $\sigma(x)$ is a solution of $D\psi = Q(x)\psi + \zeta(x)$, then all solutions of $D\psi = Q(x)\psi + \zeta(x)$ are of the form

$$\psi(x) = W(x)\{[W(a)]^{-1}[\psi(a) - \sigma(a)]\} + \sigma(x)$$

The final topic in this section is an explicit representation of a solution of $D\psi = Q(x)\psi + \zeta(x)$.

31 THEOREM

If $W(x)$ is the Wronski matrix of a solution basis for $D\psi = Q(x)\psi$ on the interval J, then

$$\sigma(x) = W(x) \int_a^x W^{-1}(t)\zeta(t) \, dt, \qquad a \in J$$

is the solution of $D\psi = Q(x)\psi + \zeta(x)$ on J satisfying $\sigma(a) = \varnothing$.

Proof: Clearly $\sigma(a) = \varnothing$. Since

$$D[\sigma(x)] = D[W(x)] \int_a^x W^{-1}(t)\zeta(t) \, dt + W(x)W^{-1}(x)\zeta(x)$$

or

$$D[\sigma(x)] = D[W(x)] \int_a^x W^{-1}(t)\zeta(t) \, dt + \zeta(x)$$

and, from Theorem 18,

$$D[W(x)] = Q(x)W(x)$$

it follows that

$$D[\sigma(x)] = Q(x)W(x) \int_a^x W^{-1}(t)\zeta(t) \, dt + \zeta(x) = Q(x)\sigma(x) + \zeta(x)$$

The form of the solution in Theorem 31 is somewhat similar to the solution of

$$Dy + p(x)y = f(x)$$

that was developed in Section 8. In particular, see Exercises 17, Section 8, Problem 4.

32 Example

Consider the linear differential system

$$D\psi = Q(x)\psi + \zeta(x)$$

on $J = (0, \infty)$, where

$$Q(x) = \begin{bmatrix} \varnothing & 1 \\ -x^{-2} & x^{-1} \end{bmatrix} \quad \text{and} \quad \zeta(x) = \begin{bmatrix} \varnothing \\ 2x^{-1} \end{bmatrix}$$

It is easy to show that the Wronski matrix of a solution basis for the homogeneous system $D\psi = Q(x)\psi$ is:

$$W(x) = \begin{bmatrix} x & x \ln(x) \\ 1 & 1 + \ln(x) \end{bmatrix}$$

and

$$W^{-1}(x) = \begin{bmatrix} x^{-1} + x^{-1} \ln(x) & -\ln(x) \\ -x^{-1} & 1 \end{bmatrix}$$

In order to use Theorem 31, we need

$$W^{-1}(x)\zeta(x) = \begin{bmatrix} x^{-1} + x^{-1} \ln(x) & -\ln(x) \\ -x^{-1} & 1 \end{bmatrix} \begin{bmatrix} \varnothing \\ 2x^{-1} \end{bmatrix}$$

$$= \begin{bmatrix} -2x^{-1} \ln(x) \\ 2x^{-1} \end{bmatrix}$$

so that

$$\int_1^x W^{-1}(t)\zeta(t)\, dt = \int_1^x \begin{bmatrix} -2t^{-1} \ln(t) \\ 2t^{-1} \end{bmatrix} dt = 2 \int_1^x \begin{bmatrix} -t^{-1} \ln(t) \\ t^{-1} \end{bmatrix} dt$$

$$= 2 \begin{bmatrix} -\int_1^x t^{-1} \ln(t)\, dt \\ \int_1^x t^{-1}\, dt \end{bmatrix} = \ln(x) \begin{bmatrix} -\ln(x) \\ 2 \end{bmatrix}$$

Thus

$$\sigma(x) = \ln(x) \begin{bmatrix} x & x \ln(x) \\ 1 & 1 + \ln(x) \end{bmatrix} \begin{bmatrix} -\ln(x) \\ 2 \end{bmatrix} = \ln(x) \begin{bmatrix} x \ln(x) \\ 2 + \ln(x) \end{bmatrix}$$

is a solution of $D\psi = Q(x)\psi + \zeta(x)$, and, therefore, all solutions of this equation are:

$$\psi(x) = W(x)\gamma + \sigma(x)$$

Finally, if the initial condition associated with this non-homogeneous system is $\psi(1) = (0, 1)^T$, then Corollary 30 can be used to complete the calculations. Since $\sigma(1) = \varnothing$, we have that

$$[W(1)]^{-1}[\psi(1) - \sigma(1)] = \begin{bmatrix} 1 & 0 \\ -1 & 1 \end{bmatrix} \begin{bmatrix} 0 \\ 1 \end{bmatrix} = \begin{bmatrix} 0 \\ 1 \end{bmatrix}$$

so that

$$\psi(x) = \begin{bmatrix} x \ln(x) \\ 1 + \ln(x) \end{bmatrix} + \ln(x) \begin{bmatrix} x \ln(x) \\ 2 + \ln(x) \end{bmatrix}$$

or

$$\psi(x) = \begin{bmatrix} x \ln(x) + x[\ln(x)]^2 \\ 1 + 3 \ln(x) + [\ln(x)]^2 \end{bmatrix}$$

is the solution of the initial-value problem.

This last example should be compared with Example 10, Section 12, since Example 32 and Example 10, Section 12, amount to two apparently different methods to solve an equivalent initial-value problem. In this sense, the result in Theorem 31 is an alternative derivation of the method of variation of parameters.

33 EXERCISES

Solve the initial-value problems of the form $D\psi = Q(x)\psi + \zeta(x)$; $\psi(a) = \alpha$. Each of the matrices listed below appeared in previous problems.

1. $Q(x) = \begin{bmatrix} 3 & 1 \\ 1 & 3 \end{bmatrix}$, $\quad \zeta(x) = (1 - 4x)\begin{bmatrix} 1 \\ 1 \end{bmatrix}$, $\quad \psi(0) = \begin{bmatrix} 0 \\ 0 \end{bmatrix}$

2. $Q(x) = \begin{bmatrix} 3x^2 & x \\ 0 & x^{-1} \end{bmatrix}$, $\quad \zeta(x) = \begin{bmatrix} 4x^2 \\ 1 \end{bmatrix}$, $\quad \psi(2) = \begin{bmatrix} 1 \\ 2 \end{bmatrix}$

3. $Q(x) = \begin{bmatrix} -2 & 0 & -5 \\ -3 & 1 & 1 \\ 0 & 0 & 3 \end{bmatrix}$, $\quad \zeta(x) = 4e^{-x}\begin{bmatrix} -1 \\ 1 \\ 1 \end{bmatrix} + 10\begin{bmatrix} 0 \\ 1 \\ 0 \end{bmatrix}$

$\psi(0) = \begin{bmatrix} 1 \\ -10 \\ -1 \end{bmatrix}$

4. Prove Theorem 27.
5. Prove Corollary 30.

Section 40 Differential Systems with Constant Coefficients

In this section, we study an important special class of linear differential systems, namely, those in which the coefficient matrix has only constants as entries. *Throughout this section Q will denote an $n \times n$ matrix of real constants.*

A linear differential system with constant coefficient matrix is written as

(1) $D\psi = Q\psi + \zeta(x)$

on J, where J is the domain of $\zeta(x)$.

It is sufficient to consider the homogeneous system

(2) $D\psi = Q\psi$

on $(-\infty, \infty)$, and apply the results of the previous section to characterize all solutions of equation (1). Many of the results of this section are similar to those of Chapter 3. Some of these points of similarity as well as much of the motivation for the approach taken in this section are illustrated in the first example.

3 Example

The second-order homogeneous equation

$$P(D)y = \varnothing$$

where $P(t) = t^2 + t - 6 = (t + 3)(t - 2)$, is equivalent to the linear differential system

$$D\psi = Q\psi$$

where

$$Q = \begin{bmatrix} 0 & 1 \\ 6 & -1 \end{bmatrix}$$

This equation appeared in Example 4, Section 14.

Since $y_1(x) = e^{-3x}$ and $y_2(x) = e^{2x}$ are two solutions of $P(D)y = \varnothing$, it follows that

$$\psi_1(x) = \begin{bmatrix} y_1(x) \\ Dy_1(x) \end{bmatrix} = \begin{bmatrix} e^{-3x} \\ -3e^{-3x} \end{bmatrix} = e^{-3x} \begin{bmatrix} 1 \\ -3 \end{bmatrix}$$

and

$$\psi_2(x) = \begin{bmatrix} y_2(x) \\ Dy_2(x) \end{bmatrix} = \begin{bmatrix} e^{2x} \\ 2e^{2x} \end{bmatrix} = e^{2x} \begin{bmatrix} 1 \\ 2 \end{bmatrix}$$

are solutions of $D\psi = Q\psi$. In fact, this pair of solutions is a solution basis for $D\psi = Q\psi$ since the Wronski matrix is:

$$W(x) = \begin{bmatrix} e^{-3x} & e^{2x} \\ -3e^{-3x} & 2e^{2x} \end{bmatrix}$$

and clearly $\det[W(x)] = 5e^{-x} > 0$.

Of course, $\psi_1(x)$ and $\psi_2(x)$ were easy to determine since the solutions of the equivalent equation $P(D)y = \varnothing$ were known. However, by substituting $\psi_1(x)$ into $D\psi = Q\psi$, we get the equation

$$-3e^{-3x} \begin{bmatrix} 1 \\ -3 \end{bmatrix} = \begin{bmatrix} 0 & 1 \\ 6 & -1 \end{bmatrix} e^{-3x} \begin{bmatrix} 1 \\ -3 \end{bmatrix}$$

and since $e^{-3x} \neq 0$ on $(-\infty, \infty)$ this equation is equivalent to

(A) $$\begin{bmatrix} 0 & 1 \\ 6 & -1 \end{bmatrix} \begin{bmatrix} 1 \\ -3 \end{bmatrix} = -3 \begin{bmatrix} 1 \\ -3 \end{bmatrix}$$

Not only is equation (A) correct, but, more importantly, it is an eigenvalue-eigenvector equation for the matrix Q. By noticing that equation (A) is of the form

$$Q\beta = r\beta$$

we have a clue as to how to determine $\psi_1(x)$ and $\psi_2(x)$ directly.

The eigenvalues of Q are the solutions of the characteristic equation $\det(rI - Q) = 0$, and in this example, the characteristic polynomial of Q is

$$\det(rI - Q) = \det \begin{bmatrix} r & -1 \\ -6 & r+1 \end{bmatrix} = r(r+1) - 6 = r^2 + r - 6$$

Notice that $\det(rI - Q) = P(r)$, where P is the operator polynomial in $P(D)y = \varnothing$.

The main results of this section are generalizations of the observations made in Example 3. As indicated in the example, the nature of the solutions of $D\psi = Q\psi$ is related to the eigenvalues and eigenvectors of the coefficient matrix Q. The essential background on the eigenvalue-eigenvector problem for a matrix is contained in Section 5, and frequent use of this material is made throughout the remainder of this chapter.

4 THEOREM

The vector-valued function $\psi(x) = e^{rx}\beta$ is a solution of $D\psi = Q\psi$ if and only if r is an eigenvalue and β is an associated eigenvector of Q, that is, $Q\beta = r\beta$.

Proof: If $\psi(x) = e^{rx}\beta$, then $D\psi(x) = re^{rx}\beta$ and $Q\psi(x) = e^{rx}Q\beta$. By putting these results in equation (2), we have

$$re^{rx}\beta = e^{rx}Q\beta$$

and since $e^{rx} > 0$, this is equivalent to

$$Q\beta = r\beta$$

5 Example

Consider the equation $D\psi = Q\psi$, where

$$Q = \begin{bmatrix} -3 & 2 \\ -2 & 2 \end{bmatrix}$$

In order to use Theorem 4, we need to know the eigenvalues and eigenvectors of Q. Since

$$\det(rI - Q) = (r+3)(r-2) + 4 = r^2 + r - 2 = (r+2)(r-1)$$

it follows that $r_1 = -2$ and $r_2 = 1$ are the eigenvalues of Q.

An eigenvector associated with $r_1 = -2$ is any nonzero vector β that is a solution of

$$Q\beta = -2\beta$$

This eigenvector equation can also be written as

$$[-2I - Q]\beta = \varnothing$$

By using this last form and taking b_1 and b_2 to be the components of β, we have to solve

$$\begin{bmatrix} 1 & -2 \\ 2 & -4 \end{bmatrix} \begin{bmatrix} b_1 \\ b_2 \end{bmatrix} = \begin{bmatrix} 0 \\ 0 \end{bmatrix}$$

In terms of the components of β, this equation can be written as the pair of equations

$$b_1 - 2b_2 = 0$$
$$2b_1 - 4b_2 = 0$$

However, the second of these equations is merely a multiple of the first equation, so the components of β must satisfy

$$b_1 = 2b_2$$

or

$$\beta = \begin{bmatrix} b_1 \\ b_2 \end{bmatrix} = \begin{bmatrix} 2b_2 \\ b_2 \end{bmatrix} = b_2 \begin{bmatrix} 2 \\ 1 \end{bmatrix}$$

is an eigenvector of Q associated with the eigenvalue $r_1 = -2$ provided $b_2 \neq 0$. Since any choice of $b_2 \neq 0$ provides an eigenvector, $b_2 = 1$ is a convenient choice and thus, by Theorem 4,

$$\psi_1(x) = e^{-2x} \begin{bmatrix} 2 \\ 1 \end{bmatrix}$$

is a solution of $D\psi = Q\psi$.

      Similarly, it can be shown that

$$\begin{bmatrix} -3 & 2 \\ -2 & 2 \end{bmatrix} \begin{bmatrix} 1 \\ 2 \end{bmatrix} = 1 \begin{bmatrix} 1 \\ 2 \end{bmatrix}$$

which implies that

$$\psi_2(x) = e^x \begin{bmatrix} 1 \\ 2 \end{bmatrix}$$

is a solution of $D\psi = Q\psi$.

6 EXERCISES

Continue the analysis of $D\psi = Q\psi$ of Example 5; in particular, do the following.

1. Show that $\psi_1(x)$ and $\psi_2(x)$ form a solution basis.
2. Show that another solution basis consists of

$$\psi_1(x) = e^x \begin{bmatrix} -1 \\ -2 \end{bmatrix}, \qquad \psi_2(x) = e^{-2x} \begin{bmatrix} 1 \\ \frac{1}{2} \end{bmatrix}$$

3. Form the Wronski matrix of the above solution basis and show that

$$D[W(x)] = QW(x)$$

4. Solve the initial-value problem $D\psi = Q\psi$, with

$$\psi(0) = \begin{bmatrix} -1 \\ 1 \end{bmatrix}$$

Transform the following linear differential equations of the form $P(D)y = \varnothing$ into a differential system $D\psi = Q\psi$ and compare the roots of $P(t)$ with the eigenvalues of Q.

5. $P(t) = t^2 - 6t - 7$
6. $P(t) = 4t^2 - 1$
7. $P(t) = t^2 + 1$
8. $P(t) = t^2 + 2t + 1$
9. Let $P(t)$ be a polynomial of degree n. Show that if $D\psi = Q\psi$ is the differential system form of $P(D)[y] = \varnothing$, then the roots of $P(t)$ are the eigenvalues of Q.

One central fact about the eigenvectors of an $n \times n$ matrix Q is reviewed here. The two possibilities are these: (i) Q has n linearly independent eigenvectors, and (ii) Q fails to have n linearly independent eigenvectors. The first case is easier to treat and is more common.

Suppose that

(7) $Q\beta_i = r_i\beta_i$

$i = 1, 2, \ldots, n$, and that the vectors (eigenvectors) β_i are linearly independent. It is easy to verify that the n equations in (7) can be written more compactly as

(8) $QB = BR$

where B is the matrix of eigenvectors; that is, the columns of B are β_1, $\beta_2, \ldots, \beta_n$, and R is a diagonal matrix of eigenvalues, $R = \text{diag}(r_1, r_2, \ldots, r_n)$. Since the columns of B are linearly independent, it follows from Theorem 16, Section 5, that B has an inverse. Thus equation (8) can be rewritten as

(9) $Q = BRB^{-1}$

Conversely, if $Q = BRB^{-1}$, where R is a diagonal matrix, then it follows that Q has n linearly independent eigenvectors, namely, the columns of B. Thus equations (7), (8), and (9) are all equivalent.

If the eigenvalues and eigenvectors of Q are as given in (7), then it follows from Theorem 4 that each of the vector-valued functions

(10) $\psi_i(x) = e^{r_i x}\beta_i$

$i = 1, 2, \ldots, n$, is a solution of equation (2). The Wronski matrix of this set of n solutions of (2) can be expressed as

(11) $W(x) = BU(x)$

where B is the matrix whose columns are the eigenvectors β_i, and

(12) $U(x) = \text{diag}(e^{r_1 x}, e^{r_2 x}, \ldots, e^{r_n x})$

Thus the Wronskian of the n solutions in (10) is:

$$\det[W(x)] = \det(B) \exp[(r_1 + r_2 + \cdots + r_n)x]$$

Therefore, $\det[W(x)] \neq 0$ if and only if $\det(B) \neq 0$. Thus the n solutions in (10) are linearly independent if and only if the eigenvectors β_i are linearly independent. These are the essential steps needed to characterize the solutions of $D\psi = Q\psi$ in the case under consideration.

13 THEOREM

If $Q = BRB^{-1}$, where $R = \text{diag}(r_1, r_2, \ldots r_n)$, then all solutions of $D\psi = Q\psi$ are of the form

$$\psi(x) = BU(x)\gamma$$

where γ is a vector of constants and $U(x)$ is defined in (12).

The representation of the solutions of $D\psi = Q\psi$ in Theorem 13 is the equivalent in matrix-vector notation of

$$(14) \qquad \psi(x) = c_1 e^{r_1 x}\beta_1 + c_2 e^{r_2 x}\beta_2 + \cdots + c_n e^{r_n x}\beta_n$$

Since these vector-valued functions are all defined on $J = (-\infty, \infty)$, any initial condition is appropriate for equation (2). However, there is a notational advantage to having the initial condition at $x = 0$. Notice that $U(0) = I$, and this fact provides another compact representation of the solutions of (2).

15 COROLLARY

Given the assumptions of Theorem 13, all solutions of $D\psi = Q\psi$ can be expressed as

$$\psi(x) = BU(x)B^{-1}\psi(0)$$

Proof: Using the form of the solutions in Theorem 13 gives

$$\psi(0) = BU(0)\gamma = BI\gamma = B\gamma$$

Thus $\gamma = B^{-1}\psi(0)$ and the result follows.

---

16 Example

Let

$$Q = \begin{bmatrix} 2 & -1 & -1 \\ 1 & 0 & -1 \\ -1 & 1 & 2 \end{bmatrix}$$

The characteristic polynomial of Q is:

$$\det(rI - Q) = r^3 - 4r^2 + 5r - 2 = (r - 1)^2(r - 2)$$

and the eigenvectors of Q are the solutions of

$$(rI - Q)\beta = \varnothing$$

With $r = 1$, the equation $(I - Q)\beta = \varnothing$ is:

$$\begin{bmatrix} -1 & 1 & 1 \\ -1 & 1 & 1 \\ 1 & -1 & -1 \end{bmatrix} \begin{bmatrix} b_1 \\ b_2 \\ b_3 \end{bmatrix} = \begin{bmatrix} 0 \\ 0 \\ 0 \end{bmatrix}$$

which is equivalent to the single equation $-b_1 + b_2 + b_3 = 0$. Thus $b_1 = 1$, $b_2 = 0$, $b_3 = 1$, and $b_1 = 1$, $b_2 = 1$, $b_3 = 0$ are two solutions of this equation, or

$$\beta_1 = \begin{bmatrix} 1 \\ 0 \\ 1 \end{bmatrix} \quad \text{and} \quad \beta_2 = \begin{bmatrix} 1 \\ 1 \\ 0 \end{bmatrix}$$

are two linearly independent eigenvectors of Q associated with the eigenvalue $r = 1$. Similarly, it is easy to show that if $r = 2$, then

$$\beta_3 = \begin{bmatrix} 1 \\ 1 \\ -1 \end{bmatrix}$$

is an eigenvector. Thus

$$B = \begin{bmatrix} 1 & 1 & 1 \\ 0 & 1 & 1 \\ 1 & 0 & -1 \end{bmatrix}$$

is a matrix of linearly independent eigenvectors of Q and $Q = BRB^{-1}$, where $R = \text{diag}(1, 1, 2)$. Since

$$B^{-1} = \begin{bmatrix} 1 & -1 & 0 \\ -1 & 2 & 1 \\ 1 & -1 & -1 \end{bmatrix} \quad \text{and} \quad U(x) = \begin{bmatrix} e^x & 0 & 0 \\ 0 & e^x & 0 \\ 0 & 0 & e^{2x} \end{bmatrix}$$

it follows from Corollary 15 that each solution of $D\psi = Q\psi$ is of the form

$$\psi(x) = BU(x)B^{-1}\psi(0)$$

Before considering some exercises, we need to investigate the consequences of having complex eigenvalues and eigenvectors of a real matrix Q. This can, and does, occur since the eigenvalues are roots of a polynomial with real coefficients. However, since these complex roots occur in conjugate pairs, the treatment of this case is very similar to that in Chapter 3.

Suppose that $a + ib$ is a complex eigenvalue of Q, that is, $b \neq 0$. Since Q is real, it follows that an associated eigenvector is also complex, or that

(17) $Q(\alpha + i\beta) = (a + ib)(\alpha + i\beta)$

where α and β are real vectors and β is not the zero vector. By taking the conjugate of both sides of equation (17), it follows that

(18) $\quad Q(\alpha - i\beta) = (a - ib)(\alpha - i\beta)$

or that $a - ib$ is an eigenvalue and that $\alpha - i\beta$ is an associated eigenvector. Thus

$$\theta_1(x) = \exp[(a + ib)x](\alpha + i\beta)$$

and

$$\theta_2(x) = \exp[(a - ib)x](\alpha - i\beta)$$

are two complex vector-valued solutions of $D\psi = Q\psi$. Since

$$\exp[(a + ib)x] = e^{ax}(\cos bx + i \sin bx)$$

and

$$\exp[(a - ib)x] = e^{ax}(\cos bx - i \sin bx)$$

these two complex solutions can be expressed as

$$\theta_1(x) = e^{ax}[(\cos bx)\alpha - (\sin bx)\beta] + ie^{ax}[(\sin bx)\alpha + (\cos bx)\beta]$$
$$\theta_2(x) = e^{ax}[(\cos bx)\alpha - (\sin bx)\beta] - ie^{ax}[(\sin bx)\alpha + (\cos bx)\beta]$$

Since

$$\tfrac{1}{2}\theta_1(x) + \tfrac{1}{2}\theta_2(x) = e^{ax}[(\cos bx)\alpha - (\sin bx)\beta]$$

and

$$-\tfrac{1}{2}i\theta_1(x) + \tfrac{1}{2}i\theta_2(x) = e^{ax}[(\sin bx)\alpha + \cos bx)\beta]$$

are two linear combinations of solutions of $D\psi = Q\psi$, it follows that

(19)
$$\psi_1(x) = e^{ax}[(\cos bx)\alpha - (\sin bx)\beta]$$
$$\psi_2(x) = e^{ax}[(\sin bx)\alpha + (\cos bx)\beta]$$

are two real-valued solutions of $D\psi = Q\psi$. It remains to be shown that $\psi_1(x)$ and $\psi_2(x)$ are linearly independent. The proof of this result is an exercise.

20 THEOREM

If $Q(\alpha + i\beta) = (a + ib)(\alpha + i\beta)$, where $b \neq 0$, then $\psi_1(x)$ and $\psi_2(x)$ as defined in (19) are two linearly independent solutions of $D\psi = Q\psi$.

The problem of determining a complex eigenvector associated with a complex eigenvalue can be reduced to a form that involves real vectors only. In order to determine the real vectors α and β in the pair of solutions (19), we first separate the eigenvalue-eigenvector equation (17) into real and imaginary parts. Thus (17) can be written as

$$Q\alpha + iQ\beta = (a\alpha - b\beta) + i(b\alpha + a\beta)$$

By equating real and imaginary parts, it follows that this equation is equivalent to the pair of equations

(21)
$$(Q - aI)\alpha + b\beta = \varnothing$$
$$-b\alpha + (Q - aI)\beta = \varnothing$$

By multiplying the first of the equations in (21) by b, the second equation by $(Q - aI)$, and then adding the equations, it follows that

$$[(Q - aI)^2 + b^2 I]\beta = \varnothing$$

By a similar reduction, it can be shown that

$$[(Q - aI)^2 + b^2 I]\alpha = \varnothing$$

or that the real vectors α and β are solutions of the same matrix equation.

22 THEOREM

If $Q(\alpha + i\beta) = (a + ib)(\alpha + i\beta)$, where $b \neq 0$, then α and β are any two linearly independent solutions of

$$[(Q - aI)^2 + b^2 I]\xi = \varnothing$$

23 Example

If

$$Q = \begin{bmatrix} -2 & 1 & 0 \\ -5 & 2 & 1 \\ -2 & -2 & 1 \end{bmatrix}$$

then $\det(rI - Q) = r^3 - r^2 + 3r + 5$, so that $r_1 = -1$, $r_2 = 1 + 2i$, and $r_3 = 1 - 2i$ are the eigenvalues of Q. We first apply Theorem 22 to obtain an eigenvector associated with $r_2 = 1 + 2i$. In this case,

$$(Q - aI)^2 + b^2 I = (Q - I)^2 + 4I = \begin{bmatrix} 8 & -2 & 1 \\ 8 & -2 & 1 \\ 16 & -4 & 2 \end{bmatrix}$$

Thus the equation $[(Q - I)^2 + 4I]\xi = \varnothing$ is easy to solve and

$$\alpha = \begin{bmatrix} 1 \\ 3 \\ -2 \end{bmatrix}, \qquad \beta = \begin{bmatrix} 0 \\ 2 \\ 4 \end{bmatrix}$$

are two linearly independent solutions of this equation. Thus, by using (19), we have that

$$\psi_1(x) = e^x[(\cos 2x)\alpha - (\sin 2x)\beta]$$

$$\psi_2(x) = e^x[(\sin 2x)\alpha + (\cos 2x)\beta]$$

are two (real) linearly independent solutions of $D\psi = Q\psi$.

Some additional questions about this equation are in the exercises.

This completes the study of $D\psi = Q\psi$, where Q has n linearly independent eigenvectors. The compact forms for the solutions of this equation, as given in Theorem 13 and Corollary 15, are especially useful in problems where the eigenvalues of Q are all real numbers. In the case of complex eigenvalues, a modified version of the form (14) can be used. However, in general, the problem of computing the eigenvalues and eigenvectors of a real matrix Q is of the same order of difficulty as that of finding the roots of a polynomial.

24 EXERCISES

If

$$Q = \begin{bmatrix} 24 & 10 \\ -65 & -27 \end{bmatrix}, \quad \alpha = \begin{bmatrix} 1 \\ -2 \end{bmatrix}, \quad \beta = \begin{bmatrix} 0 \\ -1 \end{bmatrix}$$

find the following:

1. all solutions of $D\psi = Q\psi$ on $J = (-\infty, \infty)$
2. the solution of $D\psi = Q\psi$; $\psi(0) = \alpha$
3. the solution of $D\psi = Q\psi$; $\psi(0) = \beta$
4. the solution of $D\psi = Q\psi$; $\psi(0) = \alpha - \beta$
5. the solution of $D\psi = Q\psi$; $\psi(0) = \varnothing$
6. all solutions of $D\psi = Q\psi$ on $J = (-\infty, \infty)$ which have α in their range

If

$$Q = \begin{bmatrix} -37 & -12 \\ 114 & 37 \end{bmatrix}, \quad \alpha = \begin{bmatrix} -6 \\ 19 \end{bmatrix}, \quad \beta = \begin{bmatrix} 1 \\ -1 \end{bmatrix}$$

find the following:

7. all solutions of $D\psi = Q\psi$ on $J = (-\infty, \infty)$
8. the solution of $D\psi = Q\psi$; $\psi(0) = \alpha$
9. the solution of $D\psi = Q\psi$; $\psi(0) = \beta$

If

$$Q = \begin{bmatrix} 1 & 4 & 4 \\ 0 & -7 & -10 \\ -2 & 4 & 7 \end{bmatrix}, \quad \alpha = e^{-\pi} \begin{bmatrix} -1 \\ 3 \\ -1 \end{bmatrix}, \quad \beta = \begin{bmatrix} 3 \\ 0 \\ 1 \end{bmatrix}$$

find the following:

10. all solutions of $D\psi = Q\psi$ on $J = (-\infty, \infty)$
11. the solution of $D\psi = Q\psi$; $\psi(\pi) = \alpha$
12. the solution of $D\psi = Q\psi$; $\psi(0) = \beta$

If

$$Q = \frac{1}{4} \begin{bmatrix} 17 & -3 & -12 \\ 16 & -2 & -12 \\ 3 & -3 & 2 \end{bmatrix}, \quad \alpha = \begin{bmatrix} 1 \\ 2 \\ 1 \end{bmatrix}, \quad \beta = \frac{1}{4} \begin{bmatrix} 2 \\ 1 \\ -2 \end{bmatrix}$$

find the following:

13. all solutions of $D\psi = Q\psi$ on $J = (-\infty, \infty)$
14. the solution of $D\psi = Q\psi$; $\psi(0) = \alpha$
15. the solution of $D\psi = Q\psi$; $\psi(0) = \beta$

The following problems pertain to Example 23.

16. Find an eigenvector associated with the eigenvalue $r_1 = -1$.
17. Let B be a matrix whose columns are (in any order) an eigenvector associated with $r_1 = -1$ along with α and β. Is $B^{-1}QB$ a diagonal matrix?
18. Find the solution of $D\psi = Q\psi; \psi(0) = \alpha$.

The following problems are of a general nature.

19. If Q_1 and Q_2 are $n \times n$ matrices, how do the solutions of

$$D\psi = Q_1\psi$$

$$D\psi = Q_2\psi$$

$$D\psi = (Q_1 + Q_2)\psi$$

compare?
20. Prove Theorem 20.
21. Prove Theorem 22.

To complete the characterization of the solutions of $D\psi = Q\psi$ it is necessary to consider the case in which the $n \times n$ matrix Q does not have n linearly independent eigenvectors. Various approaches to this problem are possible, but each of them depends upon advanced topics in the theory of matrices. Instead of attempting a general analysis of this case, we present a comprehensive treatment of a specific example.

25 Example

The third-order homogeneous equation

$$P(D)y = \varnothing$$

where

$$P(t) = (t + 2)^2(t - 3)$$

can be transformed into the linear differential system

$$D\psi = Q\psi$$

where

$$Q = \begin{bmatrix} 0 & 1 & 0 \\ 0 & 0 & 1 \\ 12 & 8 & -1 \end{bmatrix}$$

As expected, $\det(rI - Q) = r^3 + r^2 - 8r - 12 = P(r)$, so that the eigenvalues of Q are $-2, -2$, and 3.

It is easy to verify that a solution of

$$(3I - Q)\beta_1 = \varnothing$$

is $\beta_1 = (1, 3, 9)^T$, so that $\psi_1(x) = e^{-3x}\beta_1$ is a solution of $D\psi = Q\psi$.

An eigenvector associated with the multiple eigenvalue $r = -2$ is a solution of

$$(-2I - Q)\beta = \begin{bmatrix} -2 & -1 & 0 \\ 0 & -2 & -1 \\ -12 & -8 & -1 \end{bmatrix} \begin{bmatrix} b_1 \\ b_2 \\ b_3 \end{bmatrix} = \begin{bmatrix} 0 \\ 0 \\ 0 \end{bmatrix}$$

It is easy to show that all solutions of this eigenvector equation are of the form

$$\beta = b \begin{bmatrix} 1 \\ -2 \\ 4 \end{bmatrix}$$

where b is any constant. Thus there cannot be two linearly independent eigenvectors of Q associated with the multiple eigenvalue $r = -2$. Nonetheless $\beta_2 = (1, -2, 4)^T$ is an eigenvector, and thus $\psi_2(x) = e^{-2x}\beta_2$ is a solution of $D\psi = Q\psi$. However, none of the techniques discussed previously are useful in determining a third solution of the differential system.

We do know from Chapter 3 that $y(x) = xe^{-2x}$ is a solution of $P(D)y = \varnothing$, and a review of the transformation from $P(D)y = \varnothing$ into $D\psi = Q\psi$ reveals that

$$\psi_3(x) = \begin{bmatrix} y(x) \\ Dy(x) \\ D^2 y(x) \end{bmatrix} = \begin{bmatrix} xe^{-2x} \\ e^{-2x} - 2xe^{-2x} \\ -4e^{-2x} + 4xe^{-2x} \end{bmatrix}$$

$$= e^{-2x} \begin{bmatrix} 0 \\ 1 \\ -4 \end{bmatrix} + xe^{-2x} \begin{bmatrix} 1 \\ -2 \\ 4 \end{bmatrix}$$

is a solution of $D\psi = Q\psi$.

Let $\beta_3 = (0, 1, -4)^T$, and notice that

$$\psi_3(x) = e^{-2x}\beta_3 + xe^{-2x}\beta_2$$

$$\psi_2(x) = e^{-2x}\beta_2$$

$$\psi_1(x) = e^{3x}\beta_1$$

While β_1 and β_2 are linearly independent eigenvectors of Q, β_3 is not an eigenvector of Q. However, β_2 and β_3 are linearly independent vectors.

The relationship between β_2 and β_3 is given by

$$(Q + 2I)\beta_3 = \beta_2$$

In the terminology of matrix theory, β_3 is called a *principal vector*.

The observation made in Example 25 can be generalized.

26 THEOREM

If $Q\alpha = r\alpha$, then $\psi(x) = e^{rx}\beta + xe^{rx}\alpha$ is a solution of $D\psi = Q\psi$ if and only if $(Q - rI)\beta = \alpha$.

Proof: By direct calculations:

$$D\psi(x) = re^{rx}\beta + e^{rx}\alpha + rxe^{rx}\alpha$$

and

$$Q\psi(x) = e^{rx}Q\beta + xe^{rx}Q\alpha = e^{rx}Q\beta + rxe^{rx}\alpha$$

Thus $D\psi(x) = Q\psi(x)$ if and only if

$$e^{rx}(r\beta + \alpha) = e^{rx}Q\beta$$

and this equation is equivalent to

$$(Q - rI)\beta = \alpha$$

Theorem 26 shows how to determine an additional solution of $D\psi = Q\psi$ in the case where Q has an eigenvalue of multiplicity two for which there is only a single linearly independent eigenvector. Comments about ways to generalize this result are contained in the exercises.

27 EXERCISES

If

$$Q = \begin{bmatrix} 2 & 1 \\ 0 & 2 \end{bmatrix}, \qquad \alpha = \begin{bmatrix} 1 \\ 1 \end{bmatrix}, \qquad \beta = e^2 \begin{bmatrix} 2 \\ 1 \end{bmatrix}$$

find the following:

1. all solutions of $D\psi = Q\psi$ on $J = (-\infty, \infty)$
2. the solution of $D\psi = Q\psi$; $\psi(0) = \alpha$
3. the solution of $D\psi = Q\psi$; $\psi(1) = \beta$

If

$$Q = \begin{bmatrix} 12 & 4 \\ -25 & -8 \end{bmatrix}, \qquad \alpha = \begin{bmatrix} 1 \\ -2 \end{bmatrix}, \qquad \beta = \begin{bmatrix} -7 \\ 18 \end{bmatrix}$$

find the following:

4. all solutions of $D\psi = Q\psi$ on $J = (-\infty, \infty)$
5. the solution of $D\psi = Q\psi$; $\psi(0) = \alpha$
6. the solution of $D\psi = Q\psi$; $\psi(0) = \beta$

If

$$Q = \begin{bmatrix} 1 & 0 & 1 \\ 2 & 2 & -2 \\ -1 & 0 & 3 \end{bmatrix}, \qquad \alpha = \begin{bmatrix} -1 \\ 6 \\ -4 \end{bmatrix}, \qquad \beta = \begin{bmatrix} 1 \\ 2 \\ 0 \end{bmatrix}$$

find the following:

7. all solutions of $D\psi = Q\psi$ on $J = (-\infty, \infty)$
8. the solution of $D\psi = Q\psi$; $\psi(0) = \alpha$

9. the solution of $D\psi = Q\psi$; $\psi(0) = \beta$
10. Write the linear system $D\psi = Q\psi$ as

$$Dy_1 = y_1 + y_3$$

$$Dy_2 = 2y_1 + 2y_2 - 2y_3$$

$$Dy_3 = -y_1 + 3y_3$$

and first solve

$$D\begin{bmatrix} y_1 \\ y_3 \end{bmatrix} = \begin{bmatrix} 1 & 1 \\ -1 & 3 \end{bmatrix}\begin{bmatrix} y_1 \\ y_3 \end{bmatrix}$$

and then solve $Dy_2 = 2y_1 + 2y_2 - 2y_3$. This is another type of uncoupled system.

Let

$$Q_1 = \begin{bmatrix} 2 & 0 & 0 \\ 0 & 2 & 0 \\ 0 & 0 & 2 \end{bmatrix}, \qquad Q_2 = \begin{bmatrix} 2 & 0 & 0 \\ 0 & 2 & 1 \\ 0 & 0 & 2 \end{bmatrix}, \qquad Q_3 = \begin{bmatrix} 2 & 1 & 0 \\ 0 & 2 & 1 \\ 0 & 0 & 2 \end{bmatrix}$$

11. Compare the eigenvalues and eigenvectors of Q_1, Q_2, and Q_3.
12. Compare the solutions of $D\psi = Q_1\psi$, $D\psi = Q_2\psi$, and $D\psi = Q_3\psi$.
13. Which, if any, of the linear differential systems in Problem 12 can be transformed into a third-order linear differential equation?

Let $P(D)y = \varnothing$ be an nth-order linear differential equation, where $P(t) = (t - b)^n$, and let $D\psi = Q\psi$ be an equivalent linear differential system.

14. Show that Q has only one linearly independent eigenvector.
15. Use the solutions of $P(D)y = \varnothing$ to find n linearly independent solutions of $D\psi = Q\psi$.
16. Compute the solutions of $(bI - Q)^k\alpha = \varnothing$, where $k = 1, 2, \ldots, n$.
17. Generalize Theorem 26.

     In certain cases, the particular properties of a matrix Q are useful in studying the solutions of $D\psi = Q\psi$. We consider one such class of matrices as an example.

     As was noted in Section 5, a real $n \times n$ matrix Q is called *symmetric* if $Q = Q^T$. The important fact about symmetric matrices is the following. If Q is symmetric, then there exist real matrices R and B such that

$$R = \text{diag}(r_1, r_2, \ldots, r_n), \qquad BB^T = I, \qquad Q = BRB^T$$

An $n \times n$ matrix B which satisfies $BB^T = I$ is called an *orthogonal matrix*. Evidently, if B is an orthogonal matrix, then $B^{-1} = B^T$. This fact, together with the real eigenvalues, makes the treatment of $D\psi = Q\psi$, where Q is symmetric, particularly simple.

28 THEOREM

If Q is a symmetric matrix, then all solutions of $D\psi = Q\psi$ can be expressed as

$$\psi(x) = BU(x)B^T\psi(0)$$

where

$$Q = BRB^T$$

$$R = \text{diag}(r_1, r_2, \ldots, r_n)$$

$$BB^T = I$$

$$U(x) = \text{diag}(e^{r_1 x}, e^{r_2 x}, \ldots, e^{r_n x})$$

Because Q is assumed to be symmetric, the form of the solutions of $D\psi = Q\psi$ in Corollary 15 is easier to calculate since it is not necessary to determine B^{-1}. However, in order to exploit this advantage, it is necessary to arrange the calculations so that the matrix of eigenvectors of Q form an orthogonal matrix. This will not happen automatically, but, as illustrated in the next example, it is easy to arrange.

29 Example

If

$$Q = \begin{bmatrix} 7 & 3 \\ 3 & -1 \end{bmatrix}$$

then $Q = Q^T$, and so it is possible to factor Q as

$$Q = BRB^T$$

as indicated above.

Let $\alpha_1 = (3, 1)^T$ and $\alpha_2 = (2, -6)^T$ and notice that $Q\alpha_1 = 8\alpha_1$, while $Q\alpha_2 = -2\alpha_2$. Thus α_1 is an eigenvector associated with the eigenvalue $r_1 = 8$, and α_2 is an eigenvector associated with the eigenvalue $r_2 = -2$. Let

$$A = \begin{bmatrix} 3 & 2 \\ 1 & -6 \end{bmatrix}$$

be the matrix whose columns are these eigenvectors of Q. Since

$$AA^T = \begin{bmatrix} 3 & 2 \\ 1 & -6 \end{bmatrix}\begin{bmatrix} 3 & 1 \\ 2 & -6 \end{bmatrix} = \begin{bmatrix} 13 & -9 \\ -9 & 37 \end{bmatrix} \neq \begin{bmatrix} 1 & 0 \\ 0 & 1 \end{bmatrix}$$

evidently A is not an orthogonal matrix.

However, A is not *the* matrix of eigenvectors of Q, it is a particular choice. In fact,

$$B = \begin{bmatrix} 3b_1 & 2b_2 \\ b_1 & -6b_2 \end{bmatrix}$$

is also a matrix of eigenvectors of Q, where b_1 and b_2 are any nonzero real numbers. This follows from the fact that

$$\beta_1 = b_1\alpha_1 \quad \text{and} \quad \beta_2 = b_2\alpha_2$$

are also eigenvectors of Q.

Since

$$BB^T = \begin{bmatrix} 3b_1 & 2b_2 \\ b_1 & -6b_2 \end{bmatrix} \begin{bmatrix} 3b_1 & b_1 \\ 2b_2 & -6b_2 \end{bmatrix}$$

$$= \begin{bmatrix} 9b_1^2 + 4b_2^2 & 3b_1^2 - 12b_2^2 \\ 3b_1^2 - 12b_2^2 & b_1^2 + 36b_2^2 \end{bmatrix}$$

it follows that B is orthogonal if and only if

$$3b_1^2 - 12b_2^2 = 0$$

$$9b_1^2 + 4b_2^2 = 1$$

$$b_1^2 + 36b_2^2 = 1$$

These equations have the solution

$$b_1^2 = \tfrac{1}{10} \qquad \text{and} \qquad b_2^2 = \tfrac{1}{40}$$

Thus

$$\beta_1 = \frac{1}{\sqrt{10}} \alpha_1 \qquad \text{and} \qquad \beta_2 = \frac{1}{2\sqrt{10}} \alpha_2$$

are a pair of eigenvectors of Q which will result in an orthogonal matrix B. This illustrates the fact that in order to take full advantage of the symmetry of Q, an arbitrary set of linearly independent eigenvectors cannot be used.

Finally, there is no need to imitate these calculations in working another problem since, after an examination of the conclusion, it should be clear that the important feature of

$$\beta_1 = \begin{bmatrix} \dfrac{3}{\sqrt{10}} \\ \dfrac{1}{\sqrt{10}} \end{bmatrix} \qquad \text{and} \qquad \beta_2 = \begin{bmatrix} \dfrac{1}{\sqrt{10}} \\ \dfrac{-3}{\sqrt{10}} \end{bmatrix}$$

is that

$$\left[\frac{3}{\sqrt{10}}\right]^2 + \left[\frac{1}{\sqrt{10}}\right]^2 = \left[\frac{1}{\sqrt{10}}\right]^2 + \left[\frac{-3}{\sqrt{10}}\right]^2 = 1$$

that is, that the sum of the squares of the components of each eigenvector is one.

If $\alpha = (a_1, a_2, \ldots, a_n)^T$ is a nonzero vector and $a^2 = a_1^2 + a_2^2 + \cdots + a_n^2$, then $\beta = (1/a)\alpha$ is called a *normalized* vector. It is illustrated in Example 29 that, after finding an eigenvector of a symmetric matrix, it is necessary to replace it with a normalized vector in order that the matrix of eigenvectors be orthogonal.

30 EXERCISES

If

$$Q = \frac{1}{9}\begin{bmatrix} 25 & 2 & -28 \\ 2 & -2 & -26 \\ -28 & -26 & 13 \end{bmatrix}, \qquad \alpha = \begin{bmatrix} 3 \\ 3 \\ 3 \end{bmatrix}, \qquad \beta = \begin{bmatrix} 1 \\ -1 \\ 5 \end{bmatrix}$$

find the following:

1. an orthogonal matrix of eigenvectors of Q
2. all solutions of $D\psi = Q\psi$ on $J = (-\infty, \infty)$
3. the solution of $D\psi = Q\psi$; $\psi(0) = \alpha$
4. the solution of $D\psi = Q\psi$; $\psi(0) = \beta$

If

$$Q = \begin{bmatrix} -\frac{3}{5} & \frac{14}{5} \\ \frac{14}{5} & \frac{18}{5} \end{bmatrix}, \qquad \alpha = \begin{bmatrix} -2 \\ 1 \end{bmatrix}, \qquad \beta = \begin{bmatrix} 1 \\ -1 \end{bmatrix}$$

find the following:

5. all solutions of $D\psi = Q\psi$ on $J = (-\infty, \infty)$
6. the solution of $D\psi = Q\psi$; $\psi(0) = \alpha$
7. the solution of $D\psi = Q\psi$; $\psi(0) = \beta$
8. the solution of $D\psi = (I + Q)\psi$; $\psi(0) = \alpha + \beta$
9. the solution of $D\psi = (Q^2 + 18Q - 7I)\psi$; $\psi(0) = \emptyset$

Let

$$Q = \begin{bmatrix} a & b \\ b & -a \end{bmatrix}$$

where $a^2 + b^2 = 1$.

10. Show that Q is both symmetric and orthogonal.
11. Find the eigenvalues and eigenvectors of Q.
12. If $b = -\frac{1}{2}$, what is the solution of $D\psi = Q\psi$;

$$\psi(0) = \begin{bmatrix} 1 \\ 0 \end{bmatrix}$$

13. Are there any 2×2 matrices not of the form of Q that are symmetric and orthogonal?

The final topic in this section is a characterization of all solutions of the nonhomogeneous linear differential system

(31) $$D\psi = Q\psi + \zeta(x)$$

on J, where

$$Q = BRB^{-1}$$

$$R = \text{diag}(r_1, r_2, \ldots, r_n)$$

Although a method that is similar to the method of undetermined coefficients, (14) of Section 16, can be developed for equation (31), the derivation of the method is quite involved. However, Theorem 31, Section 39, provides a computational method that is satisfactory in many cases.

Recall that the Wronski matrix of the reduced equation

$$D\psi = Q\psi$$

as given in (11) is:

$$W(x) = BU(x)$$

where

$$U(x) = \text{diag}(e^{r_1x}, e^{r_2x}, \dots, e^{r_nx})$$

Thus, by using Theorem 31, Section 39, we have that a solution of (31) is:

(32) $$\sigma(x) = BU(x) \int_a^x [U(t)]^{-1} B^{-1} \zeta(t)\, dt$$

where $a \in J$. Notice that $\sigma(a) = \varnothing$, and thus the solutions of (31) as given in Corollary 30, Section 39, are:

$$\psi(x) = BU(x)\{[BU(a)]^{-1}[\psi(a) - \sigma(a)]\}$$
$$+ BU(x) \int_a^x [U(t)]^{-1} B^{-1} \zeta(t)\, dt$$

By using the fact that $U(a)$ is a diagonal matrix and $\sigma(a) = \varnothing$, the final result can be simplified into a useful formula.

33 THEOREM

All solutions of the linear system (31) are of the form

$$\psi(x) = BU(x - a)B^{-1}\psi(a) + BU(x) \int_a^x U(-t)B^{-1}\zeta(t)\, dt$$

where $a \in J$.

Since $[U(t)]^{-1} = \text{diag}[e^{-r_1t}, e^{-r_2t}, \dots, e^{-r_nt}] = U(-t)$, the difficulties in the integration problem depend upon how the components of $\zeta(t)$ combine with the exponential functions. Also, some or all of the eigenvalues of Q could be complex, so a modified form of Theorem 33 could be developed so that only real vector- and matrix-valued functions are used.

34 EXERCISES

1. Solve:

$$D\psi = \begin{bmatrix} 24 & 10 \\ -65 & 27 \end{bmatrix} \psi + (1 + x)\begin{bmatrix} 2 \\ -5 \end{bmatrix}; \qquad \psi(0) = \begin{bmatrix} 1 \\ -2 \end{bmatrix}$$

2. Solve:

$$D\psi = \begin{bmatrix} 1 & 4 & 4 \\ 0 & -7 & -10 \\ -2 & 4 & 7 \end{bmatrix} \psi + x \begin{bmatrix} 3 \\ -10 \\ 9 \end{bmatrix} + \begin{bmatrix} 5 \\ -7 \\ 3 \end{bmatrix}; \quad \psi(0) = \begin{bmatrix} 0 \\ 0 \\ -1 \end{bmatrix}$$

3. Solve:

$$D\psi = \begin{bmatrix} 1 & 0 & -1 \\ 0 & 1 & 1 \\ 1 & -1 & 0 \end{bmatrix} \psi + \begin{bmatrix} xe^x + e^x \\ -xe^x - 1 \\ e^x + 1 \end{bmatrix}; \quad \psi(0) = \begin{bmatrix} 0 \\ 1 \\ 0 \end{bmatrix}$$

4. Solve:

$$D\psi = \begin{bmatrix} 2 & 1 \\ 0 & 2 \end{bmatrix} \psi - \begin{bmatrix} 2 \\ -8 \end{bmatrix}; \quad \psi(0) = \begin{bmatrix} 0 \\ 0 \end{bmatrix}$$

5. Solve:

$$D\psi = \begin{bmatrix} 1 & 1 \\ 1 & 1 \end{bmatrix} \psi + \begin{bmatrix} 1 \\ 2x \end{bmatrix} - x \begin{bmatrix} 1 \\ 1 \end{bmatrix} - x^2 \begin{bmatrix} 1 \\ 1 \end{bmatrix}; \quad \psi(0) = \begin{bmatrix} 1 \\ -1 \end{bmatrix}$$

6. Solve:

$$D\psi = \begin{bmatrix} 1 & 1 \\ 1 & 1 \end{bmatrix} \psi + \begin{bmatrix} \sin 2x \\ \sin 2x \end{bmatrix} + \begin{bmatrix} 1 \\ 1 \end{bmatrix}; \quad \psi(0) = \begin{bmatrix} 0 \\ 1 \end{bmatrix}$$

7. Solve:

$$D\psi = \begin{bmatrix} 0 & 1 \\ 1 & 0 \end{bmatrix} \psi - \begin{bmatrix} 0 \\ 2 \sin x \end{bmatrix}; \quad \psi(0) = \begin{bmatrix} 4 \\ 1 \end{bmatrix}$$

8. Solve:

$$D\psi = \begin{bmatrix} 1 & 2 & 3 \\ 3 & 1 & 2 \\ 2 & 3 & 1 \end{bmatrix} \psi - 2 \begin{bmatrix} 0 \\ 1 \\ 2 \end{bmatrix}; \quad \psi(0) = \begin{bmatrix} 1 \\ 1 \\ -1 \end{bmatrix}$$

9. Find a differential system of the form $D\psi = Q\psi + \zeta(x)$ such that

$$\psi(x) = 2 \begin{bmatrix} 7 \\ -1 \\ -1 \end{bmatrix} + e^{-x} \begin{bmatrix} -3 \\ 1 \\ 0 \end{bmatrix} + e^x \begin{bmatrix} -3 \\ 0 \\ 1 \end{bmatrix} + \begin{bmatrix} 1 \\ 0 \\ 0 \end{bmatrix}$$

is a solution.

Section 41 Linear Difference Systems

This section contains a very brief treatment of systems of linear difference equations. A detailed treatment of this topic would be a duplication of the results in Sections 38, 39, and 40. This should not be necessary since the many analogies between linear differential equations and linear difference equations that are noted in Chapter 5 can be extended to include both types of linear systems. Some of the missing details appear in the exercises.

The linear difference equation

(1) $$E^n y + p_1(k)E^{n-1}y + p_2(k)E^{n-2}y + \cdots + p_n(k)y = f(k)$$

can be transformed into a system of first-order equations by defining the new variables

$$y_1 = y$$

$$y_2 = Ey$$

$$\vdots$$

$$y_n = E^{n-1}y$$

Thus

(2)
$$Ey_1 = \quad 0 \cdot y_1 \quad + y_2 \qquad + 0 \cdot y_3 \qquad + \cdots + 0 \cdot y_n$$
$$Ey_2 = \quad 0 \cdot y_1 \quad + 0 \cdot y_2 \quad + y_3 \qquad + \cdots + 0 \cdot y_n$$
$$\vdots \qquad \vdots \qquad \vdots \qquad \vdots \qquad \qquad \vdots$$
$$Ey_n = \; -p_n(k)y_1 - p_{n-1}(k)y_2 - p_{n-2}(k)y_3 - \cdots - p_1(k)y_n + f(k)$$

is the resultant linear system.

The form of the system (2) suggests the obvious choices of matrix- and vector-valued functions, namely,

$$\psi = \begin{bmatrix} y_1 \\ y_2 \\ \vdots \\ y_n \end{bmatrix}, \qquad E\psi = \begin{bmatrix} Ey_1 \\ Ey_2 \\ \vdots \\ Ey_n \end{bmatrix}, \qquad \zeta(k) = \begin{bmatrix} \varnothing \\ \varnothing \\ \vdots \\ f(k) \end{bmatrix}$$

$$Q(k) = \begin{bmatrix} \varnothing & 1 & \varnothing & \cdots & \varnothing \\ \varnothing & \varnothing & 1 & \cdots & \varnothing \\ \vdots & \vdots & \vdots & & \vdots \\ -p_n(k) & -p_{n-1}(k) & -p_{n-2}(k) & \cdots & -p_1(k) \end{bmatrix}$$

With these new functions it is clear that the system (2) can be written more compactly as

(3) $$E\psi = Q(k)\psi + \zeta(k)$$

The matrix-valued function $Q(k)$ is called a *companion matrix* of the nth-order linear equation (1). Since $p_n(k) \neq 0$, for all $k \in N$, is an essential part of the definition of equation (1), that is, Definition 6, Section 22, it follows that $Q(k)$ is nonsingular for all $k \in N$. Verification of this is left as an exercise.

Also, each entry in $Q(k)$ and $\zeta(k)$ is a member of V_∞. Thus equation (3) represents a relationship between an unknown vector-valued function ψ with domain N and its first displacement. This is, with proper modification, essentially the same way that an nth-order differential equation is transformed into a linear system. Equation (3) is a particular type of the general family of equations defined below.

4 DEFINITION

Let $Q(k)$ be an $n \times n$ matrix-valued function whose entries are members of V_∞ such that $\det[Q(k)] \neq 0$ for all $k \in N$, and let $\zeta(k)$ be a vector-valued function with domain N. The equation

$$E\psi = Q(k)\psi + \zeta(k)$$

is a *linear difference system*. If $\zeta(k) = \varnothing$, the system is called *homogeneous*; otherwise it is called *nonhomogeneous*. The matrix-valued function $Q(k)$ is called the *coefficient matrix*.

5 Example

Let

$$Q(k) = \begin{bmatrix} 2 & k+1 \\ 2^{-k} & 2 \end{bmatrix} \quad \text{and} \quad \zeta(k) = \begin{bmatrix} -(k+1)^{(2)}2^k \\ 2^{k+1} - 1 \end{bmatrix}$$

Since $\det[Q(k)] = 4 - 2^{-k}(k+1) \neq 0$ for all $k \in N$, it follows that

$$E\psi = Q(k)\psi + \zeta(k)$$

is a linear difference system. The vector-valued function

$$\psi(k) = \begin{bmatrix} 2^k \\ k2^k \end{bmatrix} = 2^k \begin{bmatrix} 1 \\ k \end{bmatrix}$$

is a solution of this linear system since

$$E\psi(k) = 2^{k+1} \begin{bmatrix} 1 \\ k+1 \end{bmatrix}$$

and

$$Q(k)\psi(k) + \zeta(k) = \begin{bmatrix} 2 & k+1 \\ 2^{-k} & 2 \end{bmatrix} \begin{bmatrix} 2^k \\ k2^k \end{bmatrix} + \begin{bmatrix} -(k+1)^{(2)}2^k \\ 2^{k+1} - 1 \end{bmatrix}$$

$$= \begin{bmatrix} 2^{k+1} + (k+1)^{(2)}2^k - (k+1)^{(2)}2^k \\ 1 + k2^{k+1} + 2^{k+1} - 1 \end{bmatrix}$$

$$= 2^{k+1} \begin{bmatrix} 1 \\ k+1 \end{bmatrix}$$

Also, this solution satisfies the initial condition

$$\psi(0) = \begin{bmatrix} 1 \\ 0 \end{bmatrix}$$

The transformation of a linear difference system into a single equation can, in general, be done in a number of ways. To illustrate this, consider the homogeneous system

(6)
$$\begin{bmatrix} Ey_1 \\ Ey_2 \end{bmatrix} = \begin{bmatrix} a(k) & b(k) \\ c(k) & d(k) \end{bmatrix} \begin{bmatrix} y_1 \\ y_2 \end{bmatrix}$$

This system is equivalent to the two equations

(7)
$$Ey_1 = a(k)y_1 + b(k)y_2$$
$$Ey_2 = c(k)y_1 + d(k)y_2$$

and, by taking the first displacement of both sides of one of these, we have

(8) $E^2y_1 = a(k + 1)Ey_1 + b(k + 1)Ey_2$

By combining (7) and (8), we have

$$E^2y_1 = a(k + 1)Ey_1 + b(k + 1)c(k)y_1 + b(k + 1)d(k)y_2$$

Finally, the first equation in (7) can be used to eliminate y_2 and give

(9)
$$b(k)E^2y_1 = [a(k + 1)b(k) + b(k + 1)d(k)]Ey_1$$
$$+ b(k + 1)[b(k)c(k) - a(k)d(k)]y_1$$

Notice that the determinant of the coefficient matrix appears as part of the coefficient of y_1 in equation (9). Also, if $b(k) = \varnothing$, then it would be possible to obtain a single first-order nonhomogeneous equation in y_2. If both $b(k) = c(k) = \varnothing$, then the system is really two *uncoupled* first-order equations which can be solved separately.

A solution of a system of the form

(10) $E\psi = Q(k)\psi + \gamma(k)$

can be obtained in a step-by-step calculation. If $\psi(0) = \beta$, where β is a vector of constants, then by using equation (10) in the form

$$\psi(k + 1) = Q(k)\psi(k) + \gamma(k)$$

$k = 0, 1, 2, \ldots$, the step-by-step solution can be formed. Notice that

$$\psi(1) = Q(0)\psi(0) + \gamma(0) = Q(0)\beta + \gamma(0)$$
$$\psi(2) = Q(1)\psi(1) + \gamma(1) = Q(1)[Q(0)\beta + \gamma(0)] + \gamma(1)$$

or

$$\psi(2) = Q(1)Q(0)\beta + Q(1)\gamma(0) + \gamma(1)$$

By continuing in this manner, we can show that each $\psi(k)$ is uniquely determined by $\psi(0) = \beta$, and thus the initial-value problem has a unique solution. The basic idea as well as the method of proof are in Chapter 5.

Each solution of equation (10) can be represented as

$$\psi(k) = c_1\psi_1(k) + c_2\psi_2(k) + \cdots + c_n\psi_n(k) + \alpha(k)$$

where $\psi_1(k), \psi_2(k), \ldots, \psi_n(k)$ are linearly independent solutions of the homogeneous system

$$E\psi = Q(k)\psi$$

and $\alpha(k)$ is a solution of the nonhomogeneous system (10). The verification of this follows the line of the analogous results in Section 39.

11 EXERCISES

1. Verify that the companion matrix $Q(k)$ that is used to transform the linear equation (1) into the linear system (3) is nonsingular for all $k \in N$.
2. Show that there is a unique vector-valued function $\psi(k)$ that is a solution of the linear system $E\psi = Q(k)\psi + \zeta(k)$ and also satisfies the initial condition $\psi(0) = \beta$.
3. Let $V_\infty{}^n$ be the set of all vector-valued functions with n components, each of which is a member of V_∞. Show that the solutions of $E\psi = Q(k)\psi$ is a subspace of $V_\infty{}^n$.
4. Let $\psi_1(k), \psi_2(k), \ldots, \psi_n(k)$ be n solutions of $E\psi = Q(k)\psi$ and $C(k)$ be a matrix-valued function whose columns are the vector-valued functions $\psi_i(k), i = 1, 2, \ldots, n$. Show that $\psi_1(k), \psi_2(k), \ldots, \psi_n(k)$ are linearly independent if and only if $\det[C(k)] \neq 0$ for all $k \in N$.
5. Compare $C(k)$ above with the Casorati matrix of a solution basis of an nth-order linear homogeneous difference equation.
6. Show that the solutions of $E\psi = Q(k)\psi$ form an n-dimensional subspace of $V_\infty{}^n$.
7. Show that if $\sigma_1(k)$ and $\sigma_2(k)$ are solutions of $E\psi = Q(k)\psi + \zeta(k)$, then $\psi(k) = \sigma_1(k) - \sigma_2(k)$ is a solution of $E\psi = Q(k)\psi$.
8. Characterize the set of all solutions of a nonhomogeneous linear difference system.

The main topic in this section is linear difference systems in which the coefficient matrix is a constant matrix. A homogeneous system of this type can be represented as

$$(12) \qquad E\psi = Q\psi$$

where Q is an $n \times n$ matrix of constants and $\det[Q] \neq 0$. It is easy to verify that equation (12) has a solution of the form

$$(13) \qquad \psi(k) = r^k\beta$$

where r is a constant and $\beta \neq \varnothing$ is a constant vector, if and only if

$$(14) \qquad r^{k+1}\beta = r^k Q\beta$$

Since $\psi(k) = \varnothing$ is always a solution of equation (12), we can assume that $r \neq 0$, and thus equation (14) is equivalent to

$$(15) \qquad Q\beta = r\beta$$

The solutions of equation (15) are the eigenvalues and eigenvectors of Q. Since $\det[Q] \neq 0$ is a requirement for (12) to be a linear difference system, it follows that if r is an eigenvalue of Q, then $r \neq 0$.

These considerations lead to the main theorem for linear homogeneous difference systems with constant coefficient matrices.

16 THEOREM

If Q is an $n \times n$ nonsingular matrix, then $\psi(k) = r^k \beta$ is a nontrivial solution of $E\psi = Q\psi$ if and only if $Q\beta = r\beta$, that is, β is an eigenvector of Q and r is an associated eigenvalue. If Q has n linearly independent eigenvectors $\beta_1, \beta_2, \ldots, \beta_n$, then each solution of $E\psi = Q\psi$ is of the form

$$\psi(k) = c_1 r_1{}^k \beta_1 + c_2 r_2{}^k \beta_2 + \cdots + c_n r_n{}^k \beta_n$$

where $Q\beta_i = r_i \beta_i$, $i = 1, 2, \ldots, n$.

Of course, a matrix Q need not have n linearly independent eigenvectors, but this case is not too difficult to analyze. The case in which Q has n linearly independent eigenvectors is the easiest to treat and is more common in applications. Therefore, the remainder of this section is devoted to the difference system (12) in which Q has n linearly independent eigenvectors.

If Q has n linearly independent eigenvectors, then we saw in Section 40 that

$$Q = BRB^{-1}$$

where $R = \text{diag}(r_1, r_2, \ldots, r_n)$ and the columns of B are eigenvectors of Q. Following the ideas of Section 40, we define

$$V(k) = \text{diag}(r_1{}^k, r_2{}^k, \ldots, r_n{}^k)$$

so that we have the following theorem.

17 THEOREM

All solutions of $E\psi = Q\psi$, where $Q = BRB^{-1}$, are of the form

$$\psi(k) = BV(k)\gamma$$

where γ is a constant vector.

Also, there is a convenient form for initial-value problems. Since $\psi(0) = BV(0)\gamma$ and $V(0) = I$, the following is an obvious corollary.

18 COROLLARY

All solutions of $E\psi = Q\psi$, where $Q = BRB^{-1}$, are of the form

$$\psi(k) = BV(k)B^{-1}\psi(0)$$

---

19 Example

The matrix used in Example 16, Section 40, is:

$$Q = \begin{bmatrix} 2 & -1 & -1 \\ 1 & 0 & -1 \\ -1 & 1 & 2 \end{bmatrix}$$

It is easy to verify that

$$Q = BRB^{-1} = \begin{bmatrix} 1 & 1 & 1 \\ 0 & 1 & 1 \\ 1 & 0 & -1 \end{bmatrix} \begin{bmatrix} 1 & 0 & 0 \\ 0 & 1 & 0 \\ 0 & 0 & 2 \end{bmatrix} \begin{bmatrix} 1 & -1 & 0 \\ -1 & 2 & 1 \\ 1 & -1 & -1 \end{bmatrix}$$

Thus

$$V(k) = \begin{bmatrix} 1^k & 0 & 0 \\ 0 & 1^k & 0 \\ 0 & 0 & 2^k \end{bmatrix} = \begin{bmatrix} 1 & 0 & 0 \\ 0 & 1 & 0 \\ 0 & 0 & 2^k \end{bmatrix}$$

and all solutions of $E\psi = Q\psi$ are of the form

$$\psi(k) = \begin{bmatrix} 1 & 1 & 1 \\ 0 & 1 & 1 \\ 1 & 0 & -1 \end{bmatrix} \begin{bmatrix} 1 & 0 & 0 \\ 0 & 1 & 0 \\ 0 & 0 & 2^k \end{bmatrix} \begin{bmatrix} 1 & -1 & 0 \\ -1 & 2 & 1 \\ 1 & -1 & -1 \end{bmatrix} \psi(0)$$

In particular, if

$$\psi(0) = \begin{bmatrix} 1 \\ 1 \\ -1 \end{bmatrix}$$

then

$$\psi(k) = 2^k \begin{bmatrix} 1 \\ 1 \\ -1 \end{bmatrix}$$

20 EXERCISES

Let Q be the matrix in Example 19.

1. Show that the solution of the initial-value problem $E\psi = Q\psi$; $\psi(0) = \beta$, is $\psi(k) = \beta$ if and only if

$$\beta = c_1 \begin{bmatrix} 1 \\ 0 \\ 1 \end{bmatrix} + c_2 \begin{bmatrix} 1 \\ 1 \\ 0 \end{bmatrix}$$

where c_1 and c_2 are constants.

2. Find all solutions of $E\psi = Q\psi$ which have the vector $\beta^T = (8, 0, 0)$ in their range.

3. Find a vector that is not in the range of any solution of $E\psi = Q\psi$.

Let

$$Q = \begin{bmatrix} 0 & 2 \\ 2 & 0 \end{bmatrix}, \quad \alpha = \begin{bmatrix} 10 \\ -8 \end{bmatrix}, \quad \beta = \begin{bmatrix} 0 \\ 2 \end{bmatrix}$$

4. Find all solutions of $E\psi = Q\psi$.
5. Find the solution of $E\psi = Q\psi$; $\psi(0) = \alpha$.
6. Find the solution of $E\psi = Q\psi$; $\psi(0) = \beta$.

Let

$$Q = \begin{bmatrix} 2 & 1 & 1 \\ 0 & 0 & -2 \\ 0 & 1 & 3 \end{bmatrix}, \quad \alpha = \begin{bmatrix} 0 \\ 2 \\ -1 \end{bmatrix}, \quad \beta = \begin{bmatrix} 3 \\ 0 \\ -1 \end{bmatrix}$$

7. Find all solutions of $E\psi = Q\psi$.
8. Find the solution of $E\psi = Q\psi$; $\psi(0) = \alpha$.
9. Find all solutions of $E\psi = Q\psi$; $\psi(0) = \beta$.

Let

$$Q = \begin{bmatrix} 0 & 1 & 0 \\ 0 & 0 & 1 \\ -4 & -8 & -5 \end{bmatrix}$$

10. Show that Q does not have three linearly independent eigenvectors.
11. Solve $E\psi = Q\psi$ by first transforming this linear homogeneous system into a third-order linear equation. Find a solution basis for the third-order linear equation and use it to obtain a solution basis for $E\psi = Q\psi$.
12. State and prove an analogue of Theorem 26, Section 40.

Since a real matrix Q need not have real eigenvalues, the problem of associating real solutions of the homogeneous system

(21) $E\psi = \psi Q$

with complex eigenvalues of Q is essentially the same as in Section 25.

Suppose that $a + ib$ is a complex eigenvalue of Q. Then, since Q is a real matrix, it follows that an associated eigenvector is also complex. Thus

(22) $Q(\alpha + i\beta) = (a + ib)(\alpha + i\beta)$

where $\alpha + i\beta$ is a complex eigenvector with α and β real vectors. By examining the conjugate of equation (22), namely,

(23) $Q(\alpha - i\beta) = (a - ib)(\alpha - i\beta)$

it follows that $a - ib$ is also an eigenvalue of Q and that $\alpha - i\beta$ is an associated eigenvector. Thus (22) and (23) imply that

(24) $\theta_1(k) = (a + ib)^k(\alpha + i\beta)$ and $\theta_2(k) = (a - ib)^k(\alpha - i\beta)$

are two linearly independent, but complex-valued, solutions of (21).

By using the polar form of the eigenvalues, namely,

$$a + ib = r(\cos s + i \sin s)$$

and
$$a - ib = r(\cos s - i \sin s)$$

it follows that

$$\tfrac{1}{2}\theta_1(k) + \tfrac{1}{2}\theta_2(k) = r^k[\cos(ks)\,\alpha - \sin(ks)\,\beta]$$
$$-\tfrac{1}{2}i\theta_1(k) + \tfrac{1}{2}i\theta_2(k) = r^k[\sin(ks)\,\alpha + \cos(ks)\,\beta]$$

and thus

$$\psi_1(k) = r^k[\cos(ks)\,\alpha - \sin(ks)\,\beta]$$
$$\psi_2(k) = r^k[\sin(ks)\,\alpha + \cos(ks)\,\beta]$$

are two real-valued solutions of equation (21). It can be shown that this pair of solutions is linearly independent. This result is summarized below.

25 THEOREM

If Q is a real matrix and $Q(\alpha + i\beta) = (a + ib)(\alpha + i\beta)$ defines a complex eigenvalue-eigenvector pair for Q, where $b \neq 0$, then

$$\psi_1(k) = r^k[\cos(ks)\,\alpha - \sin(ks)\,\beta]$$

and

$$\psi_2(k) = r^k[\sin(ks)\,\alpha + \cos(ks)\,\beta]$$

are a pair of linearly independent solutions of $E\psi = Q\psi$, where

$$r = [a^2 + b^2]^{1/2} \qquad \text{and} \qquad \cos s = \frac{a}{r}$$

26 Example

It can be verified that

$$Q(\alpha + i\beta) = (a + ib)(\alpha + i\beta)$$

where

$$Q = \begin{bmatrix} -2 & 1 & 0 \\ -5 & 2 & 1 \\ -2 & -2 & 1 \end{bmatrix}, \qquad \alpha + i\beta = \begin{bmatrix} 1 \\ 3 \\ -2 \end{bmatrix} + i \begin{bmatrix} 0 \\ 2 \\ 4 \end{bmatrix}$$

and $a + ib = 1 + 2i$. Thus Theorem 25 can be used to obtain two real solutions of $E\psi = Q\psi$ since

$$r = \sqrt{5} \qquad \text{and} \qquad \cos s = \frac{\sqrt{5}}{5}$$

About the only convenient representation of s is:

$$s = \text{arc}\cos \frac{\sqrt{5}}{5}$$

However, $1.07 < s < 1.08$, and a more accurate decimal approximation is possible. The two solutions are:

$$\psi_1(k) = 5^{k/2} \cos(ks) \begin{bmatrix} 1 \\ 3 \\ -2 \end{bmatrix} - 5^{k/2} \sin(ks) \begin{bmatrix} 0 \\ 2 \\ 4 \end{bmatrix}$$

and

$$\psi_2(k) = 5^{k/2} \sin(ks) \begin{bmatrix} 1 \\ 3 \\ -2 \end{bmatrix} + 5^{k/2} \cos(ks) \begin{bmatrix} 0 \\ 2 \\ 4 \end{bmatrix}$$

Some special results concerning a particular solution of the non-homogeneous system

(27) $E\psi = Q\psi + \zeta(k)$

can be derived. Many of these require a sophisticated use of matrix theory and are beyond the scope of this book.

In general, only one solution of the nonhomogeneous system is required, together with all solutions of the associated homogeneous system, as described in Theorem 16. There is always the step-by-step approach to producing a solution of (27) and in some cases this can be useful.

By writing equation (27) as

$$\psi(k + 1) = Q\psi(k) + \zeta(k)$$

and, for convenience, by using the initial condition $\psi(0) = \varnothing$, we have

$$\psi(1) = Q\psi(0) + \zeta(0) = \zeta(0)$$
$$\psi(2) = Q\psi(1) + \zeta(1) = Q\zeta(0) + \zeta(1)$$
$$\psi(3) = Q\psi(2) + \zeta(2) = Q^2\zeta(0) + Q\zeta(1) + \zeta(2)$$

In general, it follows by an inductive argument that

(28) $\sigma(k) = Q^{k-1}\zeta(0) + Q^{k-2}\zeta(1) + \cdots + Q\zeta(k - 2) + \zeta(k - 1)$

is a solution of

$$E\psi = Q\psi + \zeta(k); \qquad \psi(0) = \varnothing$$

Now if the factorization $Q = BRB^{-1}$, where R is diagonal, is known, then

$$Q^2 = BRB^{-1}BRB^{-1} = BR^2B^{-1}$$
$$Q^3 = BRB^{-1}BR^2B^{-1} = BR^3B^{-1}$$

and so on. Thus (28) can be expressed as

(29)
$$\sigma(k) = B[R^{k-1}B^{-1}\zeta(0) + R^{k-2}B^{-1}\zeta(1)$$
$$+ \cdots + RB^{-1}\zeta(k - 2) + B^{-1}\zeta(k - 1)]$$

This last form has the computational advantages associated with calculating powers of a diagonal matrix, R, instead of the more cumbersome form in (28).

These observations can be used in conjunction with Corollary 18 to obtain one characterization of the solutions of the linear system (27).

30 THEOREM

If $Q = BRB^{-1}$, where $R = \operatorname{diag}(r_1, \ldots, r_n)$, and $V(k) = \operatorname{diag}(r_1{}^k, \ldots, r_n{}^k)$, then all solutions of the linear nonhomogeneous difference system $E\psi = Q\psi + \zeta(k)$ are of the form

$$\psi(k) = BV(k)B^{-1}\psi(0) + \sigma(k)$$

where $\sigma(0) = 0$ and $\sigma(k)$, $k \geq 1$, is defined in (29).

31 Example

Consider $E\psi = Q\psi + \zeta(k)$, where

$$Q = \begin{bmatrix} 2 & -1 & -1 \\ 1 & 0 & -1 \\ -1 & 1 & 2 \end{bmatrix} \quad \text{and} \quad \zeta(k) = \begin{bmatrix} 1 \\ 1 \\ 0 \end{bmatrix}$$

This is the same matrix that appeared in Example 19, along with the factorization $Q = BRB^{-1}$. To determine the solution $\sigma(k)$ in (29), notice first that

$$B^{-1}\zeta(k) = \begin{bmatrix} 1 & -1 & 0 \\ -1 & 2 & 1 \\ 1 & -1 & -1 \end{bmatrix}\begin{bmatrix} 1 \\ 1 \\ 0 \end{bmatrix} = \begin{bmatrix} 0 \\ 1 \\ 0 \end{bmatrix}$$

and since $\zeta(k)$ is a constant vector, the above expression holds for all $k \in N$.

Next, notice that

$$R^j = \begin{bmatrix} 1^j & 0 & 0 \\ 0 & 1^j & 0 \\ 0 & 0 & 2^j \end{bmatrix}$$

for all $j \in N$, so that

$$R^j B^{-1}\zeta(k) = \begin{bmatrix} 0 \\ 1 \\ 0 \end{bmatrix}$$

Thus

$$R^{k-1}B^{-1}\zeta(0) + R^{k-2}B^{-1}\zeta(1) + \cdots + RB^{-1}\zeta(k-2)$$

$$+ B^{-1}\zeta(k-1) = \begin{bmatrix} 0 \\ k \\ 0 \end{bmatrix}$$

Finally,

$$\sigma(k) = \begin{bmatrix} 1 & 1 & 1 \\ 0 & 1 & 1 \\ 1 & 0 & -1 \end{bmatrix} \begin{bmatrix} 0 \\ k \\ 0 \end{bmatrix} = \begin{bmatrix} k \\ k \\ 0 \end{bmatrix} = k \begin{bmatrix} 1 \\ 1 \\ 0 \end{bmatrix}$$

for all $k \in N$. Thus by using the same initial condition as in Example 19, we have that

$$\psi(k) = 2^k \begin{bmatrix} 1 \\ 1 \\ -1 \end{bmatrix} + k \begin{bmatrix} 1 \\ 1 \\ 0 \end{bmatrix}$$

is the solution of

$$E\psi = \begin{bmatrix} 2 & -1 & -1 \\ 1 & 0 & -1 \\ -1 & 1 & 2 \end{bmatrix} \psi + \begin{bmatrix} 1 \\ 1 \\ 0 \end{bmatrix}; \qquad \psi(0) = \begin{bmatrix} 1 \\ 1 \\ -1 \end{bmatrix}$$

and this claim can be verified directly.

32 EXERCISES

1. Let Q be the matrix in Examples 19 and 31, and consider the systems $E\psi = Q\psi + \zeta(k)$ for various choices of $\zeta(k)$.
   a. Determine the solution (29), where $\zeta(k) = 2^k(0, -1, 1)^T$.
   b. Determine the solution (29), where

   $$\zeta(k) = \begin{bmatrix} 1 \\ 1 \\ -1 \end{bmatrix}$$

   c. Solve the initial-value problem $E\psi = Q\psi + \zeta(k)$, with

   $$\psi(0) = \begin{bmatrix} 0 \\ 0 \\ 0 \end{bmatrix}$$

   and

   $$\zeta(k) = k^2 \begin{bmatrix} -1 \\ -1 \\ 1 \end{bmatrix} + k \begin{bmatrix} 1 \\ 0 \\ 0 \end{bmatrix} + \begin{bmatrix} 1 \\ 1 \\ -1 \end{bmatrix}$$

   (Hint: Perhaps the superposition principle might simplify the calculations.)

2. If $R = \text{diag}(r_1, r_2, \ldots, r_n)$, use the summation calculus to obtain a representation of

   $$U(k) = R^{k-1} + R^{k-2} + \cdots + R + I$$

   and compare $U(k)$ with $V(k) = \text{diag}(r_1{}^k, r_2{}^k, \ldots, r_n{}^k)$.

3. Let Q be a nonsingular matrix and let β be a constant vector. Show that a solution of $E\psi = Q\psi + \beta$ is $\psi(k) = \alpha$ (a constant vector) if and only if α is a solution of

   $$(I - Q)\alpha = \beta$$

4. Suppose Q is nonsingular and 1 is not an eigenvalue of Q. What is the solution of the initial-value problem $E\psi = Q\psi + (I - Q)\alpha; \psi(0) = \alpha$?

5. Find all constant vectors that are solutions of

$$9E\psi = \begin{bmatrix} 25 & 2 & -28 \\ 2 & -2 & -26 \\ -28 & -26 & 13 \end{bmatrix} \psi + 3 \begin{bmatrix} 2 \\ -2 \\ 1 \end{bmatrix}$$

6. If $\beta \neq \emptyset$, show that there is no solution of

$$E\psi = Q\psi + k\beta$$

of the form $\psi(k) = k\alpha$, where α is a constant vector.

7. Find necessary and sufficient conditions for $\psi(k) = k\alpha_1 + \alpha_2$ to be a solution of $E\psi = Q\psi + k\beta$.

8. Let

$$Q = \frac{1}{4} \begin{bmatrix} 17 & -3 & -12 \\ 16 & -2 & -12 \\ 3 & -3 & 2 \end{bmatrix} \quad \text{and} \quad \zeta(k) = k \begin{bmatrix} 16 \\ 16 \\ 1 \end{bmatrix}$$

a. Solve $E\psi = Q\psi - \zeta(k)$;

$$\psi(0) = \begin{bmatrix} 1 \\ 1 \\ 1 \end{bmatrix}$$

b. Solve $E\psi = Q\psi - \zeta(k); \psi(0) = \emptyset$.

c. Solve $E\psi = Q\psi + \zeta(k); \psi(0) = \emptyset$.

9. Let

$$Q = \begin{bmatrix} 2 & -1 \\ -4 & 2 \end{bmatrix}$$

a. Show that $E\psi = Q\psi$ is not a linear difference system.

b. Review the discussion at the end of Section 25, and discuss the solutions of $E\psi = Q\psi$.

10. Transform $E^2 y - 6Ey + 12y = 2^{k+2}$ into a linear difference system and find all solutions of the system.

11. Find all solutions of

$$E\psi = \frac{1}{8} \begin{bmatrix} -3 & -15 & -14 \\ -3 & 1 & 2 \\ 5 & -15 & 6 \end{bmatrix} \psi + 2k \begin{bmatrix} -5 \\ 3 \\ 3 \end{bmatrix} - 8 \begin{bmatrix} 1 \\ -1 \\ 1 \end{bmatrix}$$

Section 42 Numerical Analysis

The procedure used in Section 38 to transform a single nth-order linear differential equation into a system of n first-order equations suggests a way to treat approximate solutions of higher-order linear differential equations. This amounts to extending the results of Chapter 6 from a single first-order equation to a system of equations of the form

(1) $\qquad D\psi = Q(x)\psi + \zeta(x)$

In particular, if $\sigma(k)$ is used to denote an approximate value of the vector $\psi(x_0 + kh)$, where $\psi(x)$ is a solution of (1), $h > 0$, and $k \in N$, then the modifications in Definition 11, Section 27, of Euler's method are obvious.

2 DEFINITION

Euler's method as applied to equation (1) is:

$$\sigma(k + 1) = \sigma(k) + h\, D\sigma(k)$$

where $\sigma(0) = \psi(x_0)$, $h > 0$, and

$$D\sigma(k) = Q(x_0 + kh)\sigma(k) + \zeta(x_0 + kh)$$

The *error function* is the vector-valued function

$$\varepsilon(k, h) = \psi(x_0 + kh) - \sigma(k)$$

Thus Euler's method amounts to transforming a linear differential system into a linear difference system. In a similar way, the other numerical methods of Chapter 6 can be extended to include linear systems of the form (1).

3 Example

The second-order equation

(A) $\qquad D^2 y + Dy - 6y = \varnothing$

can be transformed into the linear differential system

(B) $\qquad D\psi = Q\psi$

where

$$\psi = \begin{bmatrix} y \\ Dy \end{bmatrix} \quad \text{and} \quad Q = \begin{bmatrix} 0 & 1 \\ 6 & -1 \end{bmatrix}$$

Euler's method as applied to system (B) is:

$$\sigma(k + 1) = \sigma(k) + hQ\sigma(k)$$

or

(C) $\qquad \sigma(k + 1) = (I + hQ)\sigma(k)$

For each $h > 0$, system (C) is a linear difference system with coefficient matrix $(I + hQ)$. Thus the methods of Section 41 can be used to investigate system (C).

To illustrate the results of a step-by-step calculation, suppose $y(0) = 1$ and $Dy(0) = -3$ are the initial conditions imposed on equation (A). Then

$$\psi(0) = \begin{bmatrix} 1 \\ -3 \end{bmatrix} \quad \text{and} \quad \sigma(0) = \begin{bmatrix} 1 \\ -3 \end{bmatrix}$$

are the initial vectors associated with equations (B) and (C), respectively. Next, suppose $h = \frac{1}{2}$ is selected, in which case

$$(I + hQ) = \begin{bmatrix} 1 & 0 \\ 0 & 1 \end{bmatrix} + \frac{1}{2}\begin{bmatrix} 0 & 1 \\ 6 & -1 \end{bmatrix} = \frac{1}{2}\begin{bmatrix} 2 & 1 \\ 6 & 1 \end{bmatrix}$$

Thus from equation (C), we have that

$$\sigma(1) = \frac{1}{2}\begin{bmatrix} 2 & 1 \\ 6 & 1 \end{bmatrix}\begin{bmatrix} 1 \\ -3 \end{bmatrix} = \frac{1}{2}\begin{bmatrix} -1 \\ 3 \end{bmatrix}$$

$$\sigma(2) = \frac{1}{2}\begin{bmatrix} 2 & 1 \\ 6 & 1 \end{bmatrix}\frac{1}{2}\begin{bmatrix} -1 \\ 3 \end{bmatrix} = \frac{1}{4}\begin{bmatrix} 1 \\ -3 \end{bmatrix}$$

and

$$\sigma(3) = (-1)^3 2^{-3}\begin{bmatrix} 1 \\ -3 \end{bmatrix}$$

$$\sigma(4) = (-1)^4 2^{-4}\begin{bmatrix} 1 \\ -3 \end{bmatrix}$$

and so on. Here, $\sigma(k)$ is an approximation of $\psi(kh) = \psi(\frac{1}{2}k)$, and

$$\psi(\tfrac{1}{2}k) = \begin{bmatrix} y(\tfrac{1}{2}k) \\ Dy(\tfrac{1}{2}k) \end{bmatrix}$$

where $y(x)$ is the solution of (A) that satisfies $y(0) = 1$, $Dy(0) = -3$. It follows that the first component of $\sigma(k)$ is an approximate value of $y(\frac{1}{2}k)$. However, $y(x) = e^{-3x}$, so that

$$-\tfrac{1}{2}, \tfrac{1}{4}, -\tfrac{1}{8}, \tfrac{1}{16}, -\tfrac{1}{32}$$

and so on, are the approximations for

$$e^{-3/2}, e^{-3}, e^{-9/2}, e^{-6}, e^{-15/2}$$

and so on.

This example illustrates that problems similar to those discussed in Section 27 arise in connection with numerical methods for linear differential systems. However, some new problems are encountered with systems that do not occur with a single first-order equation. The results in this section are aimed toward exposing these new problems. This can be done by considering the linear differential system

(4) $\qquad D\psi = Q\psi$

where $Q = BRB^{-1}$ and $R = \mathrm{diag}(r_1, r_2, \ldots, r_n)$.

If Euler's method is applied to (4), then the resultant difference system is:

(5) $\qquad E\sigma = [I + hQ]\sigma$

Notice that the matrix in (5) has the same eigenvectors as the matrix in (4), since

$$I + hQ = BB^{-1} + hBRB^{-1} = B(I + hR)B^{-1}$$

and that

$$I + hR = \mathrm{diag}[(1 + r_1 h), (1 + r_2 h), \ldots, (1 + r_n h)]$$

Thus the methods of the previous sections can be used to characterize and compare the solutions of (4) and (5). This requires the two matrix-valued functions

$$U(x) = \mathrm{diag}(e^{r_1 x}, e^{r_2 x}, \ldots, e^{r_n x}), \qquad x \in [0, \infty)$$

and

$$V(k) = \mathrm{diag}[(1 + r_1 h)^k, (1 + r_2 h)^k, \ldots, (1 + r_n h)^k], \qquad k \in N$$

It follows from Section 40 that

(6) $\qquad \psi(x) = BU(x)B^{-1}\psi(0)$

is the solution of (4), and from Section 41, we have that

(7) $\qquad \sigma(k) = BV(k)B^{-1}\sigma(0)$

is the solution of (5).

If the initial condition $\psi(0) = \alpha$ is used with equation (4), then $\sigma(0) = \psi(0) = \alpha$ is the appropriate initial vector in the step-by-step approximate solution. Thus the error function

$$\varepsilon(k, h) = \psi(kh) - \sigma(k)$$

is, in this case,

(8) $\qquad \varepsilon(k, h) = B[U(kh) - V(k)]B^{-1}\alpha$

While $\varepsilon(k, h)$ is a vector-valued function, it is, in one important aspect, analogous to the error function (21), Section 27. Notice that $U(kh) - V(k)$ is a diagonal matrix with entries of the form

(9) $\qquad g_i(k, h) = e^{r_i h k} - (1 + r_i h)^k$

$i = 1, 2, \ldots, n$, and this is exactly the same functional form that appears in (21) of Section 27.

Evidently, it is desirable to have an error (vector) $\varepsilon(k, h)$ that is "small" in the sense of being close to the zero vector. In order to make this requirement more precise, a new concept is needed.

10 DEFINITION

Let $\alpha = (a_1, a_2, \ldots, a_n)^T$ be a vector with components a_i that are real or complex. The *norm* of α is the real-valued function defined by

$$\|\alpha\| = \max\{|a_i| : i = 1, 2, \ldots, n\}$$

11 Examples

a. If $\alpha = (2, -7, 0, 1)$, then $\|\alpha\| = 7$.
b. If $\alpha(k) = [\cos k, (-3)^k, 7]^T$, then

$$\|\alpha(k)\| = \max\{|\cos k|, 3^k, 7\}$$

or

$$\|\alpha(k)\| = 7$$

if $k = 0, 1$ and

$$\|\alpha(k)\| = 3^k$$

if $k \geq 2$.
c. If $\alpha(x) = [6e^{-x}, -2, -4]^T$, where $x \in [0, \infty)$, then

$$\|\alpha(x)\| = 6e^{-x}$$

if $0 \leq x \leq \ln 3 - \ln 2$ and

$$\|\alpha(x)\| = 4$$

if $x > \ln 3 - \ln 2$.

This definition of vector norm is just one of many ways to extend the concept of absolute value to higher dimensional spaces. Notice that if a number, -3 for example, is viewed as a vector with one component, then $\|-3\| = |-3| = 3$, that is, the norm agrees with the absolute value. In fact, the basic properties of the norm function are the same as those for the absolute value function. Even the proof of the following theorem is virtually identical with the same result for the absolute value function.

12 THEOREM

If α and β are two vectors of the same order and c is a constant, then

(i) $\|\alpha\| > 0$ if $\alpha \neq \varnothing$ and $\|\varnothing\| = 0$
(ii) $\|c\alpha\| = |c| \, \|\alpha\|$
(iii) $\|\alpha + \beta\| \leq \|\alpha\| + \|\beta\|$.

Although our main application of the norm is in the study of the error function (8), this concept is also useful in extending the notion of boundedness to vector-valued functions.

13 DEFINITION

If $\alpha(x)$ is a vector-valued function with domain J, and there is a constant c such that $\|\alpha(x)\| \leq c$ for all $x \in J$, then $\alpha(x)$ is said to be *bounded* on J.

14 Examples

    a. The function $\alpha(x)$ of Example 11c is bounded on $[0, \infty)$ since $\|\alpha(x)\| \leq 6$ for all $x \in [0, \infty)$.

    b. The function $\alpha(k)$ of Example 11b is unbounded on N since $\|\alpha(k)\| \geq 3^k$ for all $k \in N$, and 3^k is clearly an unbounded function on N.

15 EXERCISES

1. Transform the linear difference system (C) in Example 3 into a single second-order equation.

2. Apply Euler's method with $h = \frac{1}{4}$ to

$$D\psi = \begin{bmatrix} 4 & -8 \\ -8 & 4 \end{bmatrix} \psi; \qquad \psi(0) = \begin{bmatrix} 1 \\ 1 \end{bmatrix}$$

and compute $\sigma(2)$.

3. Determine the norm of the following.

    a. $\alpha = \begin{bmatrix} 3 \\ i \\ -1 \end{bmatrix}$
    b. $\alpha = \begin{bmatrix} 1 + 4i \\ -4.2 \\ 4 - 2i \end{bmatrix}$

    c. $\alpha(x) = \begin{bmatrix} \cos x \\ 1 - e^x \\ x^2 \end{bmatrix}, x \in [0, \infty)$
    d. $\alpha(k) = \begin{bmatrix} (\frac{1}{2})^k \\ -3 \\ k^{(2)} \end{bmatrix}, k \in N$

4. Prove Theorem 12.

5. Show that $\alpha(k) = [t_1{}^k, t_2{}^k, t_3{}^k, \ldots, t_n{}^k]^T$ is bounded on N if and only if $|t_i| \leq 1$ for $i = 1, 2, \ldots, n$.

6. Show that $\alpha(x) = [e^{t_1 x}, e^{t_2 x}, \ldots, e^{t_n x}]^T$ is bounded on $[0, \infty)$ if and only if $Re(t_i) \leq 0$ for $i = 1, 2, \ldots, n$. (Note: $Re(t)$ is the real part of t.)

7. Show that $\big| \|\alpha\| - \|\beta\| \big| \leq \|\alpha - \beta\|$.

8. Let A be an $n \times n$ matrix whose columns are the vectors $\alpha_1, \alpha_2, \ldots, \alpha_n$; and let β be a vector whose components are $b_1, b_2, \ldots, b_n$, respectively. Show that

$$\|A\beta\| \leq |b_1| \|\alpha_1\| + |b_2| \|\alpha_2\| + \cdots + |b_n| \|\alpha_n\|$$

and

$$\|A\beta\| \leq (\max\|\alpha_i\|)\|\beta\|$$

9. Let Q be an $n \times n$ matrix that is similar to a diagonal matrix. Show that all solutions of

$$E\sigma = Q\sigma$$

are bounded if and only if each eigenvalue of Q is less than or equal to one in absolute value.

The concept of a norm serves to transform the problem of studying the error function (8) into one of analyzing a real-valued function. This is done by defining the *normed error function* as

(16) $e(k, h) = \|\varepsilon(k, h)\|$

Notice that $e(k, h) \geq 0$ and $e(k, h) = 0$ if and only if $\varepsilon(k, h) = \varnothing$, that is, if and only if the solution of the differential equation (4) agrees exactly with the approximate solution by Euler's method. Thus all statements concerning the error in Euler's method as applied to a system of differential equations can be made in terms of $e(k, h)$.

In order to obtain an upper bound for $e(k, h)$, it is convenient to adopt a special form for the initial condition

$$\psi(0) = \sigma(0) = \alpha$$

By taking $\alpha = B\gamma$, the error function becomes

(17) $\varepsilon(k, h) = B[U(kh) - V(k)]\gamma$

Of course, for each vector γ, there is one and only one choice of α and conversely, since $\gamma = B^{-1}\alpha$. Next, let $\gamma = (c_1, c_2, \ldots, c_n)^T$, and let the columns of B (which are eigenvectors of Q) be denoted by $\beta_1, \beta_2, \ldots, \beta_n$. It is a simple matter to show that

(18) $\varepsilon(k, h) = c_1 g_1(k, h)\beta_1 + c_2 g_2(k, h)\beta_2 + \cdots + c_n g_n(k, h)\beta_n$

where $g_i(k, h)$ is defined in (9). With this representation, an upper bound for $e(k, h)$ can be obtained by taking norms of both sides of equation (18). Thus

(19)
$$e(k, h) \leq |c_1| \, \|\beta_1\| \, |g_1(k, h)|$$
$$+ |c_2| \, \|\beta_2\| \, |g_2(k, h)| + \cdots + |c_n| \, \|\beta_n\| \, |g_n(k, h)|$$

The bound in (19) depends upon an obvious generalization of Theorem 12. Notice that the bound in (19) depends upon the initial condition $\psi(0) = \alpha$, as well as the eigenvectors of Q, which are the columns of B, and the eigenvalues of Q, which appear in $g_i(k, h)$. A useful consequence of (19) is the following theorem.

20 THEOREM

If $D\psi = Q\psi$, where $Q = BRB^{-1}$ and $R = \text{diag}(r_1, r_2, \ldots, r_n)$, and $\psi(0) = \alpha$, then there exists a constant c such that the normed error in Euler's method satisfies

$$e(k, h) \leq c\{\max|e^{r_i hk} - (1 + r_i h)^k| : i = 1, 2, \ldots, n\}$$

Proof: Exercise.

The result in Theorem 20 is the basis for extending the theorems in Section 27 to systems. While in Section 27 an error analysis involved but a single function of the form

$$|e^{ahk} - (1 + ah)^k|$$

n terms of this type, namely,

$$|g_i(k, h)| = |e^{r_ihk} - (1 + r_ih)^k|$$

must be considered here.

21 Example

Suppose that

$$Q = \begin{bmatrix} 1 & 1 & 1 \\ 0 & 1 & 1 \\ 1 & 0 & -1 \end{bmatrix} \begin{bmatrix} -1 & 0 & 0 \\ 0 & -4 & 0 \\ 0 & 0 & -3 \end{bmatrix} \begin{bmatrix} 1 & -1 & 0 \\ -1 & 2 & 1 \\ 1 & -1 & -1 \end{bmatrix}$$

It is easy to verify that all solutions of $D\psi = Q\psi$ are of the form

$$\psi(x) = c_1 e^{-x} \begin{bmatrix} 1 \\ 0 \\ 1 \end{bmatrix} + c_2 e^{-4x} \begin{bmatrix} 1 \\ 1 \\ 0 \end{bmatrix} + c_3 e^{-3x} \begin{bmatrix} 1 \\ 1 \\ -1 \end{bmatrix}$$

Since each of the above eigenvectors of Q has norm one, it follows that

$$\|\psi(x)\| \le |c_1| e^{-x} + |c_2| e^{-4x} + |c_3| e^{-3x}$$

or for $x \in [0, \infty)$,

$$\|\psi(x)\| \le |c_1| + |c_2| + |c_3|$$

Thus all solutions of $D\psi = Q\psi$ are bounded on $[0, \infty)$.

The bound for the normed error function is:

$$e(k, h) \le |c_1| \, |e^{-hk} - (1 - h)^k| + |c_2| \, |e^{-4hk} - (1 - 4h)^k| + |c_3| \, |e^{-3hk} - (1 - 3h)^k|$$

Thus if

$$0 < h < \frac{2}{|-4|} = \tfrac{1}{2}$$

then $|1 - h| < 1$, $|1 - 4h| < 1$, and $|1 - 3h| < 1$, so that $e(k, h)$ and hence $\sigma(k)$ are bounded on N.

This example suggests an extension of Corollary 24, Section 27, in which $|a|$ is replaced by the eigenvalues of Q. There is a concept in matrix theory that is helpful in stating this extended result.

22 DEFINITION

Let Q be an $n \times n$ matrix with eigenvalues $r_1, r_2, \ldots, r_n$. The *spectral radius* of Q is denoted by $s(Q)$ and defined by

$$s(Q) = \max\{|r_i|: i = 1, 2, \ldots, n\}$$

Evidently, the matrix Q in Example 21 has a spectral radius of 4; that is, $s(Q) = \max\{|-1|, |-4|, |-3|\} = 4$. Notice that in this example the restriction $0 < h < \frac{1}{2}$ is equivalent to

$$0 < h < \frac{2}{s(Q)}$$

We can now state a result on boundedness of approximate solutions that includes Corollary 24, Section 27, as a special case. The proof of this theorem is left as an exercise.

23 THEOREM

Let $Q = BRB^{-1}$, where R is a diagonal matrix, and for each eigenvalue r of Q, let $Re(r) < 0$. Then all solutions of $D\psi = Q\psi$ are bounded on $[0, \infty)$, and if $0 < h < 2/s(Q)$, all approximate solutions obtained by Euler's method are bounded on N.

24 EXERCISES

1. Verify the upper bound for $e(k, h)$ in (19).

2. Prove Theorem 20.

3. Determine $s(Q)$ for the following matrices Q.

   a. $Q = \begin{bmatrix} 1 & 4 & 4 \\ 0 & -7 & -10 \\ -2 & 4 & 7 \end{bmatrix}$
   b. $Q = \frac{1}{4}\begin{bmatrix} 17 & -3 & -12 \\ 16 & -2 & -12 \\ 3 & -3 & 2 \end{bmatrix}$

   c. $Q = \frac{1}{3}\begin{bmatrix} 1 & 2 & -2 \\ 2 & -2 & -1 \\ 2 & 1 & 2 \end{bmatrix}$
   d. $Q = \begin{bmatrix} 1 & 0 & -1 \\ 0 & 1 & 1 \\ 1 & -1 & 0 \end{bmatrix}$

4. Prove Theorem 23.

5. If A is an $n \times n$ matrix whose columns are the vectors $\alpha_1, \alpha_2, \ldots, \alpha_n$, let $\|A\| = n \max\{\|\alpha_i\|: i = 1, 2, \ldots, n\}$. Show that for each matrix in Problem 3 that $s(Q) \leq \|Q\|$.

6. The function $\|A\|$ defined in Problem 5 is called a *matrix norm*. Show that this function has the three properties in Theorem 12.

7. Assume that Q has n linearly independent eigenvectors and show that if Euler's method is applied to $D\psi = Q\psi$ with $0 < h < 2/\|Q\|$ and the approximate solution is not bounded, then $D\psi = Q\psi$ has unbounded solutions. (Hint: Show that $s(Q) \leq \|Q\|$.)

8. Show that the hypotheses of Theorem 23 not only imply boundedness, but also $\lim_{x \to \infty} \|\psi(x)\| = 0$ for each solution $\psi(x)$ of $D\psi = Q\psi$, and $\lim_{k \to \infty} \|\sigma(k)\| = 0$ for each approximate solution.

   In the actual step-by-step calculation of an approximate solution of a linear differential system, the effects of rounding errors can be far more important than in the single equation case. A step-by-step calculation of an approximate solution of $D\psi = Q\psi$ amounts to solving equation (5), and here a convenient form of this solution is:

(25) $\qquad \sigma(k) = [I + hQ]^k \sigma(0)$

Notice that the calculation of $\sigma(1)$, $\sigma(2)$, and so on, amounts to raising a matrix to a power. Whether this is done using a pocket calculator or a large computer, eventually some rounding errors can be expected due to the finite number of digits used in the calculations.

   Suppose that at some step k an error is made, and the incorrect value of $\sigma(k)$ is:

(26) $\qquad \bar{\sigma}(k) = (I + hQ)^k \sigma(0) + \delta$

where δ is a nonzero vector. Under some mild restrictions which are mentioned in the exercises to follow, the error δ can be represented as

(27) $\qquad \delta = (I + hQ)^k \delta_1$

for some unique choice of δ_1. By substituting (27) into (26), we have

(28) $\qquad \bar{\sigma}(k) = [I + hQ]^k (\sigma(0) + \delta_1)$

Evidently, (28) represents a solution of the difference system (5) but with the incorrect initial condition $\bar{\sigma}(0) = \sigma(0) + \delta_1$. Thus mistakes due to rounding errors amount to producing approximations that satisfy incorrect initial conditions. The next example indicates what effect this could have.

29 Example

Let

$$Q = \begin{bmatrix} 6 & 1 \\ 20 & -2 \end{bmatrix}$$

Since

$$Q \begin{bmatrix} -1 \\ 10 \end{bmatrix} = -4 \begin{bmatrix} -1 \\ 10 \end{bmatrix} \quad \text{and} \quad Q \begin{bmatrix} 2 \\ 4 \end{bmatrix} = 8 \begin{bmatrix} 2 \\ 4 \end{bmatrix}$$

this matrix has one negative and one positive eigenvalue. The solution of $D\psi = Q\psi$, $\psi(0) = (-1, 10)^T$ is the bounded function

$$\psi(x) = e^{-4x} \begin{bmatrix} -1 \\ 10 \end{bmatrix}$$

However, all solutions of $\sigma(k + 1) = (I + hQ)\sigma(k)$ are of the form

$$\sigma(k) = c_1(1 - 4h)^k \begin{bmatrix} -1 \\ 10 \end{bmatrix} + c_2(1 + 8h)^k \begin{bmatrix} 2 \\ 4 \end{bmatrix}$$

Clearly,

$$\sigma(0) = \begin{bmatrix} -1 \\ 10 \end{bmatrix}$$

implies that $c_1 = 1$ and $c_2 = 0$, but a change in initial condition, or, equivalently, an arithmetic error in a step-by-step calculation will provide a $c_2 \neq 0$. If this happens, and it usually does, then $\sigma(k)$ is an unbounded function on N independent of the choice of h.

      There is no remedy for this situation. This particular differential system is called *unstable*. Notice that if the initial condition is changed to

$$\psi(0) = \begin{bmatrix} -1 \\ 10 \end{bmatrix} + c \begin{bmatrix} 2 \\ 4 \end{bmatrix}$$

then $c \neq 0$ implies that $\psi(x)$ is unbounded on $[0, \infty)$.

30 EXERCISES

1. Show that $(I + hQ)^k$ has in inverse except for at most n values of h, and thus equation (27) has a unique solution except for at most n values of h.
2. Let $Q = BRB^{-1}$, where $R = \text{diag}(r_1, r_2, \ldots, r_n)$ and

$$r_1 < r_2 < \cdots < r_j < 0 < r_{j+1} < \cdots < r_n$$

Show that $\psi(x)$ is a bounded solution of $D\psi = Q\psi$ if and only if $\psi(0) = B\gamma$, where at most the first j components of γ are nonzero numbers.
3. Let

$$Q = \begin{bmatrix} 24 & 10 \\ -65 & 27 \end{bmatrix}$$

Show that all solutions of $D\psi = Q\psi$ are bounded, and find an upper bound for h so that all approximations by Euler's method are bounded.

      Each of the numerical methods of Chapter 6 can be used to approximate solutions of linear differential systems. For example, if

$$D\psi = Q\psi$$

then the two-step method becomes

(31) $\sigma(k + 2) = a_1\sigma(k + 1) + a_2\sigma(k) + h[b_1 Q\sigma(k + 1) + b_2 Q\sigma(k)]$

Here, as before, $\sigma(k)$ is an approximate value of $\psi(kh)$.

      However, equation (31) is a system of second-order difference equations. This can be transformed into a standard form of a linear difference system by the change of variables

(32)
$$\sigma_1(k) = \sigma(k)$$
$$\sigma_2(k) = \sigma(k + 1)$$

The result is the following pair of matrix-vector equations:

(33)
$$\sigma_1(k + 1) = \sigma_2(k)$$
$$\sigma_2(k + 1) = (a_2 I + h b_2 Q)\sigma_1(k) + (a_1 I + h b_1 Q)\sigma_2(k)$$

By using a "partitioned form" for matrices, the equations of (33) can be expressed compactly as

(34)
$$\begin{bmatrix} \sigma_1(k + 1) \\ \sigma_2(k + 1) \end{bmatrix} = \begin{bmatrix} 0 & I \\ a_2 I + h b_2 Q & a_1 I + h b_1 Q \end{bmatrix} \begin{bmatrix} \sigma_1(k) \\ \sigma_2(k) \end{bmatrix}$$

where

$$G = \begin{bmatrix} 0 & I \\ a_2 I + h b_2 Q & a_1 I + h b_1 Q \end{bmatrix}$$

is a $2n \times 2n$ matrix in which 0 is the $n \times n$ zero matrix and I is the $n \times n$ identity matrix.

Although the details are somewhat tedious, it can be shown that if t is an eigenvalue of G, then

(35) $$t^2 - (a_1 + r_i h b_1)t - (a_2 + r_i h b_2) = 0$$

where r_i is an eigenvalue of Q.

By noticing that the polynomials in (35) are of the same form as those in (42) of Section 28, an analysis can be modeled after the results in Section 28.

FIRST-ORDER DIFFERENTIAL EQUATIONS*

Section 43 Introduction

If F is a function of three variables, then the equation

(1) $F(x, y, Dy) = \varnothing$

is a first-order differential equation. Conversely, any first-order differential equation can be written in the form (1). For example, the first-order linear equation

(2) $Dy + p(x)y = f(x)$

can be put in the form (1) by letting $F(x, y, Dy) = Dy + p(x)y - f(x)$. In Section 8, techniques were developed for finding the one-parameter family of solutions of equation (2), and for finding the unique solutions of associated initial-value problems. The purpose of this chapter is to study additional special cases of equation (1) for which there exist elementary techniques for obtaining a one-parameter family of solutions and for solving initial-value problems.

In considering equations of the form (1), a basic assumption is that the derivative, Dy, of the unknown function appears in the equation in a nontrivial manner. For example, we would not consider the equation

$$Dy + ye^x = Dy$$

---

* Most of the material in this chapter depends upon the material in Chapter 2. The exception is Section 47, which is an extension of the results in Chapter 6.

to be a first-order differential equation. Similarly, the equation

$$\frac{y \, Dy + x^2 \, Dy}{Dy} = \varnothing$$

is not a first-order differential equation.

    A first-order differential equation of the form (1), in which Dy appears with exponent 1 only, is called a *first-degree equation*. The linear equation (2) is an example of a first-order first-degree equation. On the other hand, the first-order equations

$$x \, Dy + (Dy)^{3/2} - y = \varnothing$$

and

$$[Dy]^2 + xe^x y = \varnothing$$

are not of first degree.

    Almost all of the work in this chapter is on first-order first-degree equations, and from equation (1), it is easy to verify that such an equation must, in fact, have the more special form

(3) $Q(x, y) \, Dy + P(x, y) = \varnothing$

where $Q(x, y) \neq \varnothing$. In particular, equation (2) can be written in the form (3) by letting $Q(x, y) = 1$ and $P(x, y) = p(x)y - f(x)$. There are three basic types of first-order first-degree equations which can be treated by elementary methods: linear equations, studied in detail in Section 8; separable equations (Section 44); and exact equations (Section 45). There are, in addition, certain classes of equations of the form (3) which are not linear, separable, or exact, but which can be transformed into one of these basic types either by some prescribed change of variable or by an algebraic manipulation. These additional cases of equation (3) are considered in the appropriate sections. First-order equations which are related to linear equations are treated in Section 46. The chapter is concluded with an extension of a numerical method developed in Chapter 6.

    Before proceeding to the solution techniques, it is necessary to make some preliminary remarks concerning the concepts of solution and one-parameter families of solutions, and the questions of existence and uniqueness of solutions of first-order initial-value problems. At this point, it would be worthwhile for the reader to review Section 7 where these ideas were first introduced.

    We shall assume throughout the chapter that the functions $Q(x, y)$ and $P(x, y)$ in equation (3) are continuous real-valued functions on some region $\mathscr{R}$ of the x-y plane. As should be expected from the study of linear differential equations, the points (x, y) in $\mathscr{R}$, if any, where $Q(x, y) = 0$, cause difficulties with respect to existence of solutions. In fact, these difficulties are generally more complicated in the case of nonlinear equations than they were for linear equations. In order to maintain an appropriate level of discussion, and to emphasize the solution techniques, we shall consider equation (3) only on the subsets of $\mathscr{R}$ where $Q(x, y) \neq 0$.

4 DEFINITION

Given the first-order differential equation (3), where P and Q are continuous on the region $\mathscr{R}$. The subset $\mathscr{D}$ of $\mathscr{R}$ defined by

$$\mathscr{D} = \{(x, y) \in \mathscr{R}: Q(x, y) \neq 0\}$$

is called the *domain* of the equation.

5 Examples

    a. Consider the equation

$$y\, Dy + x = \varnothing$$

        The functions $P(x, y) = x$ and $Q(x, y) = y$ are continuous on the whole plane. The domain of the equation is the set $\mathscr{D} = \{(x, y): y \neq 0\}$, that is, the plane with the x-axis deleted.

    b. Consider the equation

$$(y - x)\, Dy + e^y \tan x = \varnothing$$

        The region $\mathscr{R}$ where each of the functions $P(x, y) = e^y \tan x$ and $Q(x, y) = y - x$ is continuous is the plane with the vertical lines $x = (2n + 1)\pi/2$, $n = 0, \pm 1, \pm 2, \ldots$, deleted. The domain of the equation is the set $\mathscr{D} = \{(x, y) \in \mathscr{R}: y \neq x\}$, that is, $\mathscr{R}$ with the line $y = x$ deleted.

    c. In the form (3), the linear equation is:

$$Dy + p(x)y - f(x) = \varnothing$$

        If $p(x)$ and $f(x)$ are continuous on an interval J, then the domain of the equation coincides with the region on which the functions $P(x, y) = p(x)y - f(x)$ and $Q(x, y) = 1$ are continuous, namely, the infinite strip $\mathscr{R} = \mathscr{D} = \{(x, y): x \in J \text{ and } y \in (-\infty, \infty)\}$.

    In most of the examples and exercises in this chapter, the domains of the first-order differential equations will not be given explicitly, but we shall follow the usual convention of assuming that the appropriate domain is clear from the context.

    On the domain $\mathscr{D}$ of equation (3), we may divide by $Q(x, y)$ to obtain the equivalent equation

$$(6) \qquad Dy = \frac{-P(x, y)}{Q(x, y)}$$

which is an equation of the form

$$(7) \qquad Dy = g(x, y)$$

where g is a continuous function on $\mathscr{D}$. Equations (3) and (7) are equivalent in the sense that they have exactly the same set of solutions. Of course, equation (7) can be written in the form (3) simply by letting $P(x, y) = -g(x, y)$ and $Q(x, y) = 1$. The form (7) is normally used in discussing such theoretical aspects of first-order first-degree equations as the concept of a solution, and existence and uniqueness theory. Also, in treating numerical methods, form (7) is found to be preferable to form (3). We shall consider the two forms as being completely interchangeable.

8 DEFINITION

Let $\mathscr{D}$ be the domain of the differential equation (7). A function $y = y(x)$ is a solution of (7) [or, equivalently, of (3)] on an interval I if for all $x \in I$:

(i) the point $(x, y(x)) \in \mathscr{D}$; and

(ii) $Dy(x) = g(x, y(x))$.

In contrast to first-order linear differential equations, first-order nonlinear equations are quite complicated and difficult to analyze. The most important and immediate distinction between the linear and nonlinear cases is the fact that there is no general formula for the solutions of a nonlinear equation. This lack of a general formula, as an analogue of

$$(9) \qquad y(x) = \exp\left\{-\int_a^x p(t)\, dt\right\}\left[c + \int_a^x \exp\left\{\int_a^t p(s)\, ds\right\} f(t)\, dt\right]$$

(see Corollary 15, Section 8) for the one-parameter family of solutions of equation (2), greatly increases the importance of techniques for obtaining approximate solutions. Also, because there is no general solution method, the problem of establishing existence and uniqueness of solutions of non-linear initial-value problems is considerably more difficult. The following theorem is an extension of Theorem 14, Section 8. The proof is omitted since it involves methods which are beyond the scope of this text.

10 THEOREM

Let (x_0, y_0) be a point in the domain $\mathscr{D}$ of equation (7), and let $R = \{(x, y): |x - x_0| \leq a, |y - y_0| \leq b\}$ be a closed rectangle in $\mathscr{D}$. If $\partial g/\partial y$ is continuous on R, then there exists an interval $I = (x_0 - \delta, x_0 + \delta), 0 < \delta \leq a$, and a function $y = y(x)$ defined on I such that y is the unique solution of (7) satisfying

$$y(x_0) = y_0$$

A second major distinction between linear and nonlinear equations concerns the interval of existence of solutions. Note that all the solutions (9) of the linear equation (2) are continuously differentiable throughout the entire interval J on which the functions p and f are continuous. The following examples show that in the nonlinear case, the interval on which a solution exists does not have any simple relationship to the function g in equation (7) or to the domain of the differential equation.

11 Examples

a. Consider the initial-value problem

$$Dy = 1 + y^2; \qquad y(0) = 0$$

The domain of the equation is the entire plane, and on any closed rectangle containing $(0, 0)$, each of $g(x, y) = 1 + y^2$ and $\partial g/\partial y = 2y$ is continuous. Thus, by Theorem 10, the initial-value problem has a unique solution valid on some interval containing $x = 0$. It is easy to verify that $y(x) = \tan x$ is the solution. Clearly, this function is unbounded as $x \to \pi/2$ and as $x \to -\pi/2$. Note that the differential equation itself does not give any indication that the points $x = \pi/2$ and $x = -\pi/2$ are significant.

b. Consider the initial-value problem

$$Dy = -y^2; \qquad y(1) = 1$$

The domain of this equation is the entire plane, and on any closed rectangle containing $(1, 1)$, both $g(x, y) = -y^2$ and $\partial g/\partial y = -2y$ are continuous. The unique solution of this initial-value problem is $y(x) = 1/x$. Obviously, this function is not defined at 0. This could not have been forecast from the function g. Since the initial value is at $x = 1$, the interval I associated with $y(x) = 1/x$ could be taken to be $I = (0, \infty)$. Consider the same differential equation together with the initial condition $y(0) = -1$. The unique solution is $y(x) = 1/(x - 1)$, which is defined on $(-\infty, 1)$. Thus the interval of existence depends not only on the equation but on the initial condition as well.

These examples also illustrate that the existence and uniqueness theorem, Theorem 10, is a "local" theorem; the hypotheses guarantee the existence and uniqueness of a solution in some neighborhood $(x_0 - \delta, x_0 + \delta)$ containing the initial point x_0.

A remark concerning uniqueness of solutions is appropriate here. Although it can be established that the continuity of g in a closed rectangle containing the initial point (x_0, y_0) is sufficient for the existence of a solution, the continuity of g does not imply uniqueness.

12 Example

Consider the differential equation

$$Dy = 3y^{2/3}$$

whose domain is the entire plane. Note that $g(x, y) = 3y^{2/3}$ is continuous for all x, y, but $\partial g/\partial y = 2y^{-1/3}$ is not continuous at $y = 0$. This differential equation has the one-parameter family of solutions $y(x) = (x + c)^3$. Thus the function $y(x) = x^3$ is a member of this family which satisfies the initial condition $y(0) = 0$. However, the function $y = \emptyset$ is also a solution of the initial-value problem

$$Dy = 3y^{2/3}; \qquad y(0) = 0$$

and $y = \emptyset$ is not a member of the one-parameter family of solutions (there is no value of c such that $(x + c)^3 = \emptyset$). Even more generally, for any pair of numbers a and b, $a < 0 \leq b$, the function y_{ab} defined by

$$y_{ab}(x) = \begin{cases} (x - a)^3 & x \leq a \\ 0 & a < x < b \\ (x - b)^3 & x \geq b \end{cases}$$

is a solution of the initial-value problem, and these functions are not members of the one-parameter family. (See the graphs shown in the figure here.)

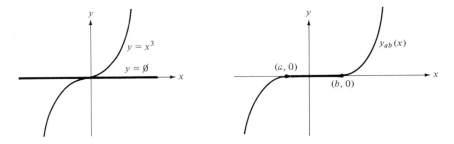

On the other hand, if (x_0, y_0) is a point in the plane and $y_0 \neq 0$, then there is a closed rectangle R containing (x_0, y_0) such that $y \neq 0$ for all $(x, y) \in R$. Thus $\partial g / \partial y$ is continuous on R and the initial-value problem

$$Dy = 3y^{2/3}; \qquad y(x_0) = y_0$$

will have a unique solution.

The reader can also refer to Example 11, Section 7, for a second example of an initial-value problem with an infinite number of solutions.

13 EXERCISES

1. Determine the domain of each of the following differential equations.
   a. $Dy + 2xy^2 = \emptyset$
   b. $y \, Dy = 2e^{4x}$
   c. $2x^3 \, Dy - y(y^2 + 3x^2) = \emptyset$
   d. $(x^2 - y^2) \, Dy + e^{xy} = \emptyset$
   e. $Dy = x/(y^2 - x^2 - 1)$
   f. $(xy - x^2) \, Dy = y^2$
2. In each of the following, verify that the given function (or family of functions) is a solution of the corresponding differential equation.
   a. $y = (c - x^2)^{1/2}, c > 0; y \, Dy + x = \emptyset$
   b. $y = (e^{4x} + 1)^{1/2}; y \, Dy = 2e^{4x}$
   c. $y = \sec x; Dy + 1 = y^2$

  d. $y = cx^2 + c^2$, c a constant; $(Dy)^2 + 2x^3 \, Dy - 4x^2y = \varnothing$

  e. $y = \sqrt{Cx - 1}$, $C > 0$; $2xy \, Dy = 1 + y^2$

  f. $y = x \tan x$; $x \, Dy = x^2 + y^2 + y$

3. For each of the following differential equations, determine the region in the x-y plane where the existence of a unique solution through any specified point is guaranteed by Theorem 10.

  a. $Dy = y^{1/3}$ b. $Dy = xy/(1 + y^2)$

  c. $(y^2 - x^2 - 1) \, Dy = x$ d. $Dy = (4 - x^2 - y^2)^{1/2}$

  e. $Dy = (x + y)^{3/2}$ f. $(x^2 - y^2) \, Dy = \ln(x - y)$

4. Consider the differential equation $Dy + 2xy^2 = \varnothing$.

  a. Show that $y(x) = 1/(x^2 - c)$, where c is any constant, is a one-parameter family of solutions.

  b. Determine a solution of the equation satisfying $y(0) = 1$. On what interval is your solution valid?

  c. Determine a solution of the equation satisfying $y(0) = -1$. On what interval is this solution valid?

5. Consider the differential equation $x^2 - y^2 + 2xy \, Dy = \varnothing$.

  a. Show that function $y = (2x - x^2)^{1/2}$ is a solution of the equation satisfying $y(1) = 1$. On what interval is this solution valid?

  b. Verify that

$$y(x) = (2cx - x^2)^{1/2} \qquad \text{and} \qquad y(x) = -(2cx - x^2)^{1/2}$$

  for any constant c, $c \neq 0$, are solutions of the equation. Where are these one-parameter families of solutions valid? Show that no member crosses the x-axis.

  c. Describe the members of the families in part (b) geometrically.

6. Consider the differential equation

$$4 \, Dy = x - (x^2 - 8y)^{1/2}$$

  a. Show that $y_1(x) = x - 2$ and $y_2(x) = x^2/8$ are each solutions of the equation satisfying $y(4) = 2$.

  b. Explain why the existence of two solutions of the equation satisfying the same initial condition does not contradict Theorem 10.

  c. Show that every member of the one-parameter family $y(x) = cx - 2c^2$ is a solution of the differential equation. Is the solution $y_2(x) = x^2/8$ a member of this one-parameter family?

   The next examples illustrate a third point of departure between linear and nonlinear equations.

14 Examples

   a. Consider the differential equation

$$Dy = \frac{-x}{y}$$

   whose domain is the plane with the x-axis deleted. It can be verified that any function which satisfies this equation also satisfies the implicit relation

$$x^2 + y^2 = c^2$$

for some value of c. This one-parameter family of curves can be found using the methods of the next section. We will regard this family of circles as a one-parameter family of solutions, with the solutions $y = y(x)$ being defined implicitly by the relation

$$G(x, y, c) = x^2 + y^2 - c^2 = \varnothing$$

for each value of the parameter c. In this example, we can solve the implicit relation to obtain the *two* one-parameter families

$$y(x) = \sqrt{c^2 - x^2} \quad \text{and} \quad y(x) = -\sqrt{c^2 - x^2}$$

The first is a family of functions whose graphs are semicircles in the upper half plane, and the second is a family of semicircles in the lower half plane. To obtain a solution of the initial-value problem

$$Dy = \frac{-x}{y}; \qquad y(0) = 1$$

we would obviously select the first family $y = \sqrt{c^2 - x^2}$ to evaluate c. A solution is $y(x) = \sqrt{1 - x^2}$.

b. Consider the differential equation

$$Dy = \frac{y + x}{y - x}, \qquad \mathscr{D} = \{(x, y): y \neq x\}$$

Using the methods of the next section, it can be shown that a one-parameter family of solutions is:

$$(x^2 + y^2)^{1/2} = c \cdot \exp\left[\tan^{-1}\left(\frac{y}{x}\right) \right]$$

In contrast to part (a) of this example, where the implicit relation could be solved for y, this relation cannot be solved for y explicitly as a function of x, although, in theory, y can be regarded as a function of x in appropriately determined regions of the plane.

As seen in these examples, solutions of nonlinear equations occur in the form of implicit functions. Also, Example 12 illustrates that a one-parameter family of solutions of a nonlinear equation need not contain all solutions of the equation. In contrast, all solutions of the linear equation (2) are given by the explicit one-parameter family of functions in (9).

We conclude this section with some remarks concerning the form of the first-order equations under consideration and the solution methods which will be developed. In Sections 44 and 45, it will be convenient to use differential notation and write equation (3) in the form

(15) $P(x, y) \, dx + Q(x, y) \, dy = \varnothing$

An advantage of the differential form (15) over the forms (3) and (7) is that this notation does not commit one in advance to a choice of independent variable. That is, in (15), we could just as well regard x as an unknown function of y and pass to the differential equation

$$\frac{dx}{dy} = \frac{-Q(x, y)}{P(x, y)}$$

whose domain would be $\mathscr{D} = \{(x, y) : P(x, y) \neq 0\}$. However, for the purpose of developing solution techniques, we can, without loss of generality, regard y as an unknown function of x, and we will continue with this convention throughout the chapter.

Recall that the solution technique associated with the first-order linear equation (2) involved multiplication of the equation by the function

$$h(x) = \exp\left\{\int_a^x p(t)\, dt\right\}$$

Since h is continuous and nonzero on the domain of equation (2), this multiplication results in a new equation with exactly the same set of solutions, that is, an equivalent equation. In the case of nonlinear equations, some of the solution techniques involve multiplication of the equation by a function $u = u(x, y)$, and this multiplication will often have the effect of either adding solutions or deleting solutions, or even both. In particular, the curves $\{(x, y) : u(x, y) = 0\}$, which are not solutions of equation (15), are "added" solutions of

(16) $u(x, y)P(x, y)\, dx + u(x, y)Q(x, y)\, dx = \varnothing$

On the other hand, the set of points in $\mathscr{D}$, the domain of equation (15), at which $u(x, y)$ is not defined may determine solutions of (15), and these would be "lost" in solving (16). Solutions which are lost because of algebraic manipulations are called *suppressed solutions*. An examination of the steps used in solving an equation will usually reveal added or suppressed solutions.

17 EXERCISES

1. Verify that the given function (or family of functions), defined implicitly, is a solution of the corresponding differential equation.
   a. $(x - c)^2 + y^2 = c^2$; $x^2 - y^2 + 2xy\, Dy = \varnothing$
   b. $x^2 y + (y^3/3) = c$; $2xy\, dx + (y^2 + x^2)\, dy = \varnothing$
   c. $y^2 = cx - 1$; $2xy\, Dy = 1 + y^2$
   d. $x^2 + \ln(y^2 - 1) = 5$; $y\, Dy + xy^2 - x = \varnothing$
   e. $y^2 + 2e^{-x}\sin(y) = 4$; $e^{-x}\sin(y)\, dx - [y + e^{-x}\cos(y)]\, dy = \varnothing$
   f. $y = \sin^{-1}(xy)$; $x\, Dy + y = Dy(1 - x^2 y^2)^{1/2}$
   g. $\sin(x)\cos(y) = c$; $\cos(x)\cos(y)\, dx - \sin(x)\sin(y)\, dy = \varnothing$
   h. $y + x = \ln[cx(y + 1)]$; $xy\, dy = (y + 1)(1 - x)\, dx$
   i. $y = x\sin(y + c)$; $x\, dy - y\, dx = x(x^2 - y^2)^{1/2}\, dy$
   j. $(x/y) + \ln(y^3/x^2) = 10$; $xy - 2y^2 - (x^2 - 3xy)\, Dy = \varnothing$

2. Consider the differential equation

$$y = x\,Dy - 2[Dy]^2$$

   a. Show that every member of the one-parameter family $y(x) = cx - 2c^2$ is a solution of the equation.

   b. Show that the function $y(x) = x^2/8$ is a solution of the equation which is not contained in the one-parameter family in (a). Such solutions are often called *singular solutions*.

   c. Fix any point x_0. Show that there is a member of the one-parameter family in (a) which is tangent to the singular solution $y(x) = x^2/8$ at the point where $x = x_0$.

   d. Sketch the singular solution and several members of the one-parameter family on the same graph.

3. The differential equation in Problem 2 is a special case of the general class of first-order equations

(A) $y = x\,Dy + f(Dy)$

where f is a given function (e.g., $f(t) = -2t^2$ in Problem 2). An equation of the form (A) is known as a Clairaut equation.

   a. Let $p = Dy$ in (A) and differentiate with respect to x. Show that this yields the equation

(B) $$\dfrac{dp}{dx}\left[x + \dfrac{df}{dp}\right] = \varnothing$$

which is equivalent to the two equations

(C) $$\dfrac{dp}{dx} = \varnothing$$

and

(D) $$x + \dfrac{df}{dp} = \varnothing$$

   b. From (C), $p = c$, where c is an arbitrary constant. Show that substituting $p = Dy = c$ in (A) yields the one-parameter family of solutions

$$y = cx + f(c)$$

   c. A singular solution of (A) can be obtained by using equations (A) and (D) to eliminate p (if possible). Verify this in the case of the differential equation in Problem 2.

4. Find the one-parameter family of solutions and the singular solutions (if any) of each of the following Clairaut equations.

   a. $y = x\,Dy + [Dy]^2$ b. $y = x\,Dy + (1/Dy)$

   c. $y = x\,Dy + ([Dy]^2 - 1)^{1/2}$ d. $y = x\,Dy + (4 + [Dy]^2)^{1/2}$

5. Consider the general Clairaut equation

$$y = x\,Dy + f(Dy)$$

Prove that the lines forming the one-parameter family of solutions are the tangents to the curve which represents its singular solution provided $d^2f/dp^2 \neq 0$, where $p = Dy$.

Section 44 Separable Equations

Consider the first-order equation

(1) $P(x, y)\, dx + Q(x, y)\, dy = \varnothing$

If $P(x, y)$ is a function of x only and $Q(x, y)$ is a function of y only, that is, if $P(x, y) = p(x)$ and $Q(x, y) = q(y)$, then (1) actually has the form

(2) $p(x)\, dx + q(y)\, dy = \varnothing$

and the variables are said to be *separated*. This equation can be integrated to yield the one-parameter family

(3) $\displaystyle\int p(x)\, dx + \int q(y)\, dy = c$

This expression determines a one-parameter family of solutions of (2), with y being defined implicitly as a function of x.

4 Example

The differential equation

$$x\, dx + y\, dy = \varnothing$$

appeared in Example 14a, Section 43. By integration, we obtain the equation

$$\frac{x^2}{2} + \frac{y^2}{2} = k$$

where k is an arbitrary constant, or parameter. Since this equation does not represent a curve if $k \le 0$, we assume $k > 0$, and we replace $2k$ by c^2 to obtain the one-parameter family of circles

$$x^2 + y^2 = c^2$$

This preliminary observation leads to the characterization of separable equations and their solution method.

5 DEFINITION

The differential equation (1) is *separable* if the functions P and Q have the forms

$$P(x, y) = p_1(x)p_2(y)$$

$$Q(x, y) = q_1(x)q_2(y)$$

respectively.

___

6 SOLUTION METHOD FOR SEPARABLE EQUATIONS

Suppose (1) is separable. Then (1) has the form

$$p_1(x)p_2(y)\,dx + q_1(x)q_2(y)\,dy = \emptyset$$

Recall that the domain of this equation is the set

$$\mathcal{D} = \{(x, y): q_1(x)q_2(y) \neq 0\}$$

Put $u(x, y) = 1/q_1(x)p_2(y)$ and multiply the equation by u. Of course, u is not defined if $q_1(x)p_2(y) = 0$, and so this step may result in lost solutions. In particular, $q_1(x) \neq 0$, but the curves $p_2(y) = 0$ should be considered for suppressed solutions. Multiplying by u yields the equation

$$\frac{p_1(x)}{q_1(x)}\,dx + \frac{q_2(y)}{p_2(y)}\,dy = \emptyset$$

in which the variables are separated. Thus a one-parameter family of solution is:

(7)
$$\int \frac{p_1(x)}{q_1(x)}\,dx + \int \frac{q_2(y)}{p_2(y)}\,dy = c$$

which, in general, determines y implicitly as a function of x for each value of the parameter c. The curves $p_2(y) = 0$ should be checked for possible solutions, and it should be determined whether any such solutions are included in the one-parameter family (7).

---

8 Example

Consider the differential equation

$$(x^2y + x^2)\,dx + (xy^2 - y^2)\,dy = \emptyset$$

The domain $\mathcal{D} = \{(x, y): y \neq 0 \text{ and } x \neq 1\}$. This equation is separable since

$$P(x, y) = x^2y + x^2 = x^2(y + 1)$$

and

$$Q(x, y) = xy^2 - y^2 = y^2(x - 1)$$

Separating the variables through multiplication by

$$u(x, y) = \frac{1}{(x - 1)(y + 1)}$$

we obtain

$$\frac{x^2}{x - 1}\,dx + \frac{y^2}{y + 1}\,dy = \emptyset$$

or

$$\left(x + 1 + \frac{1}{x - 1}\right) dx + \left(y - 1 + \frac{1}{y + 1}\right) dy = \varnothing$$

Thus the one-parameter family

$$\frac{x^2}{2} + x + \ln|x - 1| + \frac{y^2}{2} - y + \ln|y + 1| = c$$

defines y implicitly as a function of x for each value of the parameter c. By algebraic manipulation and the properties of the log function, the one-parameter family can also be written as

$$(x + 1)^2 + (y - 1)^2 + \ln[(x - 1)^2(y + 1)^2] = k$$

Since $\ln(0)$ is not defined, the function $y = -1$ is a solution of the differential equation which is not included in the one-parameter family of solutions.

9 EXERCISES

Find the one-parameter family of solutions of each of the following.

1. $Dy = y^2 + 1$
2. $xy\, dx + (1 + x^2)\, dy = \varnothing$
3. $(xy - x)\, dx + (xy + y)\, dy = \varnothing$
4. $(x^2 - 1)\, Dy + 2xy^2 = \varnothing$
5. $x \cos x \cos y + \sin y\, Dy = \varnothing$
6. $x(1 + y^2)\, dx + y(1 + x^2)\, dy = \varnothing$
7. $x(1 - y)\, dx + (1 + y^2)(x - 1)\, dy = \varnothing$
8. $e^{(x + y)} \sin x\, dx + (2y + 1)e^{-y^2}\, dy = \varnothing$
9. $(y^2 - 1)x\, dx + xy(1 - x)\, dy = \varnothing$
10. $Dy = e^{x - y}$

Solve these initial-value problems.

11. $(y + 2)\, Dy = \sin(x);\ y(0) = 0$
12. $\cos y\, dx + (1 + e^{-x}) \sin y\, dy = \varnothing;\ y(0) = \pi/4$
13. $2y\, Dy - \sin^2 x = \varnothing;\ y(0) = 1$
14. $Dy = x/(y + x^2 y);\ y(1) = 0$
15. $Dy = y^2 - 1;\ y(0) = 0$
16. $(xy^2 + y^2 + x + 1)\, dx + (y - 1)\, dy = \varnothing;\ y(2) = 0$

We now consider a class of differential equations of the form (1) which has the property that each member of the class can be converted into a separable equation by means of a change of variable.

10 DEFINITION

A function $h = h(x, y)$ is *homogeneous of degree r* if, for every $t > 0$,

$$h(tx, ty) = t^r h(x, y)$$

where r is a real number.

11 Examples

a. The function $h(x, y) = x^2 + y^2 - 2xy$ is homogeneous of degree 2 since

$$h(tx, ty) = (tx)^2 + (ty)^2 - 2(tx)(ty) = t^2[x^2 + y^2 - 2xy]$$
$$= t^2 h(x, y)$$

b. The function $g(x, y) = 2x + ye^{x/y} + \sqrt{x^2 + y^2}$ is homogeneous of degree 1 since

$$g(tx, ty) = 2(tx) + tye^{tx/ty} + \sqrt{(tx)^2 + (ty)^2}$$
$$= t(2x + ye^{x/y}) + \sqrt{t^2(x^2 + y^2)}$$
$$= t(2x + ye^{x/y} + \sqrt{x^2 + y^2}) = tg(x, y)$$

c. The function $f(x, y) = x^2 + xy^{1/2} + 3y$ is not homogeneous since

$$f(tx, ty) = t^2 x^2 + t^{3/2} xy^{1/2} + 3ty \neq t^r f(x, y)$$

for any value of r.

d. The functions $u(x, y) = x - 3y$, $v(x, y) = 2 + \sin(y/x)$, and $w(x, y) = 1/\sqrt{x + y}$ are homogeneous of degree 1, 0, and $-\frac{1}{2}$, respectively.

12 DEFINITION

The differential equation (1) is *homogeneous* if $P(x, y)$ and $Q(x, y)$ are homogeneous functions of the same degree.

Note that this use of the term "homogeneous" in describing a differential equation is much different from its use in discussing linear differential equations. The proper interpretation of the term will always be clear from the context.

13 THEOREM

If (1) is a homogeneous differential equation, then the change of dependent variable defined by

$$y = vx, \qquad dy = v\,dx + x\,dv$$

transforms (1) into a separable equation in x and v.

Proof: Assume equation (1) is homogeneous. Then each of $P(x, y)$ and $Q(x, y)$ is a homogeneous function, and their degrees are equal. Let r be the degree of P and Q. Now introduce the new variable v by means of the equations

$$y = vx, \qquad dy = v\, dx + x\, dv$$

Substituting into (1), we have

$$P(x, vx)\, dx + Q(x, vx)[v\, dx + x\, dv] = \varnothing$$

By using the homogeneity of P and Q, we obtain the equation

$$x^r P(1, v)\, dx + x^r Q(1, v)[v\, dx + x\, dv] = \varnothing$$

which can be written as

(14) $x^r[p_1(v) + vq_1(v)]\, dx + x^{r+1}q_1(v)\, dv = \varnothing$

where $p_1(v) = P(1, v)$ and $q_1(v) = Q(1, v)$ are functions of v only. Clearly, (14) is a separable equation in x and v.

Since a homogeneous equation can be transformed into a separable equation, this transformation also provides a solution method for homogeneous equations.

15 SOLUTION METHOD FOR HOMOGENEOUS EQUATIONS

Suppose (1) is a homogeneous equation, so that P and Q are homogeneous functions of the same degree, say r. Then the change of variable $y = vx$ transforms (1) into (14), which becomes

$$\frac{dx}{x} + \frac{q_1(v)\, dv}{p_1(v) + vq_1(v)} = \varnothing$$

with the variables separated. Integrating, we obtain the one-parameter family

$$\ln|x| + H(v) = c$$

where

$$\frac{dH}{dv} = \frac{q_1(v)}{p_1(v) + vq_1(v)}$$

Finally, replacing v by y/x yields the one-parameter family

(16) $\ln|x| + H\left(\dfrac{y}{x}\right) = c$

of solutions of (1). Of course, the steps in this procedure should be considered for possible suppressed solutions.

---

17 Example

In the equation

$(x^2 + y^2)\, dx - 2xy\, dy = \varnothing$

$P(x, y) = x^2 + y^2$ and $Q(x, y) = -2xy$ are each homogeneous of degree 2. Thus the equation is homogeneous. By letting $y = vx$, $dy = v\, dx + x\, dv$, and substituting into the equation, we get

$(x^2 + x^2 v^2)\, dx - 2x^2 v(v\, dx + x\, dv) = \varnothing$

or

$x^2(1 + v^2)\, dx - 2x^2 v^2\, dx - 2x^3 v\, dv = \varnothing$

Therefore,

$x^2(1 - v^2)\, dx - 2x^3 v\, dv = \varnothing$

which is a separable equation whose domain is:

$\mathscr{D} = \{(x, v) : x \neq 0, v \neq 0\}$

Dividing by $x^3(1 - v^2)$ [assuming $1 - v^2 \neq 0$], we have

$\dfrac{dx}{x} - \dfrac{2v}{1 - v^2}\, dv = \varnothing$

Integration yields the one-parameter family

$\ln|x| + \ln|1 - v^2| = c$

or

$x(1 - v^2) = k$

We note that the curves (straight lines) $v = \pm 1$ are contained in this one-parameter family, and thus no solutions were lost. Finally, replacing v by y/x, we get the one-parameter family of solutions

$x^2 - y^2 = kx$

For $k \neq 0$, this is a family of hyperbolas. The value $k = 0$ yields the pair of straight lines $y = \pm x$. We can solve this implicit relation for y to obtain the two families of functions

$y(x) = \sqrt{x^2 - kx}$ and $y(x) = -\sqrt{x^2 - kx}$

To solve the initial-value problem

$(x^2 + y^2)\, dx - 2xy\, dy = \varnothing; \qquad y(1) = 2$

we select the first family of functions to evaluate k. Using the initial condition, we find

$2 = \sqrt{1 - k}$

which yields $k = -3$. Since the functions

$$g(x, y) = \frac{-P(x, y)}{Q(x, y)} = \frac{-(x^2 + y^2)}{2xy}$$

and $\partial g/\partial y$ are continuous on a closed rectangle containing the point $(1, 2)$, we conclude, from Theorem 10, Section 43, that

$$y(x) = \sqrt{x^2 + 3x}$$

is the unique solution of the initial-value problem.

18 EXERCISES

Which of the following functions are homogeneous and of what degree?

1. $f(x, y) = x^2 + 2xy + y^2$
2. $f(x, y) = \arctan(y/x)[(x^2 + 2y^2)^{-1/2}]$
3. $f(x, y) = (x^2 + 2xy)/(x + y + 7)$
4. $f(x, y) = (x^3 + 4xy^2 + 7x^2y - 9y^3)^{1/4}$

Find a one-parameter family of solutions of the following.

5. $Dy = 2xy/(x^2 - y^2)$
6. $(xe^{y/x} + y)\, dx - x\, dy = \varnothing$
7. $x\, dy = [1 + \ln(y/x)]y\, dx$
8. $Dy = (y/x) + \sin(y/x)$
9. $(y + \sqrt{x^2 - y^2})\, dx - x\, dy = \varnothing$
10. $Dy = (x^3 + y^3)/3xy^2$

Solve the following initial-value problems and comment on the uniqueness of the solutions.

11. $x^2\, dy = (3xy + 2y^2)\, dx; \; y(3) = -2$
12. $(x + y)\, dy = (y - x)\, dx; \; y(1) = 0$
13. $x\sin(y/x)\, dy = [x + y\sin(y/x)]\, dx; \; y(1) = 0$

The remainder of this section can be omitted without loss of continuity.

We have seen that a homogeneous equation can be transformed into a separable equation by means of a prescribed change of variable. In an analogous manner, it might seem reasonable to expect that there are classes of equations of the form (1) which are not homogeneous, but which can be transformed into a homogeneous equation. Of course, solving any such equation will be a two-step procedure: first, the transformation into a homogeneous equation; and second, the transformation of the homogeneous equation into a separable equation. We conclude this section with a brief discussion of two such types of differential equations.

First, consider the differential equation

(19) $(a_1x + b_1y) \, dx + (a_2x + b_2y) \, dy = \emptyset$

Since $P(x, y) = a_1x + b_1y$ and $Q(x, y) = a_2x + b_2y$ are each homogeneous functions of degree 1, (19) is a homogeneous equation. However, the following equation, which is similar in form to (19),

(20) $(a_1x + b_1y + c_1) \, dx + (a_2x + b_2y + c_2) \, dy = \emptyset$

is not homogeneous. The form of the functions $P(x, y) = a_1x + b_1y + c_1$ and $Q(x, y) = a_2x + b_2y + c_2$ in (20) suggests the possibility of a change of variable of the type

(21) $x = u + h \qquad y = v + k$

where h and k are numbers which are to be selected, if possible, so that the constants c_1 and c_2 are eliminated. Geometrically, this amounts to considering the equations $a_1x + b_1y + c_1 = 0$ and $a_2x + b_2y + c_2 = 0$ as a pair of lines in the plane, and translating the coordinate axes to the point of intersection (if there is one). With this idea in mind, there are two cases to consider.

CASE I.   The lines have a unique point of intersection (i.e., $a_1b_2 - a_2b_1 \neq 0$). Let the ordered pair (h, k) be the unique solution of the pair of equations

$$a_1x + b_1y + c_1 = 0$$
$$a_2x + b_2y + c_2 = 0$$

and define the new variables u and v by means of the equations in (21). Then $dx = du$, $dy = dv$, and substituting into (20) yields

$$\left[a_1(u + h) + b_1(v + k) + c_1\right] du$$
$$+ \left[a_2(u + h) + b_2(v + k) + c_2\right] dv = \emptyset$$

which simplifies to

(22) $(a_1u + b_1v) \, du + (a_2u + b_2v) \, dv = \emptyset$

Equation (22) can now be solved using the solution method (15).

23 Example

Consider the differential equation

$$(x - y + 3) \, dx + (x + 2y - 3) \, dy = \emptyset$$

The lines $x - y + 3 = 0$ and $x + 2y - 3 = 0$ have slopes 1 and $-\frac{1}{2}$, respectively. Their unique point of intersection is $(h, k) = (-1, 2)$. Defining u and v by

$$x = u - 1, \qquad y = v + 2$$

we obtain the equation

$$(u - v)\,du + (u + 2v)\,dv = \varnothing$$

which is homogeneous in u and v. Introducing the new variable z by means of the equations

$$v = zu, \qquad dv = u\,dz + z\,du$$

we have

$$(u - zu)\,du + (u + 2zu)(u\,dz + z\,du) = \varnothing$$

Separating the variables leads to

$$\frac{du}{u} + \frac{1 + 2z}{1 + 2z^2}\,dz = \varnothing$$

which has the one-parameter family of solutions

$$\ln|u| + \tfrac{1}{2}\ln(1 + 2z^2) + \frac{1}{\sqrt{2}}\tan^{-1}\sqrt{2}\,z = c$$

In terms of the original variables x and y, this one-parameter family is:

$$\ln|x + 1| + \tfrac{1}{2}\ln\left[1 + 2\left(\frac{y - 2}{x + 1}\right)^2\right] + \frac{1}{\sqrt{2}}\tan^{-1}\sqrt{2}\left(\frac{y - 2}{x + 1}\right) = c$$

CASE II.   The lines do not have a unique point of intersection (i.e., $a_1b_2 - a_2b_1 = 0$). The assumption $a_1b_2 - b_2a_1 = 0$ implies that there exists a non-zero constant k such that $a_2x + b_2y = k(a_1x + b_1y)$. Thus in this case, (20) can be written as

(24) $[a_1x + b_1y + c_1]\,dx + [k(a_1x + b_1y) + c_2]\,dy = \varnothing$

Introduce a new variable v by means of the equation $v = a_1x + b_1y$. Then $dv = a_1\,dx + b_1\,dy$, so that

$$dy = \frac{1}{b_1}(dv - a_1\,dx)$$

provided $b_1 \ne 0$. Note that if $b_1 = 0$, then (24) has a particularly simple form; it is, for example, separable. Assuming, therefore, that $b_1 \ne 0$, the change of variable produces the equation

$$(v + c_1)\,dx + (kv + c_2)\left(\frac{1}{b_1}\,dv - \frac{a_1}{b_1}\,dx\right) = \varnothing$$

or

$$\left[v + c_1 - \frac{a_1}{b_1}v - \frac{c_2 a_1}{b_1}\right]dx + \frac{1}{b_1}(kv + c_2)\,dv = \varnothing$$

which is a separable equation in x and v.

25 Example

Consider the differential equation

$$(x + y)\, dx + (2x + 2y + 3)\, dy = \varnothing$$

Here the lines $x + y = 0$ and $2x + 2y + 3 = 0$ are parallel. Put $v = x + y$. Then $dv = dx + dy$ and $dy = dv - dx$. Substituting into the equation yields

$$v\, dx + (2v + 3)(dv - dx) = \varnothing$$

or

$$-(v + 3)\, dx + (2v + 3)\, dv = \varnothing$$

which is separable. Separating the variables, we obtain

$$dx - \frac{2v + 3}{v + 3}\, dv = \varnothing$$

or

$$dx - \left(2 - \frac{3}{v + 3}\right) dv = \varnothing$$

This equation has the one-parameter family of solutions

$$x - 2v + 3\ln|v + 3| = c$$

In terms of x and y, we have

$$\ln|x + y + 3|^3 - x - 2y = c$$

When each of $P(x, y)$ and $Q(x, y)$ in equation (1) is a linear combination of terms of the form $x^r y^s$, where r and s are real numbers, it is sometimes possible to transform the equation into a homogeneous equation by means of the change of variable

(26) $y = z^k, \qquad dy = kz^{k-1}\, dz$

Here, k is a real number which is to be determined, if possible, such that the transformed equation is homogeneous. We illustrate the technique with an example.

27 Example

Consider the differential equation

$$(2x - y^2)\, dx + (y^3 + 2xy)\, dy = \varnothing$$

This equation is neither separable nor homogeneous. Introducing the new variable z by means of the equations in (26), we have that

$$(2x - z^{2k})\, dx + (z^{3k} + 2xz^k)kx^{k-1}\, dz = \varnothing$$

or

$$(2x - z^{2k})\, dx + k(z^{4k-1} + 2xz^{2k-1})\, dz = \varnothing$$

For this equation to be homogeneous, $P(x, z) = 2x - z^{2k}$ and $Q(x, z) = z^{4k-1} + 2xz^{2k-1}$ must be homogeneous functions, and their degrees must be equal. Now

$$P(tx, tz) = t2x - t^{2k}z^{2k}$$

and we conclude that $2k = 1$, or $k = \frac{1}{2}$, in order for $P(x, z)$ to be homogeneous. Similarly,

$$Q(tx, tz) = t^{4k-1}z^{4k-1} + t^{2k}2xz^{2k-1}$$

so that $4k - 1 = 2k$, or $k = \frac{1}{2}$, for $Q(x, z)$ to be homogeneous. Thus the change of variable $y = z^{1/2}$ transforms the equation into

$$(2x - z)\, dx + (z^{3/2} + 2xz^{1/2})\tfrac{1}{2}z^{-1/2}\, dz = \varnothing$$

or

$$2(2x - z)\, dx + (z + 2x)\, dz = \varnothing$$

which is homogeneous in x and z. Following the methods of this section, it can be verified that a one-parameter family of solutions is:

$$\ln|x| + \tfrac{1}{2}\ln\left(4 + \frac{y^2}{x}\right) + \tan^{-1}\left(\frac{y^2}{2x}\right) = c$$

28 EXERCISES

Find a one-parameter family of solutions of each of the following.

1. a. $(2x + y + 2)\, dy + (2y - x - 6)\, dx = \varnothing$
    b. $(x + y - 1)\, dx + (x - y - 3)\, dy = \varnothing$
    c. $(3x + 2y + 1)\, dx = (3x + 2y - 1)\, dy$
    d. $Dy = (x + y + 1)/(-2x - 2y + 1)$
    e. $Dy = [(x - y + 3)/(1 - y + x)]^2$ (Hint: Try the technique used on the above problems.)

2. Investigate solution methods for equations of the form

$$Dy = F\left[\frac{ax + by + c}{dx + ey + f}\right]$$

3. Which of the following equations can be transformed into homogeneous equations by the change of variable (26)?
    a. $(7x^{-3}y^3 - 4x^3)\, dx = (2y + 3x^2)\, dy$
    b. $[x^2y^2 - 2(x^4/y^2)]\, dy = \sqrt{x}\, dx$
    c. $[(1/x) + (2/x^2y)]\, dy + [(y^4/x) - x^2y]\, dx = \varnothing$

4. Find a one-parameter family of solutions of each of the following equations.
    a. $(x^4y^2 - 3y)\, dx + x\, dy = \varnothing$ b. $2xy\, dx + (2x^2 - 3y)\, dy = \varnothing$

Section 45 Exact Equations

The treatment of exact differential equations requires some basic facts about functions of several variables. We state without proof the necessary properties here. If $G = G(x, y)$ is a function of two variables defined on the region R, and if the first partial derivatives of G exist throughout R, then the expression

(1) $\qquad dG = \dfrac{\partial G}{\partial x}\, dx + \dfrac{\partial G}{\partial y}\, dy$

is called the *total differential* of G. Since $\partial G/\partial x$ and $\partial G/\partial y$ are themselves functions of x and y, (1) can also be written in the form

(2) $\qquad dG = P(x, y)\, dx + Q(x, y)\, dy$

where $P(x, y) = \partial G/\partial x$ and $Q(x, y) = \partial G/\partial y$. If G has first partial derivatives on R, and if the mixed second partial derivatives of G, $\partial^2 G/\partial y\, \partial x$ and $\partial^2 G/\partial x\, \partial y$, are continuous on R, then it can be shown that

(3) $\qquad \dfrac{\partial^2 G}{\partial y\, \partial x} = \dfrac{\partial^2 G}{\partial x\, \partial y}$

Finally, if $G(x, y)$ and $\partial G/\partial y$ are continuous on R, then

$$\frac{\partial\left[\int_{x_0}^{x} G(t, y)\, dt\right]}{\partial y} = \int_{x_0}^{x} \frac{\partial G(t, y)}{\partial y}\, dx$$

Now, suppose $G = G(x, y)$ is a function of two variables, and suppose G has continuous first partial derivatives, $\partial G/\partial x = P(x, y)$ and $\partial G/\partial y = Q(x, y)$, on a region R. Then the first-order differential equation

$$P(x, y)\, dx + Q(x, y)\, dy = \varnothing$$

which is the same as

$$\left(\frac{\partial G}{\partial x}\right) dx + \left(\frac{\partial G}{\partial y}\right) dy = dG = \varnothing$$

has the one-parameter family of solutions $G(x, y) = c$.

This observation leads to the characterization of another class of equations of the form

(4) $\qquad P(x, y)\, dx + Q(x, y)\, dy = \varnothing$

5 DEFINITION

The differential equation (4) on the domain $\mathcal{D}$ is *exact* if and only if there is a function $G = G(x, y)$ defined on a region $R \supseteq \mathcal{D}$ such that $\partial G/\partial x = P(x, y)$ and $\partial G/\partial y = Q(x, y)$, where $Q(x, y) \neq 0$ for all $(x, y) \in \mathcal{D}$.

The following theorem gives a method for testing for exactness, and its proof provides a solution technique for exact equations.

6 THEOREM

Let the functions $P(x, y)$ and $Q(x, y)$ in equation (4) have continuous first partial derivatives on $\mathscr{D}$. Then (4) is exact if and only if

(A) $$\frac{\partial P}{\partial y} = \frac{\partial Q}{\partial x}$$

on $\mathscr{D}$.

Proof: Suppose (4) is exact. Then, according to the definition, there is a function $G = G(x, y)$ such that $\partial G/\partial x = P$ and $\partial G/\partial y = Q$. Since P and Q have continuous first derivatives, G has continuous second derivatives, and so

$$\frac{\partial^2 G}{\partial y\, \partial x} = \frac{\partial^2 G}{\partial x\, \partial y}$$

But $\partial^2 G/\partial y\, \partial x = \partial P/\partial y$ and $\partial^2 G/\partial x\, \partial y = \partial Q/\partial x$. Thus (A) holds if (4) is exact.

Suppose now that (A) holds. To show that (4) is exact, we must construct a function $G = G(x, y)$ such that $\partial G/\partial x = P$ and $\partial G/\partial y = Q$. Choose any point $(x_0, y_0) \in \mathscr{D}$ and integrate P with respect to x to obtain

$$G(x, y) = \int_{x_0}^{x} P(t, y)\, dt + g(y)$$

where $g = g(y)$ is an unknown function of y which is to be determined so that $\partial G/\partial y = Q$. Clearly, $\partial G/\partial x = P$. Calculating the derivative of G with respect to y, we have

$$\frac{\partial G}{\partial y} = \frac{\partial \int_{x_0}^{x} P(t, y)\, dt}{\partial y} + Dg(y)$$

$$= \int_{x_0}^{x} \frac{\partial P(t, y)}{\partial y}\, dt + Dg(y)$$

By setting $\partial G/\partial y = Q$ and solving for $Dg(y)$, we get

$$Dg(y) = Q(x, y) - \int_{x_0}^{x} \frac{\partial P(t, y)\, dt}{\partial y}$$

Now, from (A), $\partial P/\partial y = \partial Q/\partial x$. Thus

$$Dg(y) = Q(x, y) - \int_{x_0}^{x} \frac{\partial Q(t, y)}{\partial t}\, dt$$

$$= Q(x, y) - [Q(x, y) - Q(x_0, y)]$$

$$= Q(x_0, y)$$

Therefore,

$$g(y) = \int_{y_0}^{y} Q(x_0, s)\, ds$$

and

(7) $$G(x, y) = \int_{x_0}^{x} P(t, y)\, dt + \int_{y_0}^{y} Q(x_0, s)\, ds$$

By our construction of G, $\partial G/\partial x = P$ and $\partial G/\partial y = Q$, so that (4) is exact if (A) holds. This completes the proof of the theorem.

_____

Some remarks concerning the theorem and its proof are appropriate here. In the second part of the proof, we selected, arbitrarily, a point $(x_0, y_0) \in \mathcal{D}$, integrated P with respect to x, added a function $g = g(y)$, and called the sum $G = G(x, y)$. The function g was to be determined so that $\partial G/\partial y = Q$. Because of symmetry, it should be clear that we could have integrated Q with respect to y, added a function $h = h(x)$, and called that sum $G = G(x, y)$, where h is to be determined so that $\partial G/\partial x = P$. By proceeding in this manner, the result is:

(8) $$G(x, y) = \int_{y_0}^{y} Q(x, s)\, ds + \int_{x_0}^{x} P(t, y_0)\, dt$$

This alternative procedure will be illustrated below. If the differential equation (4) is exact, then the equation $G(x, y) = c$, where G is either of the expressions (7) or (8), is a one-parameter family of solutions which determines y implicitly as a function of x for each value of the parameter c.

9 Examples

a. Consider the differential equation

$$(e^{-x} + 2ye^{2x} + 2x \cos y)\, dx + \left(2y - \frac{1}{y} + e^{2x} - x^2 \sin y\right) dy = \varnothing$$

Here,

$$P(x, y) = e^{-x} + 2ye^{2x} + 2x \cos y$$

and

$$Q(x, y) = 2y - \frac{1}{y} + e^{2x} - x^2 \sin y$$

and calculating the appropriate partial derivatives gives

$$\frac{\partial P}{\partial y} = 2e^{2x} - 2x \sin y = \frac{\partial Q}{\partial x}$$

Thus the equation is exact. To find G, either of the expressions (7) or (8) could be used. However, the following technique, based upon the second part of the proof of Theorem 6, is usually simpler. We know that there is a function $G = G(x, y)$ such that

$$\frac{\partial G}{\partial x} = e^{-x} + 2ye^{2x} + 2x \cos y$$

$$\frac{\partial G}{\partial y} = 2y - \frac{1}{y} + e^{2x} - x^2 \sin x$$

Thus, from the first of these equations, we set

$$G(x, y) = \int \left[e^{-x} + 2ye^{2x} + 2x \cos y \right] dx + g(y)$$

or

(A) $G(x, y) = -e^{-x} + ye^{2x} + x^2 \cos y + g(y)$

where $g = g(y)$ is a function of y which is to be determined so that $\partial G/\partial y = Q = 2y - (1/y) + e^{2x} - x^2 \sin y$. Now, from (A),

$$\frac{\partial G}{\partial y} = e^{2x} - x^2 \sin y + Dg(y)$$

and setting this equal to Q, we have

$$e^{2x} - x^2 \sin y + Dg(y) = 2y - \frac{1}{y} + e^{2x} - x^2 \sin y$$

Thus

$$Dg(y) = 2y - \frac{1}{y}$$

and

$$g(y) = y^2 - \ln|y|$$

The constant of integration here is omitted since we are seeking only one function $g = g(y)$ such that $\partial G/\partial y = Q$, not all functions with this property. It now follows that

$$G(x, y) = -e^{-x} + ye^{2x} + x^2 \cos y + y^2 - \ln|y|$$

and

$$-e^{-x} + ye^{2x} + x^2 \cos y + y^2 - \ln|y| = c$$

is a one-parameter family of solutions defining y implicitly as a function of x.

b. Using the alternative procedure suggested in the remarks following the proof of the theorem, we have

$$G(x, y) = \int \left[2y - \frac{1}{y} + e^{2x} - x^2 \sin y \right] dy + h(x)$$

$$= y^2 - \ln|y| + ye^{2x} + x^2 \cos y + h(x)$$

where $h = h(x)$ is to be determined so that $\partial G/\partial x = P$. By calculating $\partial G/\partial x$ and setting this derivative equal to P, we obtain

$$2ye^{2x} + 2x \cos y + Dh(x) = e^{-x} + 2ye^{2x} + 2x \cos y$$

It follows that

$$Dh(x) = e^{-x}$$

The obvious choice is $h(x) = -e^{-x}$. Thus

$$G(x, y) = y^2 - \ln|y| + ye^{2x} + x^2 \cos y - e^{-x}$$

and

$$y^2 - \ln|y| + ye^{2x} + x^2 \cos y - e^{-x} = c$$

is a one-parameter family of solutions, which agrees with our result in part (a).

10 EXERCISES

Find a one-parameter family of solutions of the following.

1. $\cos(x + y^2)\,dx + 2y\cos(x + y^2)\,dy = \varnothing$
2. $(\cos 2y - 3x^2y^2)\,dx + (\cos 2y - 2x\sin 2y - 2x^3y)\,dy = \varnothing$
3. $[(x + y)^{-1} + y^2]dx + [(x + y)^{-1} + 2xy]\,dy = \varnothing$
4. $e^{-x}\sin y\,dx - (y + e^{-x}\cos y)\,dy = \varnothing$
5. $(x + y\cos x)\,dx + \sin x\,dy = \varnothing$
6. $[(y/x) + 6x)]dx + (\ln(x) - 2)\,dy = \varnothing$
7. $(3y^2 + y\sin 2xy)\,dx + (6xy + x\sin 2xy + 3y^2)\,dy = \varnothing$
8. $(\cos 2y - 3x^2y^2)\,dx + (y + e^y - 2x\sin 2y - 2x^3y)\,dy = \varnothing$
9. $(ye^x + 2x\cos y)\,dx + (e^x - x^2\sin y)\,dy = \varnothing$
10. $(2\cos y - y\cos x + e^x)\,dx + (\sin x - 2x\sin y + y^2 - 1)\,dy = \varnothing$
11. $[(e^x + \ln(y) + (y/x)]\,dx + [(x/y) + \ln(x) + \sin y]\,dy = \varnothing$

Find a solution of each of the following initial-value problems.

12. $[2xy - \sin(x - 3)]\,dx = (2y - x^2)\,dy; \, y(3) = 1$
13. $[x(x^2 + 2y)^{-1/2} + \ln(1 + y)]\,dx + [(x^2 + 2y)^{-1/2} + x(1 + y)^{-1}]\,dy = \varnothing;$ $y(-2) = 0$
14. $(ye^{xy} + 2x)\,dx + (xe^{xy} - 2y)\,dy = \varnothing; \, y(0) = 2$
15. $\cosh(x - y^2)\,dx = 2y\cosh(x - y^2)\,dy; \, y(2) = \sqrt{2}$
16. $(1 + y^2 + xy^2)\,dx + (x^2y + y + 2xy)\,dy = \varnothing; \, y(1) = 1$
17. Show that any differential equation in which the variables are separated is exact. See equation (2), Section 44.
18. Can you find a function $f(x)$ and a function $g(y)$ such that the equation

$$g(y)\sin x\,dx + y^2f(x)\,dy = \varnothing$$

is exact?

19. Determine all functions $g(y)$ such that the differential equation

$$g(y)e^y \, dx + xy \, dy = \varnothing$$

is exact.

Continuing with the discussion of exact differential equations, it often happens that

(11) $P(x, y) \, dx + Q(x, y) \, dx = \varnothing$

is not exact, but a function $\mu = \mu(x, y)$ can be determined so that the equation

(12) $\mu(x, y)P(x, y) \, dx + \mu(x, y)Q(x, y) \, dy = \varnothing$

is exact. For all $(x, y) \in \mathscr{D}$ such that $\mu(x, y) \neq 0$, equations (11) and (12) are equivalent in the sense that they have the same set of solutions.

13 DEFINITION

A function $\mu = \mu(x, y)$ is an *integrating factor* for the equation (11) if the differential equation (12) is exact.

In the discussion of separable equations

$$p_1(x)p_2(y) \, dx + q_1(x)q_2(y) \, dy = \varnothing$$

the solution method consisted of multiplying the equation by $\mu(x, y) = [q_1(x)p_2(y)]^{-1}$. It is easy to see that the resulting equation

$$\frac{p_1(x)}{q_1(x)} \, dx + \frac{q_2(y)}{p_2(y)} \, dy = \varnothing$$

is exact. Thus μ is an integrating factor for the separable equation. As another example, the function $\mu(x) = \exp\{\int_a^x p(t) \, dt\}$ is an integrating factor for the linear equation

$$Dy + p(x)y = f(x)$$

In (11), assume that the functions P and Q have continuous first derivatives and that $\mu = \mu(x, y)$ is an integrating factor. Assume also that μ has continuous first derivatives. Then the equation

$$\mu(x, y)P(x, y) \, dx + \mu(x, y)Q(x, y) \, dy = \varnothing$$

is exact, and by Theorem 6,

$$\frac{\partial[\mu P]}{\partial y} = \frac{\partial[\mu Q]}{\partial x}$$

which implies

$$\mu \frac{\partial P}{\partial y} + P \frac{\partial u}{\partial y} = \mu \frac{\partial Q}{\partial x} + Q \frac{\partial u}{\partial x}$$

or

(14) $\quad \mu\left(\dfrac{\partial P}{\partial y} - \dfrac{\partial Q}{\partial x}\right) + P\dfrac{\partial \mu}{\partial y} - Q\dfrac{\partial \mu}{\partial x} = \varnothing$

Thus it follows that an integrating factor for (11) must satisfy the first-order partial differential equation (14). Unfortunately, (14) is usually at least as difficult to analyze as the original equation, so that, in practice, integrating factors are difficult to find. The following theorem specifies two important special cases in which an integrating factor for (11) can be determined.

15 THEOREM

Given the differential equation (11). If the quotient

$$\frac{\dfrac{\partial P}{\partial y} - \dfrac{\partial Q}{\partial x}}{Q} = v$$

is a function of x only, then

$$\mu(x) = \exp\left[\int_{x_0}^{x} v(t)\, dt\right]$$

is an integrating factor. If the quotient

$$\frac{\dfrac{\partial P}{\partial y} - \dfrac{\partial Q}{\partial x}}{P} = w$$

is a function of y only, then the function

$$\mu(y) = \exp\left[-\int_{y_0}^{y} w(t)\, dt\right]$$

is an integrating factor.

Proof: We prove the first part of the theorem and leave the second part as an exercise. Assume that

$$\frac{\dfrac{\partial P}{\partial y} - \dfrac{\partial Q}{\partial x}}{Q} = v(x)$$

is a function of x only, and multiply (12) by $\mu(x) = \exp\{\int_{x_0}^{x} v(t)\, dt\}$ to obtain

(B) $\quad \mu(x)P(x, y)\, dx + \mu(x)Q(x, y)\, dy = \varnothing$

Checking for exactness, we have

$$\frac{\partial[\mu P]}{\partial y} = \mu\frac{\partial P}{\partial y}$$

and

$$\frac{\partial[\mu Q]}{\partial x} = v\mu Q + \mu \frac{\partial Q}{\partial x}$$

$$= \frac{\dfrac{\partial P}{\partial y} - \dfrac{\partial Q}{\partial x}}{Q} \mu Q + \mu \frac{\partial Q}{\partial x}$$

$$= \mu \frac{\partial P}{\partial y}$$

Thus (B) is exact.

---

16 Example

Consider the equation

$$2xy \, dx + (y^2 - 3x^2) \, dy = \varnothing$$

on $\mathscr{D} = \{(x, y) : y \neq \pm\sqrt{3}x\}$. Since $\partial P/\partial y = 2x$ and $\partial Q/\partial x = -6x$, the equation is not exact. However,

$$\frac{\dfrac{\partial P}{\partial y} - \dfrac{\partial Q}{\partial x}}{P} = \frac{2x - (-6x)}{2xy} = \frac{4}{y}$$

so that, by Theorem 15, $\mu(y) = \exp\{-\int (4/y) \, dy\}$ is an integrating factor. Multiplying the equation by $\mu(y) = y^{-4}$, assuming $y \neq \varnothing$, we have

$$2xy^{-3} \, dx + (y^{-2} - 3x^2 y^{-4}) \, dy = \varnothing$$

which is exact. To solve this equation, put

$$G(x, y) = \int 2xy^{-3} \, dx + g(y)$$

$$= x^2 y^{-3} + g(y)$$

Now $\partial G/\partial y = -3x^2 y^{-4} + Dg(y) = y^{-2} - 3x^2 y^4$. Therefore,

$$Dg(y) = y^{-2}$$

and

$$g(y) = -y^{-1}$$

Thus

$$G(x, y) = x^2 y^{-3} - y^{-1}$$

and

$$x^2 y^{-3} - y^{-1} = c$$

or

$$x^2 - y^2 = cy^3$$

is a one-parameter family of solutions. Note that the function $y = \varnothing$ is a solution of the equation which is not contained in this one-parameter family.

17 EXERCISES

1. Find a one-parameter family of solutions of each of the following differential equations.

   a. $4x^3 y\, dx - (x^4 + y^2)\, dy = \varnothing$

   b. $(2y + 3x)\, dx + (x - 18yx^{-1})\, dy = \varnothing$

   c. $(4x^4 y^3 + 1)\, dx + (3x^5 y^2 - xy^{-1})\, dy = \varnothing$

   d. $(2xy^4 e^y + 2xy^3 + y)\, dx + (x^2 y^4 e^y - x^2 y^2 - 3x)\, dy = \varnothing$

   e. $xy\, dx + (x^2 + y^2 + y)\, dy = \varnothing$

   f. $(2xy + 1)\, dx + x^2\, dy = \varnothing$

2. Find a solution of the initial-value problem

   $$[y^{-3} \cos(x - y) + y]\, dx + [4x - y^{-3} \cos(x - y)]\, dy = \varnothing; \qquad y(1) = 1$$

3. Show that $\mu(x, y) = 1/2xy(1 - xy)$ is an integrating factor for the equation

   $$(xy^2 + y)\, dx + 3x^2 y\, dy = \varnothing$$

4. Notice that the equation in Problem 3 can be written as

   $$y[xy + 2]\, dx + x(3xy)\, dy = \varnothing$$

   and that the integrating factor is:

   $$\mu(x, y) = \frac{1}{xy[(xy + 2) - 3xy]}$$

   In general, verify that an integrating factor for an equation of the form

   $$yf(x \cdot y)\, dx + xg(x \cdot y)\, dy = \varnothing$$

   is:

   $$\mu(x, y) = \frac{1}{xy[f(x \cdot y) - g(x \cdot y)]}$$

5. Complete the proof of Theorem 15.

6. Solve the equation in Example 16 by the method (15), Section 44.

The remainder of this section can be omitted without loss of continuity.

If each of P and Q in (12) involves only terms of the form $x^r y^s$, then the equation may admit an integrating factor of the form $\mu(x, y) = x^m y^n$. Equations of this special type were also considered at the end of the preceding section, and so the technique illustrated here offers a second possible solution procedure for equations of this form.

18 Example

Consider the equation

$$(2xy + y^3)\, dx + (3x^2 + xy^2)\, dy = \varnothing$$

It is easy to verify that this equation is not exact and that Theorem 15 does not apply. Since the functions P and Q have the suggested form, we multiply the equation by $\mu(x, y) = x^m y^n$, where m and n are to be determined, if possible, so that μ is an integrating factor. We obtain

$$(2x^{m+1}y^{n+1} + x^m y^{n+3})\, dx + (3x^{m+2}y^n + x^{m+1}y^{n+2})\, dy = \varnothing$$

Using Theorem 6 to test for exactness yields

$$2(n + 1)x^{m+1}y^n + (n + 3)x^m y^{n+2}$$
$$= 3(m + 2)x^{m+1}y^n + (m + 1)x^m y^{n+2}$$

which leads to the pair of equations

$$2(n + 1) = 3(m + 2)$$
$$n + 3 = m + 1$$

This pair of equations has the solution $m = -8$, $n = -10$, and so $\mu(x, y) = x^{-8}y^{-10}$ is an integrating factor. Multiplying the given equation by μ yields the exact equation

$$(2x^{-7}y^{-9} + x^{-8}y^{-7})\, dx + (3x^{-6}y^{-10} + x^{-7}y^{-8})\, dy = \varnothing$$

A one-parameter family of solutions is:

$$7x + 3y^2 = cx^7 y^9$$

The function $y = \varnothing$ is also a solution of the equation which is not contained in this one-parameter family. In conclusion, note that if we attempt to transform this equation into a homogeneous equation using the method described at the end of the preceding section, then we set $y = z^k$, $dy = kz^{k-1}\, dz$, and substitute into the differential equation. We get

$$(2xz^k + z^{3k})\, dx + (3x^2 + xz^{2k})kz^{k-1}\, dz = \varnothing$$

and in order for this to be a homogeneous differential equation, k must satisfy the two equations

$$k + 1 = 3k \qquad \text{and} \qquad k + 1 = 3k - 1$$

Clearly, this is impossible, so that method cannot be applied here.

19 EXERCISES

1. Find a one-parameter family of solutions of the differential equation

$$(x^2 y + 2y^3)\, dx - (2x^3 + 3xy^2)\, dy = \varnothing$$

2. Find an integrating factor of the form $\mu(x, y) = x^m y^n$ for

$$(4x^{-4}y^2 - 2x^{-2}y)\,dx + (3x^{-3}y + x^{-1})\,dy = \varnothing$$

3. Find an integrating factor of the form $\mu(x, y) = x^m y^n$ for

$$(3y^{-1} + x^{-3/2}y^{-2})\,dx + (2xy^{-2} - x^{-1/2}y^{-3})\,dy = \varnothing$$

4. For what values of a, b, c, and d (if any) will the equation

$$(3x^a y^b - x^c y^d)\,dx + (2x^2 y^{-1} - x^{-3}y)\,dy = \varnothing$$

have an integrating factor of the form $\mu(x, y) = x^m y^n$?

5. Find a one-parameter family of solutions of the differential equation

$$(4xy + 3y^3)\,dx + (2x^2 + 5xy^3)\,dy = \varnothing$$

Section 46 Equations Related to Linear Equations

The first-order linear differential equation

(1) $Dy + p(x)y = f(x)$

where p and f are continuous on an interval J, was studied in detail in Section 8. We can use the approach taken in this chapter to reestablish the results of that section. Writing (1) in the form

$$P(x, y)\,dx + Q(x, y)\,dy = \varnothing$$

we have

(2) $[p(x)y - f(x)]\,dx + dy = \varnothing$

Since $\partial P/\partial y = p(x)$ and $\partial Q/\partial y = 0$, (2) is not exact. However,

$$\frac{\dfrac{\partial P}{\partial y} - \dfrac{\partial Q}{\partial x}}{Q} = p(x)$$

is a function of x only, so that $\mu(x) = \exp\{\int_{x_0}^x p(t)\,dt\}$, $x_0 \in J$, is an integrating factor. Multiplying (2) by μ and solving as an exact equation yields the one-parameter family of solutions

(3) $$y(x) = \exp\left\{-\int_{x_0}^x p(t)\,dt\right\}\left[c + \int_{x_0}^x f(t)\exp\left\{\int_{x_0}^t p(s)\,ds\right\}dt\right]$$

Thus the integrating factor approach is an example of a technique which is applicable to first-order linear equations and which can be extended to include a larger class of first-order differential equations.

In this section, we consider certain classes of first-order equations which are related to linear equations either by a prescribed transformation or by a solution method which has been developed for the linear case.

The first such class of equations is characterized by the following definition.

4 DEFINITION

A *Bernoulli equation* is a first-order differential equation which has the form

(5) $Dy + p(x)y = f(x)y^r$

where p and f are continuous functions on an interval J and r is a real number, $r \neq 0, 1$.

The reason for the restrictions $r \neq 0$ and $r \neq 1$ is that in each of these cases, (5) reduces to a first-order linear equation.

6 THEOREM

The change of dependent variable defined by the equations

$$v = y^{1-r}, \qquad Dv = (1-r)y^{-r}\,Dy$$

transforms (5) into a linear equation in x and v.

Proof: Multiply (5) by y^{-r} to obtain

$$y^{-r}\,Dy + p(x)y^{1-r} = f(x)$$

Introducing the new dependent variable v as specified in the hypothesis yields the equation

$$\frac{1}{1-r}\,Dv + p(x)v = f(x)$$

or

$$Dv + (1-r)p(x)v = (1-r)f(x)$$

which is a linear equation in x and v.

The fact that a Bernoulli equation can be transformed into a linear equation provides a solution method for the class of Bernoulli equations.

7 Example

The differential equation

$$Dy + \frac{1}{x-2}\,y = 5(x-2)y^{1/2}$$

is a Bernoulli equation, with $r = \frac{1}{2}$. Multiply by $y^{-1/2}$, $y \neq \varnothing$, to obtain

(A) $y^{-1/2}\,Dy + \dfrac{1}{x-2}\,y^{1/2} = 5(x-2)$

Define v by the equations $v = y^{1/2}$, $Dv = \frac{1}{2}y^{-1/2}\,Dy$, and substitute into (A). We have

$$2\,Dv + \frac{1}{x-2}\,v = 5(x-2)$$

or

$$Dv + \frac{1}{2(x - 2)} \, v = \tfrac{5}{2}(x - 2)$$

The one-parameter family of solutions of this equation is:

$$v(x) = (x - 2)^2 + c(x - 2)^{-1/2}$$

To determine y as a function of x, we have

$$[y(x)]^{1/2} = (x - 2)^2 + c(x - 2)^{-1/2}$$

or

$$y(x) = [(x - 2)^2 + c(x - 2)^{-1/2}]^2$$

as a one-parameter family of solutions. Note that $y = \varnothing$ is a solution of the equation which is not contained in the one-parameter family.

8 EXERCISES

1. Find a one-parameter family of solutions of each of the following differential equations.
   a. $Dy + xy = xy^3$
   b. $Dy + y^2(x^2 + x + 1) = y$
   c. $Dy = 4y + 2e^x\sqrt{y}$
   d. $2xy \, Dy = 1 + y^2$
   e. $(x - 2) Dy + y = 5(x - 2)^2 y^{1/2}$
   f. $y \, Dy - xy^2 + x = \varnothing$
2. Find a solution of each of the following initial-value problems.
   a. $Dy + xy - y^3 e^{x^2} = \varnothing$; $y(0) = \tfrac{1}{2}$
   b. $x \, Dy + y - y^2 \ln x = \varnothing$; $y(1) = 1$
   c. $2x^3 \, Dy = y(y^2 + 3x^2)$; $y(1) = 1$
   d. $Dy + y \tan x - y^2 \sec^3 x = \varnothing$; $y(0) = 3$
3. Show that the change of variable $u = \ln y$ transforms the equation

   $$Dy - \left(\frac{y}{x}\right) \ln y = xy$$

   into a linear equation. Find a one-parameter family of solutions.
4. Show that the change of variable indicated in Problem 3 transforms

   $$Dy + yf(x) \ln y = g(x)y$$

   into a first-order linear equation.
5. Can you determine a change of variable which will transform

   $$\cos y \, Dy + g(x) \sin y = f(x)$$

   into a linear equation?

We now consider another class of first-order differential equations, the members of which are closely related to linear equations.

9 DEFINITION

A *Riccati equation* is a first-order differential equation having the form

(10) $Dy + p(x)y^2 + q(x)y + r(x) = \varnothing$

where p, q, and r are continuous functions on an interval J.

There is no general solution method for the Riccati equation. However, if one solution, $y = y_1(x)$, of the equation is known, then, using y_1, a one-parameter family of solutions for the equation can be constructed.

11 THEOREM

If $y = y_1(x)$ is a solution of (10), and if $z = z(x, c)$ is a one-parameter family of solutions of the Bernoulli equation

(B) $Dz + [q(x) + 2p(x)y_1(x)]z = -p(x)z^2$

then $y(x) = y_1(x) + z(x, c)$ is a one-parameter family of solutions of (10).

Proof: Let $y = y_1(x)$ be a solution of (10). Then

$$Dy_1(x) + p(x)y_1{}^2(x) + q(x)y_1(x) + r(x) = \varnothing$$

Subtracting this equation from (10) yields

(A) $D(y - y_1) + p(y^2 - y_1{}^2) + q(y - y_1) = \varnothing$

By algebraic manipulation, (A) can be written as

$$D(y - y_1) + (2py_1 + q)(y - y_1) + p(y - y_1)^2 = \varnothing$$

Letting $z = y - y_1$, we have

$$Dz + (2py_1 + q)z = -pz^2$$

which is a Bernoulli equation in x and z. Therefore, if $z = z(x, c)$ is a one-parameter family of solutions of (B), then

$$y(x) = y_1(x) + z(x, c)$$

is a one-parameter family of solutions of (10).

12 Example

Consider the Riccati equation

$$Dy - x^3y^2 - \frac{1}{x}y + x^5 = \varnothing$$

It is easy to see that $y_1(x) = x$ is a solution of the equation. We now consider the Bernoulli equation

$$Dz + \left[\frac{-1}{x} + 2(-x^3)x\right]z = -(-x^3)z^2$$

or

$$Dz - \left(2x^4 + \frac{1}{x}\right) z = x^3 z^2$$

This equation is transformed into the linear equation

$$Dv + \left(2x^4 + \frac{1}{x}\right) v = -x^3$$

by means of the change of variable

$$v = z^{-1}, \qquad Dv = -z^{-2} Dz$$

The one-parameter family of solutions of the linear equation is:

$$v(x) = \frac{c - e^{(2/5)x^5}}{2xe^{(2/5)x^5}}$$

Therefore,

$$z(x) = \frac{2xe^{(2/5)x^5}}{c - e^{(2/5)x^5}}$$

and

$$y(x) = x + \frac{2xe^{(2/5)x^5}}{c - e^{(2/5)x^5}}$$

is a one-parameter family of solutions of the given equation.

13 EXERCISES

1. In each of the following Riccati equations, one solution is specified. Use the given solution to determine a one-parameter family of solutions.
   a. $Dy - xy^2 - (1/x)y + x^3 = \varnothing$, $y_1(x) = x$
   b. $Dy + e^{-x}y^2 - y - e^x = \varnothing$, $y_1(x) = -e^x$
   c. $Dy + y^2 = 1 + x^2$, $y_1(x) = x$
2. Verify that $y_1(x) = x$ is a solution of

$$Dy + \left(\frac{y}{x}\right)^2 + \left(\frac{y}{x}\right) = 3$$

   Find a solution of the equation satisfying the initial condition $y(1) = 2$.
3. Show that the Riccati equation

$$Dy + by^2 = 2x^{-2}$$

   has a real solution of the form $y(x) = a/x$ if and only if $b \geq -\frac{1}{8}$.
4. Show that the constant function $y(x) = r$ is a solution of the constant-coefficient Riccati equation

$$\frac{dy}{dx} + ay^2 + by + c = \varnothing$$

   if and only if r is a real root of the quadratic polynomial $at^2 + bt + c$.

5. Use the result in Problem 4 to find a one-parameter family of solutions of

$$Dy + y^2 - 3y + 2 = \varnothing$$

The special Riccati equation

(14) $Dy + y^2 + Q(x)y + R(x) = \varnothing$

on J, is closely related to the second-order linear equation

(15) $D^2y + Q(x)\, Dy + R(x)y = \varnothing$

on J, and, as a consequence, (14) occurs frequently in a variety of applications. Before investigating the relationship between (14) and (15), we remark that if the coefficient function p in the general Riccati equation (10) is nonzero and differentiable on J, then the change of variable defined by $y = u/p$ transforms (10) into

(16) $Du + u^2 + \left(q - \dfrac{Dp}{p}\right)u + pr = \varnothing$

which has the form (14). The verification is left as an exercise.

17 THEOREM

If $z = z(x)$ is a solution of (15) such that $z(x) \neq 0$ for all $x \in J$, then $v(x) = Dz(x)/z(x)$ is a solution of (14). Conversely, if $v = v(x)$ is a solution of (14), then

$$z(x) = \exp\left[\int_{x_0}^x v(t)\, dt\right], \qquad x_0 \in J$$

is a solution of (15).

Proof: Suppose $z = z(x)$ is a solution of (15), and $z(x) \neq 0$ for all $x \in J$. Put $v = Dz/z$. Then

$$Dv = \frac{D^2z}{z} - \frac{[Dz]^2}{z^2} = \frac{D^2z}{z} - v^2$$

or

$$Dv + v^2 = \frac{D^2z}{z}$$

Since z is a solution of (15), $D^2z = -Q\, Dz - Rz$. Thus

$$Dv + v^2 = \frac{-Q\, Dz - Rz}{z} = -Qv - R$$

and $v = Dz/z$ is a solution of (14).
   Now assume that $v = v(x)$ is a solution of (14), and let

$$z(x) = \exp\left\{\int_{x_0}^x v(t)\, dt\right\}, \qquad x_0 \in J$$

Then

$$Dz = v \exp\left\{\int_{x_0}^{x} v(t)\,dt\right\} = vz$$

and

$$D^2z = v\,Dz + z\,Dv$$
$$= v^2z + z\,Dv$$
$$= v^2z + z(-v^2 - Qv - R)$$

since v satisfies (14). Simplifying and using the fact that $Dz = vz$, we have

$$D^2z = -Q\,Dz - Rz$$

Thus

$$D^2z + Q\,Dz + Rz = \varnothing$$

and $z(x) = \exp\{\int_{x_0}^{x} v(t)\,dt\}$ is a solution of (15).

From Theorem 11, if we have one solution of a Riccati equation, then a one-parameter family of solutions can be constructed. By analyzing the associated second-order linear equation, it is sometimes possible to find a solution of a given Riccati equation. We illustrate with an example.

18 Example

Consider the Riccati equation

(A) $Dy + e^{-x}y^2 - y - 4e^x = \varnothing$

Here, $p(x) = e^{-x}$, $q(x) = -1$, and $r(x) = -4e^x$. Since p is nonzero and differentiable on $J = (-\infty, \infty)$, put $y = z/e^{-x} = ze^x$. Then $Dy = e^x\,Dz + e^xz$, and substituting into the equation yields

$$e^x\,Dz + e^xz + e^{-x}(e^xz)^2 - e^xz - 4e^x = \varnothing$$

or

$$e^x\,Dz + e^xz + e^xz^2 - e^xz - 4e^x = \varnothing$$

Thus after multiplying by e^{-x}, we have

(B) $Dz + z^2 - 4 = \varnothing$

which has the form (14). The associated second-order linear equation is:

(C) $D^2u - 4u = \varnothing$

Each of the functions $u_1(x) = e^{2x}$ and $u_2(x) = e^{-2x}$ is a solution of (C). Thus by Theorem 17,

$$z(x) = \frac{Du_1}{u_1} = \frac{2e^{2x}}{e^{2x}} = 2$$

is a solution of (B). Of course, any linear combination of u_1 and u_2 could have been used to construct a solution z of (B). Finally, $y_1(x) = 2/e^{-x} = 2e^x$ is a solution of (A). Using y_1, the Bernoulli equation associated with (A) is:

$$Dz + [2(e^{-x})2e^x + (-1)]z = -e^{-x}z^2$$

or

(D) $Dz + 3z = -e^{-x}z^2$

By letting $v = z^{-1}$, this equation is transformed into the linear equation

$$Dv - 3v = e^{-x}$$

whose one-parameter family of solutions is:

$$v(x) = \frac{ce^{3x} - e^{-x}}{4}$$

Since $[z(x)]^{-1} = v(x)$, we have

$$z(x) = \frac{4}{ce^{3x} - e^{-x}}$$

and

$$y(x) = 2e^x + \frac{4}{ce^{3x} - e^{-x}}$$

is a one-parameter family of solutions of the given equation.

19 EXERCISES

1. As observed in Example 6, Section 12, the function $y_1(x) = x$ is a nonzero solution of

$$D^2y - \frac{1}{x}Dy + \frac{1}{x^2}y = \varnothing$$

on $[1, \infty)$. Determine a one-parameter family of solutions of the associated Riccati equation

$$Dy + y^2 - \frac{1}{x}y + \frac{1}{x^2} = \varnothing$$

2. Choose a member from the one-parameter family obtained in Problem 1 and use Theorem 11 to determine another solution of the second-order equation

$$D^2y - \frac{1}{x}Dy + \frac{1}{x^2}y = \varnothing$$

Compare your result with Example 6, Section 12.

3. Verify that the change of variable $y = u/p$ transforms (10) into (16).

4. Consider the equation

$$Dy + 2y^2 + \tfrac{1}{2} = \varnothing$$

Show that the method suggested by Problem 4 in Exercises 13, does not yield a *real* solution of the equation. Transform this equation into the form (14) and determine a solution by applying Theorem 17. On what interval is your solution valid?

5. Consider the equation

$$Dy - 2y^2 + x^{-2} = \varnothing$$

Put the equation in the form (14) and apply Theorem 17 to find a solution of the equation. [Hint: The related second-order equation is a Cauchy equation: see Section 20.]

6. Continue Problem 5 by finding three different solutions (not necessarily linearly independent) of the Riccati equation. Show that

$$\frac{y - y_1}{y - y_2} = c\,\frac{y_3 - y_1}{y_3 - y_2}$$

where y_1, y_2, and y_3 are the three solutions, is a one-parameter family of solutions of the equation.

   As in the case of linear equations, the Taylor series method is useful for finding a series representation of a solution of a differential equation. The following example illustrates the Taylor series method applied to a nonlinear initial-value problem. See Section 30 for additional details.

20 Example

Consider the initial-value problem

$$Dy = y^2 + 1; \qquad y(0) = 0$$

of Example 11a, Section 43.

   By the existence and uniqueness theorem, Theorem 10, Section 43, this initial-value problem has a unique solution $y = y(x)$ valid on some interval I containing $x = 0$. Actually, from Example 11a, Section 43, I can be taken to be $(-\pi/2, \pi/2)$. From the differential equation

(A) $$Dy(x) = y^2(x) + 1$$

so that

$$Dy(0) = 0 + 1 = 1$$

Since $y(x)$ is continuously differentiable, $Dy(x)$ is continuously differentiable, and differentiating (A), we get

(B) $$D^2y(x) = 2y(x)\,Dy(x)$$

and

$$D^2y(0) = 0$$

Continuing in this manner, we obtain

$$D^3y(0) = 2; \qquad D^4y(0) = 0; \qquad D^5y(0) = 16$$

and so on. Of course, these calculations become increasingly more complicated. Using the results above, we have the first five terms of the Taylor series expansion of $y(x)$, that is

$$y(x) = \sum_{n=0}^{\infty} \frac{D^n y(0)}{n!} x^n = x + \frac{1}{3}x^3 + \frac{16}{5!}x^5 + \cdots$$

$$= x + \tfrac{1}{3}x^3 + \tfrac{2}{15}x^5 + \cdots$$

and this is the series representation of tan x.

21 EXERCISES

Use the Taylor series method to find the first five terms of a series solution of the following initial-value problems.

1. $Dy = x^2 + y^2; y(1) = 1$
2. $Dy = 1 - x^2 - y^2; y(0) = 0$
3. $Dy = \sin(y); y(1) = \pi/2$
4. $Dy = \ln(y - x); y(1) = 2$
5. Obtain the series solution of the initial-value problem

$$y \, Dy = 1; \qquad y(1) = 1$$

Also, solve this initial-value problem using the method in (6), Section 44. Compare the two solutions.

Section 47 Numerical Analysis

All of the numerical methods of Chapter 6 can be applied to the general first-order equation

(1) $Dy = g(x, y)$

However, in the case where (1) is a nonlinear differential equation, the analysis of the numerical approximations becomes less precise.

Euler's method as applied to (1) is:

(2) $v(k + 1) = v(k) + hg[x_0 + kh, v(k)]$

The first-order initial-value problem

(3) $Dy = y^2 + 1; \qquad y(0) = 0$

was mentioned in Sections 43 and 46. The solution is:

$$y(x) = \tan(x)$$

Here, $x_0 = 0$ and $g[x_0 + kh, v(k)] = [v(k)]^2 + 1$. Thus Euler's method yields the difference equation

(4) $\qquad v(k + 1) = v(k) + h\{[v(k)]^2 + 1\}; \qquad v(0) = 0$

This is a nonlinear difference equation. In general, it is most unlikely that a convenient explicit representation of the solutions of a nonlinear difference equation can be determined. However, a number of potentially useful facts about the initial-value problem (4) can be obtained.

The following table contains the results (to five digits) of a sample calculation, with $h = 0.3$, of the difference equation (4). Notice that the solution of the difference equation is an increasing function, whereas the solution of the differential equation has a discontinuity at $x = \pi/2$. Thus while $\lim\limits_{x \to \pi/2} |y(x)| = \infty$, it appears that $\lim\limits_{k \to \infty} v(k) = \infty$. In any case, Euler's method results in a very poor approximation with this choice of h.

k	$y(k)$	$y(0.3k)$
0	0	0
1	0.30000	0.30934
2	0.62700	0.68416
3	1.0449	1.2602
4	1.6725	2.5722
5	2.8117	14.101
6	5.4834	-4.2863

In fact, it is not too difficult to begin a step-by-step solution of (4) with an unspecified value of $h > 0$. Some of the details are:

$$v(1) = v(0) + h\{[v(0)]^2 + 1\} = 0 + h(0 + 1) = h$$
$$v(2) = h + h[h^2 + 1] = h^3 + 2h$$

and

$$v(3) = h^7 + 4h^5 + 5h^3 + 3h$$

Notice that $v(1)$, $v(2)$, and $v(3)$ are all polynomials in h and that each is an odd function. Also, it seems clear that for $h > 0$, $v(1) < v(2) < v(3)$, and $v(3) > 3v(1)$. While further calculations of $v(k)$ will get progressively more complicated, there are enough clues here to suggest some general properties of $v(k)$ which can be proved by induction.

5 EXERCISES

1. Show by induction that $v(k)$ is a polynomial of degree $2^k - 1$, that only odd powers of h are present in $v(k)$, and that all coefficients are positive integers.

2. Show that $v(k) > kv(1)$, and thus that $\lim\limits_{k \to \infty} v(k) = \infty$ for all $h > 0$.

3. Compare $v(3)$ with the series representation of $y(3h) = \tan(3h)$, and show that $v(3) < y(3h)$ for all $h \in (0, \pi/6)$.

4. Solve $Dy = y^2 - 1$; $y(0) = 0$, and show that the solution is bounded on $[0, \infty)$.

5. Apply Euler's method to $Dy = y^2 - 1$; $y(0) = 0$, and compare $v(3)$ with $y(3h)$.

6. Compare $\lim_{n \to \infty} v(n)$ with $\lim_{n \to \infty} y(nh)$, where y is the solution of $Dy = y^2 - 1$; $y(0) = 0$.

The evidence, both experimental and theoretical, seems to be that Euler's method provides a poor approximate solution of the initial-value problem (3). Also, if, in an effort to obtain more satisfactory results, a relatively small value of h is used, then the effects of rounding errors in the step-by-step solution of (4) could be quite substantial. For example, if $h = 10^{-4}$ is used to determine an approximate value of $y(1.0)$, then since $1.0 = 10^4 10^{-4}$, the value of $v(10,000)$ is required. Notice from Problem 1, Exercises 5, that $v(10,000)$ is a polynomial in h of quite a high degree. Thus for at least this reason, some rounding errors should be expected, although precise information on the nature of these errors is difficult to determine.

In order to obtain a better understanding of Euler's method as applied to the general first-order equation (1), it is necessary to make some assumption about $g(x, y)$. The examples in this chapter and especially initial-value problem (3) illustrate that a seemingly mild $g(x, y)$ can lead to the potentially troublesome situation of having an unbounded solution.

The basic assumption in this chapter is that $g(x, y)$ is continuous in some region. This in turn implies that all solutions of (1) are members of $\mathscr{C}^1(J)$ for some interval J. The additional assumptions that are needed in order to analyze Euler's method, as well as other numerical methods, are these:

(6) $\begin{cases} \text{(a) } J = [0, b] \text{ for some } b > 0 \\ \text{(b) } y \in \mathscr{C}^2(J), \text{ and} \\ \text{(c) } \partial g/\partial y \text{ is continuous and bounded for } x \in J \text{ and } y \in (-\infty, \infty). \end{cases}$

The choice of $J = [0, b]$ is merely for convenience and simplicity of notation. Any other closed interval would do just as well and merely amounts to a trivial change of variables. Next, condition (6b) holds for the vast majority of linear first-order equations in Chapters 2 and 3. In fact, if

$$Dy = q(x)y + f(x)$$

then $q(x) \in \mathscr{C}^1(J)$ and $f(x) \in \mathscr{C}^1(J)$ implies that

$$D^2 y = q(x) \, Dy + y \, Dq(x) + Df(x)$$

or that $y \in \mathscr{C}^2(J)$. Finally, the third restriction in (6) is true for linear equations since if $g(x, y) = q(x)y + f(x)$, then

$$\frac{\partial g}{\partial y} = q(x)$$

and $q(x) \in \mathscr{C}(J)$ is clearly a bounded function on J.

Thus the assumptions made in (6) amount to assuming that the nonlinear differential equation (1) is (very roughly speaking) close to a linear equation. Notice that whether or not a first-order equation satisfies (6) could be difficult to determine. Also, it is easy to consider a nonlinear equation that does not satisfy all parts of (6). For example, (6c) fails to hold for equation (3).

7 Example

The first-order nonlinear equation

$$Dy = 2(x + 1) \cos^2 y$$

satisfies the conditions in (6). In particular,

$$\frac{\partial [2(x + 1) \cos^2 y]}{\partial y} = -4(x + 1) \sin y \cos y$$

and for $x \in [0, 1]$ and $y \in \mathscr{C}(-\infty, \infty)$,

$$|-4(x + 1) \sin y \cos y| < 8$$

8 THEOREM

If $g(x, y)$ satisfies (6c), then there is a number $L > 0$ such that

$$|g(x, u) - g(x, v)| \le L|u - v|$$

for all $x \in J$ and arbitrary u and v.

Proof: It is a consequence of the mean value theorem (Theorem 2, Section 2) that if $u > v$, then for $x \in J$,

$$g(x, u) = g(x, v) + \frac{\partial g(x, c)}{\partial y}(v - u)$$

where $c \in (v, u)$.
Thus

$$|g(x, u) - g(x, v)| \le \left| \frac{\partial g(x, c)}{\partial y} \right| |u - v|$$

and since $\partial g/\partial y$ is bounded, there is some number $L > 0$ such that

$$\left| \frac{\partial g(x, c)}{\partial y} \right| \le L$$

The error function that arises when Euler's method is applied to the differential equation (1) with initial condition $y(0) = y_0$ is:

$$E(k, h) = y(kh) - v(k)$$

Since it is assumed in (6) that $y \in \mathscr{C}^2(J)$, we can use Theorem 2, Section 2, to get that

$$y((k + 1)h) = y(kh + h) = y(kh) + h\,Dy(kh) + \frac{h^2}{2} D^2 y(c_k)$$

where

$$kh < c_k < (k + 1)h$$

Also from Euler's method (2),

$$v(k + 1) = v(k) + hg[kh, v(k)]$$

Thus

(9) $E(k + 1, h) = y(kh) + h\,Dy(kh) + \dfrac{h^2}{2} D^2 y(c_k) - v(k) - hg[kh, v(k)]$

Next, from the differential equation (1), we have that

$$Dy(kh) = g[kh, y(kh)]$$

so that (9) can be rewritten as

(10) $E(k + 1, h) = E(k, h) + h\{g[kh, y(kh)] - g[kh, v(k)]\}$

$$+ \dfrac{h^2}{2} D^2 y(c_k)$$

Thus for a given choice of h, the error function is a solution of a first-order linear difference equation, namely, equation (10). The assumption in (6b) resulted in this linear difference equation, and thus a more precise investigation of this special case is possible. Notice also that an initial condition is implicit with equation (10), since clearly

(11) $E(0, h) = 0$

However, it is not possible to solve this initial-value problem since the numbers c_k are not known.

It is possible to obtain an upper bound for $E(k, h)$, and this is done by taking absolute values of both sides of (10) and using the triangle inequality. Thus

$$|E(k + 1, h)| \le |E(k, h)| + h|g[kh, y(kh)] - g[kh, v(k)]|$$

$$+ h^2 \frac{|D^2 y(c_k)|}{2}$$

By applying Theorem 8 and condition (6b), we get

$$|E(k + 1, h)| \le |E(k, h)| + Lh\,|y(kh) - v(k)| + h^2 M$$

where $|\partial g/\partial y| \le L$ and $|D^2 y/2| \le M$. Since $y(kh) - v(k) = E(k, h)$, this simplifies to

(12) $|E(k + 1, h)| \le (1 + Lh)|E(k, h)| + h^2 M$

The form of the inequality (12) together with the initial condition (11) suggest a general result that is applicable here.

13 THEOREM

If $\mu \in V_\infty$, $\mu(0) = 0$, and if for all $k \in N$,

$$|\mu(k + 1)| \leq a|\mu(k)| + b$$

where $a > 1$ and $b > 0$, then

$$|\mu(k)| \leq \left(\frac{a^k - 1}{a - 1}\right) b$$

for all $k \in N$.

Proof: By observing that

$$|\mu(1)| \leq a|\mu(0)| + b = b = \left(\frac{a - 1}{a - 1}\right) b$$

and

$$|\mu(2)| \leq a|\mu(1)| + b \leq ab + b = (a + 1)b = \left(\frac{a^2 - 1}{a - 1}\right) b$$

the result follows by induction.

By applying Theorem 13 to the inequality (12), we get

$$|E(k, h)| \leq \left[\frac{(1 + Lh)^k - 1}{1 + Lh - 1}\right] h^2 M$$

or

(14) $$|E(k, h)| \leq [(1 + Lh)^k - 1]\frac{hM}{L}$$

for all $k \in N$.

In general, it is not practical to determine the upper bounds M and L, and even when they can be determined, the upper bound in (14) tends to be a gross overestimate of $|E(k, h)|$. The main reason for working out (14) is to use it to prove a convergence theorem, which is the next topic.

15 THEOREM

If equation (1) satisfies the conditions in (6), $t \in J$ and $\varepsilon > 0$, then there is a positive integer k and a number $h > 0$ such that $kh = t$ and $|E(k, h)| < \varepsilon$.

Proof: Let $t = kh$, or, equivalently, $h = t/k$, where k is to be determined. With this choice of h, the upper bound in (14) becomes

(A) $$\left|E\left(k, \frac{t}{k}\right)\right| \leq \left[\left(1 + \frac{Lt}{k}\right)^k - 1\right]\frac{Mt}{kL} \leq (e^{Lt} - 1)\frac{Mt}{kL}$$

This last estimate comes from the fact that $(1 + [Lt/k])^k$ is an increasing function of k whose limit is e^{Lt}. Thus if k is an integer such that

$$k > \frac{(e^{Lt} - 1)Mt}{\varepsilon L}$$

then

$$(e^{Lt} - 1)\frac{Mt}{kL} < \varepsilon$$

and therefore, using (A),

$$\left| E\left(k, \frac{t}{k}\right) \right| < \varepsilon$$

Also,

$$h = \frac{t}{k} < \frac{\varepsilon L}{M(e^{Lt} - 1)}$$

implies that $|E(k, h)| < \varepsilon$, where $kh = t$.

The proof of this theorem actually contains a rule for picking the stepsize h. If

(16) $$h < \frac{\varepsilon L}{M(e^{Lt} - 1)}$$

then Euler's method will produce an approximate value of $y(t)$ that has an error of at most ε. Once again, in most examples, the right side of (16) is much smaller than it need be.

17 Example

We continue with Example 7, where

$Dy = 2(x + 1) \cos^2 y$

for $x \in J = [0, 1)$.

$D^2 y = 2 \cos^2 y - 4(x + 1) \sin y \cos y[2(x + 1) \cos^2 y]$

or

$D^2 y = 2 \cos^2 y[1 - 4(x + 1)^2 \sin y \cos y]$

Thus

$|D^2 y| < 34 = M$

for $x \in J$ and $y \in (-\infty, \infty)$. Also, from Example 7,

$$\left| \frac{\partial g}{\partial y} \right| < 8 = L$$

Suppose an approximation for $y(1)$ is required with an error of less than 10^{-4}. Then $t = 1$ and $\varepsilon = 10^{-4}$ so that

$$h < \frac{10^{-4}8}{34(e^8 - 1)}$$

Since $h = 10^{-9}$ meets this error bound, this means that $k = 10^9$. With this choice of h and t, 10^9 steps in the calculation of Euler's method would be required. If this many calculations were done, for example, on a ten-digit calculator, one could expect a substantial amount of rounding errors.

As was noted earlier, conditions (6) hold for a large class of linear differential equations, and thus the convergence theorem, Theorem 15, is applicable. However, even for these equations, a determination of the upper bound for h in (16) could be difficult. Several associated problems are suggested in the exercises.

18 EXERCISES

1. Find a better bound L for the equation in Example 17.
2. Consider $Dy = q(x)y + f(x)$, where $q(x) \in \mathscr{C}^1[0, b]$ and $f(x) \in \mathscr{C}^1[0, b]$. Show that M must satisfy

$$\left| \frac{Df - qf + (q^2 - Dq)y}{2} \right| \le M$$

3. If $Dy = g(x, y)$ and each of the functions

$$g, \quad \frac{\partial g}{\partial x}, \quad \text{and} \quad \frac{\partial g}{\partial y}$$

are continuous and bounded for $x \in J$ and $y \in (-\infty, \infty)$, show that conditions (6) are satisfied.

4. What conditions can be imposed on f_1 and f_2 so that the separable equation

$$Dy = f_1(x)f_2(y)$$

satisfies conditions (6)?

5. Show that no Bernoulli equation satisfies conditions (6).
6. Consider $Dy = g(x, y)$ on the interval $[a, b]$, where $a \ne 0$, and assume that the modified conditions (6) hold. What is the form of the bound for h needed to approximate $y(b)$?
7. Suppose $f \in \mathscr{C}^1(-\infty, \infty)$ and f and Df are bounded on $(-\infty, \infty)$. Does $Dy = f(x \cdot y)$ satisfy conditions (6)? What about $Dy = f(x - y)$?

This analysis of Euler's method is merely an example of the kind of results that are important considerations in computational mathematics. The basic ideas used in this section can be extended to include a more general class of numerical methods. Naturally, the analysis becomes progressively more complicated as more involved computational methods are considered. Since the purpose of this book is to provide an introductory level survey of differential equations, we must leave the advanced topics to books devoted to numerical analysis.

The topics and examples in Chapters 6, 9, and 10 are intended to illustrate that the matter of finding approximate solutions of differential equations is hardly a routine activity. A good understanding of the classical theory of differential equations is essential. As a final topic in this section, we consider an approach to approximating a solution of a Riccati equation.

The specific nature of $g(x, y)$ in the general first-order equation (1) can sometimes be used in the formulation of a numerical method. Since the Riccati equation is a nonlinear equation that has been studied extensively, we will use it as an illustration.

If $g(x, y) = -[y^2 + q(x)y + r(x)]$, then $Dy = g(x, y)$ can be written as the Riccati equation (14) of Section 46, namely,

(19) $Dy + y^2 + q(x)y + r(x) = \emptyset$

Notice that this particular choice of $g(x, y)$ does not satisfy condition (6c). This does not mean that Euler's method would fail to yield useful approximate values, since the conditions (6) are merely sufficient for convergence of Euler's method. We consider the nonlinear equation (19) along with the initial condition

(20) $y(0) = b \neq 0$

The related second-order linear equation is, as in (15), Section 46,

(21) $D^2 y + q(x)\, Dy + r(x)y = \emptyset$

This second-order equation can be transformed into a system of first-order equations

(22) $D\begin{bmatrix} y_1 \\ y_2 \end{bmatrix} = \begin{bmatrix} \emptyset & 1 \\ -r(x) & -q(x) \end{bmatrix}\begin{bmatrix} y_1 \\ y_2 \end{bmatrix}$

by the methods of Section 38. Of course, if

$$\begin{bmatrix} y_1 \\ y_2 \end{bmatrix}$$

is a solution of (22), then $y = y_1$ is a solution of (21) and $Dy = y_2$. Thus any solution of equation (22) can be assumed to be of the form

(23) $\sigma = \begin{bmatrix} u \\ Du \end{bmatrix}$

where u is a solution of equation (21).

Theorem 17, Section 46, contains the connection between the solutions of equations (19) and (21). All that is needed is a suitable initial condition for equation (22) to produce a solution of (22), and hence a solution of (21). Consider the initial condition

(24) $\sigma(0) = \begin{bmatrix} 1 \\ b \end{bmatrix}$

From (23), this initial condition implies that

(25)
$$u(0) = 1$$
$$Du(0) = b$$

are the initial conditions associated with the second-order equation (22). It follows from Theorem 17, Section 46, that

(26) $y(x) = \dfrac{Du(x)}{u(x)}$

is a solution of the Riccati equation (19) and, using the initial condition (24), we have that

$$y(0) = \frac{Du(0)}{u(0)} = \frac{b}{1} = b$$

and this agrees with initial condition (20).

Thus an approximate solution of the linear homogeneous system (22) with initial condition (24) will give, by the use of (26), an approximate solution of the Riccati equation (19) with initial condition (20). Numerical methods for equations of the form (22) were studied in Section 42.

11

APPLICATIONS

This chapter contains some illustrations of how the theory of differential equations and difference equations can be used in various applications. These particular illustrations have been selected because they cover most of the topics that are treated in the text.

In the very recent past, the bulk of the applications of differential equations was found in the various branches of engineering and some of the physical sciences. Now, however, it is very common to see differential equations and difference equations appear in the literature of business, economics, political science, medicine, and many newer areas of specialization. Naturally, the number of illustrations here must be limited, but it is quite common for equations of essentially the same form to occur in many unrelated applications. The best source of examples in a particular area of specialization is a book or journal devoted to that field.

There is no ideal way to arrange these illustrations since some of them suggest how various topics in the text can be applied to a particular problem. For the most part, practical problems do not exactly fit into one of the special categories studied in the text. Thus a satisfactory (and satisfying) solution of a nontrivial practical problem usually results from a combination of approaches. This is, or is close to, an art form that is acquired through experience.

A numbers of questions are raised within the illustrations and these questions may be considered as exercises. In addition, the treatment of each illustration is by no means exhaustive, so it is quite appropriate to ignore the approach to a particular problem and develop your own.

Finally, it is hoped that these few illustrations will help to provide some motivation to read and study the various topics in the text.

1 Vibrating Mechanical Systems

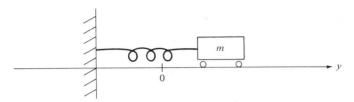

The diagram shown here represents a simple mechanical system consisting of a mass m connected to a wall by a spring. The mass moves on a smooth surface, and y (a function of time) represents its displacement from the wall when the mass is displaced from its equilibrium position 0. It is assumed that the spring exerts a force (restoration) that is proportional to the displacement, and that the surface exerts a resistance (damping) to the motion of the mass that is proportional to the velocity. This resistance, for example, could be due to friction. From Newton's law, the motion is described by

(A) $m\,D^2 y = -a\,Dy - by, \qquad D = \dfrac{d}{dt}$

where $a \geq 0$ and $b \geq 0$ are the damping and restoration proportionality constants, respectively. More generally, if an external force is applied to this system, say $Q(t)$, that is dependent on time only, then

(B) $m\,D^2 y = -a\,Dy - by + Q(t)$

Note that each of these equations is a second-order linear differential equation with constant coefficients. Thus the results in Chapter 3 apply.

We begin an analysis of vibrating mechanical systems by considering the case where there is no damping and no external forces. In this case, the system is in a state of *free vibration*, and from (A) the equation describing the system can be written as

$m\,D^2 y + by = \varnothing, \qquad b > 0$

or

(C) $D^2 y + k^2 y = \varnothing$

where $k^2 = b/m$. The two-parameter family of solutions is:

$y(t) = A \cos kt + B \sin kt$

In a particular problem, the constants A and B are determined from prescribed initial conditions. For example, if the mass is displaced one unit and released with an initial velocity of one unit per second in the direction of the equilibrium position at time $t = 0$, then the initial conditions are $y(0) = 1$, $Dy(0) = -1$, and the corresponding solution is:

$$y(t) = \cos kt - \frac{1}{k} \sin kt$$

By letting $R = (A^2 + B^2)^{1/2}$ and δ the angle such that $A = R \cos \delta$ and $B = R \sin \delta$, the two-parameter family can be written in the form

(D) $y(t) = R \cos(kt - \delta)$

It is clear from (D) that the solutions of (C) are periodic functions of period $2\pi/k$ and frequency k. These positive numbers are called the *natural period* and the *natural frequency* of the system. The numbers R and δ are called the *amplitude* and *phase shift*. The motion represented by (D) is usually called *simple harmonic motion*. The graph is shown in the figure here.

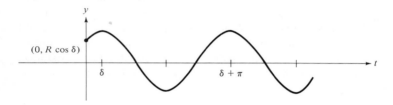

If we include the effects of damping, then the differential equation describing the motion is:

(E) $m D^2 y + a Dy + by = \varnothing$, $a > 0$

and the system is in a state of *damped free vibration*. The polynomial associated with equation (E) is:

$$P(s) = ms^2 + as + b$$

and its two roots are:

$$r_1, r_2 = \frac{-a \pm \sqrt{a^2 - 4bm}}{2m} = \frac{-a}{2m} \pm \left(\frac{a^2}{4m^2} - \frac{b}{m}\right)^{1/2}$$

It is convenient to define new constants in terms of a, b, and m. Let $r = a/2m > 0$, where r is called the *damping factor*; let $k^2 = b/m$, where k is called the *frequency*; and let $q = (k^2 - r^2)^{1/2}$, where q is called the *oscillation factor*. In terms of these constants, we have

$$r_1 = -r + iq \qquad \text{and} \qquad r_2 = -r - iq$$

and the three possible forms of the solutions of (E) are:

(i) $q^2 = k^2 - r^2 > 0$, $y(t) = e^{-rt}(A \cos qt + B \sin qt)$

(ii) $q^2 = k^2 - r^2 = 0,$ $y(t) = e^{-rt}(At + B)$

(iii) $q^2 = k^2 - r^2 < 0,$

$$y(t) = Ae^{(-r+q)t} + Be^{(-r-q)t}, \quad -r + q < 0, \quad -r - q < 0.$$

It is important to note that in all three cases $\lim\limits_{t \to \infty} y(t) = 0$, regardless of the values of the constants A and B. Comparing these solutions with (D), we see that without damping the motion continues in a "free" periodic manner, while with damping the motion decays to zero as time increases.

In cases (ii) and (iii), there is no "oscillatory" behavior. These cases are referred to as *critical damping* and *overdamping*, respectively. Depending upon the initial conditions, the graph of a solution must look like one of the curves shown in the figure here.

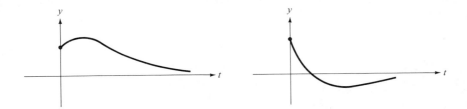

Case (i) is called *underdamped* motion. By letting $R = (A^2 + B^2)^{1/2}$ and $\sin(\delta) = (B/R)$, as in the case of free vibrations, we get

$$y(t) = Re^{-rt} \cos(qt - \delta)$$

The graph is shown in the figure here.

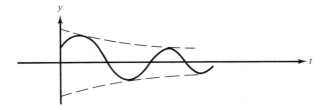

We now examine the effect of an externally applied force $Q(t)$ on the spring-mass system. The equation of motion in this case is given by (B), and the system is said to be in a state of *forced vibration*. As a particular example, suppose that there is no damping, that is, $a = 0$, and that the applied force is periodic and proportional to $\cos ht$. The equation of motion in this case can be written as

(F) $D^2 y + k^2 y = G \cos ht$

where $k^2 = b/m$ and G is the constant of proportionality. The two-parameter family of solutions of (F) is:

$$y(t) = A \cos kt + B \sin kt + \frac{G}{k^2 - h^2} \cos ht$$

provided $h \neq k$. As in the previous examples, this expression can also be written in the form

(G) $y(t) = R \cos(kt - \delta) + \dfrac{G}{k^2 - h^2} \cos ht$

where the constants R and δ are determined by the initial conditions. The resulting motion is the sum of two periodic functions of different frequencies and, in general, different amplitudes. A typical graph of $y(t)$ is shown in the figure here.

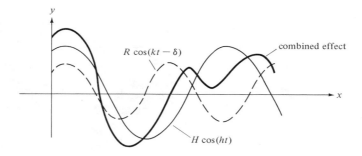

An interesting phenomenon known as *resonance* occurs when the period of the applied force equals the natural period of the system. In (G), note that if h is "close to" k, then the amplitude of y is "large." If $h = k$ in (F), then the two-parameter family of solutions is:

(H) $y(t) = A \cos kt + B \sin kt + \dfrac{Gt}{2k} \sin kt$

and it is clear that the motion is unbounded as $t \to \infty$. Of course, in actual practice, the spring would break at some time $t > 0$. The graph of a typical member of the family is shown in the figure here.

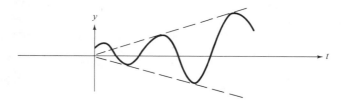

In general, the motion of the spring-mass system, with damping and the forcing function $G \cos ht$, is described by

(I) $\qquad D^2 y + \dfrac{a}{m} Dy + \dfrac{b}{m} y = G \cos ht$

The two-parameter family of solutions of (I) has the form

(J) $\qquad y(t) = Ae^{r_1 t} + Be^{r_2 t} + F \cos(ht - \beta)$

where r_1 and r_2 are the roots of the polynomial associated with the reduced equation of (I), and F and β are constants which are determined explicitly from the coefficients m, a, b, G, and h. As shown in the case of damped free vibrations, the function

$$y_c(t) = Ae^{r_1 t} + Be^{r_2 t}$$

has limit zero as $t \to \infty$. Thus, as $t \to \infty$,

$$y(t) \to z(t) = F \cos(ht - \beta)$$

For this reason, $y_c(t)$ is called the *transient solution* and $z(t)$ is called the *steady state* solution.

Some related exercises are given in the following list.

a. Equation (E) can be written $D^2 y + 2r\, Dy + k^2 y = \varnothing$ by letting $r = a/2m$ and $k^2 = b/m$. In each of the cases (i) $r = 1$, $k = \sqrt{2}$; (ii) $r = 1$, $k = 1$; and (iii) $r = 1$, $k = \sqrt{3}/2$; graph the solution that satisfies the initial condition $y(0) = 1$; $Dy(0) = -1$.

b. In the case of underdamping, show that if t_1 and t_2 are successive relative maxima of a solution, then $y(t_2)/y(t_1) = \exp[-2\pi(r/q)]$, where $r = a/2m$ and $q = (k^2 - r^2)^{1/2}$.

c. Show that if the damping constant a in equation (E) is such that the motion is either critically damped or overdamped, then the mass passes through 0 at most once.

d. In the case of critical damping, determine a condition on $Dy(0)$ such that a solution satisfying $y(0) = 1$ will pass through zero.

e. Consider equation (F), that is, forced vibration without damping, and with $h \neq k$, Show that the solution of (F) satisfying $y(0) = Dy(0) = 0$ can be written in the form

$$y(t) = \frac{2G}{(k^2 - h^2)} \sin \frac{(k - h)t}{2} \sin \frac{(k + h)t}{2}$$

If $|k - h|$ is small, then the resulting motion can be used to illustrate the phenomenon known as a *beat*.

f. Complete the analysis of the spring-mass system in the case of forced vibrations by considering equation (I) in the three cases $q > 0$, $q = 0$, and $q < 0$, where $q = (k^2 - r^2)^{1/2}$.

g. Show that a steady state solution $z(t)$ of (I) is given by

$$z(t) = \frac{G \cos(ht - \delta)}{[(k^2 - h^2)^2 + 4r^2 h^2]^{1/2}}$$

where

$$\delta = \frac{k^2 - h^2}{[(k^2 - h^2)^2 + 4r^2 h^2]^{1/2}}$$

2 An Electrical Circuit

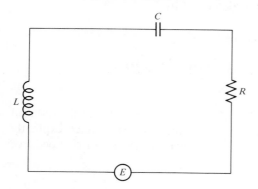

The diagram shown here represents an electrical circuit consisting of a resistance of R ohms, an inductance of L henries, a capacitance of C farads, and an electromotive force of $E(t) = E \sin wt$, where E and w are positive constants. The equation that describes the current, $i(t)$, in the circuit is:

(A) $LD^2 i + R\,Di + \left(\dfrac{1}{C}\right) i = wE \cos(wt)$

Equation (A) is of the same form as the equation of motion of the vibrating mechanical system (1) under forced vibrations. Thus an analogy can be made between the mechanical and the electrical quantities.

Equation (A) can be written in operator form as

$P(D)i = wE \cos(wt)$

where $P(s) = Ls^2 + Rs + (1/C)$. If L, R, and C are each positive, then the roots of $P(s)$ are:

$$s_1 = \frac{-R + \sqrt{R^2 - 4(L/C)}}{2L} \qquad s_2 = \frac{-R - \sqrt{R^2 - 4(L/C)}}{2L}$$

Evidently, both of these numbers (whether real or complex) have a negative real part, so that all solutions of the homogeneous equation

$P(D)i = \varnothing$

are such that $\lim\limits_{t \to \infty} i(t) = 0$. Therefore, regardless of the initial conditions, the solution of equation (A) approaches a "steady state" as time increases, and this steady state solution is a particular solution of equation (A).

The steady state solution is in a particularly simple form and somewhat easier to interpret if, as was the case of the vibrating mechanical system, new constants are defined. The following defines these constants:

$$x = Lw - \frac{1}{Cw} \qquad\qquad \text{(reactance)}$$

$$z = \sqrt{x^2 + R^2} \qquad\qquad \text{(impedance)}$$

$$\theta = \arc \sin\left(\frac{x}{z}\right) = \arc \cos\left(\frac{R}{z}\right) \qquad \text{(phase angle)}$$

In terms of these new constants, it is fairly easy to show that if $i(t)$ is a solution of equation (A), then

(B) $\qquad \lim_{t \to \infty} i(t) = \frac{E}{z} \sin(wt - \theta)$

and this is the steady state solution.

Some related questions are given in the following list.

a. Verify that (B) is the steady state solution of equation (A) with L, R, and C positive.
b. If $R = 0$, what conditions on x will cause equation (A) to have unbounded solutions?
c. If $q(t)$ is the charge (in coulombs) on capacitor C and $Dq = i(t)$, what is $q(t)$ in the steady state condition?
d. Investigate the electrical circuit in the case $L = 0$. Is there a steady state solution in this case?

3 Commercial Applications

Suppose the amount of a loan L is to be repaid by equal payments P. The payments are made periodically with a portion of each payment representing the interest i on the balance due $B(k)$ after the kth payment is made. Thus

$$B(k + 1) = B(k) + iB(k) - P$$

or

$$B(k + 1) - (1 + i)B(k) = -P$$

is a first-order linear nonhomogeneous difference equation that describes this method of repayment.

Using the methods of Section 23, we have that the solutions of this difference equation are of the form

$$B(k) = C(1 + i)^k + \frac{P}{i}$$

and since $B(0) = L$, the unique choice of C is found by solving

$$B(0) = L = C + \frac{P}{i}$$

Therefore,

$$B(k) = L(1 + i)^k - P\left[\frac{(1 + i)^k - 1}{i}\right]$$

is a solution of this problem.

If a total of n payments are to be made, then $B(n) = 0$, in which case

$$P = L\left[\frac{i(1 + i)^n}{(1 + i)^n - 1}\right]$$

As an example, suppose a person wishes to repay a loan of $1000.00 by making four equal payments. The payments are to be made semiannually, and the annual interest rate is 8%. In this case, $L = \$1000.00$, $n = 4$, and $i = 0.08/2 = 0.04$, so that

$$P = 1000 \left[\frac{0.04(1.04)^4}{(1.04)^4 - 1} \right] = \$275.49$$

Some related exercises are given in the following list.

a. If a person can afford to make monthly payments of $300.00 for a period of ten years, and the current annual interest rate is 9%, show how to determine how much money he can borrow.

b. If M dollars are deposited in a savings bank whose annual interest rate is i, what is the balance in the account after k years?

c. If a person's beginning salary is $10,000 per year, he receives an annual salary increase of 5% and invests 10% of his total annual income each year in bonds that pay 6% interest, how many years will it be until his annual income from investments only exceeds his beginning salary?

d. A seller asks $5500.00 as the cash price of a building lot. A buyer offers to make a down payment of $2000.00 and pay an additional $2000.00 each year for two years. What is the effective annual interest rate that the buyer is offering the seller?

4 Calculating Determinants

In some special cases, it is possible to calculate determinants of matrices by solving difference equations. One such case is the *Jacobi matrices* which have the form

$$J_5 = \begin{bmatrix} a & b & 0 & 0 & 0 \\ c & a & b & 0 & 0 \\ 0 & c & a & b & 0 \\ 0 & 0 & c & a & b \\ 0 & 0 & 0 & c & a \end{bmatrix}$$

Although this is J_5, there are matrices of this form for all positive integers; it should be clear from this example how J_k is defined. These matrices are a special case of the *tridiagonal* matrices.

In order to calculate $\det(J_k)$ for $k = 1, 2, \ldots$, we use the elementary rule for evaluating a determinant on the first row of J_{k+3}. Thus

$$\det(J_{k+3}) = a \det(J_{k+2}) - bc \det(J_{k+1})$$

Evidently, this is a second-order linear homogeneous difference equation, so that the methods of Section 25 can be applied.

In order to put the problem into a standard form, let

$$V(k) = \det(J_{k+1})$$

Thus the equation can be expressed as

$$V(k + 2) = aV(k + 1) - bcV(k)$$

or

$$P(E)V(k) = \varnothing$$

where

$$P(t) = t^2 - at + bc$$

The roots of $P(t)$ are:

$$t = \frac{a \pm \sqrt{a^2 - 4bc}}{2}$$

and the initial conditions are:

$$V(0) = \det J_1 = \det(a) = a$$

$$V(1) = \det J_2 = \det \begin{bmatrix} a & b \\ c & a \end{bmatrix} = a^2 - bc$$

This gives enough information to break the problem into cases depending on the three constants that define J_k. For example, suppose that

$$a^2 = 4bc$$

In this case, $P(t)$ has a single root of multiplicity two, namely $t = a/2$. Thus

$$V(k) = C_1 \left(\frac{a}{2}\right)^k + C_2 k \left(\frac{a}{2}\right)^k$$

is the two-parameter family of solutions of the difference equation. Also, the initial conditions, in this case, are:

$$V(0) = a$$

and

$$V(1) = a^2 - bc = a^2 - (\tfrac{1}{4})a^2 = \frac{3a^2}{4}$$

so that

$$a = V(0) = C_1$$

and

$$\frac{3a^2}{4} = V(1) = C_1 \left(\frac{a}{2}\right) + C_2 \left(\frac{a}{2}\right)$$

or $C_2 = a/2$. Therefore,

$$V(k) = a \left(\frac{a}{2}\right)^k + \left(\frac{a}{2}\right) k \left(\frac{a}{2}\right)^k = \det(J_{k+1})$$

or

$$\det(J_{k+1}) = \left(\frac{a}{2}\right)^{k+1} (2 + k) = \left(\frac{a}{2}\right)^{k+1} (1 + k + 1)$$

or

$$\det(J_k) = \left(\frac{a}{2}\right)^k (1 + k)$$

$$k = 1, 2, \ldots .$$

Some related exercises are given in the following list.

a. Find a formula for $\det(J_k)$ if $b = c$.

b. Find a formula for $\det(J_k)$ if $b = -c$.

c. A somewhat related matrix form is a type of *band matrix* that also comes up in applications. One of these is:

$$B_5 = \begin{bmatrix} a & 0 & b & 0 & 0 \\ 0 & a & 0 & b & 0 \\ c & 0 & a & 0 & b \\ 0 & c & 0 & a & 0 \\ 0 & 0 & c & 0 & a \end{bmatrix}$$

Can you find a difference equation for B_k?

5 Electrical Networks

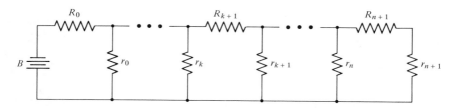

The diagram shown here depicts an electrical network that consists of $2(n + 2)$ resistors denoted by R_0, $R_1, \ldots, R_{n+1}$ and r_0, $r_1, \ldots, r_{n+1}$, together with a battery whose voltage is B. The current flowing through resistor R_j is denoted by $I(j)$, and with the help of Kirchhoff's law, we can write the equations of the voltages in each of the loops of this network.

We consider the simple case of $R = r_j = R_j$, where $j = 0$, 1, 2, $\ldots$, $n + 1$, that is, where all resistors have the same value. In the first, or left-most loop, the sum of the voltages is:

(A) $I(0)R + [I(0) - I(1)]R = B$

while in the last loop,

(B) $I(n + 1)R + I(n + 1)R = [I(n) - I(n + 1)]R$

For the intermediate loops, we have that

(C) $I(k + 1)R + [I(k + 1) - I(k + 2)]R = [I(k) - I(k + 1)]R$

Equations (A) and (B) consist of boundary conditions, while equation (C) is a second-order linear homogeneous difference equation whose solutions can be characterized using the methods of Section 25.

By eliminating the common factor R, equation (C) can be expressed as

(D) $P(E)I(k) = \varnothing$

where $P(t) = t^2 - 3t + 1$. Since $P(t) = 0$ if and only if t is one of the two numbers

$$t_1 = \left(\tfrac{3}{2}\right) + \left(\tfrac{5}{4}\right)^{1/2} \qquad t_2 = \left(\tfrac{3}{2}\right) - \left(\tfrac{5}{4}\right)^{1/2}$$

it follows that the solutions of equation (D) are of the form

(E) $I(k) = c_1(t_1)^k + c_2(t_2)^k$

The boundary conditions in equations (A) and (B) can be expressed in the form

(F)
$$I(1) = 2I(0) - \left(\frac{B}{R}\right)$$

$$I(n) = 3I(n+1)$$

Also, by taking $k = 1$ and $k = n$ in equation (E), we get another pair of equations for $I(1)$ and $I(n)$ which can be used together with (F) to obtain values for c_1 and c_2 in terms of $I(0)$ and $I(n + 1)$.

Some related exercises are given in the following list.

a. Obtain c_1 and c_2 in equation (E).
b. Consider the network where $R = R_j$ and $r_j = cR$ for all j, where c is a positive constant.
c. Consider the network where $R_{j+1} = 2R_j$ for $j = 0, 1, \ldots, n$, and $R = R_0 = r_0 = r_1 = \cdots = r_{n+1}$.
d. Consider the problem of determining the voltage drops across R_k in networks of this type.

6 Circuit Response

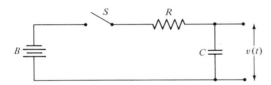

The diagram shown here represents a simple *R-C* circuit consisting of a battery (*B* volts), a resistor, a capacitor, and a switch (shown in the open position). It is desired to know the voltage $V(t)$ across the capacitor that results from closing the switch for a time $\bar{t}$, and then opening it.

Assuming an idealized switch, the voltage across the circuit is defined by a function whose graph is shown here.

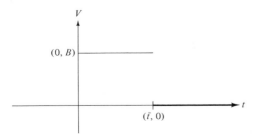

This is a piecewise continuous function defined in terms of the Heaviside function (see Section 37) as

$$BH(t) - BH(t - \bar{t})$$

Thus the voltage, $V(t)$, across the capacitor satisfies the equation

$$(RC)DV + V = BH(t) - BH(t - \bar{t})$$

and the Laplace transform of this equation is:

$$RC\mathscr{L}(DV) + \mathscr{L}(V) = B\mathscr{L}[H(t)] - B\mathscr{L}[H(t - \bar{t})]$$

This in turn becomes

$$RC[s\mathscr{L}(V) - V(0)] + \mathscr{L}(V) = \frac{B}{s} - \frac{Be^{-st}}{s}$$

Assuming that $V(0) = 0$, that is, there was no charge on C prior to the closing of the switch, then

$$\mathscr{L}(V) = \frac{B\left(\dfrac{1}{s} - \dfrac{e^{-ts}}{s}\right)}{RCs + 1} = \frac{B}{RC}\left[\frac{1}{s[s + (RC)^{-1}]} - \frac{e^{-\bar{t}s}}{s[s + (RC)^{-1}]}\right]$$

$$= B\left[\frac{1}{s} - \frac{1}{s + (RC)^{-1}} - \frac{e^{-\bar{t}s}}{s} + \frac{e^{-\bar{t}s}}{s + (RC)^{-1}}\right]$$

Thus

$$V(t) = B[1 - e^{-t/RC} - H(t - \bar{t}) + e^{-((t-\bar{t})/RC)}H(t - \bar{t})]$$

Evidently, $V(t)$ is maximal when $t = \bar{t}$, $V(\bar{t}) < B$, and $\lim_{t \to \infty} V(t) = 0$. The graph of $V(t)$ is shown here.

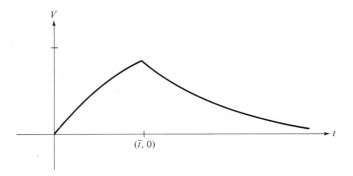

Some related questions are given in the following list.

a. Assume that $\bar{t} > RC \ln(2)$. Is there a value of $t > \bar{t}$ such that $V(t) = B/2$?
b. Suppose that $V(0) = B/2$. What is the graph of $V(t)$?
c. Suppose that $V(0) = 0$ and that the switch is closed but is not opened again. What is $V(t)$ and what is $\lim_{t \to \infty} V(t)$?

7 Mixing Problems

A number of applications amount to what are called mixing problems. The material being mixed might be chemical compounds, voters, parts on an assembly line, and so forth.

As an example of a simple mixing problem, suppose that a tank contains 40 gallons of a salt solution, the total amount of the salt being 10 pounds. At a certain time, a salt solution with a concentration of 2 pounds/gallon is pumped into the tank at a rate of 3 gallons/minute, while liquid is being drained from the tank at the same 3-gallon/minute rate. Assume that during this time the solution in the tank is being stirred to keep the concentration of salt uniform throughout the tank.

Let $x(t)$ be the total amount of salt in the tank at any time t. Then $x(0) = 10$, and salt is entering the tank at a rate of 6 pounds/minute. The rate at which salt is leaving the tank is $(\frac{3}{40})x(t)$ since the amount of solution in the tank remains fixed at 40 gallons. Thus

$$Dx = 6 - \tfrac{3}{40}x; \qquad x(0) = 10$$

is the initial-value problem that describes this mixing problem.

The methods of Section 8 or Section 13 can be used to solve this first-order linear initial-value problem, and the solution is:

$$x(t) = 80 - 70 \exp\left[\frac{-3t}{40}\right]$$

To find, for example, how long it will take to double the amount of salt initially in the tank, we could solve

$$x(t) = 20$$

or

$$80 - 70 \exp\left[\frac{-3t}{40}\right] = 20$$

Thus

$$t = (\tfrac{40}{3})(\ln 7 - \ln 6)$$

and with the help of a table, we have that $2.06 > t > 2.00$, so that it takes about two minutes.

Clearly there will always be less than 80 pounds of salt in the tank and

$$\lim_{t \to \infty} x(t) = 80$$

Some related questions about this mixing problem are these: How will the problem be changed if the solution is being drained from the tank at a rate of 4 gallons/minute? Will it take more or less time until there are 20 pounds of salt in the tank?

For a slightly more involved mixing problem, consider the situation depicted in the diagram shown here.

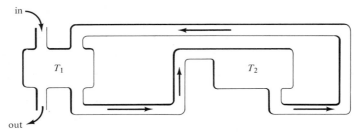

Here the two tanks T_1 and T_2 each contain a salt solution, and there is a flow of 4 gallons/minute circulating through the piping from T_1 to T_2. Suppose that, in addition to this flow between the tanks, plain water is flowing into T_1 at 3 gallons/minute while the salt solution is flowing out of T_1 at the same rate.

We make two simplifying assumptions, namely, that the solutions in the tanks are kept uniform by stirring, and that the volume of the solution in the connecting pipes is negligible. Notice that in this problem the volume of the solutions in T_1 and T_2 remains fixed. Suppose that T_1 has 100 gallons of solution and T_2 has 200 gallons of solution, and let $s_1(t)$ and $s_2(t)$ be the amount of salt in the respective tanks at time t. Evidently,

$$Ds_1 = -\tfrac{3}{100}s_1 - \tfrac{4}{100}s_1 + \tfrac{4}{200}s_2 = \tfrac{-7}{100}s_1 + \tfrac{2}{100}s_2$$

$$Ds_2 = \tfrac{4}{100}s_1 - \tfrac{4}{200}s_2$$

These equations can be expressed as the linear system

(A) $$D\begin{bmatrix} s_1 \\ s_2 \end{bmatrix} = \left(\frac{1}{100}\right)\begin{bmatrix} -7 & 2 \\ 4 & -2 \end{bmatrix}\begin{bmatrix} s_1 \\ s_2 \end{bmatrix}$$

and this system can be solved using the methods of Section 40.

Some related exercises are given in the following list.

a. Find all solutions of equation (A).

b. Let $s(t) = s_1(t) + s_2(t)$ be the total amount of salt in the system at time t. Find t such that $s(t) = \tfrac{1}{2}s(0)$.

c. Show that if a salt solution is flowing into T_1, then the resultant linear system is nonhomogeneous.

d. Can you formulate a mixing problem of this type which leads to a linear system with a symmetric matrix?

8 Epidemics

Suppose that in a total population of organisms:

$x(t)$ is the number of members who are susceptible to, but are not infected by, a certain disease;

$y(t)$ is the number of members who are infected; and

$z(t)$ is the number of members who either expired, recovered, or are immune.

Each of these are functions of time, and it is assumed that for each t the sum of these functions is a constant n (the total population). One theory to explain the spread of an epidemic is based upon the assumptions that for some positive constants a and b,

$$Dx = -axy$$

$$Dy = -Dx - by$$

This is a pair of (coupled) first-order nonlinear equations. While equations of this type were not covered explicitly in the text, it is possible to obtain certain information about the solutions of these equations by using elementary methods. In terms of differentials, the pair of equations become

$$dx = (-axy)\, dt$$

$$dy = (axy - by)\, dt$$

Thus

$$\frac{dy}{dx} = \frac{(axy - by)}{(-axy)} = -1 + \left(\frac{b}{a}\right) x^{-1}$$

which is a separable equation. Therefore,

(A) $$y = -x + \left(\frac{b}{a}\right) \ln(x) + c$$

is a one-parameter family that gives the relationship between y and x. It is easy to show that the absolute maximum of this function on the interval $(0, n)$ occurs at $x = (b/a)$.

By using the assumption that $x + y + z = n$, together with equation (A), we have that

$$z = n - \left(\frac{b}{a}\right) \ln(x) - c$$

Thus it is fairly easy to get relationships for y and z in terms of x.

Finally, yet another possibility, in so far as studying this proposed model of the spread of an epidemic, is to use equation (A) in the first equation

$$Dx = -axy$$

and thereby get the relationship between x and t. This equation is:

(B) $$Dx = ax^2 - bx \ln(x) - acx$$

The constant c can be determined from appropriate initial conditions. For example, if $z(0) = n/2$ and $y(0) = 0$, that is, if half of the population has been inoculated and are therefore immune, and no member of the population has the disease, then $x(0) = n/2$ and

$$c = \left(\frac{n}{2}\right) - \left(\frac{b}{a}\right) \ln\left(\frac{n}{2}\right)$$

Some related questions are given in the following list.

a. Equation (B) is a nonlinear equation of a type studied in Chapter 10. Identify it.
b. Use the Taylor series method to determine the first three terms of the series solution of equation (B) with initial conditions $y(0) = 0$; $z(0) = (n/2)$. Is the disease spreading through the population?
c. What is a physical interpretation of the initial conditions $z(0) = 0$; $y(0) = (n/2)$? Determine the constant c in this case.
d. Use the initial conditions in Problem c to determine $Dx(0)$, and give an interpretation of this result.
e. Assume that $a > (2b)/n$. Use the initial conditions in Problem c to show that $Dy(0)$ is positive. What is the meaning of this?

9 A Simple Pendulum

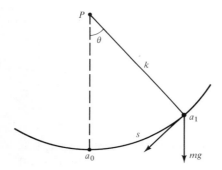

The diagram shown here represents the path of motion of a pendulum that consists of a wire of length k supported from a pivot at P and having a weight of mass m at the point a_1. The mass moves along the circular arc and s (a function of time) is the distance along the arc from a_1 to a_0.

From elementary mechanics, we have that the tangential force due to gravity is $mg \sin \Theta$. Also, since $s = k\Theta$, it follows that

$$D^2 s = k\, D^2 \Theta$$

so that the force $m\, D^2 s$ can be expressed as $mk\, D^2 \Theta$. Thus the equation

$$mk\, D^2 \Theta = -mg \sin \Theta$$

or

(A) $D^2 \Theta + r \sin \Theta = \varnothing$

where $r = (g/k) > 0$, describes the motion of this pendulum.

Quite a number of simplifying assumptions are implicit in equation (A). Some are the following: the wire is rigid and weightless; there is no friction at the pivot point; and no air friction is encountered. Thus this is a highly idealized model of the true physical situation.

Equation (A) is a second-order nonlinear differential equation, but it is related to a first-order nonlinear equation as well as a second-order linear equation. Because of the importance of equation (A) in applications, much is known and has been written about it. We discuss three approaches to an analysis of a particular initial-value problem associated with equation (A).

In order to simplify matters, we select the special case of $r = 1$; this amounts to making a specific choice for the length k, namely, $k = g$. Thus in our special case, the equation is:

(B) $D^2\Theta + \sin \Theta = \varnothing$

The initial conditions to be used in the illustrations are:

(C)
$$\Theta(0) = \frac{\pi}{12}$$

$$D\Theta(0) = -\tfrac{3}{2}$$

The above initial conditions are consistent with the diagram since they describe the case where the weight is moving along the arc from a_1 towards a_0, and thus $\Theta(t)$ is a decreasing function of time. We consider the particular problem of finding the time t at which the weight reaches the point a_0. This amounts to solving $\Theta(t) = 0$.

The first approach involves "linearizing" equation (B). Since $\lim_{\Theta \to 0} (\sin \Theta / \Theta) = 1$, it follows that for "small" values, Θ is approximately equal to $\sin \Theta$. The initial value of Θ in (C) is $\pi/12$ (or, equivalently, $15°$), which may not be considered "small." Nonetheless, we consider the "linearized" equation

(D) $D^2\Theta + \Theta = \varnothing$

together with the initial conditions (C). The solution of this initial-value problem is:

$$\Theta(t) = \left(\frac{\pi}{12}\right) \cos t - (\tfrac{3}{2}) \sin t$$

Thus $\Theta(t) = 0$ if and only if

$$(\tfrac{3}{2}) \sin t = \left(\frac{\pi}{12}\right) \cos t$$

or

$$\tan t = \frac{\pi}{18}$$

or

$$t = \operatorname{arc\,tan}\left(\frac{\pi}{18}\right)$$

By use of a table, we find that

(E) $t = 0.172$

is the approximate time for the pendulum to reach the point a_0.

A second approach involves a formal power series solution of equation (B) rewritten in the form

$$D^2\Theta = -\sin\Theta$$

Thus $D^3\Theta = -\cos\Theta\, D\Theta$, and so on, so that the initial conditions (C) can be used to determine $D^n\Theta(0)$ for $n = 2, 3, 4, 5, \ldots$. Therefore, a power series representation of $\Theta(t)$ is:

$$\Theta(t) = \Theta(0) + D\Theta(0)t + \frac{D^2\Theta(0)t^2}{2} + \cdots$$

By using the facts that

$$\sin\frac{\pi}{12} = \frac{\sqrt{2}}{4}(\sqrt{3} - 1)$$

and

$$\cos\frac{\pi}{12} = \frac{\sqrt{2}}{4}(\sqrt{3} + 1)$$

we have that

$$\Theta(t) = \frac{\pi}{12} - \tfrac{3}{2}t - \frac{\sqrt{2}}{8}(\sqrt{3} - 1)t^2 + \frac{\sqrt{2}(\sqrt{3} + 1)}{16}t^3 + \cdots$$

If only the first three terms in the series representation of $\Theta(t)$ are used, then the solution of $\Theta(t) = 0$ amounts to solving a quadratic equation. The approximate value

(F) $t = 0.164$

is the positive solution. If four terms are used to approximate $\Theta(t)$, then

$$\Theta(t) = \frac{\pi}{12} - \tfrac{3}{2}t - \frac{\sqrt{2}}{8}(\sqrt{3} - 1)t^2 + \frac{\sqrt{2}}{16}(\sqrt{3} + 1)t^3$$

and for this choice we find that $\Theta(0.172) > 0$, whereas $\Theta(0.173) < 0$. This agrees quite well with the approximate value in (E).

Another approach to the problem begins by noticing that $2\, D\Theta$ is an integrating factor for equation (B). On multiplying through by $2\, D\Theta$, equation (B) becomes

$$2\, D\Theta\, D^2\Theta + 2\, D\Theta \sin\Theta = \emptyset$$

which implies that

(G) $(D\Theta)^2 - 2\cos\Theta = b$

where b is a constant. A choice of b that is consistent with the initial conditions (C) is found by solving

$$[D\Theta(0)]^2 - 2\cos[\Theta(0)] = b$$

or

$$b = \frac{9}{4} - \frac{2\sqrt{2}}{4}(\sqrt{3} + 1) = \frac{9 - 2\sqrt{2}(\sqrt{3} + 1)}{4}$$

While equation (G) is first order, it is a second-degree equation. However, it is equivalent to

$$(D\Theta)^2 = 2\cos\Theta + b$$

and although there are two possibilities in solving for $D\Theta$, evidently the choice

(H) $$D\Theta = -\sqrt{2\cos\Theta + b}$$

is consistent with the problem of finding a solution of $\Theta(t) = 0$ since Θ is clearly a decreasing function as the pendulum moves from a_1 to a_0.

The first-order nonlinear equation (H) is separable, that is, it could be written as

$$\frac{d\Theta}{\sqrt{2\cos\Theta + b}} = -dt$$

However,

$$\int_{\pi/12}^{\Theta} \frac{dx}{\sqrt{2\cos x + b}}$$

is quite a problem in its own right since there is no elementary function whose derivative is the integrand. This is a so-called *elliptic integral*. While the study of elliptic integrals is interesting, it is not an elementary topic.

We conclude by considering the application of Euler's method to equation (H). To do this, we take a decimal approximation for the constant b. The initial-value problem that arises from Euler's method is:

(I)
$$V(k + 1) = V(k) - h\{2\cos[V(k)] + 0.318148\}^{1/2}$$
$$V(0) = 0.261799$$

where $V(0)$ is a decimal approximation of $\Theta(0) = \pi/12$.

The solution of (I) was calculated using eight-decimal precision. Four choices for h were used, namely $h_i = 10^{-(1+i)}$, where $i = 1, 2, 3, 4$. In each case, the value of j was determined such that $V(j) \geq 0$ and $V(j + 1) \leq 0$. Since $V(k)$ is an approximation of $\Theta(hk)$, the approximate solution of $\Theta(t) = 0$ was computed by linear interpolation, namely,

$$t_i = h_i\left[j + \frac{V(j)}{V(j) - V(j + 1)}\right]$$

$i = 1, 2, 3, 4$. The results of these calculations are:

$t_1 = 0.17288$ where $h = 0.01$

$t_2 = 0.17281$ $h = 0.001$

$t_3 = 0.17280$ $h = 0.0001$

$t_4 = 0.17281$ $h = 0.00001$

Notice that $t_1 > t_2 > t_3$ but that $t_4 > t_3$. Evidently, in changing from $h = 0.001$ to $h = 0.00001$, the theoretical advantages were offset by the additional rounding errors.

10 Curves of Pursuit

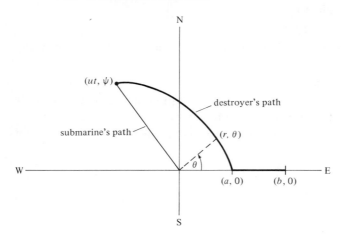

The graph shown here represents, in polar coordinates, a curve of pursuit. The object giving chase follows the dark curve with the intention of intercepting the fleeing object which is following the straight line path.

For example, suppose a destroyer at the point $(b, 0)$ sights a submerging submarine at $(0, 0)$. The destroyer's captain assumes that the submarine follows a straight line path of unknown direction but at a known speed of u miles per hour. What path should the destroyer follow in order to assure that it intercepts the submarine?

In order to consider a specific illustration, suppose that the submarine's speed is $u = 10$ while the destroyer's speed is $u = 20$. Thus the position of the submarine at time t is $(10t, \psi)$, where ψ is unknown.

If at the time of sighting $(t = 0)$, the destroyer is at the point $(6, 0)$, then the destroyer might proceed to the point $(2, 0)$ since because of the relative speeds the destroyer can travel 4 miles while the submarine travels 2 miles. At this time, the destroyer and the submarine are both 2 miles from $(0, 0)$. Also, had the submarine gone due east after submerging, then the destroyer would have intercepted it at $(2, 0)$.

Next, suppose the destroyer turns north and follows the dark curve. Let s be the distance traveled along this path from the point $(2, 0)$ to the point (r, θ). By using the polar form of the distance formula, we have that

(A) $(Ds)^2 = r^2(D\theta)^2 + (Dr)^2$

By assumption, $Ds(t) = 20$, which is the speed of the destroyer. Also, in order for interception to take place, it is necessary that

(B) $r(t) = 10t$

that is, the destroyer should be the same distance from the origin as the submarine. Thus substitution of (B) into (A) gives that

(C) $400 = 100t^2(D\theta)^2 + 100$

since $Ds(t) = 20$.

Equation (C) simplifies to

$$(D\theta)^2 = 3t^{-2}$$

and since it was assumed that the destroyer turned north, it follows that $D\theta > 0$. Therefore, $\theta(t)$ is a solution of

$$D\theta = \sqrt{3}\, t^{-1}$$

or

$$\theta(t) = \sqrt{3}\, \ln(t) + c$$

for some constant c.

We have from (B) that $t = r/10$, so that

$$\theta(t) = \sqrt{3}\, \ln\left(\frac{r}{10}\right) + c$$

Since $\theta = 0$ when $r = 2$, it follows, by solving for c, that

$$\theta = \sqrt{3}\, \ln\left(\frac{r}{2}\right)$$

or

(D) $r = 2\exp[3^{-1/2}\theta]$

is the path that the destroyer follows.

Some related exercises are given in the following list.

a. Use equation (D) to show that the time (in hours) from sighting to interception is less than

$$\frac{\exp[3^{-1/2}(2\pi)] + 1}{5}$$

b. Suppose that the destroyer turns south at $(2, 0)$. Derive the curve of pursuit in this case.

c. Suppose that the destroyer captain assumes that the submarine went due west. Find a curve of pursuit for this case.

d. Find a sharp upper bound for the total number of miles that the destroyer might travel before intercepting the submarine. If the destroyer's speed is only 15 miles per hour, what is the upper bound for the total miles traveled?

11 Families of Curves and Orthogonal Trajectories

The equation

(A) $y - 1 = c(x - 2)^2$

where c is an arbitrary constant, describes a one-parameter family of curves. Each member of the family is a parabola whose vertex is at $(2, 1)$ and whose axis is the line $x = 2$. The graphs of several members of this family are shown in the figure.

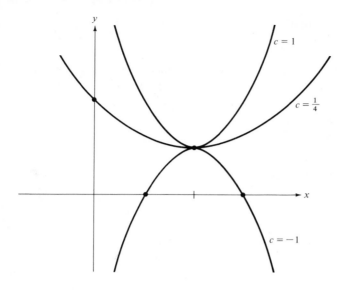

Through each point (x_0, y_0) in the plane, $x_0 \neq 2$, there passes exactly one member of the family. For if we fix (x_0, y_0), $x_0 \neq 2$, then c is uniquely determined by the equation $y_0 - 1 = c(x_0 - 2)^2$, or

$$c = \frac{(y_0 - 1)}{(x_0 - 2)^2}$$

Fix any point (x, y), $x \neq 2$. By using equation (A), the slope of the parabola at (x, y) is $Dy = 2c(x - 2)$. But from (A), $c = (y - 1)/(x - 2)^2$, and we conclude that

(B) $$Dy = \frac{2(y - 1)}{(x - 2)}$$

Thus every member of the one-parameter family of parabolas is a solution of the first-order differential equation (B). Equivalently, equation (A) is a one-parameter family of solutions of the differential equation (B).

In a similar manner, the equation

(C) $$x^2 - y^2 = 2cx$$

where c is an arbitrary constant, is a one-parameter family of hyperbolas. To find a differential equation which has (C) as a one-parameter family of solutions, we differentiate (C) to obtain

(D) $$2x - 2y\,Dy = 2c$$

and then eliminate the constant c using the two equations (C) and (D). The result is:

(E) $$2xy\,Dy - x^2 - y^2 = \emptyset$$

Consider the one-parameter family of ellipses

(F) $\left(\dfrac{x-2}{2}\right)^2 + (y-1)^2 = c$

each member of which has its center at $(2, 1)$ and its major axis on the line $y = 1$. By differentiating (F) and eliminating c as illustrated above, we find that

(G) $Dy = -\dfrac{(x-2)}{2(y-1)}$

is a first-order differential equation having (F) as a one-parameter family of solutions. Compare the differential equations (B) and (G), and note that Dy in (B) is the negative reciprocal of Dy in (G). The geometric interpretation of this is that if (x, y) is a point of intersection of a parabola from (A) and an ellipse from (G), then their respective tangents at (x, y) are perpendicular, or *orthogonal*. The two curves are said to be orthogonal to each other at (x, y). See the figure.

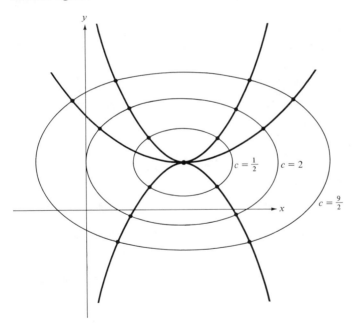

In general, if each of F_1 and F_2 is a one-parameter family of plane curves, then F_1 and F_2 are said to be *orthogonal trajectories* if each member of F_1 is orthogonal to each member of F_2.

Examples of orthogonal trajectories occur naturally in a variety of physical situations. Iron filings sprinkled on a pane of glass atop a bar magnet arrange themselves along a family of curves F_1 indicating the direction of magnetic force. The set of curves F_2 in the plane of the glass of equal potential energy is orthogonal to the members of F_1. In hydro-dynamics, the flow of a fluid from a large reservoir into a narrow channel produces a family F_1 of streamlines. The curves of equal velocity potential are orthogonal to the family of streamlines. See the figure.

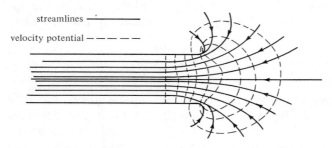

A procedure for finding a one-parameter family F_2 which is or-thogonal to a given family F_1 is:

(i)   determine a first-order differential equation having the F_1 as a family of solutions

(ii) replace Dy in the differential equation by $-1/Dy$, producing the differ-ential equation for the orthogonal family F_2

(iii) determine the one-parameter family of solutions of the differential equation produced in (ii).

For example, if we apply this procedure to the family of hyperbolas (C), then we obtain from (E) the differential equation

(H) $2xy + (x^2 + y^2)\, Dy = \varnothing$

which can be solved using the methods in Chapter 10.

Some related questions are given in the following list.

a. Determine the orthogonal trajectories of (C) by finding a one-parameter family of solutions of (H).

b. Determine a family of orthogonal trajectories for each of the following families.

(1) all circles passing through the origin with centers on the x-axis

(2) $xy = c$

(3) $(x^2/c^2) + [y^2/(c^2 - y)] = 1$

c. Determine a second-order differential equation having the family of curves

$$y(x) = c_1 x + c_2 x^{1/2}$$

as a two-parameter family of solutions.

d. Determine a second-order differential equation having the family of curves

$$(x - c_1)^2 + (y - c)^2 - 1 = \varnothing$$

as a two-parameter family of solutions.

Answers
to Selected
Exercises

Please note that the first digit indicates the section, and the second digit indicates the Exercise number in that section.

(7.4) 1a. first-order ordinary; 1c. second-order partial; 1e. third-order ordinary; 2a. y_1 is a solution and y_2 is not a solution; 2c. y_1 is a solution and y_2 is not a solution; 2f. u is not a solution; 2h. y_1 is a solution; 2j. y_1 and y_2 are solutions; 4c. $r = 0, 1, 2$; 5a. $r = 0$; 5c. $r = \frac{1}{2}, \frac{3}{2}$.

(7.12) 1. $y(x) = 2e^{5x}$; 3. $y(x) = (1 - 2e^{x-1})^{-1}$; 5. $y(x) = 3^{-1}\cos(3x) - \sin(3x)$; 7. $y(x) = 9x^{1/2} - (\frac{17}{4})x$; 10c. $y(10) = y_0 e^{-(1/10)\ln 2}$ or approximately $(0.9)y_0$; 11b. $y(1) = 1000e^{0.06}$ or approximately 1061.80; 11c. 7% simple interest is better.

(8.6) 1a. nonhomogeneous with singular points at $x = 1, -1$; 1c. nonhomogeneous with no singular points; 1e. nonhomogeneous with singular points at $x = n\pi$, n an integer; 3a. $y(x) = 6x^{-1} + \frac{1}{2}x$; 3c. $y(x) = \emptyset$; 4b. $y(x) = x^2 + 1 + c(x^2 + 1)^{-1}$.

(8.11) 1a. $y(x) = ce^{3x}$; 1c. $y(x) = ce^{-x^2}$; 1e. $y(x) = c/x$; 1g. $y(x) = c(1 - x^2)^{-1}$; 2b. $y(x) = (\frac{2}{3})(x + 1)$; 2d. $y(x) = x^{-2}$; 2f. $y(x) = -(x^2 + 1)^{-1}$; 6c. $y(x) = c$; 7b. $\int_0^\infty q(x)\,dx = \infty$.

(8.17) 1a. $y(x) = ce^{x^2/2} - 1$; 1c. $y(x) = ce^x + 2xe^x$; 1e. $y(x) = \sin x + c\cos x$; 1f. $y(x) = e^{-x^3}(c - \int_0^x e^{t^3}\,dt)$; 1g. $y(x) = ce^{-x^2} + x^2 - 1$; 1i. $y(x) = \csc(x)[c + \frac{1}{2}x - \frac{1}{4}\sin(2x)]$; 2a. $y(x) = 2e^{-2x} + 2x - 1$; 2c. $y(x) = e^{1-x^2} + \frac{1}{2}(x^2 - 1)$; 2e. $y(x) = x^2\cos(x) - \frac{1}{4}\pi x\cos(x)$; 3b. $y(x) = cx^{-1} + (\frac{1}{3})x^2 + (1 - x^{-1})e^x$; 6b. $(i)y^{-1}(x) = cx - (\frac{1}{3})x^4$; 7a. $b = a^{-1}, c = a^{-2}(a - 1)$; 8a. $y(x) = [b\sin(ax) - a\cos(ax)]/(a^2 + b^2)$; 8b. $y(x) = \lim_{x\to\infty} z(x)$; 9a. $y(x) = 3x^{-4} + x^5$; 9b. $y(1) = 1$.

(9.8) 1a. nonhomogeneous with singular point at $x = -1$; 1c. homogeneous with singular point at $x = 0$; 1e. homogeneous with no singular points; 2c. $p(x) = ax(x - 1)$ for any $a \neq 0$; 3c. $y(x) = e^{-x^2}$; 4c. there is not.

(9.15) 2b. $y(x) = x$;   2c. there is not;   3c. $y(x) = x \ln x$;
4c. $y(x) = (\frac{8}{5})x^3 + (\frac{2}{5})x^{-2} + x^3 \ln(x) - 2x^2$.

(10.7) 1a. and 1c. linearly independent;   1d., 1f., and 1g. linearly dependent.

(10.14) 1b. $y(x) = e^{4x} + e^{-2x}$;   2b. $y(x) = \frac{1}{2}x^3 + (\frac{5}{2})x^{-1}$;   5b. dependent;
5c. independent;   9a. independent;   9c. dependent;   9f. independent.

(10.21) 2b. $y(x) = c_1 x^4 + c_2 x^{-1}$;   2c. $y(x) = \varnothing$;   3b. $y(x) = c_1 e^x + c_2 e^{-x} + c_3 e^{2x}$;
3c. $y(x) = -\frac{1}{2}e^x + (\frac{1}{6})e^{-x} + (\frac{1}{3})e^{2x}$;   4a. $p(x) = ax + a, a \neq 0$,
$q(x) = bx^2 + cx + (b + c), b \neq 0$;   4b. $y(x) = c_1(x + 1) + c_2(x^2 + 1)$;
7a. $\det[W(x)] = \det[W(a)] \exp(-\int_a^x p_1(t)\, dt)$;   7b. $\det[W(x)] = k$ (constant);
9a. $D^2 y = \varnothing$;   9b. $D^2 y - 4y = \varnothing$;   9d. $D^2 y + 4y = \varnothing$;
10. $b_1 c_2 - b_2 c_1 \neq 0$.

(11.9) 1c. $y(x) = 3\cos(x) - \sin(x) + (3x^2 - 2)$;   2c. $y(x) = c_1 + c_2 \ln(x) + \frac{1}{4}x^2 + x$;
4c. $y(x) = \frac{1}{2}(-7e^x + 4xe^x + 3e^{-x} + 4 + 4x + 2xe^{-x})$;
5b. $y(x) = c_1 x + c_2(x^2 - 1) + x \ln(x)$;
7d. $y(x) = c_1 e^x + c_2 e^{3x} + (\frac{1}{10})\sin(x) + (\frac{1}{5})\cos(x) - \frac{1}{2}xe^x$.

(12.7) 1. $\{e^{-2x}, xe^{-2x}\}$;   3. $\{x, x^2 + 1\}$;   5. $\{e^{x^2}, e^{-x^2}\}$;   7. $\{e^x - 1, \frac{1}{2}(e^x + 1)^{-1}\}$;
9. $y(x) = c_1 x^{-1/2} \sin(x) + c_2 x^{-1/2} \cos(x)$;
10. $y(x) = c_1 x + c_2\{x \ln[(1 + x)/(1 - x)]^{1/2} - 1\}$.

(12.11) 1. $z(x) = \frac{1}{2}\sec(x)$;   3. $z(x) = x^2 + 1$;   5. $z(x) = x \sin(x) + \cos(x) \ln(\cos x)$;
7. $z(x) = (\frac{1}{10})e^x[\cos(x) - 3\sin(x)]$;   9. $z(x) = -1 - x^2$;
11. $z(x) = (\frac{1}{6})x^2(x^2 - 3)$;   13. $y(x) = c_1 x + c_2 x^{-1} + x \ln(x)$;
15. $y(x) = x + (\frac{1}{6})x^4 + \frac{1}{2}x^2$.

(12.16) 1. $y(x) = c_1 x^2 + c_2 x + c_3 - \sin(x)$;   2. $y(x) = c_1 x + c_2 x^2 + c_3 x^3$;
4. $y(x) = c_1 e^x + c_2 e^{2x} + c_3 e^{-x} - xe^x$;   5b. $y(x) = c_1 + c_2 x^{-2} + c_3 x^3 + x^4$.

(13.13) 1c. $y(x) = c_1 e^{-x} + c_2 e^{4x}$;   3c. $y_2(x) = xe^{3x}$;   5. $\{e^x, xe^x, x^2 e^x\}$;
6b. $P(t) = t^3 - (\frac{5}{2})t^2 - 7t + 4$.

(14.8) 1. $y(x) = c_1 e^{-x} + c_2 e^{-3x}$;   3. $y(x) = c_1 + c_2 e^{2x}$;   5. $y(x) = c_1 e^{x/2} + c_2 e^{2x}$;
7. $y(x) = c_1 e^{-2x} + c_2 xe^{-2x}$;   9. $y(x) = c_1 e^x + c_2 xe^x$;
11. $y(x) = c_1 e^{-x} + c_2 e^x + xe^x + x^2 + 2$;
13. $y(x) = c_1 \sin(x) + c_2 \cos(x) + 2e^x(x - 1)$.

(15.9) 1. $y(x) = c_1 e^x + c_2 e^{2x} + c_3 e^{3x}$;
3. $y(x) = e^x[c_1 \cos(2x) + c_2 \sin(2x)] + c_3 e^{2x} + c_4 e^{-2x}$;
5. $y(x) = c_1 \sin(x) + c_2 \cos(x) + e^{2x}[c_3 \sin(3x) + c_4 \cos(3x)]$;
9c. $e^{-x}(-12x^2 + 32x - 14)$;   10c. $e^{2x}(x^3 + 6x^2 + 6x)$.

(15.16) 1a. $y(x) = c_1 e^x + e^{-(1/2)x}[c_2 \sin(\frac{3}{4})^{1/2}x + c_3 \cos(\frac{3}{4})^{1/2}x]$;
1c. $y(x) = c_1 + e^{-3x}[c_2 \cos(2x) + c_3 \sin(2x)]$;
1e. $y(x) = c_1 \sin(x) + c_2 \cos(x) + c_3 \sin(2^{1/2}x) + c_4 \cos(2^{1/2}x)$;
1g. $y(x) = c_1 e^{-x} + c_2 xe^{-x} + c_3 x^2 e^{-x} + c_4 e^x + e^{-x}[c_5 \sin(x) + c_6 \cos(x)]$;
1i. $y(x) = c_1 e^x + e^{(1/2)x}[c_2 \sin(\frac{7}{4})^{1/2}x + c_3 \cos(\frac{7}{4})^{1/2}x]$;
1k. $y(x) = c_1 e^{2x} + c_2 xe^{2x} + c_3 \sin(x) + c_4 \cos(x)$;
2b. $y(x) = 2(e^x - 2e^{2x} + e^{3x})$;   2d. $y(x) = (\frac{1}{15})[3e^x - \sin(3x) - 3\cos(3x)]$;
3b. $D^2 y - 10\, Dy + 25y = \varnothing$;   4a. $D^3 y - D^2 y + Dy - y = \varnothing$;
4c. $D^4 y + 5\, D^2 y + 4y = \varnothing$;   7a. $D^2 y(0) = 4e$;   7b. $y(0) = k, Dy(0) = -k$,
$D^2 y(0) = k$ (constant);   9b. (ii) $y(x) = x[c_1 \sin(\ln x^2) + c_2 \cos(\ln x^2)]$.

(16.7) 1c. $D^5 y + 2\, D^4 y + D^3 y + 2\, D^2 y = \varnothing$;   2b. $P(t) = t^3(t^2 + 4)(t + 1)$;
2d. there is no polynomial;   3a. $P(t) = (t - 1)^2(t^2 + 1)$;
3c. $P(t) = t^3(t + 1)^2(t^2 + 4t + 5)$;   3e. $P(t) = t(t^2 + 4)(t^2 - 2t + 5)$.

(16.15) 1. $y(x) = c_1 e^x + c_2 e^{2x} + (\frac{1}{10})[\sin(x) + 3\cos(x)]$;
3. $y(x) = c_1 e^{-x} + c_2 xe^{-x} + c_3 e^{-4x} + (\frac{1}{6})x^2 e^{-x}$;
5. $y(x) = c_1 e^{-2x} + c_2 e^x - (\frac{1}{3})xe^x + \frac{1}{2}x^2 e^x$;

7. $y(x) = c_1 e^x + c_2 x e^x - \frac{1}{2}[\sin(x) + \cos(x) + x \sin(x)]$;

9. $y(x) = c_1 e^{-2x} + c_2 x e^{-2x} + (\frac{1}{6})x^3 e^{-2x}$;

11. $y(x) = c_1 e^{-2x} + c_2 e^x + \frac{1}{4}e^{2x} - e^{-x}$;

13. $y(x) = e^x[c_1 \sin(x) + c_2 \cos(x) + \frac{1}{2}x \sin(x)]$;   15. $y(x) = e^x - 1$;

17. $y(x) = \varnothing$;   19. $y(x) = (\frac{1}{12})e^{2x} + (\frac{13}{15})e^{-x} - (\frac{3}{20})\sin(2x) + (\frac{1}{20})\cos(2x)$;

21c. all solutions bounded for $b \neq r$ and all solutions unbounded for $b = r$;

22b. $y(0) = k, Dy(0) = -(k + \frac{4}{3})$.

(16.19)   1. $y(x) = c_1 e^x + c_2 x e^x + x^2 e^x(2 \ln(x) - 3)$;   3. $z(x) = x e^{-x}$;

5. $z(x) = (\frac{1}{4}) \sin(2x)$;   6. $z(x) = (\frac{1}{6})(x^2 e^x - 3x e^{-x})$.

(17.7)   1a. 18;   1c. 3;   1e. $6(-1)^{k+1}$;   1h. $(-1)^{k+1}(k + 2)k!$;   2a. 2;

2c. $(k + 1)!$;   6. Ef is increasing and Δf is positive;   8. yes.

(18.8)   1a. $k!/(k^2 + k + 1)$;   1c. $7^k 6^3$;   1e. $k!(k - 1)2^{-(k+1)}$;   1g. $2^k(1 - k)/(k + 1)!$;

4. $\Delta^3 f(k) = -8(-1)^k - 27(-2)^k - 64(-3)^k$;   6a. $x = -\frac{1}{2}$;   6c. no solution;

6f. $x = -3, -5$.

(18.14)   6. $\Delta g^3 = (\Delta g)^3 + 3g(\Delta g)^2 + 3g^2(\Delta g)$;

10. $\Delta^n[(-1)^k f(k)] = (-1)^{k+n} \displaystyle\sum_{i=0}^{n} \binom{n}{i} 2^i \Delta^{n-i} f(k)$.

(18.21)   3. yes;   6. $x^2 + p(x)$ where p is periodic;   8. $b = 1$.

(19.9)   1a. 60;   1c. $\frac{1}{2}$;   1f. $-(\frac{1}{24})$;   4b. $x = -1, -2, \ldots, -n$;   5b. $x^{(2)} - x$;

5d. $x^{(-1)}(x + 3)^{(4)}$;   5f. $x^{(-4)}(x + 3)^{(2)}$;   6c. $4x^{(-3)}x$;   7b. 3;   7d. $-\frac{5}{16}$.

(19.21)   2b. $f(x) = x^3 - 6x^2$;   2c. $f(x) = x^4 - 3x^3 + 2x^2 + 7$;

5b. $f(x) = x^{(3)} - 2x^{(1)}$;   10. $a_0 = 4, a_1 = 1, a_2 = 4$;

11. $Ef(x) = x^4 + 4x^3 + (\frac{7}{2})x^2 + (\frac{3}{2})x - 8$,

$Ef(x) = x^{(4)} + 10x^{(3)} + (\frac{45}{2})x^{(2)} + 10x^{(1)} - 8$,

$Ef(x) = 24\dbinom{x}{4} + 60\dbinom{x}{3} + 45\dbinom{x}{2} + 10\dbinom{x}{1} - 8$.

(19.27)   4. polynomial of degree seven.

(20.12)   4b. $y(x) = 5x^{2/3} \ln(x)$;   6. $y(x) = x^{9/2} - x^{-1/2} + 4x^2 + x - 1$;

11. $L_2 = \frac{1}{4}(2x - 1)^2 D^2 + \frac{1}{2}(2x - 1) D - I$;   12. $y(x) = 5(2x - 1)^{-1} + \frac{1}{2}$.

(20.15)   1. $c_1 = 0, c_2 = 1$;   3. $\{x, x \ln(x), x^{-4}\}$;   5. $y(x) = 1 + x + x^2 + \ln(x)$;

10. $y(x) = x^3 + 3x^3 \ln(x) + x^{-1} \ln(x)$.

(21.9)   2b. $\dbinom{x}{4} + p(x)$;   2c. $x^{(2)}x^{(-2)} + p(x)$;   2d. $(-\frac{1}{8})(-7)^x + p(x)$;

5. $(r - 1)^{-2}[(r - 1)x - r]r^x + p(x)$.

(21.14)   1a. 1863;   1c. 209;   1f. $\frac{1}{4}[(-1)^n(1 - 2n) - 1]$;   3a. 2628;   3d. 2510;

4b. divergent;   4d. 2;   4e. 0.

(22.5)   2. yes;   4b. $\varnothing$;   4d. $3(2)^k$;   4g. $8k^{(2)} - 8k^{(1)} - 4$.

(22.8)   1. $[E^2 + (k - 2)E - (k + 1)I]y = 2(k + 1)$;   2b. $(E^8 + E^7 - 3^j I)y = \varnothing$.

(23.11)   1a. $y(k) = k^{(-2)}$;   1b. $y(k) = (-3)^k$;   1f. $y(k) = (k + 2)(k + 1)^{-1}$;

1g. $y(k) = (k + 1)2^k$;   2a. $y(k) = k + 3$;   2c. $y(k) = 8k$;   2f. $y(k) = k + 2$;

2g. $y(k) = \ln(k + 1) - 4$.

(23.13)   1c. $-\frac{2}{9}$;   1e. -24.

(24.14)   1a. dependent;   1f. dependent;   3. $\varnothing$.

(24.21)   4. $y(k) = -\frac{1}{2}(k + 1)^{(2)}$;   5c. $y(k) = k(k!)$;   6. yes.

(25.5)   2a. $y(k) = (-2)^k y(0)$;   4. $y(k) = 4^k - 1$;   5. $Q(-1) \neq 0$.

(25.10)   3a. $y(k) = 3 - 3(-7)^k$;   3d. $y(k) = (18)^{(1/2)k}[2 \cos(\frac{1}{4}k\pi) + \sin(\frac{1}{4}k\pi)]$;

4a. all;   4c. some;   4e. none;   5b. $y(0) = y(1) = 0$;   5d. $y(0) = c$,

$y(1) = -\frac{1}{2}c$.

(25.15)   4b. $P(t) = t^4 - 4t^3 + 16t - 16$;   6a. constant functions;

6c. $y(k) = c_1 + c_2(-\frac{1}{3})^k$;   6d. $y(k) = \varnothing$.

(25.19)   1a. $y(k) = \sin(1 - k) + \sin(k) - \sin(1)$;
1c. $y(k) = 1 - k + (18)^{(1/2)k}[2 \cos(\tfrac{1}{4}k\pi) + \sin(\tfrac{1}{4}k\pi)]$;   1e. $y(k) = 2^k - 1$;
2a. no bounded solutions;   2c. all bounded solutions;   2e. some bounded solutions;   6. $y(k) = c_1 \sin(rk) + c_2 \cos(rk) + 1$.

(27.13)   2. $v(k) = (1 + h)^k$;   3. $v(1) = 1 + h, y(h) = \exp[\sin(h)]$;   5. $v(1) = -3$, $y(4) = e^{-4}$.

(27.43)   1a. $v(k) = a^{-1}(a + b)(1 + ah)^k - (b/a) + \delta(1 + ah)^k$.

(28.26)   2. the first three terms are identical;   9a. unstable;   9b. and 9c. stable.

(28.40)   3. $T(x^4) = h^4[k^4(1 - a_1 - a_2) + 4k^3(2 - a_1 - b_1 - b_2) + 6k^2(4 - a_1 - 2b_1) + 4k(8 - a_1 - 3b_1) + (16 - a_1 - 4b_1)]$;   7. none.

(28.49)   4a. $v(k + 3) = -18v(k + 2) + 9v(k + 1) + 10v(k) + h[9\, Dv(k + 2) + 18\, Dv(k + 1) + 3\, Dv(k)]$.

(29.23)   4. $|e(k, h)| = |(1 + a^{-2})e^{ahk} - (1 + ah + \tfrac{1}{2}(ah)^2)^k|$.

(29.26)   1. $c_1 = [a(t_1 - t_2)]^{-1}\{-t^2 + [1 + ah + \tfrac{1}{2}(ah)^2]\}(a + b)$.

(29.34)   3. $-2 < ah < 0$;   4. the three terms agree with e^{ah}.

(29.38)   2. $v(2) = [v(1)]^2$.

(30.5)   1. $y(x) = a_0 \sum_{n=0}^{\infty} [(-2)^n/n!]x^n = a_0 e^{-2x}$;
3. $y(x) = a_0 \sum_{n=0}^{\infty} [(-1)^n/n!]x^n + (-x + (\tfrac{3}{2})x^2 - \tfrac{1}{2}x^3 + (\tfrac{1}{8})x^4 - (\tfrac{1}{40})x^5 + \cdots) = ce^{-x} + 2x - 3$;
5. $y(x) = a_0 \sum_{n=0}^{\infty} (x^n/n!) + \sum_{n=1}^{\infty} (nx^n/n!) = a_0 e^x + xe^x$;
8. $y(x) = 2(x - 1) - \tfrac{1}{2}(x - 1)^2 + (\tfrac{1}{6})(x - 1)^3 - (\tfrac{1}{24})(x - 1)^4 + \cdots = x - e^{-(x-1)}$;
9b. $y(x) = a_0 \sum_{n=0}^{\infty} [(-1)^n x^{2n}/n!] + \sum_{n=0}^{\infty} [(-1)^n x^{2(n+1)}/(n + 1)!] = ce^{-x^2} + 1$;
9c. $y(x) = 1$;   12. $y(x) = 1 - (x - 1) + \tfrac{1}{2}(x - 1)^2 - (\tfrac{1}{24})(x - 1)^4 + \cdots$.

(30.11)   2. $y(x) = x + (\tfrac{2}{3})x^3 + (\tfrac{8}{15})x^5 + \cdots$ on $(-1, 1)$;
3. $y(x) = 1 + 2^{-(1/2)}(x - \tfrac{1}{4}\pi)^2 - (\tfrac{1}{6})(x - \tfrac{1}{4}\pi)^3 + \cdots$ on $(-\infty, \infty)$;
5. $y(x) = \sum_{n=0}^{\infty} (-1)^n(x - 1)^n$ on $(0, 2)$;   7. $y(x) = \tfrac{1}{2} + (x - 1)$.

(31.9)   1a. no singular points;   1d. $x = 1, -1$ are singular points;
1f. $x = 0$ is a singular point;   1h. $x = 2, 3$ are singular points;
3. $y_1(x) = 1 - [(2x)^2/2!] + [(2x)^4/4!] - \cdots = \cos(2x)$, $y_2(x) = 2x - [(2x)^3/3!] + [(2x)^5/5!] - \cdots = \sin(2x)$, both converge for all x;
5. $y_1(x) = 1 - (x^3/3!) + (4x^6/6!) - \cdots$, $y_2(x) = x - (2x^4/4!) + (10x^7/7!) - \cdots$, both converge for all x;
7. $y_1(x) = \sum_{n=0}^{\infty} [(x + 1)^{2n}/2^n n!], y_2(x) = \sum_{n=0}^{\infty} [2^n n!(x + 1)^{2n+1}/(2n + 1)!]$, both converge for all x;
9. $y_1(x) = 1 - (x^2/2) - (x^3/6) + (x^4/8) + (x^5/120) + \cdots$, $y_2(x) = x + x^2 - \tfrac{1}{4}x^4 + \cdots$, on $(-1, 1)$;
11. $y_1(x) = 1 + 2(x + 2)^2 - (x + 2)^3 + (\tfrac{1}{3})(x + 2)^4 - (\tfrac{13}{20})(x + 2)^5 + \cdots$, $y_2(x) = (x + 2) + \tfrac{1}{2}(x + 2)^3 - \tfrac{1}{2}(x + 2)^4 + (\tfrac{1}{40})(x + 2)^5 + \cdots$, on $(-\infty, \infty)$;
13. $y_1(x) = 1 - 3x^2, y_2(x) = x - (x^3/3)$;
15. $y_1(x) = 1, y_2(x) = \arctan(x)$;
18. $y_1(x) = 1 + (x^3/2) + (x^6/10) + \cdots, y_2(x) = x + (x^4/3) + (x^7/18) + \cdots$, $y_3(x) = x^2 + (x^5/4) + (x^8/28) + \cdots$;
20b. $an^2 + (b - a)n + c = (n - k)(an + r)$ where k is a positive integer and $-r/a$ is not a positive integer.

(31.13)   1. $y(x) = c_1 e^x + c_2 e^{-x} + \tfrac{1}{2}xe^x$, or $y(x) = a_0(1 + [x^2/2!] + [x^4/4!] + \cdots) + a_1(x + [x^3/3!] + [x^5/5!] + \cdots) + [x^2/2!] + [x^3/3!] + [2x^4/4!] + \cdots$;
3. $y(x) = a_0[1 - (x + 1)^2] + a_1[(x + 1) - (x + 1)^2] + \sum_{n=3}^{\infty} [(x + 1)^n/n(n - 1)]$;
5. $y(x) = a_0 + a_1[(x - 1) + \{(x - 1)^3/6\} + \{(x - 1)^5/40\} + \cdots] + [(x - 1)^3/6] - [(x - 1)^4/24] + \cdots$ on $(0, 2)$;
7. $y(x) = \sin(x) + \tfrac{1}{2}x \sin(x)$;   9. $y(x) = 1 + 4x^2 + (4x^4/3)$;

12. $y(x) = 1 + (x - \frac{1}{2}\pi) - [(x - \frac{1}{2}\pi)^3/3!] + [2(x - \frac{1}{2}\pi)^4/4!] + [7(x - \frac{1}{2}\pi)^5/5!] + \cdots$;

14. $y(x) = a_0[1 + \{(x + 1)^3/3\} + \{(x + 1)^6/18\} + \cdots] + a_1[(x + 1) + \{(x + 1)^4/4\} + \{(x + 1)^7/28\} + \cdots]$.

(32.16) 1a. $x = 1$ is a regular singular point;   1e. $x = 0$ is an irregular singular point and $x = 1$ is a regular singular point;   1g. $x = 0$ is an irregular singular point; 1j. $x = 1, -1$ are regular singular points;   1m. $x = 0$ is a regular singular point and $x = -\frac{1}{2}$ is an irregular singular point;   3a. the point at infinity is a regular singular point;   3b. the point at infinity is an irregular singular point; 5. $y_1(x) = 1 - (\frac{1}{10})(x + 1)^2 + (\frac{1}{440})(x + 1)^4 - \cdots$ on $(-\infty, \infty)$, $y_2(x) = x^{1/3}[1 - (\frac{1}{14})x^2 + (\frac{1}{728})x^4 - \cdots]$ on $(0, \infty)$; 7. $y_1(x) = x \sum_{n=0}^{\infty} (n!)^{-2}x^n$ on $(-\infty, \infty)$, there is no second solution of this form; 9. $y_1(x) = (x - 2)[1 - (\frac{1}{7})(x - 2) + (\frac{1}{154})(x - 2)^2 - \cdots]$ on $(-\infty, \infty)$, $y_2(x) = (x - 2)^{1/4}[1 - (x - 2) + (\frac{1}{10})(x - 2)^2 - \cdots]$ on $(2, \infty)$; 11. $y_1(x) = x^{1/2}\sum_{n=0}^{\infty} [(-1)^n(x/2)^{2n}/n!(n + \frac{1}{2})^{(n)}]$ on $(0, \infty)$, $y_2(x) = x^{-(1/2)}\sum_{n=0}^{\infty} [(-1)^n(x/2)^{2n}/n!(n - \frac{1}{2})^{(n)}]$ on $(0, \infty)$; 13. $y_1(x) = x^2[1 - (\frac{4}{11})x + (\frac{13}{330})x^2 + \cdots]$ on $(0, \infty)$, $y_2(x) = x^{1/4}[1 + (\frac{1}{6})x - (\frac{17}{24})x^2 + \cdots]$ on $(0, \infty)$; 15. $y_1(x) = x^{3/2}[1 - (\frac{5}{14})x + (\frac{5}{72})x^2 - (\frac{5}{264})x^3 + \cdots]$, $y_2(x) = x^{-1}$.

(32.22) 1. $y_1(x) = \sum_{n=0}^{\infty} [(-1)^n(6)^{n+1}(n + 1)x^n/(2n + 3)!]$, $y_2(x) = x^{-(3/2)}[1 - 3x - (\frac{9}{2})x^2 + (\frac{3}{2})x^3 - (\frac{9}{40})x^4 + \cdots]$; 3. $y_1(x) = (x + 2)^2 \sum_{n=0}^{\infty} (-1)^n(n!)^{-2}(x + 2)^n$, $y_2(x) = (x + 2)^3 z(x) + y_1(x) \ln(x + 2)$, where $z(x)$ is a power series in $(x + 2)$; 5. $y_1(x) = x^4 \sum_{n=0}^{\infty} (n + 1)x^n$ on $(0, 1)$, $y_2(x) = (\frac{1}{3})(3 + 2x + x^2)$; 8. $y_1(x) = (x - 1)^{-1} \sum_{n=0}^{\infty} (-1)^n(n!)^{-2}(x - 1)^n$ on $(1, \infty)$, $y_2(x) = z(x) + y_1(x) \ln(x - 1)$; 11. $y_1(x) = 3x^3 \sum_{n=0}^{\infty} [2^{n+1}(n + 1)! \, x^{2n}/(2n + 3)!]$ on $(-\infty, \infty)$, $y_2(x) = z(x) + Cy_1(x) \ln(x)$ where $z(x)$ is a power series in x; 13. $y_1(x) = x^2 \sum_{n=0}^{\infty} [(-1)^n x^n/n!]$ on $(-\infty, \infty)$, $y_2(x) = x[1 + x^2 - (\frac{3}{2})x^3 + (\frac{47}{72})x^4 + \cdots] - y_1(x) \ln(x)$; 16a. $r = 1$;   16b. $y_1(x) = x$;   16c. $y_1(x) = x$, $y_2(x) = xe^{1/x}$.

(33.15) 2. $P_4(x) = (\frac{1}{8})(35x^4 - 30x^2 + 3)$;   4. $q = (\frac{1}{5})(4P_3 - 10P_2 + 31P_1 - 10P_0)$; 14b. $y(x) = 1 + [\alpha(\alpha + 1)/2(1!)^2](x - 1) + \{[\alpha^2(\alpha + 1)^2 - 2\alpha(\alpha + 1)]/2^2(2!)^2\}(x - 1)^2 + \cdots$.

(33.31) 3. $\Gamma(n + \frac{1}{2}) = (2n)! \, \pi^{1/2}/n! \, 4^n$; 5. $J_b(x) = [1/\Gamma(b + 1)](x/2)^b \sum_{k=0}^{\infty} [(-1)^k(x/2)^{2k}/k! \, (k + b)^{(k)}]$.

(33.44) 1a. $J_3 = (8x^{-2} - 1)J_1 - 4x^{-1}J_0$, $J_4 = (6x^{-2} - 1)(8x^{-1})J_1 - (24x^{-2} - 1)J_0$; 1b. $J_4 = 8x^{-1}(1 - 6x^{-2}) \, DJ_0 - (24x^{-2} - 1)J_0$.

(33.53) 5. $T_3(x) = 4x^3 - 3x$, $T_4(x) = 8x^4 - 8x^2 + 1$, $T_5(x) = 16x^5 - 20x^3 + 5x$.

(33.64) 5. $H_3(x) = 8x^3 - 12x$, $H_4(x) = 16x^4 - 48x^2 + 12$, $H_5(x) = 32x^5 - 160x^3 + 120x$.

(33.73) 6. $L_3(x) = 6^{-1}(6 - 18x + 9x^2 - x^3)$, $L_4(x) = (24)^{-1}(24 - 96x + 72x^2 - 16x^3 + x^4)$, $L_5(x) = (5!)^{-1}(120 - 600x + 600x^2 - 200x^3 + 25x^4 - x^5)$.

(34.6) 7. $Y(s) = s^{-2} - s^{-2}e^{-3s}$.

(35.14) 1. all;   2a. $Y(s) = (s - 2)^{-1}$;   2c. $Y(s) = (-s^2 + 2s - 4)/(s^3 - s^2 + s - 1)$; 3b. $Y(s) = (16s^3 - 6s^2 + 20s - 8)/(s^2 + 1)(2s - 1)^3$.

(36.12) 1a. $f(x) = 6e^{-7x}$;   1c. $f(x) = (\frac{1}{5}) \sin(5x)$;   1e. $f(x) = e^{-4x} \cos(x)$; 1g. $f(x) = x \cos(2x) + x \sin(2x)$.

(36.17) 1. $y(x) = 3e^{2x} - e^x$;   3. $y(x) = \emptyset$;   5. $y(x) = xe^{-x} + x - 2 + \frac{1}{4}e^x$;

7. $y(x) = 2e^{-x}\cos(x) + xe^{-x}\sin(x)$;   9. $y(x) = 2e^{x-2} - (x-2)e^{x-2} - e^{2(x-2)}$.

(36.19)   1a. $y(0) = 0$;   1b. $-3y(0) + Dy(0) + 1 = 0$;   2. all;   3. none.

(36.25)   4a. $f(x) = (\frac{1}{5})(e^{3x} - e^{-2x})$;   4c. $f(x) = \frac{1}{2}[\sin(x) + x\cos(x)]$;

      5. $y(x) = (1/k)\sinh(kx) * f(x)$;   7a. $y(x) = x + (\frac{1}{6})x^3$;   7c. $y(x) = 1$.

(37.9)   1. no (f_1 is piecewise continuous on $[0, b]$ for all $b < \infty$);   3. no;   5. yes;

      7. yes;   9a. 1;   9b. 1;   11a. $39\frac{1}{2}$;   11b. divergent;   13. 5.

(37.16)   2. $F(s) = s^{-2} - s^{-1}e^{-s} - s^{-1}e^{-2s} - s^{-2}e^{-2s}$;   4. $F(s) = 2s^{-1}e^{-5s}$;

      6. $F(s) = s^{-1} - s^{-1}e^{-2s} + s^{-2}e^{-2s} - s^{-2}e^{-4s} - 2s^{-1}e^{-4s} + (s+1)^{-1}e^{-4s}$;

      8. $F(s) = s^{-2} - s^{-1}e^{-s} - s^{-1}e^{-2s} - \cdots = s^{-2} - s^{-1}(e^s - 1)^{-1}$;

      10. $F(s) = 2s^{-3} - 2s^{-3}e^{-3s} - 3s^{-2}e^{-3s}$;

      12. $F(s) = s^{-1}(s^2 + 1)^{-1}(e^{-\pi s} - e^{-2\pi s})$;

      14. $F(s) = s^{-2}(1 - 2e^{-s} + 2e^{-3s} - e^{-4s})$;

      16. $f(x) = \sin(x) - \sin(x)H(x - \pi)$;

      20. $f(x) = x^2 - 4(x-2)H(x-2) - (x-2)^2H(x-2) - 4H(x-4)$.

(37.20)   1. $y(x) = 1 - \cos(x) + \sin(x) + H(x-1)[\cos(x-1) - 1]$;

      3. $y(x) = 1 - e^{-x} - xe^{-x} + H(x-2)[(x-4) + xe^{-(x-2)}]$;

      5. $y(x) = 2 + \sin(x) + H(x-2)[2\cos(x-2) - \sin(x-2) + (x-4)]$.

(38.9)   2. $Dy_1 = y_2, Dy_2 = -5y_2 - 4y_1 + \ln x$;

      4. $Dy_1 = y_2, Dy_2 = y_3, Dy_3 = 11y_2 - 2y_3 + 12$.

(38.15)   1b. $D\psi = \begin{bmatrix} 0 & 1 \\ -4 & -5 \end{bmatrix}\psi + \begin{bmatrix} 0 \\ \ln x \end{bmatrix}$ on $(0, \infty)$;

      1d. $D\psi = \begin{bmatrix} 0 & 1 & 0 \\ 0 & 0 & 1 \\ 0 & 11 & -2 \end{bmatrix}\psi + \begin{bmatrix} 0 \\ 0 \\ 12 \end{bmatrix}$ on $(-\infty, \infty)$;   4. $D[Q\psi(x)] = Q[D\psi(x)]$.

(38.17)   1. $D^3y_1 - 12\,D^2y_1 + 44\,Dy_1 - 48y_1 = \varnothing$.

(39.6)   3. $\psi_1(x) = e^{-4x}\begin{bmatrix} 1 \\ -4 \end{bmatrix}, \psi_2(x) = e^{-x}\begin{bmatrix} 1 \\ -1 \end{bmatrix}$;

      4. $\psi_1(x) = e^{-4x}\begin{bmatrix} 1 \\ -4 \\ 16 \end{bmatrix}, \psi_2(x) = e^{-x}\begin{bmatrix} 1 \\ -1 \\ 1 \end{bmatrix}, \psi_3(x) = e^{3x}\begin{bmatrix} 1 \\ 3 \\ 9 \end{bmatrix}$.

(39.21)   1a. $\begin{bmatrix} 1 & e^x \\ x & x^2 \end{bmatrix}\begin{bmatrix} 7 \\ 4 \end{bmatrix}$;   1c. $\begin{bmatrix} -1 & \sin x \\ x & x \end{bmatrix}\begin{bmatrix} 2 \\ 3 \end{bmatrix} + \begin{bmatrix} e^x & 2 \\ 1 & 6 \end{bmatrix}\begin{bmatrix} 8 \\ -4 \end{bmatrix}$;

      2a. independent;   2c. independent;   4. either $\{\psi_1, \psi_5\}$ or $\{\psi_1, \psi_6\}$.

(39.24)   1. $\psi(x) = 3e^x\begin{bmatrix} 1 \\ 0 \\ 0 \end{bmatrix} + 2e^{-2x}\begin{bmatrix} 1 \\ 1 \\ 0 \end{bmatrix} - e^{3x}\begin{bmatrix} -2 \\ 1 \\ -1 \end{bmatrix}$;

      2. $\psi(x) = 3\begin{bmatrix} 3e^{x^3} \\ -x \end{bmatrix} + \begin{bmatrix} 1 \\ 0 \end{bmatrix}$.

(39.33)   1. $\psi(x) = x\begin{bmatrix} 1 \\ 1 \end{bmatrix}$;   2. $\psi(x) = \begin{bmatrix} 1 \\ x \end{bmatrix}$;   3. $\psi(x) = \begin{bmatrix} e^{-x} \\ -10 \\ -e^{-x} \end{bmatrix}$.

(40.6)   4. $\psi(x) = e^x\begin{bmatrix} 1 \\ 2 \end{bmatrix} - e^{-2x}\begin{bmatrix} 2 \\ 1 \end{bmatrix}$.

(40.24) 1. $\psi(x) = c_1 e^{-x}\begin{bmatrix} 2 \\ -5 \end{bmatrix} + c_2 e^{-2x}\begin{bmatrix} -5 \\ 13 \end{bmatrix}$; 2. $\psi(x) = 3e^{-x}\begin{bmatrix} 2 \\ -5 \end{bmatrix} + e^{-2x}\begin{bmatrix} -5 \\ 13 \end{bmatrix}$;

5. $\psi(x) = \varnothing$; 6. $\psi(x) = 3ce^{-x}\begin{bmatrix} 2 \\ -5 \end{bmatrix} + c^2 e^{-2x}\begin{bmatrix} -5 \\ 13 \end{bmatrix}$ where $c > 0$;

7. $\psi(x) = c_1 e^{-x}\begin{bmatrix} 1 \\ -3 \end{bmatrix} + c_2 e^{x}\begin{bmatrix} -6 \\ 19 \end{bmatrix}$;

10. $\psi(x) = e^{-x}[c_1 \cos(2x) + c_2 \sin(2x)]\begin{bmatrix} 1 \\ -1 \\ 1 \end{bmatrix}$

$+ e^{-x}[c_2 \cos(2x) - c_1 \sin(2x)]\begin{bmatrix} -1 \\ 2 \\ -1 \end{bmatrix} + c_3 e^{3x}\begin{bmatrix} 0 \\ 1 \\ -1 \end{bmatrix}$;

13. $\psi(x) = c_1 e^{x/2}\begin{bmatrix} 1 \\ 1 \\ 1 \end{bmatrix} + c_2 e^{x/4}\begin{bmatrix} 3 \\ 4 \\ 3 \end{bmatrix} + c_3 e^{7x/2}\begin{bmatrix} 1 \\ 1 \\ 0 \end{bmatrix}$.

(40.27) 1. $\psi(x) = (c_1 + c_2 x)e^{2x}\begin{bmatrix} 1 \\ 0 \end{bmatrix} + c_2 e^{2x}\begin{bmatrix} 0 \\ 1 \end{bmatrix}$;

4. $\psi(x) = (c_1 + c_2 x)e^{2x}\begin{bmatrix} 2 \\ -5 \end{bmatrix} + c_2 e^{2x}\begin{bmatrix} -5 \\ 13 \end{bmatrix}$;

7. $\psi(x) = c_1 e^{2x}\begin{bmatrix} 1 \\ -1 \\ 1 \end{bmatrix} + c_2 e^{2x}\begin{bmatrix} 0 \\ 1 \\ 0 \end{bmatrix} + c_3 e^{2x}\begin{bmatrix} 1 \\ 0 \\ 2 \end{bmatrix} + c_3 x e^{2x}\begin{bmatrix} 1 \\ -2 \\ 1 \end{bmatrix}$.

(40.30) 2. $\psi(x) = c_1 e^{-3x}\begin{bmatrix} 1 \\ 2 \\ 2 \end{bmatrix} + c_2 e^{x}\begin{bmatrix} 2 \\ -2 \\ 1 \end{bmatrix} + c_3 e^{6x}\begin{bmatrix} -2 \\ -1 \\ 2 \end{bmatrix}$;

5. $\psi(x) = c_1 e^{-2x}\begin{bmatrix} 2 \\ -1 \end{bmatrix} + c_2 e^{5x}\begin{bmatrix} 1 \\ 2 \end{bmatrix}$.

(40.34) 1. $\psi(x) = 3e^{-x}\begin{bmatrix} 2 \\ -5 \end{bmatrix} + e^{-2x}\begin{bmatrix} -5 \\ 13 \end{bmatrix} + x\begin{bmatrix} 2 \\ -5 \end{bmatrix}$;

2. $\psi(x) = e^{3x}\begin{bmatrix} 0 \\ 1 \\ -1 \end{bmatrix} + \begin{bmatrix} x \\ -1 \\ -x \end{bmatrix}$; 3. $\psi(x) = \begin{bmatrix} xe^x \\ 1 \\ xe^x \end{bmatrix}$; 5. $\psi(x) = \begin{bmatrix} 1 \\ -1 \end{bmatrix} + \begin{bmatrix} x \\ x^2 \end{bmatrix}$;

7. $\psi(x) = e^{x}\begin{bmatrix} 2 \\ 2 \end{bmatrix} + e^{-x}\begin{bmatrix} 2 \\ -2 \end{bmatrix} + \begin{bmatrix} \sin x \\ \cos x \end{bmatrix}$; 8. $\psi(x) = \begin{bmatrix} 1 \\ 1 \\ -1 \end{bmatrix}$.

(41.20) 2. $\psi(k) = 2^k\begin{bmatrix} 1 \\ 1 \\ -1 \end{bmatrix} - 8\begin{bmatrix} 1 \\ 1 \\ 0 \end{bmatrix} + 8\begin{bmatrix} 1 \\ 0 \\ 1 \end{bmatrix}$; 4. $\psi(k) = c_1(-2)^k\begin{bmatrix} 1 \\ -1 \end{bmatrix} + c_2(2)^k\begin{bmatrix} 1 \\ 1 \end{bmatrix}$;

7. $\psi(k) = c_1(2)^k \begin{bmatrix} 1 \\ -1 \\ 1 \end{bmatrix} + c_2 \begin{bmatrix} -1 \\ 2 \\ -1 \end{bmatrix} + c_3(2)^k \begin{bmatrix} 0 \\ 1 \\ -1 \end{bmatrix}.$

(41.32)   1a. $\sigma(k) = 2^k \begin{bmatrix} 0 \\ -1 \\ 1 \end{bmatrix} - \begin{bmatrix} 0 \\ -1 \\ 1 \end{bmatrix}$;   1c. $\psi(k) = \begin{bmatrix} k^2 \\ k \\ -k \end{bmatrix}$;

4. $\psi(k) = \alpha$;   8a. $\psi(k) = 3(\tfrac{1}{2})^k \begin{bmatrix} 1 \\ 1 \\ 1 \end{bmatrix} + 2k \begin{bmatrix} 1 \\ 1 \\ 1 \end{bmatrix}$;   8b. $\psi(k) = 2k \begin{bmatrix} 1 \\ 1 \\ 1 \end{bmatrix}$;

11. $\psi(k) = [c_1 \cos(\tfrac{1}{2}k\pi) + c_2 \sin(\tfrac{1}{2}k\pi)] \begin{bmatrix} -2 \\ 0 \\ 1 \end{bmatrix}$

$+ [c_2 \cos(\tfrac{1}{2}k\pi) - c_1 \sin(\tfrac{1}{2}k\pi)] \begin{bmatrix} 1 \\ -1 \\ 2 \end{bmatrix} + c_3(-1)^k \begin{bmatrix} 3 \\ 1 \\ 0 \end{bmatrix} + 8k \begin{bmatrix} -1 \\ 1 \\ -1 \end{bmatrix}.$

(42.15)   2. $\sigma(2) = \varnothing$;   3a. 3;   3b. $(20)^{1/2}$;   3d. 3 if $k \le 2$ and $k^{(2)}$ if $k > 2$.

(42.24)   3a. 3;   3b. $\tfrac{7}{2}$.

(43.13)   1b. $\{(x, y) | y \ne 0\}$;   1d. $\{(x, y) | y \ne \pm x\}$;   1f. $\{(x, y) | x \ne 0 \text{ and } y \ne x\}$;
3a. $\{(x, y) | y \ne 0\}$;   3c. $\{(x, y) | y^2 - x^2 \ne 1\}$;   3e. $\{(x, y) | x + y \ge 0\}$;
4b. $y(x) = (x^2 + 1)^{-1}$ on $(-\infty, \infty)$;   4c. $y(x) = (x^2 - 1)^{-1}$ on $(-1, 1)$;
5c. families of circles, centers at $(c, 0)$.

(43.17)   4a. $y(x) = cx + c^2$, $y(x) = -x^2/4$;
4c. $y(x) = cx + (c^2 - 1)^{1/2}$, $y(x) = -(x^2 - 1)^{1/2}$.

(44.9)   1. $y(x) = \tan(x) + c$;   3. $(y - 1)\exp(x + y) = c(x + 1)$;
5. $x \sin(x) + \cos(x) - \ln|\cos(y)| = c$;
7. $x + \ln|x - 1| - (y^2/2) - y - \ln(y - 1)^2 = c$;   9. $(y^2 - 1)^{1/2} = c(1 - x)$;
11. $y^2 + 4y = 2 - 2\cos(x)$;   13. $4y^2 = 2x - \sin(2x) + 4$;
15. $y(x) = -\tanh(x)$.

(44.18)   1. homogeneous, degree 2;   3. not homogeneous;   4. homogeneous,
degree $\tfrac{3}{4}$;   5. $y = c(x^2 + y^2)$;   7. $y(x) = xe^{cx}$;   9. $\ln(x) - \sin^{-1}(y/x) = c$;
11. $y = -(\tfrac{2}{5})x^2(x + y)$;   13. $\ln(x) + \cos(y/x) = 1$, use Theorem 10,
Section 43, to comment on uniqueness.

(44.28)   1a. $12x - 4y + x^2 - 4xy - y^2 = c$;   1c. $\ln(15x + 10y - 1)^2 + 5(x - y) = c$;
1e. $(x - y)^2 + \ln(x - y + 2)^2 = -8x + c$;   3a. $y = z^2$;   3c. this equation
cannot be transformed into a homogeneous equation by means of (26);
4b. $y^2(y - x^2) = c$.

(45.10)   1. $\sin(x + y^2) = c$;   3. $\ln(x + y) + xy^2 = c$;   5. $x^2 + 2y \sin(x) = c$;
7. $6xy^2 - \cos(2xy) + 2y^3 = c$;   9. $ye^x + x^2 \cos(y) = c$;
11. $e^x + x \ln(y) + y \ln(x) - \cos(y) = c$;   13. $(x^2 + 2y)^{1/2} + x \ln(1 + y) = 2$;
15. $\sinh(x - y^2) = \varnothing$, or $y^2 = x$;   19. $g(y) = e^{-y}[(y^2/2) + c]$.

(45.17)   1a. $x^4 - y^2 = cy$;   1c. $x^4 y^3 + \ln(x/y) = c$;   1e. $6x^2 y^2 + 3y^4 + 4y^3 = c$;
2. $\sin(x - y) + xy^4 = 1$.

(45.19)   1. $x^2 y^4 + y^6 = cx^4$;   3. $\mu(x, y) = y^2 x^{1/2}$.

(46.8)   1a. $y^{-2} = 1 + ce^{x^2}$;   1c. $y(x) = (ce^{2x} - e^x)^2$;   1d. $y^2 = cx - 1$;
2b. $y(x) = [1 + \ln(x)]^{-1}$;   2d. $y(x) = 3\cos^2(x)/[\cos(x) - 3\sin(x)]$;
5. $u = \sin(y)$.

(46.13) 1b. $y(x) = (e^{3x} - ce^x)/(e^{2x} + c)$; 2. $y(x) = (5x^5 + 3x)/(5x^4 - 1)$;
  5. $y(x) = (2e^x + c)/(e^x + c)$.

(46.19) 1. $y(x) = x^{-1} + [x(c + \ln x)]^{-1}$; 4. $y(x) = -\frac{1}{2}\tan(x)$ on $(-\pi/2, \pi/2)$;
  5. $y(x) = 1/2x$.

(46.21) 1. $y(x) = 1 + 2(x - 1) + 3(x - 1)^2 + (\frac{11}{3})(x - 1)^3 + (\frac{29}{6})(x - 1)^4 + \cdots$;
  3. $y(x) = (\pi/2) + (x - 1) - (\frac{1}{6})(x - 1)^3 + \cdots$.

(47.5) 4. $y(x) = -\tanh(x)$.

(47.18) 1. 4; 4. Df_1 continuous on J and Df_2 continuous and bounded on
  $(-\infty, \infty)$; 7. yes.

Index